Tradition and Innovation in Construction Project Management

Tradition and Innovation in Construction Project Management

Editors

Yongjian Ke
Jingxiao Zhang
Simon P. Philbin

MDPI • Basel • Beijing • Wuhan • Barcelona • Belgrade • Manchester • Tokyo • Cluj • Tianjin

Editors
Yongjian Ke
University of Technology Sydney
Sydney, Australia

Jingxiao Zhang
Chang'an University
Xi'an, China

Simon P. Philbin
London South Bank University
London, UK

Editorial Office
MDPI
St. Alban-Anlage 66
4052 Basel, Switzerland

This is a reprint of articles from the Special Issue published online in the open access journal *Buildings* (ISSN 2075-5309) (available at: https://www.mdpi.com/journal/buildings/special_issues/tradition_innovation_construction_management).

For citation purposes, cite each article independently as indicated on the article page online and as indicated below:

LastName, A.A.; LastName, B.B.; LastName, C.C. Article Title. *Journal Name* **Year**, *Volume Number*, Page Range.

ISBN 978-3-0365-8114-9 (Hbk)
ISBN 978-3-0365-8115-6 (PDF)

Contents

About the Editors

Yongjian Ke

Yongjian Ke is an associate professor in the School of Built Environment, University of Technology Sydney, Australia, and is a highly acclaimed project management researcher. His research spans a range of challenges in the delivery of infrastructure projects, including risk allocation, contracting behaviours, social acceptance, and social sustainability, among others. He is particularly interested in sustainable infrastructure development and public–private partnerships (PPP) and their potential to drive new investment and collaboration opportunities between Australia and China. His current work includes two National Science Foundation of China-funded projects; one to develop a "social license for infrastructure" framework for Chinese contexts and the other to investigate value co-creation in infrastructure projects. He also looks at social sustainability within PPP projects and the use of procurement to create social value within and beyond the initial project aims. He is a Department Editor for the Project Management Journal, a Guest Professor at Chang'an University in China, and a Research Associate at the Australia–China Relations Institute. Yongjian holds a PhD in Management Science and Engineering and a Bachelor of Engineering from Tsinghua University in China.

Jingxiao Zhang

Jingxiao Zhang is currently a Professor at the School of Economics and Management, Chang'an University, China. He has served as a visiting scholar at Illinois State University's Department of Technology in the USA and at Queensland University of Technology's Department of Civil Engineering and Built Environment in Australia. Currently, he is a visiting scholar at London South Bank University and a research institute member at the Nathu Puri Institute for Engineering and Enterprise, also at London South Bank University. Prof. Zhang has an extensive publication record, with over 170 papers to his name. This includes 100 Chinese papers, 70 SCI/SSCI papers, and 2 highly cited papers within his field. He has published 42 SCI papers, 18 SSCI papers, over 50 SCI/SSCI double-search-indexed papers, and more than 20 CSSCI/EI search-indexed papers in renowned international journals such as *JCLP*, *ASCE-JPI*, and *IJPR*. Since 2017, his papers have garnered 32,939 readings and 1029 citations. His current research interests encompass various areas, including green and intelligent construction of major infrastructure, new construction industrialization, BIM education for engineering management, intelligent logistics, and emergency management, as well as engineering cost and engineering management. Prof. Zhang has been involved in over 40 projects funded by the National Natural Science Foundation of China, the National Social Science Foundation of China, and the International Program of the 2020 UK–China–BRI (Vietnam/Indonesia) Education Partnership Initiative. Currently, he is leading projects such as the China Sichuan–Tibet Railway Major Fundamental Science Problems Special Fund, the Integrated Built Environment Teaching & Learning in the Joint Curriculum Development amid Digital-Driven Industry 4.0, and the Research on the Mechanism and Implementation Path of High-Quality Development for China's Construction Industry in the Context of Digital Transformation, among others.

Simon P. Philbin

Simon P. Philbin is a Professor and Director of Engineering and Enterprise in the School of Engineering at London South Bank University (LSBU) in the United Kingdom, where he currently leads the Nathu Puri Institute for Engineering and Enterprise. His primary research interests are in the area of engineering management with a particular focus on sustainability, technology-driven innovation, and project management. He holds a PhD (Brunel University) and BSc (University of Birmingham), both in chemistry, as well as an MBA (Open University Business School). Previous research positions include Visiting Fellow at Imperial College Business School and Visiting Fellow in the School of Business, Economics and Informatics at Birkbeck, University of London. He has authored/co-authored over 100 peer-reviewed journal and conference papers, and has presented research studies at various international conferences across North America, Asia, and Europe. He previously served as the 2019/20 President of the American Society for Engineering Management (ASEM) and he is currently Co-Editor-in-Chief of the Engineering Management Journal (EMJ).

buildings

MDPI

Editorial

Tradition and Innovation in Construction Project Management

Yongjian Ke [1,*], Jingxiao Zhang [2] and Simon P. Philbin [3]

1 School of Built Environment, University of Technology Sydney, Sydney, NSW 2007, Australia
2 School of Economics and Management, Chang'an University, Xi'an 710064, China; zhangjingxiao@chd.edu.cn
3 School of Engineering, London South Bank University, London SE1 0AA, UK; philbins@lsbu.ac.uk
* Correspondence: yongjian.ke@uts.edu.au

1. Introduction and Aim of the Special Issue

Construction project management is a multidimensional discipline that requires meticulous consideration of various critical aspects, including cost, quality, schedule requirements, as well as social and environmental impacts, and broader stakeholder interests. In recent years, the field has encountered challenges and opportunities arising from the adoption of emerging technologies, novel materials, advanced methods and processes, and complex contractual arrangements. One notable example is the integration of Building Information Modelling (BIM) in project management, which has received considerable attention [1–3]. Consequently, construction project management is poised to leverage the benefits of digital transformation within the Industry 4.0 paradigm [4,5] while embracing new technologies and management systems to promote sustainable development [6,7].

The Special Issue of the *Buildings* journal, titled "Tradition and Innovation in Construction Project Management," serves as a platform to showcase cutting-edge international research and development in this important field. The issue aimed to explore the interplay between tradition and innovation in construction project management and identify new research studies and practices that harness the potential of digital technologies and drive sustainable development across the sector.

2. Commentary on the Published Papers

This Special Issue comprises 22 publications, consisting of 20 research articles and two review articles. Within a short period (as of June 2023), the issue has garnered over 30,000 views and received 45 citations, indicating its significant impact. With contributions from 101 co-authors representing 11 countries, this issue offers a global perspective on tradition and innovation in construction project management. The featured papers encompass a wide range of topics, employ diverse methodologies, and examine various geographical contexts, collectively providing valuable insights into the challenges and opportunities within the construction industry.

Table 1 presents a comprehensive summary of the 22 publications, categorizing them based on their general topic tag and specific topic tag. While some articles may address multiple topics, we have assigned a primary topic focus to each paper. Through this synthesis, we have identified five overarching topic groups: technology, management, outcome, product, and urban. This classification aids in organizing the papers and highlighting the main themes explored in this Special Issue.

The papers in this Special Issue unequivocally showcase technology's innovation and transformative potential in the construction industry. For instance, Hartmann et al. [8] utilized digitalization to optimize construction processes in a hospital construction project in Germany. B. Wang et al. [13] identified a significant difference in the efficiency of green technology innovation in the construction industry of the Chengdu-Chongqing metropolitan agglomeration, with an upward trend. H. Li et al. [18] explored the key factors influencing the digital transformation of the construction industry and developed a comprehensive

Citation: Ke, Y.; Zhang, J.; Philbin, S.P. Tradition and Innovation in Construction Project Management. *Buildings* **2023**, *13*, 1537. https://doi.org/10.3390/buildings13061537

Received: 13 June 2023
Accepted: 14 June 2023
Published: 16 June 2023

evaluation system. Doukari et al. [19] compared conventional and automated approaches to creating construction schedules using 4D BIM in real-world projects. Xu et al. [24] provided a systematic commentary on the state-of-the-art research on artificial neural networks in construction management, revealing insufficient attention and implementation challenges. Elrefaey et al. [28] investigated the impact of COVID-19 on the adoption of digital technology in construction projects in the UAE.

Table 1. The content of the Special Issue "Tradition and Innovation in Construction Project Management".

Authors	Manuscript Title	Article Type	General Topic Tag	Specific Topic Tag
Hartmann et al. [8]	Optimizing Interfaces of Construction Processes by Digitalization Using the Example of Hospital Construction in Germany	Research	Technology	Digitalization
Kumaraswamy et al. [9]	Accelerating the Delivery of Low-Carbon Buildings by Addressing Common Constraints: Perspectives from High-Rise, High-Density Cities	Research	Outcome	Low Carbon
Y. Wang et al. [10]	Risk Propagation Model and Simulation of an Assembled Building Supply Chain Network	Research	Management	Risk Management
H. Yan, Yang et al. [11]	Regret Theory and Fuzzy-DEMATEL-Based Model for Construction Program Manager Selection in China	Research	Management	Resource
Bailey et al. [12]	A Critical Scoping Review of Disability Employment Research in the Construction Industry: Driving Social Innovation through More Inclusive Pathways to Employment Opportunity	Research	Outcome	Social Equity
B. Wang et al. [13]	Spatiotemporal Differentiation and Influencing Factors of Green Technology Innovation Efficiency in the Construction Industry: A Case Study of Chengdu–Chongqing Urban Agglomeration	Research	Technology	Green Technology
Mridha et al. [14]	The Influence Innovation Has on the Visual Appearance and Aesthetic Preference of Architectural Products	Research	Product	Architecture
J. Zhang et al. [15]	Current and Future Trends of Resource Misallocation in the Construction Industry: A Bibliometric Review with Grounded Theory	Research	Management	Resource
Y. Li, Liang et al. [16]	Behavioral Research in Construction Engineering Management: A Review from a Neuropsychological Perspective	Review	Management	Behavior

Table 1. *Cont.*

Authors	Manuscript Title	Article Type	General Topic Tag	Specific Topic Tag
Huaman-Orosco et al. [17]	Barriers to Adopting Lean Construction in Small and Medium-Sized Enterprises—The Case of Peru	Research	Management	Lean Construction
H. Li et al. [18]	Systematic Identification of the Influencing Factors for the Digital Transformation of the Construction Industry Based on LDA-DEMATEL-ANP	Research	Technology	Digitalization
Doukari et al. [19]	The Creation of Construction Schedules in 4D BIM: A Comparison of Conventional and Automated Approaches	Research	Technology	Digitalization
Bahamid et al. [20]	The Current Risk Management Practices and Knowledge in the Construction Industry	Research	Management	Risk Management
Y. Li, Xie et al. [21]	Evaluation of Park Accessibility Based on Improved Gaussian Two-Step Floating Catchment Area Method: A Case Study of Xi'an City	Research	Urban	Assessment
J. Wang et al. [22]	Safety Risk Assessment of Prefabricated Buildings Hoisting Construction: Based on IHFACS-ISAM-BN	Research	Management	Risk Management
Y. Zhang et al. [23]	How Private Enterprises' Participation Behaviors Evolve with Incentive Modes in PPPs: An Evolutionary Game View	Research	Management	Behavior
Xu et al. [24]	Application of Artificial Neural Networks in Construction Management: A Scientometric Review	Review	Technology	Artificial Intelligence
L. Wang et al. [25]	A Real Estate Early Warning System Based on an Improved PSO-LSSVR Model—A Beijing Case Study	Research	Urban	Assessment
Yang et al. [26]	Organization Synchronization in Response to Complex Project Delays: Network-Based Analysis	Research	Management	Behavior
Huang et al. [27]	Safety Risk Assessment Using a BP Neural Network of High Cutting Slope Construction in High-Speed Railway	Research	Management	Risk Management
Elrefaey et al. [28]	Impacts of COVID-19 on the Use of Digital Technology in Construction Projects in the UAE	Research	Technology	Digitalization
H. Yan, Zheng et al. [29]	Risk Coupling Evaluation of Social Stability of Major Engineering Based on N-K Model	Research	Management	Risk Management

This Special Issue provides insights into the dynamic landscape of construction project management, thereby underscoring the significance of embracing innovative strategies and approaches to overcome challenges in the sector and ensure successful project delivery.

- Notably, risk management emerges as a prominent theme, with Y. Wang et al. [10] developing a risk propagation model for an assembled building supply chain network. Bahamid et al. [20] summarized risk management strategies in Yemeni building projects. J. Wang et al. [22] established a safety risk assessment framework based on Bayesian Network, improved Human Factors Analysis and Classification System, and improved the Similarity Aggregation Method. Huang et al. [27] proposed a safety risk assessment model using a BP neural network algorithm. Yan et al. [29] investigated the coupling evaluation method to analyze the influence of social stability risk factors in major projects;
- Furthermore, the inclusion of behavioral research signifies the growing recognition of understanding the behavior of contracting parties in projects. Y. Li et al. [16] employed a bibliometric analysis of 1254 studies from the Web of Science to explore behavioral research in construction engineering management. Y. Zhang et al. [23] revealed the evolutionary law of private enterprises' participation behavior under different incentive modes in Public–Private Partnership projects. Whereas, Yang et al. [26] established an index system of organization interactions to determine synchronizability and effective interaction methods;
- In the realm of resource management, Yan et al. [11] combined regret theory and the Fuzzy-DEMATEL method to develop a multi-attribute model for construction program manager selection. Additionally, J. Zhang et al. [15] analyzed the current state, consequences, and emerging trends in resource management research at the macroscopic level to explore potential solutions;
- Finally, this Special Issue delves into the domain of lean construction management, with Huaman-Orosco et al. [17] exploring barriers, challenges, and strategies for adopting lean practices in small and medium-sized enterprises.

The Special Issue sheds light on an innovative view of project outcomes, with a specific focus on sustainability, low-carbon approaches, social equity, and related factors. In this regard, Kumaraswamy et al. [9] identified potential strategies for expediting the delivery of low-carbon buildings in high-rise and high-density cities by addressing common constraints identified in recent studies. Furthermore, Bailey et al. [12] presented the findings of a critical scoping review of publications on the employment of individuals with disabilities in construction, revealing that research in this area remains nascent internationally and significant knowledge gaps exist compared to mainstream disability employment research.

Moreover, the Special Issue encompasses innovation in product management and urban-related topics, featuring research articles and review papers that contribute to our understanding of advancements in architectural products, visual aesthetics, accessibility assessment, and real estate evaluation, among others. For instance, Mridha et al. [14] investigated the connection between architectural product innovation and aesthetic preference, examining the significant role of innovation in perceiving the aesthetic preference of architectural products. Y. Li, Xie, et al. [21] developed an improved Gaussian-based two-step floating catchment area method to measure park accessibility for various travel modes in 5 min, 15 min, and 30 min scenarios. L. Wang et al. [25] developed a real estate early warning system based on an improved PSO-LSSVR model and test its effectiveness in the city of Beijing.

3. Conclusions

The papers featured in this Special Issue exemplify the various ways in which innovation can be utilized to overcome challenges, enhance efficiency, and promote sustainable development in the industry. Categorized into five broad topic groups—technology, management, outcome, product, and urban—these research papers provide comprehensive insights into the intersection of tradition and innovation in construction project management.

As the construction industry undergoes continuous evolution, embracing both tradition and innovation becomes imperative. Traditional project management principles and practices serve as a robust foundation, while innovation unlocks new possibilities and facilitates ongoing improvement. By integrating innovative technologies, materials, and methodologies into construction project management, stakeholders can boost productivity, minimize costs, mitigate environmental impacts, and deliver exceptional building outcomes for society.

Conflicts of Interest: The authors declare no conflict of interest.

References

1. Travaglini, A.; Radujković, M.; Mancini, M. Building information modelling (BIM) and project management: A stakeholders perspective. *Organ. Technol. Manag. Constr. Int. J.* **2014**, *6*, 1001–1008. [CrossRef]
2. Ma, X.; Darko, A.; Chan, A.P.; Wang, R.; Zhang, B. An empirical analysis of barriers to building information modelling (BIM) implementation in construction projects: Evidence from the Chinese context. *Int. J. Constr. Manag.* **2022**, *22*, 3119–3127. [CrossRef]
3. Olanrewaju, O.I.; Kineber, A.F.; Chileshe, N.; Edwards, D.J. Modelling the relationship between Building Information Modelling (BIM) implementation barriers, usage and awareness on building project lifecycle. *Build. Environ.* **2022**, *207*, 108556. [CrossRef]
4. Newman, C.; Edwards, D.; Martek, I.; Lai, J.; Thwala, W.D.; Rillie, I. Industry 4.0 deployment in the construction industry: A bibliometric literature review and UK-based case study. *Smart Sustain. Built Environ.* **2021**, *10*, 557–580. [CrossRef]
5. Sepasgozar, S.M. Differentiating digital twin from digital shadow: Elucidating a paradigm shift to expedite a smart, sustainable built environment. *Buildings* **2021**, *11*, 151. [CrossRef]
6. Stanitsas, M.; Kirytopoulos, K.; Leopoulos, V. Integrating sustainability indicators into project management: The case of construction industry. *J. Clean. Prod.* **2021**, *279*, 123774. [CrossRef]
7. Tushar, Q.; Bhuiyan, M.A.; Zhang, G.; Maqsood, T. An integrated approach of BIM-enabled LCA and energy simulation: The optimized solution towards sustainable development. *J. Clean. Prod.* **2021**, *289*, 125622. [CrossRef]
8. Hartmann, S.; Gossmann, D.; Kalmuk, S.; Klemt-Albert, K. Optimizing Interfaces of Construction Processes by Digitalization Using the Example of Hospital Construction in Germany. *Buildings* **2023**, *13*, 1421. [CrossRef]
9. Kumaraswamy, M.M.; Hewa Welege, N.M.; Pan, W. Accelerating the Delivery of Low-Carbon Buildings by Addressing Common Constraints: Perspectives from High-Rise, High-Density Cities. *Buildings* **2023**, *13*, 1455. [CrossRef]
10. Wang, Y.; Sun, R.; Ren, L.; Geng, X.; Wang, X.; Lv, L. Risk Propagation Model and Simulation of an Assembled Building Supply Chain Network. *Buildings* **2023**, *13*, 981. [CrossRef]
11. Yan, H.; Yang, Y.; Lei, X.; Ye, Q.; Huang, W.; Gao, C. Regret Theory and Fuzzy-DEMA^{TEL}-Based Model for Construction Program Manager Selection in China. *Buildings* **2023**, *13*, 838. [CrossRef]
12. Bailey, S.; Carnemolla, P.; Loosemore, M.; Darcy, S.; Sankaran, S. A Critical Scoping Review of Disability Employment Research in the Construction Industry: Driving Social Innovation through More Inclusive Pathways to Employment Opportunity. *Buildings* **2022**, *12*, 2196. [CrossRef]
13. Wang, B.; Chen, H.; Ao, Y.; Liao, F. Spatiotemporal Differentiation and Influencing Factors of Green Technology Innovation Efficiency in the Construction Industry: A Case Study of Chengdu-Chongqing Urban Agglomeration. *Buildings* **2023**, *13*, 73. [CrossRef]
14. Mridha, M.; Kuys, B.; Suhaimi, S.N. The Influence Innovation Has on the Visual Appearance and Aesthetic Preference of Architectural Products. *Buildings* **2023**, *13*, 19. [CrossRef]
15. Zhang, J.; Dong, F.; Ballesteros-Pérez, P.; Li, H.; Skitmore, M. Current and Future Trends of Resource Misallocation in the Construction Industry: A Bibliometric Review with Grounded Theory. *Buildings* **2022**, *12*, 1731. [CrossRef]
16. Li, Y.; Liang, J.; Huang, J.; Yang, M.; Li, R. Behavioral Research in Construction Engineering Management: A Review from a Neuropsychological Perspective. *Buildings* **2022**, *12*, 1591. [CrossRef]
17. Huaman-Orosco, C.; Erazo-Rondinel, A.A.; Herrera, R.F. Barriers to Adopting Lean Construction in Small and Medium-Sized Enterprises—The Case of Peru. *Buildings* **2022**, *12*, 1637. [CrossRef]
18. Li, H.; Han, Z.; Zhang, J.; Philbin, S.P.; Liu, D.; Ke, Y. Systematic Identification of the Influencing Factors for the Digital Transformation of the Construction Industry Based on LDA-DEMA^{TEL}-ANP. *Buildings* **2022**, *12*, 1409. [CrossRef]
19. Doukari, O.; Seck, B.; Greenwood, D. The Creation of Construction Schedules in 4D BIM: A Comparison of Conventional and Automated Approaches. *Buildings* **2022**, *12*, 1145. [CrossRef]
20. Bahamid, R.A.; Doh, S.I.; Khoiry, M.A.; Kassem, M.A.; Al-Sharafi, M.A. The Current Risk Management Practices and Knowledge in the Construction Industry. *Buildings* **2022**, *12*, 1016. [CrossRef]
21. Li, Y.; Xie, Y.; Sun, S.; Hu, L. Evaluation of Park Accessibility Based on Improved Gaussian Two-Step Floating Catchment Area Method: A Case Study of Xi'an City. *Buildings* **2022**, *12*, 871. [CrossRef]
22. Wang, J.; Guo, F.; Song, Y.; Liu, Y.; Hu, X.; Yuan, C. Safety Risk Assessment of Prefabricated Buildings Hoisting Construction: Based on IHFACS-ISAM-BN. *Buildings* **2022**, *12*, 811. [CrossRef]

23. Zhang, Y.; Yi, H.; Xie, H.; Zheng, J.; Wang, Y. How Private Enterprises' Participation Behaviors Evolve with Incentive Modes in PPPs: An Evolutionary Game View. *Buildings* **2022**, *12*, 709. [CrossRef]
24. Xu, H.; Chang, R.; Pan, M.; Li, H.; Liu, S.; Webber, R.J.; Zuo, J.; Dong, N. Application of Artificial Neural Networks in Construction Management: A Scientometric Review. *Buildings* **2022**, *12*, 952. [CrossRef]
25. Wang, L.; Rong, X.; Chen, Z.; Mu, L.; Jiang, S. A Real Estate Early Warning System Based on an Improved PSO-LSSVR Model—A Beijing Case Study. *Buildings* **2022**, *12*, 706. [CrossRef]
26. Yang, L.; Hu, X.; Zhao, X. Organization Synchronization in Response to Complex Project Delays: Network-Based Analysis. *Buildings* **2022**, *12*, 662. [CrossRef]
27. Huang, J.; Zeng, X.; Fu, J.; Han, Y.; Chen, H. Safety Risk Assessment Using a BP Neural Network of High Cutting Slope Construction in High-Speed Railway. *Buildings* **2022**, *12*, 598. [CrossRef]
28. Elrefaey, O.; Ahmed, S.; Ahmad, I.; El-Sayegh, S. Impacts of COVID-19 on the Use of Digital Technology in Construction Projects in the UAE. *Buildings* **2022**, *12*, 489. [CrossRef]
29. Yan, H.; Zheng, Z.; Huang, H.; Zhou, X.; Tang, Y.; Hu, P. Risk Coupling Evaluation of Social Stability of Major Engineering Based on N-K Model. *Buildings* **2022**, *12*, 702. [CrossRef]

buildings

MDPI

Article

Optimizing Interfaces of Construction Processes by Digitalization Using the Example of Hospital Construction in Germany

Sabine Hartmann [1,*], Dirk Gossmann [2], Suzan Kalmuk [2] and Katharina Klemt-Albert [1]

1 Institute for Construction Management, Digital Engineering and Robotics in Construction, RWTH Aachen University, 52070 Aachen, Germany
2 Faculty of Civil Engineering, RWTH Aachen University, 52062 Aachen, Germany
* Correspondence: hartmann@icom.rwth-aachen.de; Tel.: +49-2418025145

Abstract: In hospital construction, additional challenges must be considered, such as an increased number of stakeholders and building trades, such as medical and laboratory technology. Due to the increasing requirements and challenges, associated construction processes are becoming more intricate. Especially for complex building types, the effects of this development are clearly noticeable and cause considerable disruptions to the construction process. A main difficulty constitutes the missing definition of the interfaces of building trades and participants. In the present study, interfaces in hospital construction were identified and analyzed by guided interviews with experts from the health sector. The qualitative content analysis, according to Mayring, was used for the evaluation to derive appropriate solution approaches. This paper presents the interfaces using the example of hospital construction in Germany and general approaches of optimization. Hereby, the digital method Building Information Modeling (BIM) plays a decisive role in the optimization of interfaces, especially in complex buildings. Furthermore, a task and building trade control matrix is required to better coordinate the interfaces. The identified approach intends to alleviate potential disputes and misunderstandings among stakeholders, as well as to improve time and financial predictability, which are particularly valuable during inflationary periods.

Keywords: digitalization; BIM; hospital construction; interfaces; optimization; critical infrastructure; building information modeling; healthcare management; project management

Citation: Hartmann, S.; Gossmann, D.; Kalmuk, S.; Klemt-Albert, K. Optimizing Interfaces of Construction Processes by Digitalization Using the Example of Hospital Construction in Germany. *Buildings* **2023**, *13*, 1421. https://doi.org/10.3390/buildings13061421

Academic Editors: Simon P. Philbin, Yongjian Ke and Jingxiao Zhang

Received: 20 March 2023
Revised: 20 May 2023
Accepted: 21 May 2023
Published: 30 May 2023

1. Introduction

Over the past 20 years, productivity in the global economy has increased by about 2.8% annually [1]. The manufacturing industry was able to achieve an increase of 3.6%, while the construction industry could increase productivity by only 1%. In 2020, global construction recorded a volume of over $12 trillion US dollars (USD) and is expected to register a Compound Annual Growth Rate (CAGR) of about 7% until 2030 [2] (pp. 1–3). Despite positive forecasts and immense technical progress, the construction industry faces major deficits in the coordination of construction measures, resulting in longer construction times and higher costs. In order to remedy these deficits and optimize construction processes, digitalization in the construction industry is inevitable, but it poses major challenges. Nonetheless, digitalization offers great opportunities for future developments in this sector.

1.1. Issue

Due to the uniqueness of each building, the requirements for the realization of construction projects are highly complex nowadays. Therefore, construction projects require process optimization of the workflows already in the planning phase. This is necessary due to the scope and desired speed of realization of projects, as well as the number of stakeholders involved. Additional pressure on costs and deadlines, as well as unforeseen changes in conditions, increase the need for information and communication in construction projects. Otherwise, the quality of design and execution will be compromised [3] (p. 173).

A significant problem in construction projects is further planning during construction. Planning is often overtaken by reality, and coordination between the stakeholders is often difficult to reconcile with the construction process. Coordinating the provision of construction services is an additional and special challenge in large-scale projects [4]. Furthermore, there are supply bottlenecks due to impending resource constraints or logistical difficulties caused by the pandemic. In the following, three worst-case scenarios from projects in Germany are described, which, due to significant design errors, have led to a drastic increase in costs and a considerable delay in project completion.

A large number of design and construction errors were discovered at the new Berlin airport, which opened with a delay of nine years. First, construction work had already been commissioned, although the necessary building licence for the construction had not been issued by the local authorities. Secondly, the planning phase had not been completed before the start of construction, so modifications had to be made during the construction phase. The subsequent changes caused high costs that had not been included in the calculations. Thirdly, the construction was characterized by considerable deficiencies in execution. A particular difficulty was the fire protection, which was affected by the technical problems of the fire alarm systems and incorrect installation of cabling. Some of the construction deficiencies could be traced back to inadequate or even missing construction supervision, which delayed consultations and impeded proper interface coordination [5]. In total, the planning and construction costs increased from an estimated 1.9 billion euros (EUR) to 5.9 billion euros (EUR) in 2020, when the airport was finalized [6].

The Elbphilharmonie, a concert hall in Hanover, built from 2007 to 2016, is another well-known construction project due to its increased costs. At the beginning of planning, the opening was envisaged in 2010, whereby the total costs were estimated at around 186 million euros (EUR) [7]. Due to a multitude of poorly coordinated contracts and interfaces between executing companies, the project duration was delayed by about 6 years and led to construction costs of 789 million euros (EUR) [8].

The project Stuttgart 21, entailing the rearrangement of a railway junction in the city of Stuttgart, represents another example of the effects of inadequate planning and coordination. The scheduled project completion, initially planned for the end of 2019 to thebeginning of 2020, has been postponed several times. Meanwhile, the opening of Stuttgart's main station is scheduled for December 2025, and other parts of the project are scheduled for even later [9]. The costs of the project increased from 2.6 billion euros (EUR) to 9.15 billion euros (EUR) [7]. The extent of the variations in costs of the described projects can be illustrated in Figure 1.

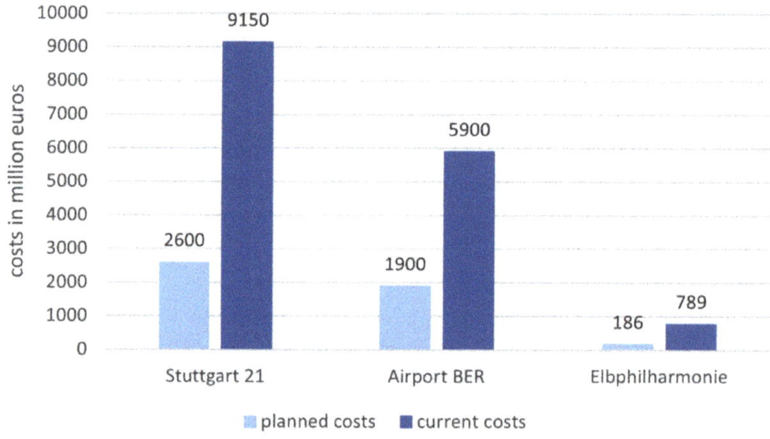

Figure 1. Planned and current costs of large-scale projects in Germany in 2017 (own illustration).

In Germany, the analyzed projects are classified in a special category for complex buildings, so-called "Sonderbauten", due to their size and type of use. Compared to residential and administrative buildings, projects in this category must fulfill particularly high requirements, e.g., in the area of fire protection. With increasing complexity of the regulatory and constructional requirements of a project, planning becomes more extensive with an increased risk of potential errors. As part of the critical infrastructure, hospitals in Germany also fall into the category of so-called "Sonderbauten" due to their special structural and regulatory requirements that need to be considered during planning and in further life cycle phases [10]. In addition, there is a multitude of use-specific requirements and stakeholders, such as doctors and the extra trade of medical technology.

The specific challenges in hospital construction can, for example, be presented by the construction project Centre for Operative Medicine 2 (ZOM II) in Düsseldorf. The project shows that hospital projects are prone to planning errors and associated execution deficiencies, mainly due to their high complexity. The ZOM II of the University Hospital Düsseldorf was mainly designed for surgically oriented hospitals. Five different clinics are integrated in ZOM II, including a neurosurgical clinic and an orthopaedic clinic. The clinics have a joint wing for surgeries with ten operating rooms. Furthermore, a central emergency department, a helicopter landing pad on the roof, as well as an intensive and intermediate care ward, are available [11]. For those patients who do not require intensive care, there are 288 beds in modern two-bed rooms [12]. Due to the organisation of these different subject-specific areas, the hospital requires distinct supply and logistics planning for the ongoing operational phase. The building of ZOM II was opened in 2014, with a delay of four years, and the additional costs amounted to about 180 million euros (EUR) [13]. University Hospital Düsseldorf has suffered a great loss due to the delayed commissioning. For example, equipment that had already been purchased could not be used and was already outdated at the time of commissioning. Additional burdens resulted from hiring new staff, which could not yet be employed due to the above-mentioned delays, but they, nevertheless, claimed their wages. The court proceedings, which have involved damages of 63 million euros (EUR), have not been concluded yet [14].

Effects of delays are particularly serious when construction work is carried out on existing buildings during the ongoing operational phase in hospitals. Construction delays can cause immense effects on the ongoing hospital operation and should be prevented to ensure the wellbeing and appropriate care of patients. In this context, the quality of the construction work should not be neglected under any circumstances, as otherwise harmful effects on the patient's health may occur (e.g., impure hospital hygiene). Furthermore, redundant protection of the functionality of vital equipment and devices in hospitals is necessary at all times to maintain supply in the event of technical malfunctions or emergencies.

It should be kept in mind that the medical field is in a constant process of change, which is reflected in even shorter innovation cycles in medical technology. This is accompanied by an increase in the necessary investment costs for the construction and operation of hospitals to be able to guarantee high-quality health care. The constantly changing medical, political, and constructional requirements often lead to hasty decisions. Therefore, hospital operators are unable to meet the demand with sustainable safeguarding. However, every renovation measure in a hospital leads to a strong restriction of the ongoing operation. The aim is to create efficient, sustainable, and goal-oriented planning by involving all necessary stakeholders at an early stage of planning to keep the costs, the effort, and the high complexity in hospital construction manageable. Thus, all relevant requirements should be considered from the very beginning, and interfaces can be defined in advance [15] (p. 3).

Renovation measures, extensions, or refurbishments can usually take place simultaneously with the ongoing operation of a hospital. By taking certain protective measures, hospitals can maintain their day-to-day operations. Therefore, the inconvenience caused by the construction work, such as noise and soiling, must be curbed and adapted to the operational conditions [16].

1.2. Purpose

Due to these comprehensive effects, not only in terms of construction, but also in terms of social relevance, there is a huge need for action in the handling of project management in construction, according to Viering et al. in 2007 [17]. In complex buildings, the planning and definition of interfaces in the construction process play a vital role. Interfaces are points of contact between different matters or objects and are also defined in conventional linguistic usage as a connection or transition point between two processes or, in more general terms, between two areas. In construction practice, this consideration cannot be limited to two units, but it must be extended to a chain of units, which occur, among other things, in the form of tasks, activities, and responsibilities of the stakeholders.

Wukonig described, in 2011, how interfaces in construction projects arise not only between the stakeholders, but generally between functionally separated areas and tasks of a project [18] (p. 6f.). The points of contact arise in the subtasks of the different design work, the execution of different trades, as well as monitoring and control tasks. In many cases, this is due to the contractually demarcated areas of liability and responsibility. For coordinating interfaces, a proper and time-critical communication between all stakeholders is indispensable, starting with the definition of uniform terms. Otherwise, technical terms or concepts can be interpreted differently and cause confusion and misunderstandings. The aim of communication is primarily the exchange in information, as well as the identification of information gaps, to link the created subtasks with each other. In order to avoid diverging interpretations, there is a need to control the flow of information in a construction project so that information gaps are kept as narrow as possible. Control is achieved by defining the availability of data for stakeholders to avoid contradictory information and to ensure that the flow of information does not become excessive, as well as that the actual activities are not neglected. Further control elements are available in the form of checklists, tables, or as plans with standardised presentation rules [18] (p. 78f.).

In 2015, Lin developed the ConBIM-Interface Management system to enhance interface information sharing and tracking efficiency. It uses three-dimensional interface maps and the BIM approach to track and manage interfaces in a graphic form [19]. In 2018, Alaloul et al. presented potential benefits, as well as challenges and strategies, for the introduction of smart technologies in the construction process [20]. Whereas, in 2022, Safikhani et al.'s paper highlighted the concrete benefits to the construction process of using VR technology in combination with the digital method Building Information Modeling (BIM) [21]. In 2022, Sun and Liu proposed a novel hybrid model of Digital Twin-Building Information Modeling. The model helps in assisting the dispatch systems in the construction projects to a greater extent than when compared to the implementation of individual technologies [22]. In an article from 2022, Li et al. proposed the novel idea of Hospital Information Modeling, an expansion of BIM for hospital settings [23]. Sepasgozar et al. presented a roadmap for developing and implementing disruptive technologies for the construction industry, in 2023 [24]. Various digital technologies, such as BIM or augmented reality (AR), are presented to optimize processes.

In this article, the process of construction measures is itemised regarding their problems concerning the reduction of the effects of planning and construction faults in the future. These difficulties are identified and analyzed using the example of hospital construction in Germany, as there is a particularly large number of interfaces to be defined due to the special characteristics of the use and requirements arising from critical infrastructure. Based on this, solution approaches are developed to optimize these interfaces and to reduce mistakes in the planning phase. In developing the solutions, digital methods are used to simplify the flow of information and communication and, thus, to intervene at the source of the interface problems. The solution approaches are transferable to other complex buildings with infrastructural, as well as social, significance.

2. Methodology

The literature research was carried out to obtain information about the status quo regarding interfaces in construction process and possibilities for optimization using digital methods for describing the research gap and the purpose of this article. The issue was dealt with using the example of hospital construction, since, due to the complexity caused by a higher number of trades, stakeholders require more interfaces than in other types of building. Since this work was developed as part of the research project KlinikBIM, whereby a guideline for the implementation of BIM in hospital construction in Germany is being created, the requirements for hospital construction in Germany were taken into account. In order to obtain background information, including the legal framework, as well as the basic requirements of construction measures in hospitals, a review of the literature was carried out.

Based on these findings, an analysis of interfaces in hospital construction processes was conducted through ten guided interviews with experts on the topic, who are from Germany. They are also part of the KlinikBIM project consortium. The interviewed experts are experienced in the field of hospital construction and belong to different subject-specific areas, such as structural design, architecture, or the construction and operations department of the operator. Thereby, the intention was to identify interfaces in hospital construction, as well as to obtain an opinion on digitization topics and their potential for optimization. The interviews were conducted by video conference with the use of a guideline. Although the questions were specified, the interviewees were still able to answer freely and give examples. The interviewer was also given the opportunity to ask questions in this regard. Mayring's qualitative content analysis was chosen for the evaluation of the guided interviews, which aim to filter out a certain structure from the material to apply it in the form of a category system.

The guided expert interviews and qualitative content evaluation offer the advantage of generating new insights while proceeding in a highly rule-governed and schematic manner. Furthermore, the described methodology enables the establishment new theoretical considerations, based on interview transcripts. Because of the underlying transparency of qualitative content analysis and its applicability to a wide variety of content, it represents an adequate tool for gaining scientific knowledge in the present case.

The evaluation method allows us to assign and to analyze the different interfaces to superordinate problems, such as errors in the planning or in the construction phase. Based on the results from the literature research and the interview evaluation, solution approaches for optimizing the interfaces could be developed. In addition, strategic recommendations for action are given.

3. Legal Background

In this research report, the legal requirements of the health care system and building law are exemplified by the German regulatory framework. Due to the federal structure of Germany, the building regulations differ in the various German states. Nevertheless, the deviations and the influence on the objective of the present research can be assessed as minor.

In Europe, the primary approach is the sovereign regulation of buildings by the state. There is a superordinate European framework of building regulations, such as EU Directive 305/2011, which regulates the marketing of construction products in the European Economic Area and defines the basic requirements for buildings as a European standard. Due to the principle of subsidiarity prevailing in the EU, the implementation of these requirements is the responsibility of the EU member states themselves. On this basis and the European interest in the exchange in goods and products, most European countries have building regulations and standards that may differ in content, but they serve to avert hazards and to ensure a functioning coexistence [25] (pp. 11 and 16).

In this article, Germany/Lower Saxony's ordinances and laws were used as an example, as this work is connected with the research project KlinikBIM, and the experts of the consortium were from Germany, predominantly from the federal state of Lower Saxony.

Besides new constructions, measures to secure and further develop existing buildings, such as reconstruction and renovation measures, were also considered. This leads to a large number of laws and regulations, which partly need to be taken into account in hospital construction and partly can be regarded as recommendations. However, since there is no clear nationwide model ordinance on the requirements for hospitals, some federal states, such as Baden-Württemberg or Brandenburg, have issued recommendations and ordinances. On the contrary, Lower Saxony follows the model hospital ordinance or works on the basis of the recommendations and ordinances of other federal states because the model hospital ordinance only serves as a guide and is thus formulated with restraint [10].

In addition to the building regulations, there is a particular fee schedule in Germany, which represents the chronological sequence of a planning process by means of service phases [26] (p. 96). In addition, the scope of services for project management is described by the German Committee of the Associations and Chambers of Engineers and Architects for the Fee Schedule (AHO) in the AHO publication series in booklet no. 9 [27]. The descriptions of the AHO and the fee schedule are not statutory requirements, but merely a standard established in the market. Furthermore, there are numerous standards that are relevant to hospital construction, for example, DIN 1946-4 for ventilation and air-conditioning systems in the healthcare sector [28], DIN 6812 for medical X-ray systems [29], or DIN EN ISO 7396-1 concerning piping systems for medical compressed gases and vacuums [30].

4. Findings from Interviews to Identify the Key Interfaces in Hospital Construction

To analyze the existing interface problems in hospital construction, the expert interviews were considered and evaluated separately. Based on the resulting findings, the interface problems can be narrowed down and analyzed. For all interface issues, the later collisions are identified, and their impact is greater. Consequently, the following evaluation is used to develop solution-based approaches that can be used to facilitate construction measures in the future.

4.1. Identification of Important Stakeholders in Hospital Construction Measures

According to the interviewed experts, the hospital is considered the building owner. Often, the hospital has its own construction department, which is responsible for planning and construction measures in the building. It represents the interface between the hospital executive board, the supervisory board, the board of directors, the physicians, and the nursing staff. Furthermore, the interests of the patients and their visitors must be taken into account. Regarding the building law, building supervision authority is the highest priority. Furthermore, municipalities, as well as representatives of public interests, participate in target planning at the state level, and they support the administrative level. They ensure that the interests of the municipality are respected, as well as that development plans are created and adapted, if necessary. Furthermore, other authorities involved, such as the nature conservation authority or the trade control, belong to the administrative level as representatives of public interests. Another representative of public interests is the fire brigade, which must be involved in fire protection measures.

The realization of the building project begins with the commissioning of the planning services. Especially, for new hospital buildings, general planners are commonly commissioned to simplify communication. However, working with architects and various specialist planners is also feasible. Due to the complexity of the planning, a consultant advises the building owner. The large number of specialized planners led to the commissioning of the executing companies, which is usually carried out as an individual contract. To assist the owner, a project manager takes over the coordination of the different stakeholders. Legal consulting can be considered for general advice or for decisive questions. The most important stakeholders in hospital construction, according to the interviewed experts, are outlined in Figure 2.

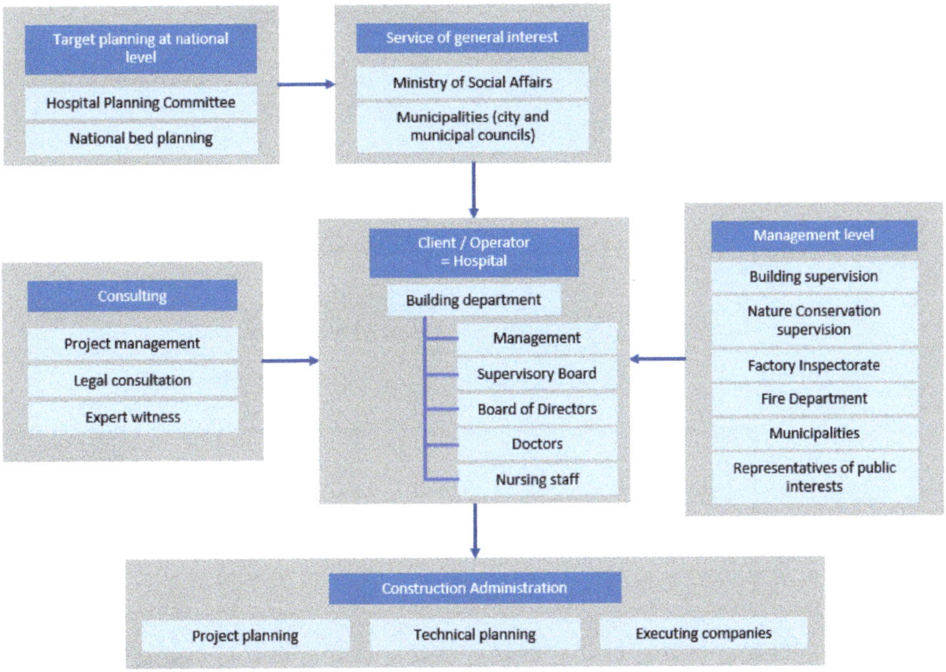

Figure 2. Stakeholders in clinic constructions (own illustration).

4.2. Main Building Trades and Interfaces in Hospital Construction

This section focuses on the various building trades and disciplines that need to be taken into account in hospital construction to avoid collisions and interface issues, according to the experts. Besides the typical collisions of the building trades of heating, ventilation and air conditioning (HVAC), as well as sanitary services and electricity, the experts have reported on many other interfaces in hospital buildings, which occur in view of high safety requirements of fire protection and hospital hygiene. Furthermore, medical and laboratory technology represent another challenge in terms of early and precise planning. In this context, it is important to ensure that reserves are planned so that dimensions and ports suffice in the event of change or replacement in equipment. Electrical engineering is one of the largest building trades involved, which requires precise planning of cable routing in coordination with the technical building equipment. A list of the most important disciplines is shown in Figure 3.

Based on the evaluation of the interviews, the main interfaces of measures in hospital construction are summarised in Figure 4, as follows.

4.3. Impact of Interfaces Using the Example of Doors

According to the experts, decisive components with high coordination efforts are the doors located in the corridors. Due to the requirements of the necessary functionality in running a hospital, these are an essential element and example of the undisturbed operational flow. Doors have high demands on fire protection and access restrictions and need to be planned and executed by different trades. For electrically operated doors, this includes the door itself, the supply line, as well as the internal wiring for controlling the door, provided by electrical engineering. An exemplary structure of an automated door system with access restriction is shown in Figure 5.

Figure 3. Integral technical planning in hospital construction (own illustration).

	Heating, ventilation, air conditioning	Sanitary	Electrical engineering	Medical technology	Fire protection	Architecture	*Doors*
Heating, ventilation, air conditioning		X	X	X	X	X	
Sanitary	X		X	X	X	X	
Electrical engineering	X	X		X	X	X	x
Medical technology	X	X	X		X	X	
Fire protection	X	X	X	X		X	x
Architecture	X	X	X	X	X		x
Doors			x		x	x	

Figure 4. Evaluation of the main interfaces and collisions, which are marked with an "×" (own illustration).

Due to the large number of planners and trades involved, this is where most of the interfaces arise, and particular attention is needed during the planning phase. According to the experts, in conventional planning, some services, such as the supply line of the doors, are forgotten to be tendered, which is only noticed during execution. Consequently, a supplement is issued, which considerably increases the additional costs due to the large number of doors in a hospital.

Another difficulty is that doors are usually components with high fire protection relevance. Doors along escape routes must be able to be opened manually in the event of fire, but they also protect against the spread of smoke in the building. Moreover, automatic opening is often required for hospital operations because patient transport in beds is part of everyday operations, especially in the corridors. Furthermore, it is necessary to consider functional areas, which may only be entered by personnel using access authorization. In this area, doors must be installed in a manner that allows them to be opened in the direction of the escape route at any time, even without the electrical presentation of authorization. In the event of fire, the doors must be able to be opened from both sides without the electrical presentation of authorization. For this reason, the wiring and the direction of installation must be considered during installation. Card readers for checking access authorization need to be installed in the immediate vicinity on the correct side and connected to the door.

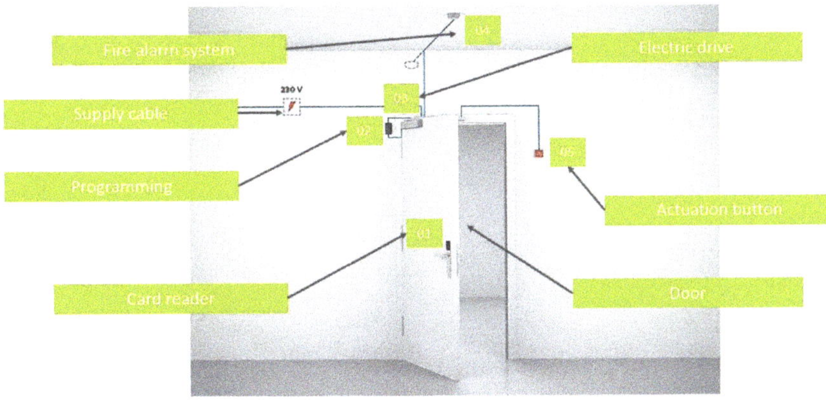

Figure 5. Structure of an automated door system with access restriction. (Adapted with permission from [31]. 2020, Dormakaba Group.)

5. Solution Approaches

5.1. Application of Digital Methods

The implementation of digitalization in the projects of the interviewed experts varies. On the basis of the expert interviews conducted, assuming that interfaces can be reduced by means of digital planning, it was analyzed whether the options listed below can lead to a reduction in interface problems.

5.1.1. Impact by Application of BIM

The BIM method is considered a pioneer of digital planning. In terms of hospital construction and the many different and complex interfaces, BIM is a suitable tool for reducing interface problems. If applied consistently, a BIM model contains all project information and keeps the data centrally in a browser- or cloud-based project space, a so-called Common Data Environment (CDE). The information attached to the individual building components simplifies the calculation of quantities and costs, e.g., by determining the number of necessary fire compartments directly from the model. This is a significant advantage, especially in hospital construction, due to the complex requirements for ventilation technology. It also simplifies the exchange in data between the stakeholders and ensures that everyone is on the same planning level to minimize the number of design errors due to a lack of communication. As a result, problematic interfaces and collisions can be identified and rectified in the early stages of planning. In a further part of this article, it is described how earlier handling of planning errors has a lower impact on costs than if it would be remedied at a later stage of planning (Figure 6).

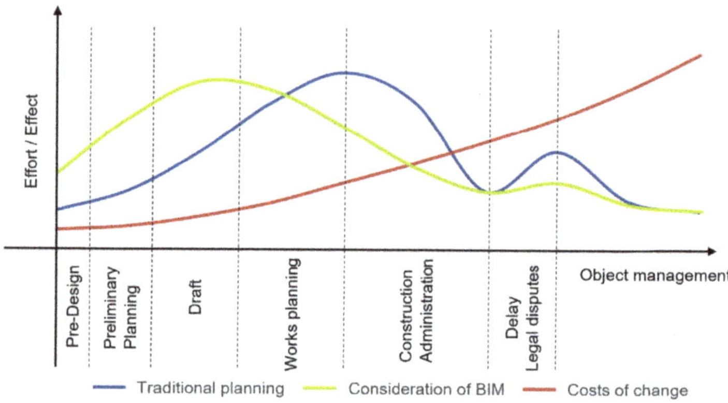

Figure 6. Effort of traditional planning and planning with BIM, according to the MacLeamy curve (own illustration based on [32] (p. 32)).

BIM also offers enormous advantages for facility management. With a digital twin, an economical and effective operation in a hospital can be facilitated. The main advantage of a digital twin can be seen in the aspect of a central data source to which everyone involved has access. With the right management, all necessary data are always available and up-to-date, so that the necessary information is available at all times in the event of repairs, maintenance, and servicing measures. However, continuous adaptation and updating of the model is necessary, already, in the operational phase. Moreover, a BIM model can only be used successfully under the condition of fixed collision rules and project-related definitions in close consultation with all stakeholders. This is performed through the elaboration of the so-called Employer's Information Requirements and the BIM Execution Planning, in which the exact procedure for using BIM in the project is described. However, these procedures are out of the scope of the present paper.

The implementation of the BIM method in Germany entails several obstacles, as revealed in the interviews. One significant problem is the lack of specifications that define the remuneration and costs of BIM services. It should also be considered that procurement costs are initially incurred for IT, as well as for the qualification of staff to use BIM. Accordingly, some hospital operators have been rather critical about the implementation of BIM so far. Furthermore, it was criticized that the actors involved in the planning process assume a higher than necessary planning effort. In contrast to the conventional method, there is no increased total planning effort with BIM, but efforts are shifted to earlier sub-phases before execution (see Figure 6). By shifting the effort into earlier project phases, the costs of unforeseen planning changes will be kept lower compared to later changes because the gradient of the cost of changes increases during the progress of the project.

Despite the positive influence of BIM on the coordination of construction projects, disruptions, such as delays due to delivery bottlenecks or unforeseen events, are still possible. Furthermore, the use of BIM creates new digital interfaces, which must also be coordinated in the already mentioned Employer's Information Requirements and the BIM Execution Planning. In the event of occurring planning errors, the stakeholders often refuse responsibility, which ultimately leads to legal disputes. These disputes can only be reduced by clarifying responsibilities at an early stage, e.g., by applying the BIM methodology.

5.1.2. BIM Use Cases

In the further course of this chapter, further digital use cases are described in order to reduce interface errors, whereby BIM constitutes a prerequisite for application. By means of checking design variants, different technical concepts are examined at the beginning of the planning regarding the investment costs. Thereby, functional influences are focused and constitute the basis of the examination for selection of the most economical variant. This is carried out in the early planning stage to determine the demand and is expected to result in an increased degree of cost and schedule certainty, as well as faster commissioning. In addition, it is possible to efficiently design and optimize the routes between the functional areas using a variant check. Operational concepts of, for instance, fire protection officers or hospital hygienists, are to be examined individually by the various professional actors regarding their feasibility and economic efficiency. Furthermore, the functional plans and designs need to be checked for compliance with current regulations in order to identify collisions or planning errors before the start of construction works.

So far, according to the interviewees, a major part of the variant checks is still carried out based on two-dimensional plans. Due to the high number of plans and concepts, as well as the complex planning requirements, it can be assumed that errors may be overlooked, despite a variant check, leading to collisions during execution. A possibility for improvement is the application of the BIM methodology, which ensures that there is only one model with all designs and information to check.

Another possibility to detect planning errors is to carry out simulations during the planning phase. The most common type of simulation involves energy considerations, e.g., to specify heating and cooling cycles. Simulations can also improve the efficiency of the planning of ventilation systems. By visualizing the architectural geometries, pipe routing can be planned and checked for economic efficiency. Another advantage of testing different geometries stands out in fire protection planning. In this way, smoke development in the event of a fire can be simulated and contained with suitable measures. If a preliminary check takes place by means of simulations, it is possible to simulate routes of insertion for medical equipment, as well as of technical systems to eliminate possible complications during planning and to adjust the dimensions. In this context, the simulative preliminary test of the so-called "active principle test", which is relevant in Germany under building regulations, is referred to. It is used to test the interaction of different technical systems in the event of a hazard.

However, up until now, simulations are often carried out with conventional software due to a lack of a BIM implementation strategy. By using the BIM methodology, there are further possibilities, in addition to the advantages mentioned above, to improve planning and construction site work processes and to represent them in an enhanced degree of detail. Virtual reality (VR) can be used to create computer-generated, interactive environments that can be displayed in real time with terminal devices, such as VR glasses, tablets, or smartphones at no great financial expense. The combination of an existing three-dimensional model with VR enables variant considerations, preliminary tests, and collisions to be displayed and checked more clearly. Special attention is paid to the flow of logistical processes. In this way, fire protection concepts, escape routes, traffic routes, or even logistical effects of construction measures can be visualized and adapted in advance.

Starting with fire protection concepts and the planning of escape routes, simulation, in combination with VR, can be useful. In the past, fire protection measures were only tested in their technical function in order to be able to exclude functional errors in the equipment. By visual inspection of the building, escape routes can be viewed in their entirety and checked regarding their stability in the event of a fire. The use of VR also provides a visualization of the development of fire and smoke, which makes necessary adjustments to the fire protection systems visible.

In addition, the traffic and transport routes must fulfill a multitude of constructional requirements. Due to the high number of these routes in a hospital, a manual check is unfeasible. Using a virtual passageway, routes, door widths, room heights, area requirements, lighting, and markings can be checked. Especially, door and corridor widths must be checked with regard to bed transport or accessibility. The introduction of new medical equipment can also be simulated at an early stage using VR, so that problems with insufficient door widths can be rectified before delivery. VR can be also used to visualize the planning for laypersons so that hospital specialists can check whether all necessary connections and equipment have been taken into account.

Construction measures during ongoing hospital operations always have an impact on the supply and functioning of a hospital. To keep these effects to a minimum, construction measures are demarcated with dust protection walls. By using virtual simulations, the effects can be determined in advance so that measures can be taken to protect patients in advance. In summary, the use of VR aims to optimize processes and to eliminate planning errors and collisions, with minimal effort at the beginning of planning, to avoid delays, to reduce additional costs, and to enforce efficient and comprehensible workflows.

While VR completely replaces reality with virtual sensory impressions, AR aims at a computer-supported extension of reality perception, where the view of the real environment is supplemented with computer-generated superimpositions. By overlaying virtual and real points, visual information, the location of components, and the positioning of reference points on the building, construction experts and laypersons can use visualization to initiate potential changes. The application of AR is not specific to a project phase and can be implemented throughout the entire building life cycle. During the planning phase, the use of AR allows representatives of different trades to view the BIM model from different perspectives. Models can be analyzed more efficiently, and potential problems can be identified by showing and hiding single units. Furthermore, the superimposition of design models, such as a technical building equipment model, with reality, enables early detection of conflict points, which can be rectified before construction work begins.

It is further possible to check and record the current status of construction by overlaying planning models with scan models of the building object. In addition, changes and construction deviations can be checked and imported back into the model to obtain an as-built model at the end of execution. In addition, the use of AR offers the option of projecting plans onto building components, e.g., in order to be able to show pipe to avoid damage during construction. Additionally, it may prove to be useful to prevent execution errors due to the attachment of components or the preparation of breakthroughs with the help of the exact visual specification.

In addition to the advantages of AR in the planning and execution phase, the possibility of virtual representation of the as-built model serves to simplify maintenance, repairs, and servicing measures in the operational phase. By feeding back the manufacturer information of maintenance-relevant components into the BIM model during the construction phase, an automatic maintenance report of the objects can be created, which reduces the workload of facility management. Maintenance technicians can then be sent to the corresponding object and quickly identify it using indoor navigation via AR glasses. This creates huge potential for reducing maintenance working hours by linking virtual step-by-step instructions to the maintenance-relevant components. In the future, there is the possibility of being able to practice standardized work steps by combining them with artificial intelligence and, thus, reducing costs with the help of standardized and routinized processes.

The digital applications mentioned above serve the communication of the project participants across the entire life cycle. Using the virtual models, location-independent meetings with a view of the construction project can be realized. Thereby, cross-trade problems can be discussed rapidly without tedious coordination effort. In addition, the use of visual forms of representation is recommended regarding visits to authorities, as well as citizens' initiatives. By overlaying the construction status with relevant information from the BIM model, use cases can be viewed together on the construction site or in the office. Particularly, with regard to the public purchasers in hospital construction, as well as the many complex requirements and conditions, the possibility of an uncomplicated joint consideration of critical points is enforced. Furthermore, due to the many necessary agreements with the various parties involved and users from other disciplines, a virtual view of the planning represents a great benefit in order to view the interactions of emerging projects in their environment and to contribute to the design.

In summary, with digital methods, such as BIM, many interface problems, such as geometric collisions, can be minimized. However, interdisciplinary tasks for the execution of construction are, at this point, difficult to coordinate.

5.2. Development of a Task and Trade Control Matrix

Due to the high level of complexity and the large number of stakeholders, but also for user-specific coordination adapted to the building, there is a need for technical and contextual coordination. Based on the findings from the expert interviews, a trade control matrix is considered a suitable tool, especially for hospital construction. Within the framework of the present study, a first draft of such a matrix was developed, which is illustrated compactly with reference to relevant information.

Up until now, control matrices have only been used for safety-related functions and their functional correlation, e.g., as a basis for the active principle test of technical equipment. The intention is to assist specialist planners, building owners, and operators, as well as executing companies, in keeping an overview and being able to present hazard scenarios. At the beginning of the planning process, a risk analysis is performed to determine possible and probable hazards in terms of their use and environment. The risk assessment must be continuously reviewed and updated throughout the entire life cycle of a building.

For the purpose of this paper, the principle of safety control matrix is transferred to the entire construction project and its stakeholders for an optimization of planning and execution. Thereby, the control matrix should be applied and updated by all stakeholders in each project phase. The associated advantages are demonstrated through clear definition of the interfaces and the corresponding responsibilities, which facilitate coordination and reduce errors in planning and execution. The precise demarcation of interfaces clarifies who is responsible for and involved in which part of the construction measures from the beginning. Nevertheless, should errors occur, those responsible can be clearly identified on the basis of the task control matrix. Another advantage of using a task and trade control matrix becomes apparent in construction supervision. Site managers and project controllers obtain an improved overview, with all the necessary information compactly summarized and organized. Furthermore, executing companies obtain more insight into their areas of responsibility and can schedule their work more easily.

A task and trade control matrix was created with specification of the building trades involved in hospital construction (compare Figure 7). In this context, it should be noted that only the main building trades are listed in this figure for simplification. As participants vary in every project, they need to be adapted specifically. In addition, heating, ventilation, and air conditioning are grouped together for a simplified illustration; likewise, the executing trades in building construction have been kept general and are to be adapted to the corresponding building component.

By specifying the trades involved and naming the subcontractors and their contact persons, an overview of the distribution of tasks and responsibilities can be created. Specifying the location of the components helps to further enable a direct assignment of the corresponding component. In this way, the division of tasks for a construction service can be applied to a large number of identical components. The matrix merely needs to be multiplied and related to its respective component. Subordinately, the construction work can be divided into its sub-services, so that an exemplary checklist for an automated door system with access restriction is created based on the fundamental requirements elaborated through the literature research and the expert interviews (see Figure 7).

Construction: Automated door system with access restriction

Location component	Building section A						…
	1. Floor Corridor-1	2. Floor Corridor-1	3. Floor Corridor-1	1. Floor Corridor-2	2. Floor Corridor-2	3. Floor Corridor-2	…

X = Responsibility
I = Indication
C = Coordination
S = Support

Task / Trade involved in construction		Client / User	Sample company & sample contact person	Technical Building Equipment - HVAC	Sample company & sample contact person	Technical Building Equipment - Sanitary	Sample company & sample contact person	Technical Building Equipment - Electricity	Sample company & sample contact person	Building construction	Sample company & sample contact person	Fire protection	Sample company & sample contact person	Medical technology	Sample company & sample contact person	Hospital hyginics	Sample company & sample contact person	Occupational safety	Sample company & sample contact person	Information (references, standards, plans, documents, etc.)
Task	**Comment**	**Responsibility**	**Status**	**Responsibility**	**Status**	**Responsibility**	**Status**	**Responsibility**	**Status**	**Responsibility**	**Status**	**Responsibility**	**Status**	**Responsibility**	**Status**	**Responsibility**	**Status**	**Responsibility**	**Status**	
Planning																				
Selection of systems		X																		
Supply cable																				
Electric drive								X												
Actuation button								X												
Card reader								X												
Fire alarm system	Cable 230V							X				C								
Delivery and installation																				
Door	Opening in direction of escape									X										
Electronic cylinder								X												
Actuation button								X												
Card reader on the wall																				
Fire alarm system								X				C								
Wiring																				
Electric drive								X												
Actuation button								X												
Card reader								X												
Fire alarm system								X				C								
Programming																				
Closing forces and speed								X										I		
Initial coding closing element								X												
Access restriction		I						X												
Fire alarm system	Automatic unlocking by alarm or power failure							X				C								
Commissioning																				
Commissioning door								X												
Signage	Note access restriction																	X		

Figure 7. Exemplary task and trade control matrix using the example of an automated door system with access restriction (own illustration).

Ultimately, the approach of a task and trade control matrix pursues the intention of developing a standardized processes for construction services and creating generally applicable interface catalogues that can be used for future construction projects. Using this task and trade control matrix, interfaces can be delineated at an early stage, thus reducing possible planning errors. By linking it to a construction schedule, supplements, disputes, and the resulting extensions to construction times to complete construction measures are reduced. In addition, it is possible to link the component-specific task and trade control matrix directly to the respective objects in the BIM model in order to visualize the state of execution. By programming the matrix of a component, it can be linked to the same component of the object and catalogued in the BIM model and identified by its location. By selecting the location of the respective component in the matrix, the correct object is highlighted in the model. The query about the status of the construction is then made based on the matrix and can be displayed in the BIM model. Thereby, a laborious search for the appropriate component in the model can be avoided, and the status of the construction work can be queried individually for each component.

Lastly, the virtual representation in the BIM model makes subtasks clearer. If this option is further combined with the possibility of VR / AR, the components can be viewed directly on the construction site or in the office. Thereby, responsibilities can be better represented and more easily documented in meetings with the stakeholders. Additionally, transparency is created regarding the construction work and its subtasks, by which the stakeholders have both a written description of the work and a visual image from every perspective. As a result, ambiguities can be minimized, and execution errors can be avoided.

6. Conclusions

The planning and execution of hospital buildings is highly complex due to the numerous requirements described. Using BIM in connection with a CDE, communication can be improved and become more transparent, which is indispensable for complex construction projects. More efficient results can be achieved through a regulated flow of information regarding the timeline and content. As a prerequisite for the application of BIM, all stakeholders need to possess the necessary software programmes and have the appropriate software knowledge and basic knowledge of the BIM application. Furthermore, the project participants should use an exchangeable format, such as Industry Foundation Classes (IFC), to fully exploit the potential for communication. Accordingly, the openBIM approach is explicitly recommended for handling.

Furthermore, the use of BIM as the basis for processing of integral building planning enables recognition and elimination of interfaces and collisions at an early stage. Due to model-based data, BIM also provides the opportunity to carry out simulations, such as the spread of fire and smoke or the evacuation of a building, to perform necessary changes early on. After all, the later interface problems are identified, the greater the impact, since the cost of change increases exponentially over the course of the project. Planning errors can be identified through automated model-based checks with BIM. However, for the identification of interdisciplinary interface errors in construction, the use of a task and trade control matrix is recommended, in which the interfaces and responsibilities are demarcated, leading to a decrease in the potential for interface errors. The intention is to link BIM models to the matrices, so that they can be assigned to the respective objects in the digital model. Although the solutions presented in this report were created based on hospital construction, they can be transferred to any other building type. Especially complex projects can benefit from this approach, as the described challenges and risks are particularly important to clearly define the interfaces. Nevertheless, an adaptation of the tasks and trades in the matrix is necessary for different building types and for each project. Furthermore, it is important to mention that the developed solution approach in this present study has not been piloted. In future, the application of the task and trade control matrix in connection with BIM needs to be examined in a pilot project, and the results need to be reflected and, if necessary, adapted. The benefits and effort are to be analyzed.

Notably, the German regulations are not yet adapted to the application of the BIM methodology. Since the introduction of the BIM method cannot be simplified by adapting a single set of rules, a considerable amount of adaptation of the legal framework in the construction industry is required. Besides the fee schedule, public procurement law, and contract design, the question of liability needs to be clarified. Especially, in hospital construction, many legal regulations must be taken into account, which are interrelated. An additional complicating factor in Germany is that many requirements must be observed and laid down in different laws and regulations, and, thus, they differ partly in the various regions of Germany. Therefore, based on the results of this work, the need for a specialized digital directive in hospital construction for national application is expressed. This should simplify the construction process regarding the merging of hospital construction guidelines and ancillary rights and present them more clearly.

To ensure the optimization of construction projects in hospitals by using BIM, strategic change management must be established in the construction departments of the hospitals in order to qualify the staff accordingly. To spread and promote the use of BIM in hospital construction and to maximally exploit the advantages of BIM, clear recommendations for the use of BIM should be formulated. Accordingly, an application-oriented guideline for the step-by-step implementation of BIM in hospital construction projects is required. The guideline should describe an adequate BIM strategy with potential use cases based on the complex requirements and framework conditions in healthcare construction. To successfully advance digitalization in the construction industry, it is vital to adapt and further develop various digital supplementary tools, e.g., for model-based simulation or the application of VR and AR. In regard to the entire life cycle of buildings, relevant data from the as-built model must be linked to the facility management software environment by digital tools. Considering that existing buildings are not designed with BIM and that there is no digital twin, difficulties emerge in the planning of reconstruction. Due to outdated or missing plans, construction measures are difficult to arrange without collisions and plan changes occurring during construction. In this respect, the elaborated solution-approaches are not always applicable to conversion measures. Further research is needed to determine the economic impact and worth of digitally scanning existing buildings for the operation phase. Thereby, it can be decided whether and in which cases digital methods are useful for renovations and extensions of existing buildings. A comparison of the economic aspects in relation to the effort involved can help to balance the disadvantages and benefits.

Author Contributions: Conceptualization, S.H., D.G. and S.K.; methodology, S.H., D.G. and S.K.; validation, S.H., D.G. and K.K.-A.; formal analysis, S.H., D.G. and S.K.; investigation, S.H. and S.K.; resources, D.G. and S.K.; data curation, S.H., D.G. and S.K.; writing—original draft preparation, S.H. and S.K.; writing—review and editing, S.H. and K.K.-A.; visualization, S.K.; supervision, K.K.-A.; project administration, S.H.; funding acquisition, K.K.-A. All authors have read and agreed to the published version of the manuscript.

Funding: This research paper was written as part of the project KlinikBIM. This project was funded by the German Federal Institute for Research on Building, Urban Affairs and Spatial Development on behalf of the German Federal Ministry of the Interior, for Building and Home Affairs with funds from the Zukunft Bau research promotion.

Data Availability Statement: The transcribed interviews are not publicly available due to privacy restrictions.

Conflicts of Interest: The authors declare no conflict of interest.

References

1. McKinsey & Company. Available online: https://www.mckinsey.com/de/news/presse/mckinsey-studie-produktivitat-der-baubranche-in-deutschland-stagniert (accessed on 29 January 2023).
2. Emergen Research (Ed.) *Global Construction Market*; Emergen Research: Surrey, BC, Canada, 2001.

3. Schönbach, R.; Klemt-Albert, K.; Aßmus, E.; Bergmann, M. Entwicklung des Masterplan BIM für Bundesbauten. *Bauingenieur* **2021**, *96*, 173–181. [CrossRef]
4. UnternehmerTUM. Available online: https://www.unternehmertum.de/themen/fokusthema-built-environment/die-zukunft-des-bauens-eine-branche-im-wandel (accessed on 29 May 2022).
5. Sieben, P. Berliner Flughafen BER vor der Eröffnung: Eine Chronik des Scheiterns. 2020. Available online: https://www.ingenieur.de/ (accessed on 21 October 2020).
6. Statista. Available online: https://de.statista.com/statistik/daten/studie/245914/umfrage/kosten-des-flughafens-berlin-brandenburg/ (accessed on 18 May 2022).
7. Statista. Available online: https://de.statista.com/statistik/daten/studie/788419/umfrage/geplante-und-aktuelle-kosten-von-grossbauprojekten-in-deutschland/ (accessed on 18 May 2022).
8. WELT (Ed.) *Etappen Beim Bau der Elbphilharmonie*; WELT: Berlin, Germany, 2017.
9. Schönhaar, L. Stuttgart 21: Aktueller Stand und Fertigstellung des Neuen Tiefbahnhofs. In BW24. 2020. Available online: https://www.bw24.de/stuttgart/stuttgart-21-aktueller-stand-fertigstellung-tiefbahnhof-bahnprojekt-stuttgart-ulm-durchgangsbahnhof-90001931.html (accessed on 10 December 2020).
10. Deja, K. *Bauvorschriften für Krankenhäuser—Sonderregelungen in Einigen Bundesländern*; Remondis Medison GmbH: Willstätt, Germany, 2020.
11. Universitätsklinikum Düsseldorf. Zentrum für Operative Medizin II. Die Einrichtungen. Available online: https://www.uniklinik-duesseldorf.de/patienten-besucher/klinikeninstitutezentren/zentrum-fuer-operative-medizin-ii/einrichtungen (accessed on 22 December 2022).
12. Universitätsklinikum Düsseldorf. Zentrum für Operative Medizin II. Die Ausstattung. Available online: https://www.uniklinik-duesseldorf.de/patienten-besucher/klinikeninstitutezentren/zentrum-fuer-operative-medizin-ii/ausstattung (accessed on 22 December 2022).
13. Sieckmeyer, D. *ZOM II: Uni-Klinik Will 27 Millionen Einklagen*; Westdeutsche Zeitung: Wuppertal, Germany, 2015.
14. Sieckmeyer, D. *Uni-Prozess: Es geht um 63 Millionen Euro*; Westdeutsche Zeitung: Wuppertal, Germany, 2016.
15. Roth, C.; Dombrowski, U.; Fisch, N.M. *Zukunft. Klinik.Bau—Strategische Planung von Krankenhäusern*; Springer Fachmedien: Wiesbaden, Germany, 2015.
16. Tabori, E.; Dettenkofer, M. *Baumaßnahmen und Krankenhaushygiene*; Springer Medizin Verlag GmbH & Springer: Berlin/Heidelberg, Germany, 2017.
17. Viering, M.G.; Liebchen, J.H.; Kochendörfer, B. (Eds.) *Managementleistungen im Lebenszyklus von Immobilien*, 1st ed.; Treubner: Wiesbaden, Germany, 2007.
18. Wukonig, T. Schnittstellenmanagement in der Bauprojektabwicklung. 2011. Available online: https://diglib.tugraz.at/download.php?id=576a71512468f&location=browse (accessed on 10 May 2023).
19. Lin, Y.-C. Use of BIM approach to enhance construction interface management: A case study. *J. Civ. Eng. Manag.* **2015**, *21*, 201–207. [CrossRef]
20. Alaloul, W.S.; Liew, M.S.; Zawawi, N.A.W.A.; Mohammed, B.S. Industry Revolution 4.0: Future Opportunities and Challenges in Construction Industry. In Proceedings of the 6th International Conference on Civil, Offshore and Environmental Engineering (ICCOEE2018), Kuala Lumpur, Malaysia, 13–14 August 2018. [CrossRef]
21. Safikhani, S.; Keller, S.; Schweiger, G.; Pirker, J. Immersive virtual reality for extending the potential of building information modeling in architecture, engineering, and construction sector: Systematic review. *Int. J. Digit. Earth* **2022**, *15*, 503–526. [CrossRef]
22. Sun, H.; Liu, Z. Research on Intelligent Dispatching System Management Platform for Construction Projects Based on Digital Twin and BIM Technology. *Adv. Civ. Eng.* **2022**, *2022*, 8273451. [CrossRef]
23. Li, Y.; Pan, X.; Han, Y.; Lavy, S. From building information modeling to hospital information modeling. In *Research Companion to Building Information Modeling*; Edward Elgar Publishing Limited: Cheltenham, UK, 2022; pp. 593–613. [CrossRef]
24. Sepasgozar, S.M.E.; Khan, A.A.; Smith, K.; Romero, J.G.; Shen, X.; Shirowzhan, S.; Li, H.; Tahmasebinia, F. BIM and Digital Twin for Developing Convergence Technologies as Future of Digital Construction. *Buildings* **2023**, *13*, 441. [CrossRef]
25. Lu, S.-H. *Bauregelwerke und Baukultur—Europa und USA—Ein kulturübergreifender Vergleich*; Transcript Publisher: Bielefeld, Germany, 2019; ISBN 978-3-8376-4824-9.
26. Friedrich, F.; Heidenreich, S. *Deutsch für Architekten und Bauingenieure*; Springer Vieweg: Berlin/Heidelberg, Germany, 2021. [CrossRef]
27. German Committee of the Associations and Chambers of Engineers and Architects for the Fee Schedule. *Booklet No. 9: Projektmanagement in der Bau- und Immobilienwirtschaft—Standards für Leistungen und Vergütung*, 5th ed.; Reguvis: Cologne, Germany, 2020; ISBN 978-3-8462-1120-5.
28. *DIN 1946-4*; Ventilation and Air Conditioning—Part 4: Ventilation in Buildings and Rooms of Health Care. German Institute for Standardization. Beuth: Berlin, Germany, 2018. [CrossRef]
29. *DIN 6812*; Medical X-ray Equipment up to 300 Kv—Rules of Construction for Structural Radiation Protection. German Institute for Standardization. Beuth: Berlin, Germany, 2021. [CrossRef]
30. *DIN EN ISO 7396-1*; Medical Gas Pipeline Systems—Part 1: Pipeline Systems for Compressed Medical Gases and Vacuum. German Institute for Standardization. Beuth: Berlin, Germany, 2019. [CrossRef]

31. Dormakaba Group. Available online: https://www.dormakaba.com/de-de/produkte-loesungen/systemloesungen (accessed on 21 June 2022).
32. Egger, M.; Hausknecht, K.; Liebich, T.; Przbylo, J. BIM-Leitfaden für Deutschland. In *Research Companion to Building Information Modeling*; Edward Elgar Publishing: Cheltenham, UK, 2013.

buildings

MDPI

Article

Systematic Identification of the Influencing Factors for the Digital Transformation of the Construction Industry Based on LDA-DEMATEL-ANP

Hui Li [1,*], Zhengji Han [1], Jingxiao Zhang [2,*], Simon P. Philbin [3], Die Liu [1] and Yongjian Ke [4]

1 School of Civil Engineering, Chang'an University, Xi'an 710061, China
2 School of Economics and Management, Chang'an University, Xi'an 710061, China
3 Nathu Puri Institute for Engineering and Enterprise, School of Engineering, London South Bank University, London SE1 0AA, UK
4 School of Built Environment, University of Technology Sydney, Sydney, NSW 2007, Australia
* Correspondence: lihui9922@126.com (H.L.); zhangjingxiao964@126.com (J.Z.)

Abstract: There is an urgent need to improve our understanding of the digital transformation of the construction in order to leverage the benefits of the wider adoption of the Industry 4.0 paradigm. However, there is a lack of systematic research on the digital transformation pathway of the construction industry. In view of this, this study uses the LDA theme model to explore the key influencing factors for the digital transformation of the construction industry and builds a digital comprehensive evaluation system of the construction industry with DEMATEL-ANP. The findings are as follows: Firstly, five elements of the construction industry, such as enterprise resources, enterprise capabilities, enterprise spirit, macro environment and industry environment, have an important impact on the digital transformation of the construction industry. Secondly, the ability of construction enterprises has the most significant influence on the digital transformation of the construction industry. This empirical study provides policy suggestions and an implementation framework for realizing high-quality development of the construction industry based on digital technological innovation. The study helps construction enterprises to understand the necessity of digital transformation and provides a theoretical basis and practical ideas for construction enterprises to formulate their own digital transformation strategies.

Keywords: construction industry; digital transformation digitalization; influencing factors; LDA thematic model; DEMATEL-ANP

Citation: Li, H.; Han, Z.; Zhang, J.; Philbin, S.P.; Liu, D.; Ke, Y. Systematic Identification of the Influencing Factors for the Digital Transformation of the Construction Industry Based on LDA-DEMATEL-ANP. *Buildings* **2022**, *12*, 1409. https://doi.org/10.3390/buildings12091409

Academic Editor: Farook Hamzeh

Received: 15 July 2022
Accepted: 6 September 2022
Published: 8 September 2022

1. Introduction

As part of the fourth industrial revolution or so called Industry 4.0, digital technologies such as big data, artificial intelligence, block chain, Internet of Things and machine learning have spread widely around the world and have the potential to transform how companies operate in the digital context [1–3]. The spread and application of digital technology has promoted the development of companies digitalization [4]. With the ongoing intensification of global competition, the digital transformation of companies has become a general trend in many countries [5–7]. This trend includes not only the application of digital technology, but also the profound change of a business model [4]. Consequently, the digital transformation of traditional industries has become key for enterprises to secure and maintain competitiveness [8–10]. Therefore, various countries have introduced measures to promote the digital development of enterprises. The United States has successively issued policies, such as 'National Artificial Intelligence Research and Development Strategic Plan' and 'Intelligent Manufacturing Revitalization Plan' to promote the integration of digital technology in traditional industries [11]. As a strong country in traditional manufacturing, Germany has gradually improved its digital transformation plan with Industry 4.0 as the core and launched 'Digital Strategy 2025' to promote the development of traditional manufacturing industries [12]. In 2021, China's 'Government Work Report' emphasized the need for accelerating digital development, creating new advantages of digital economy as well as promoting digital industrialization and industrial digital transformation [13].

As a traditional resource-intensive sector, the construction industry has made significant contributions to national economic development [14]. However, the inherent problems in the sector, such as high consumption, low efficiency and in some cases poor management have been seriously restricting the development of the construction industry. Promoting digital transformation of the construction industry has therefore become an inevitable requirement to enable the transformation and upgrading of the construction industry as part of a high-quality development strategy [15–17]. For this reason and in the field of engineering construction, developed economy countries have successively issued construction industry development strategies based on the adoption of digital technology, such as the Strategic Plan for Infrastructure Reconstruction (USA), the 'Build 2025' Strategy (UK) and the 'Construction Site Productivity Revolution' Strategy (Japan) [18].

Although the construction industry is actively promoting digital transformation, compared with other industries, the application of digital technology in the construction industry is still at a relatively low level and subject to certain challenges [19]. The overall digital development level of the construction industry ranks last among twenty-two industries, even lower than metallurgy, mining and agriculture [18]. In the case of China and to some extent, the low level of digitalization in the construction industry hinders the high-quality development of country's construction industry and the realization of the transformation goal. Therefore, it is an urgent priority to explore the development path of digital transformation of the construction industry and clarify the key factors that impact the digital development of China's construction industry. Although academic circles have recognized the importance of developing an effective digital transformation strategy for the success of construction enterprises, nevertheless certain questions remain unanswered: What are the key factors behind the digital transformation of the construction industry? What are the structural relationships among these key influencing factors? What mechanisms are needed for the digital transformation of the construction industry? Consequently, the above questions and corresponding propositions need to be explored both theoretically and empirically.

As a traditional industry with inherent problems, such as low productivity, periods of negative growth, high consumption of resources and energy, and relatively low levels of scientific and technological focus, the digital transformation of the construction industry is potentially a long and complicated process; during which the sector will face challenges in technology, resources, ability, organization, culture and management [20]. Indeed, from a practical point of view, the digital transformation of the construction industry is still in its infancy. Furthermore, a typical position of "I don't know how to do it" is an important problem faced by construction enterprises, which is an active area of investigation for various researchers.

There are two main focus areas for current research on this matter. Firstly, there is research on the impact of the digital technology application layer on the digital transformation of the construction industry. For instance, Zhen Xu et al. studied the application of 3D printing technology combined with a construction robot in the construction industry as well as integration with building materials and building environment to realize green and sustainable building construction and digital transformation [21]. Through a process of literature review and bibliometrics, Gleiliu et al. identified that the current level of digital intelligent construction is still in its infancy, and the key to digital transformation is to harness the advantages of intelligent technology to promote the productivity and management performance of construction projects [22]. Whereas Seungwon Baek et al. studied the application of NLP (natural language processing) technology in the field of architecture, and identified that NLP could have a potential role in building engineering and management [23].

Secondly, there is research using qualitative analysis to explore the impact of digital transformation on the construction industry from the macro level. For instance, Han Qiuming and others discussed the development mechanism of industrial intelligence through a method based on expert interviews and identified that strengthening policy guarantees and improving infrastructure can potentially accelerate the rate of digital transformation [24]. Whereas Chen Ke studied the participants and policy tools of construction industry transformation and identified that there is a need to consider the different characteristics of the governmental entities, construction enterprises, universities and scientific research institutions as well as and digital solution providers. This is required to promote the development of digital transformation in the construction industry through leveraging interactions between government and university-industry collaborative research, thereby injecting vitality into the digital transformation of the construction industry [25]. Further, Xie Xianqi and others studied the development path for the new generation of building quality and safety management systems, identified that digital twin technology has good potential in the construction industry and is an important driving force for the digital transformation of the construction industry [26]. On the whole, the existing research often focuses on one factor, or some set of factors, but does not necessarily build a complete system of key influencing factors for the digital transformation of construction industry. At the same time, research from extant literature also considers the influence of internal and external factors of construction enterprises but does not further explain which factors are more worthy of attention.

Therefore, this research study is based on the global perspective, and attempts to obtain more comprehensive research literature relating to the digital transformation of the construction industry, which is filtered and processed before being imported into LDA thematic model. Through the LDA theme model, the research theme of digital transformation of construction industry is excavated, and the key influencing factors of digital transformation of the construction industry are revealed. The DEMATEL method is used to analyze the mutual influence of each factor, and a comprehensive influence matrix among each factor is established. Finally, the ANP network hierarchy is established according to the influence relationship obtained by DEMATEL method. Thereafter it is imported into SD software for calculation, and the weight of each influence index is obtained, so as to quantitatively evaluate the key influencing factors for the digital transformation of the construction industry.

The research contributions of this study can be considered according to the following aspects: (1) Research on the path of digital transformation of construction industry has been improved from a global perspective. (2) An in-depth discussion on the mechanism of various factors affecting the digital transformation of the construction industry is provided. (3) The comprehensive impact index system of digital transformation of the construction industry is constructed. (4) The study provides a reference for government departments to formulate industrial policies for the transformation of the construction industry.

2. Analysis of Key Terms

2.1. Digital Transformation

In recent years, digital transformation has gradually become a core strategic direction of global technological change. Indeed, the governments of various countries have successively introduced digital strategies to guide the development of digital technology and promote digital transformation [27]. For example, in August 2015, Singapore released the "Smart Country 2025 Plan", which focuses on using artificial intelligence (AI) and data science, immersive media, Internet of Things (IOT) and network security technologies to improve social productivity, increase employment opportunities for highly skilled jobs, accommodate an aging population and cultivate social cohesion. In May 2016 and in order to build a stronger and safer digital Denmark, the Danish government jointly formulated and promulgated the national digital strategy deployment by the central, regional and local governments called "Digital Strategy 2016–2020". This strategy laid out a blueprint for the digital transformation of government departments, enterprises and individuals. In Australia's "Digital Transformation Strategy 2025", a specific roadmap for improving the digital service supply mode for individuals and enterprises is provided, and the current preparations and future acceleration plans for realizing the digital transformation strategy in Australia are described in detail. At the Fourth Plenary Session of the 19th CPC Central Committee, China identified that it was necessary to "establish and improve the use of the Internet, big data, artificial intelligence and other technical means for digital transformation, promote the construction of digital government, and strengthen the orderly sharing of data" [27].

However, the question of "what is digital transformation" has been a controversial topic among researchers and well-known organizations across the world. Some scholars prefer to define digital transformation from the technical level. Westerman et al. defined digital transformation as using technology to fundamentally improve the performance or influence of enterprises [28,29]. Whereas Fitzgerald et al. believe that digital transformation is the use of new digital technologies (such as social media, mobile, analytics or embedded devices) to achieve significant business improvements (such as enhancing customer experience, optimizing operations or creating new business models) [30,31]. Other scholars define digital transformation from the organizational level. In this regard, Demirkan et al. believe that digital transformation represents a profound and accelerated transformation of business activities, processes, capabilities and modes, making full use of the changes and opportunities brought by digital technology and its impact on the whole society according to a strategic priority perspective [32]. Whereas Haffke defines digital transformation as including the digitalization of sales and communication channels, which provides a new platform and a new way of interacting with customers as well as digitalization of company products (namely products and services), which replaces or adds physical products. Digital transformation also describes how to trigger tactical and strategic business movements through data-driven insights, and the introduction of digital business models, so as to realize new ways of value capture [33]. On the basis of previous studies, Vial summarized the definition of digital transformation, and holds that digital transformation refers to the process of triggering significant changes in entity attributes through the combination of information technology, computing technology, communication technology and connection technology, in order to improve the entity [34]. This definition includes four attributes of digital transformation, namely: target entity, means, scope and degree of change, and expected outcome.

2.2. Transformation of the Construction Industry

In the 21st century, the fourth industrial revolution has brought great technological and scientific progress, which embraces the use of computers and networked physical systems. The construction industry has also benefited from this progress, resulting in the concept of the digital transformation of the construction industry and this has attracted much attention in the past few years [35]. For instance, Sawhney et al. defined this phenomenon as a "transformative framework" in which three kinds of changes have taken place, viz. industrial production and construction, network physical system, and digital technology [36]. Some examples of digital technologies include building information modeling (BIM), public data environment (CDE), UAV (unmanned aerial vehicle) system, cloud-based project management, augmented reality/virtual reality (AR/VR), artificial intelligence (AI), network security, big data and analysis, blockchain and laser scanners. Within the scope of network physical system, there are robots and automation, sensors, the Internet of Things as well as workers with wearable sensors, actuators, additive manufacturing, off-site and on-site construction, and equipment with integrated sensors and embedded systems. Although the digital transformation of the construction industry does not only refer to the application of technology in the field of construction. Raihan Maskuriy and others suggest that it also includes the whole process from construction resettlement conditions to design and investment preparation, as well as the construction process itself and the operation and maintenance of buildings and the re-enactment of government construction legislation, which includes the standardization of new processes. From the project management perspective, there is a need for project and budget preparation, construction approval, construction management and specification of the complete electronic construction to be implemented for digital transformation projects. Furthermore, the principles of public construction contracts should be applied by law to ensure necessary deliver of the project management process [37].

3. Research Design

In this empirical study, the LDA-DEMATEL-ANP model is used to discover the research theme and analyze the key factors for the digital transformation of the construction industry (as shown in Figure 1). Latent Dirichlet Allocation (LDA) is a nonparametric hierarchical Bayesian model based on probability graph, which has become one of the mainstream topic models and is widely used in computing research, such as text mining. As an unsupervised machine learning method, LDA can accurately and effectively mine potential topic information in texts, and help researchers find potential topics in large-scale text information [38]. Compared with the traditional statistical analysis based on keywords, the LDA topic model is not characterized by a single co-occurrence word pair clustering, but by generating a series of terms related to the topic by probability method, digging the semantic information of the topic deeply, and measuring the intensity of the topic and the relationship between the topics by quantification. This approach can judge the development trend of the subject field more accurately. In this research study, we first extracted a series of topics related to all documents from the literature records of digital research in the construction industry, and then determined and identified 12 research topics as the key factors for digital transformation in the construction industry according to the intensity of topics and research needs.

Decision making trial and evaluation laboratory (DEMATEL) is a systematic analysis method that combines chart theory with a matrix. This process seeks the logical relationship among key influencing factors in the form of matrix through data, and calculates the influence degree and affected degree of each key influencing factor, which serves as the theoretical basis for constructing the causal relationship model among various factors [39]. In this study, we distributed questionnaires to ten experts to determine the interaction between these twelve topics.

The Analytic Network Process (ANP) is a decision-making method proposed by T.L. Saaty of the University of Pittsburgh in 1996, which adapts to the non-independent hierarchical structure. It is a new practical decision-making method developed on the basis of AHP [40]. In this study, the ANP network hierarchy is established through the influence relationship among twelve topics quantitatively calculated by DEMATEL method, and the index system of key influencing factors is constructed, and then the effective quantitative evaluation of the digital transformation of the construction industry is realized through the constructed index system of key influencing factors. This technical route has certain innovative significance for discovering the key influencing factors and constructing the index system of such factors for the digital transformation of the construction industry.

3.1. Data Sources

The Web of Science and Cnki are the data sources of the literature. In Cnki, the literature type is set as periodical. There is a need to set the professional search, and search with the theme of "construction industry" and "digitalization". In the web of science, literature type is set as papers, meeting and comprehensive papers. There is a need to set the basic search and search with the theme of "construction" and "digital" Thereafter, download all the literature information and export it in batch in Excel format, sift out the repeated, irrelevant and incomplete literature, and finally obtain a total of 50 literature. As shown in Table 1.

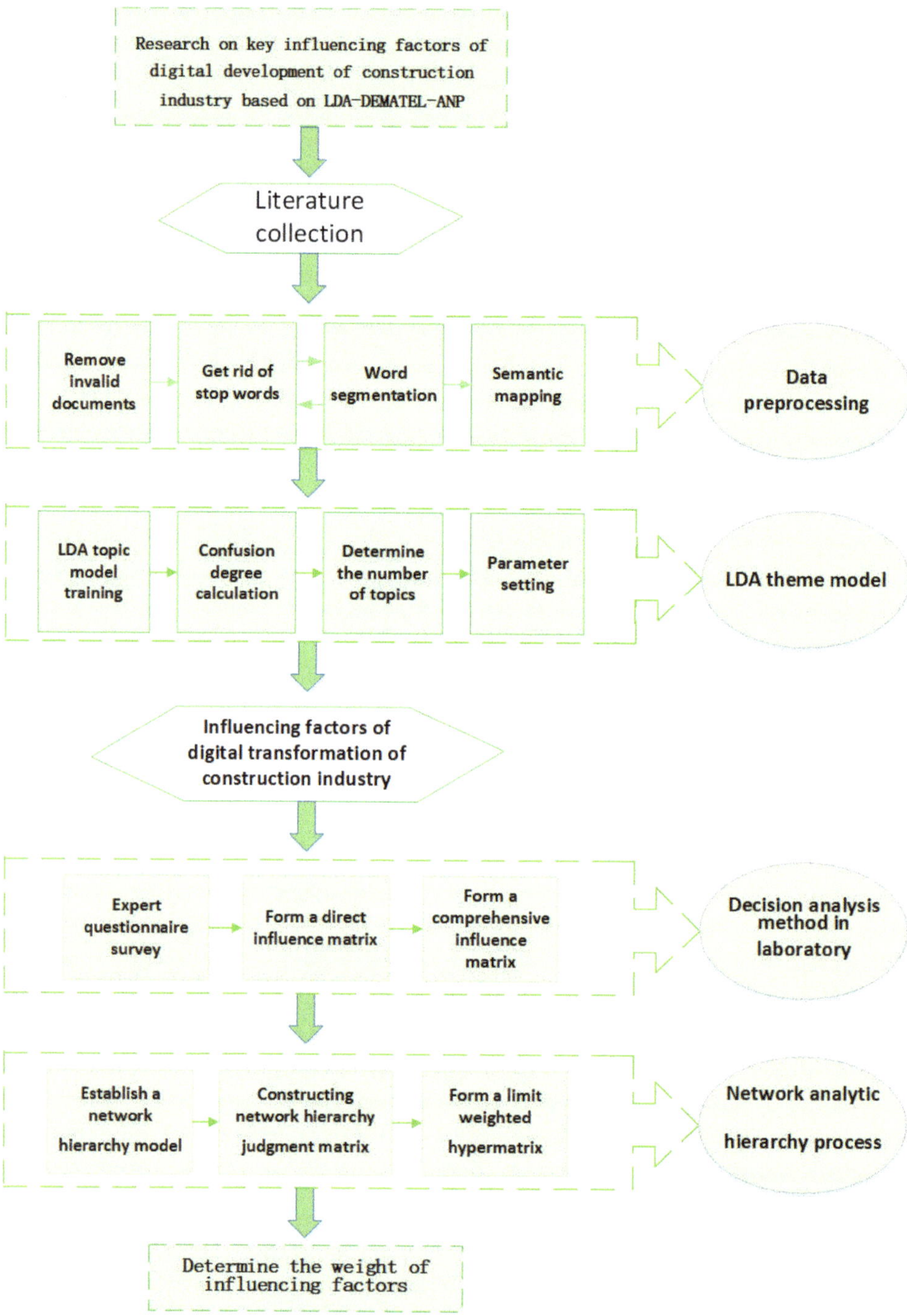

Figure 1. The research framework.

Table 1. The representative literature data.

No.	Title	Literature Content
1	A Scientometric Review of Smart Construction Site in Construction Engineering and Management: Analysis and Visualization	With the extensive development and application of information technologies in construction engineering and management (CEM), the construction site is experiencing a rapid digital revolution and transformation. Since smart construction site has become the current research trend and one of the most hot topics ...
2	Factors Influencing Adoption and Integration of Construction Robotics and Automation Technology in the US	Robotics and automation technology (RAT) has emerged as one of the most important drivers of the Industry 4.0 digital transformation of industrial operations. Although RAT has been explored i n construction over the last several decades, its applications in the field have been limited, despite its potential and rise in application opportunities ...
...	...	...
50	A critical review of text-based research in construction: Data source, analysis method, and implications	The advancement of natural language processing and text mining techniques facilitate automatic non-trivial pattern extraction and knowledge discovery from text data. However, text-based research has received less attention compared to image- and sensor-based research in the construction industry ...

3.2. Data Processing

This part of the process involves the need to extract titles, keywords, abstract and text information from literature information to form the corpus source of the LDA model. Part-of-speech analysis and part-of-speech restoration are carried out on the corpus source file with Jieba in Python, and the obtained data is preprocessed by word segmentation and stop words removal to obtain the text corpus. Secondly, subject extraction is undertaken to obtain the document-word matrix.

3.3. LDA Thematic Model Training

This involves building the LDA model with Sklearn package in Python software. Before building a model, it is necessary to determine the optimal number of topics of the model. In this study, the optimal number of topics of the model is determined by combining the model Perplexity. The calculation formula of the degree of perplexity is shown in Formula (1). Where M is the number of documents, N_d is the number of words, and $P(w_d)$ is the probability of w_d in words in documents.

$$\text{perplexity} = \exp\left\{ \frac{-\sum_{i=1}^{M} \log(P(w_d))}{\sum_{i=1}^{M} N_d} \right\} \quad (1)$$

The degree of perplexity indicates the uncertainty (i.e., information entropy) of the topic to which the document belongs. Perplexity is a standard method to measure the similarity of LDA topics [41]. When the downward trend of the degree of perplexity is no longer obvious or at the inflection point, the k value at this time is the optimal number of topics. In this study, when the number of topics is 12, 13 and 14, the perplexity of LDA model of text collection is in the lowest area. At the same time, considering the good theoretical and explanatory nature of topics, and avoiding over-classification of topics, this study set the optimal number of topics as K = 12 (as shown in Figure 2).

With regard to the setting of parameters, the Sklearn package in Python software is used to infer the distribution of topics and words. Where the number of iterations(max_iter) = 2000, algorithm for solving LDA (learning_method) is "batch", a priori parameter α(doc_topic_prior) of LDA = 50/k, k = number of topics, a priori parameter β(topic_word_prior) of LDA = 0.01 and other parameters use default values. Finally, 15 words with the highest probability under each topic are extracted, and outputs according to the word frequency from big to small, and five high-probability words under each topic are selected as the representatives of the topic meaning, which are used as the core words for topic identification.

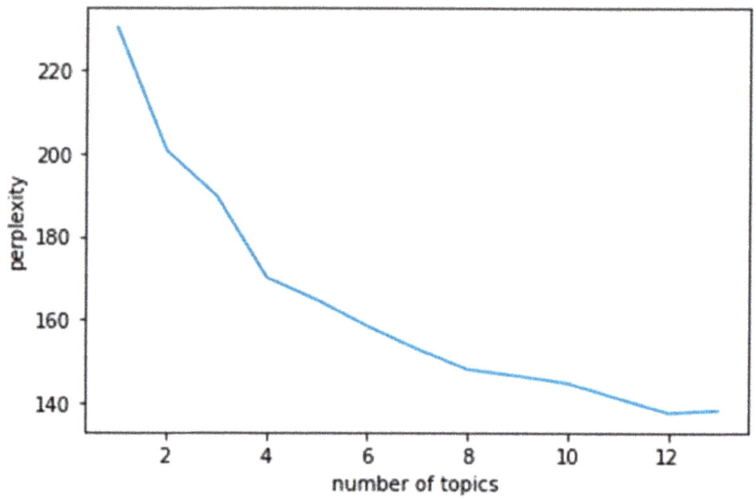

Figure 2. The trend of perplexity under different topic numbers.

3.4. Constructing Evaluation Indicators of Key Influencing Factors

Step 1: Determine the index relationship based on DEMATEL method. According to the expert level 4 DEMATEL scale, the relationship between indicators is compared pairwise, and an initial impact matrix $A = [a_{ij}]$ n × n is formed. The initial impact matrix is standardized, and a comprehensive impact matrix is formed through formula calculation.

Step 2: Establish a network hierarchy. DEMATEL method determines the significant relationship between evaluation indexes through quantitative calculation, which provides a basis for establishing network hierarchy. The network hierarchy of ANP method includes two levels, namely the control layer and network layer, and the network layer is composed of corresponding evaluation indexes and the relationship between them.

Step 3: Generate the judgment matrix. Suppose there are n indexes in the network layer, which are A1, A2, ... , An, and Ai contains secondary indexes Ai1, Ai2, ... , Aik. Then, take the element AjL in AJ as the criterion, and score the elements in the index Ai according to their action intensity on Ajl according to the nine-level scale method, so as to obtain the judgment matrix.

Step 4: Calculate the weight vector matrix. The feature vectors are transformed into standard feature vectors through standardization, and the consistency coefficient is used to test whether the judgment matrix meets the consistency requirements. If the consistency coefficient is greater than 0.1, it is considered that the judgment matrix has failed the consistency test, and it is necessary to score again to obtain new data.

Step 5: Determine the weighted hypermatrix. The standardized feature vectors of each network layer under the control of the control layer are combined together to form a super matrix W_w, and the weighted super matrix W_w is obtained by standardizing it.

Step 6: Calculate the limit hypermatrix. The Formula (2) is used to find the limit of the weighted hypermatrix W_w. If the limit is convergent and unique, the weight of each complexity index can be obtained through the limit, which also shows that the index weight can fully reflect the action relationship among the indexes.

$$\lim_{k \to \infty} w_w^k \quad (2)$$

4. Empirical Results and Analysis

4.1. Analysis of Word Frequency

Word frequency is a common technique used in text mining to evaluate the repetition of a word in a corpus. The more times a word appears in the corpus, the more likely it is to be the focus of study. Python is used to segment the text data of this study, and the words with the minimum word number of two words and the top 100 word frequency are set to make a word cloud map (as shown in Figure 3).

Figure 3. A word cloud map based on the top 100 word frequency.

In Figure 3 and according to the word size distribution, we can see that digital transformation of the construction industry has a concentration on the concept (or term) of reform, followed by administration, capital, talents and construction.

4.2. Training Results of LDA Thematic Model

After the analysis process of LDA model training and topic extraction, the optimal topics with k = 12 were finally selected, and the topic clustering and pyLDAvis visualization are carried out (as shown in Figure 4). By extracting five high-probability words under each topic, they are used as the characteristic words of this topic. It is an important process to explore the path of digital transformation of construction industry that how to effectively transform the superficial characteristic words of digital transformation of construction industry into the deep influencing factors of digital transformation of construction industry. Therefore, on the basis of synthesizing the clustering results of various topics, combining with the meanings of high-probability words, this paper consults relevant literature and consults experts to manually identify topics, and obtains the distribution of topics-terms (as shown in Table 2).

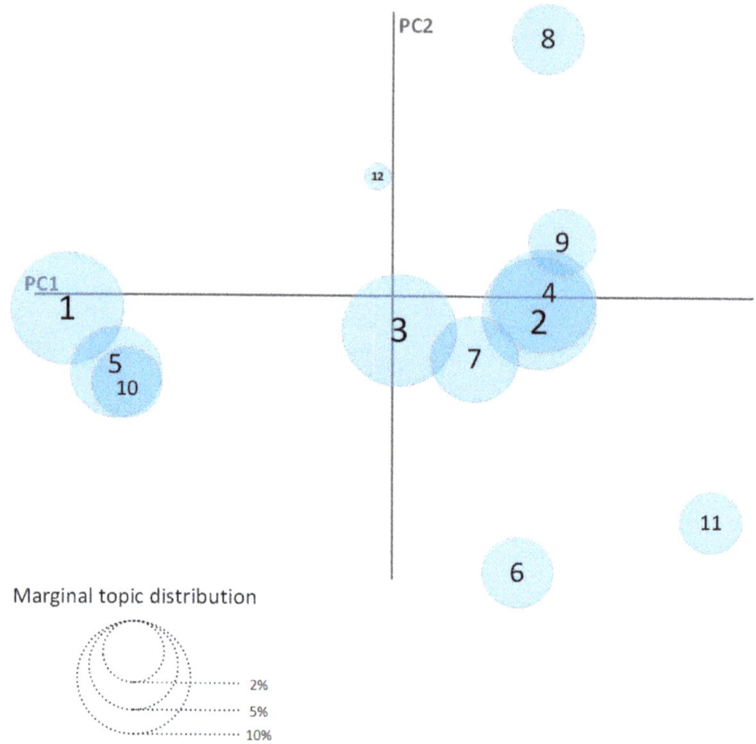

Figure 4. The visualization of Pyldavis theme-key influencing factors for the digital construction industry.

Table 2. The theme-term distribution of key influencing factors for digital transformation of the construction industry.

	Theme Identification	Document Topic Probability	Words (5 Words with High Probability Related to the Topic)
1	Digitization technology	0.128	Construction, construction site, robots, equipment, machinery
2	Policy environment	0.127	Government affairs, reform, administration, matters, cases
3	Financial strength	0.118	Investment, entity, productivity, capital, funds
4	Enterprise management level	0.105	The whole process, project management, construction site, management and control, manpower
5	International digital environment	0.105	Employees, Global, Perspective, Globalization, International
6	Hi-tech innovation capacity	0.084	Scientific research, innovation ability, core technology, vitality and competitiveness
7	Ordinary workers	0.084	Labor, labor, training, supply, skills
8	Entrepreneurship	0.067	Reform, uncertainty, managers, pressure, willingness
9	Social responsibility	0.064	Responsibility, public, benefit, public service, community
10	Green building	0.059	Green, technology, materials, ecology, construction
11	Business model	0.053	Users, partners, business processes, revenues, business models
12	Digital talents	0.006	Talent, academia, team, high quality, difficult problems.

4.3. Thematic Analysis

As can be observed from the theme identification in Table 2, are some hot topics that are emphasized in research on the digital transformation of the construction industry. Hereafter the 12 hot topics are analyzed in turn.

1. Digital technology. In recent years, digital technologies such as 3D printing technology, robotics, Internet of Things, assembly, BIM (Building Information Model) and digital twin have been widely used in construction industry [42,43]. BIM technology, in particular, has been popularized in China's construction industry, but at present, the effect of these technologies on the digital transformation of the construction industry is quite limited. Indeed, digital construction technology in China's construction industry is used for the purpose of application, but only for the bidding of engineering projects, which fails to fundamentally improve the efficiency, quality, safety and environmental protection level of engineering projects. Therefore, and in the future, it is important to promote the innovation and development of digital scientific and technological achievements through digital technology as well as enhance the core competitiveness of construction enterprises, so as to promote the structural change and digital transformation of the construction industry.

2. Policy environment. With the deepening of the digital transformation process in the construction industry, the policies and systems for the digital transformation of the construction industry have become more and more prominent, which has become a hot topic in the research and practice of the digital transformation of the construction industry. In many cases the government is the leader and regulator of the digital transformation of the construction industry. Moreover, laws and regulations have a positive impact on the digital transformation of the construction industry [44,45]. For example, the United States, Australia, countries in the European Union as well as other countries have relatively robust legal systems in terms of protecting scientific and technological innovation, and research and development of digital technologies related to the construction industry is also at the forefront of the world. Financial policies from governments, such as loan discounts, tax reduction and exemption, and financial subsidies, can also effectively stimulate the enthusiasm of construction enterprises for digital transformation [18].

3. Financial strength. In order to truly realize digital transformation, enterprises must improve and upgrade existing equipment and introduce digital talents and technologies. However, in recent years and with the global epidemic situation, rising labor costs, slowing consumer demand, and tightening of national housing policy, the continuous operating costs of construction enterprises, especially small and medium-sized construction enterprises, are constantly rising [46,47]. Faced with the natural cost of digital transformation, unpredictable transformation cycle and uncertain transformation income, the construction enterprises, which have insufficient funds, hold a cautious wait-and-see attitude towards digital transformation. Thus, capital has become a major constraint factor in the transformation of most construction enterprises.

4. Enterprise management level. Under the tide of digitalization, the market always completes self-renewal and upgrading in the process of constantly eliminating those enterprises that are inefficient and ineffective in providing value to users. The deep integration of digital technology within the construction industry is not only beneficial to the quality change, efficiency change and power change of the construction industry, but also related to the competitive advantage of enterprises in the market [34,48]. Therefore, with the full application of digitalization in the construction industry, construction enterprises are required to make adaptive management changes and adjustments to their internal organizational structure, marketing mode, product design and employment mode [49]. There is also a need to continuously improve the efficiency of value creation and supply and achieve the deep integration of digital technology and construction industry.

5. International digital environment. Nowadays, the world has entered the digital age, and the data resources provided by digital technologies, such as artificial intelligence, Internet of Things, and big data have replaced the dominant position of old production factors, such as coal and oil. The production process is more dependent on capital investment and technical support, which adversely affects the traditional resource-intensive and labor-intensive construction industry. For the traditional Chinese construction industry, this means that the low-cost dividend disappears, and the industrial space is squeezed, which potentially leads to the decline of the position of China's construction industry in the

world. On the contrary, countries with advanced technology, intensive capital and a higher education level will have more advantages over the competition [50]. Therefore, China's construction industry must seize the opportunity in the process of digital transformation and improve its position in the global value chain.

6. Scientific and technological innovation ability. With the advent of the digital age, the capabilities afforded by scientific and technological innovation has become the foundation for the survival and steady development of construction enterprises [51,52]. The economic benefit brought by scientific and technological innovation is not only a symbol to measure the survival and competitiveness of enterprises, but also a core factor to measure the market position and growth potential of enterprises. Although construction enterprises continue to increase the intensity of scientific and technological innovation, most of them are still at the level of imitation learning and technology introduction, as well as lacking core technologies with independent intellectual property rights. Indeed, the contribution rate of scientific and technological innovation has been at a low level [53], and the core technology "card neck" problem has not been alleviated. This is accompanied by the acceleration of economic globalization, the influx of internationally renowned construction contractors, the increasingly fierce competition in the construction market, and the unsustainable traditional development mode of the construction industry. Under the new situation, China's construction enterprises must adapt to the market changes and the demand of engineering science and technology and improve the level and ability of scientific and technological innovation, so as to gain the initiative in market competition and realize the high-quality development of the construction industry.

7. Ordinary labor force. In the context of an aging population and declining demographic dividend, the construction industry, as a labor-intensive industry, has received a huge impact [52]. On the one hand, existing construction workers are becoming older. On the other hand, as an important source of labor in the construction industry, the new generation of migrant workers are less and less willing to enter the construction industry [54]. As the main part of the labor force, the middle-aged and elderly workers often update their skills slowly with sometimes limited adoption of digital knowledge and technology, and often can't adapt to the mechanized and intelligent construction environment [55]. In order to alleviate the pressure of an aging construction industry, construction enterprises must speed up digital transformation, promote the transformation of production mode and the adjustment and optimization of industrial structure.

8. Entrepreneurship. Entrepreneurship is a key driving force for economic development, and it has an important influence on the practical effect of digital transformation [56,57]. Some researchers have divided entrepreneurship into two types, namely innovative entrepreneurship and arbitrage entrepreneurship [58]. Driven by innovative entrepreneurial spirit, construction enterprises actively explore the path of deep integration of digital technology and the construction industry, overcome the difficulties and obstacles in the implementation of digital transformation of construction industry, promote the cooperation efficiency of digital supply chain among construction enterprises, and continuously improve the performance of construction enterprises. In addition, by exerting the diffusion effect of entrepreneurship in management practice, it is helpful to improve the risk-taking awareness of construction enterprises, speed up the effective transformation of knowledge and information acquired by construction enterprises through the external network to the inside, and exert the knowledge spillover effect to achieve enterprise performance growth.

9. Social responsibility. As far as construction enterprises are concerned, there is a need to express a social responsibility to employees, customers, shareholders, suppliers, government and other stakeholders, which is an effective way to shape the firm's corporate image and potentially improve financial performance [59,60]. So as to better promote the digital transformation of enterprises. As far as the government is concerned, it is also willing to encourage construction enterprises to actively undertake social responsibilities and form positive feedback through financial subsidies and loan interest subsidies. In addition, enterprises' active social responsibility can not only enhance consumers' internal

perception, which has a positive effect on the digital reform of enterprises, but also enhance enterprises' external perception. Consequently, the positive response of enterprises' internal mechanism to digital reform will also be strengthened [61].

10. Green buildings. The proportion of carbon emissions in China's construction operation stage accounts for 22% of the total carbon emissions in the whole society. Therefore, energy conservation and carbon reduction in the construction sector is crucial to the realization of China's "30.60" double-carbon strategic goal [62]. In the era of the digital economy, digital technology is the best tool to achieve China's double-carbon goal [63]. In this context, it is urgent for the construction industry to step up the rate of digital transformation to achieve the goal of double carbon. In addition and according to the goal of digital transformation, green upgrading has become an important supporting force for high-quality economic development and securing people's happy and content life [64]. The demand for green upgrading of the construction industry has also forced enterprises to pursue the road of digital transformation.

11. Business model. It can be observed that innovation through digital technologies (such as the Internet of Things, big data, machine learning and artificial intelligence as well as cloud computing) continuously promote the high-quality development of China's economy, and correspondingly the traditional business model of enterprises has undergone subversive changes [65,66]. The construction industry is no exception. Facing the uncertainty and complexity of the development of the digital economy, construction enterprises must innovate their business models to identify the direction of digital technology change, secure industrial policy orientation, find competitors' threats and catch customers' demand trends, as well as quickly search for exploratory knowledge that matches the development orientation through environmental insight [67]. This is required to successfully realize the process of digital transformation, otherwise, the opportunities arising from digital transformation may not be effectively identified and eventually pursued [68].

12. Digital talents. Digital talents are an important factor of production for digital transformation. There is therefore a need to upgrade the level of talents in the construction industry, which is related to the development of digital science and technology capabilities in the construction industry [69,70]. With the development of digital theory and technology in the construction industry, the demand for talents who have a good understanding of both information and communications technology (ICT) as well as engineering technology is huge. In this context, the shortage of digital talents is one of the main factors that restricts the digital transformation of the construction industry [71]. Only by enhancing the mechanism of talent development, improving the quality of talents, and building the digital innovation mechanism of cooperation between companies and scholars under the government, equality and Industry-University-Research, can the successful digital transformation of the construction industry be realized [72].

4.4. Determination of the Relationship of Indicators Based on the DEMATEL Method

There is a mutual influence relationship among the impact indicators of the digital transformation of the construction industry. The influence relationship and degree can be determined by DEMATEL, and the indicators are recorded as Si{i =1, 2, J ... 11, 12} in order. First, 30 experts in the construction industry were given questionnaires to collect the mutual influence among the indicators of digital transformation of the construction industry judged by each expert. The 30 experts have been engaged in construction and management related work in the construction industry for many years. All of them have senior engineer and above titles, and there are many senior engineers at the level of chief engineer and professor, with rich professional experience and high credibility. The questionnaire uses the scale of 0–3 to indicate the degree of influence among indicators. See Table 3 for the scale and corresponding meanings. After calculating the arithmetic average of data from the 30 experts, the direct influence matrix A is obtained.

Table 3. Scale and meaning.

Scale a_{ij}	Meaning
0	Has no effect on Si Sj.
1	Have weak influence on Si Sj.
2	Have a moderate impact on Si Sj.
3	Have strong influence on Si Sj.

The maximum method is used to standardize the direct influence matrix *A*, and the standardized direct influence matrix X is obtained. Where a_{ij} is the value of the *i*-th row and the *j*-th column in *A*.

$$X = \frac{A}{max_{1 \le i \le 12} \sum_{i=1}^{12} a_{ij}} \quad (3)$$

The comprehensive influence matrix is calculated with the help of MATLAB software. The Formula (4) is used to calculate the comprehensive influence matrix, where T is the comprehensive influence matrix and E is the identity matrix.

$$T = X(E - X)^{-1} \quad (4)$$

In order to remove the less influential values in the digital transformation index system of the construction industry, a reasonable threshold λ is selected to process the comprehensive impact matrix T, and the processed comprehensive impact matrix D is obtained. When $a_{ij} \le \lambda$ in the comprehensive influence matrix T, $a_{ij} = 0$ is taken as the comprehensive influence matrix D after treatment, which indicates that Si has no influence on Sj or the influence degree is negligible, otherwise $a_{ij} = 1$ is taken, which indicates that Si has influence on Sj. The value of λ is usually judged according to the experience of experts, which also makes its value subjective. In order to effectively reduce the influence of subjective experience on the results, this paper chooses the sum of the mean and standard deviation of each value in the comprehensive influence matrix T as the value of λ.

$$D = \begin{pmatrix} 0 & 1 & 1 & 1 & 1 & 0 & 0 & 0 & 0 & 0 & 0 & 0 \\ 0 & 0 & 1 & 1 & 1 & 0 & 0 & 0 & 0 & 1 & 0 & 0 \\ 0 & 1 & 0 & 1 & 1 & 0 & 0 & 0 & 0 & 1 & 0 & 0 \\ 0 & 0 & 0 & 0 & 0 & 0 & 0 & 1 & 1 & 1 & 0 & 0 \\ 0 & 0 & 1 & 1 & 0 & 0 & 0 & 0 & 0 & 1 & 0 & 0 \\ 0 & 0 & 1 & 1 & 1 & 0 & 1 & 0 & 0 & 0 & 0 & 0 \\ 0 & 0 & 0 & 0 & 0 & 1 & 0 & 0 & 0 & 0 & 0 & 0 \\ 1 & 1 & 0 & 0 & 0 & 0 & 0 & 0 & 1 & 1 & 0 & 1 \\ 0 & 0 & 0 & 1 & 1 & 1 & 0 & 1 & 0 & 1 & 0 & 1 \\ 1 & 1 & 0 & 1 & 0 & 0 & 0 & 0 & 0 & 0 & 0 & 0 \\ 0 & 0 & 0 & 0 & 0 & 0 & 0 & 1 & 0 & 1 & 0 & 0 \\ 0 & 0 & 0 & 0 & 0 & 0 & 0 & 0 & 0 & 1 & 0 & 0 \end{pmatrix} \quad (5)$$

4.5. ANP Calculation Index Weight

The process of calculating index weight by ANP is extremely complicated, and it is difficult to calculate by manual method. Therefore, this study uses ANP programming calculation software Super Decision (SD for short) to calculate the index weight by ANP.

The twelve key influencing factors from the LDA theme model are summarized as follows:

- Enterprise resources (A1): capital resources (financial strength) and human resources (digital talents);
- Enterprise capability (A2): management capability (enterprise management level), scientific and technological innovation capability, organizational change capability (business model);

- Enterprise spirit (A3): entrepreneurship and social responsibility;
- Macro environment (A4): policy environment, technical environment (digital technology);
- Industry environment (A5): market competition (international digital environment), aging labor force (ordinary workers), green requirements (green buildings).

According to the comprehensive influence matrix d obtained by DEMATEL, the ANP network structure for digital transformation of the construction industry is established, as shown in Figure 5. In the figure, the two-way arrow indicates that the indexes in two element groups influence each other, the one-way arrow indicates that the indexes in the tail element group influence the indexes in the head element group, and the circular arrow indicates that the indexes in the element group influence each other. Among them, the comprehensive evaluation system of the key influencing factors of digital transformation of construction industry is the control layer; Enterprise resources A1, enterprise capabilities A2, enterprise spirit A3, macro environment A4 and industry environment A5 are the first-level indicators in the network layer; Resources A_{11}, human resources A_{12}, management ability A_{21}, scientific and technological innovation ability A_{22}, organizational change ability A_{23}, entrepreneurship A_{31}, social responsibility A_{32}, policy environment A_{41}, technical environment A_{42}, market competition A_{51}, aging workforce A_{52} and green requirement A_{53} are secondary indicators, that is, A = ($A_1A_2A_3A_4A_5$); A1 = ($A_{11}A_{12}$); A2 = ($A_{21}A_{22}A_{23}$); A3 = ($A_{31}A_{32}$); A4 = ($A_{41}A_{42}$); A5 = ($A_{51}A_{52}A_{53}$).

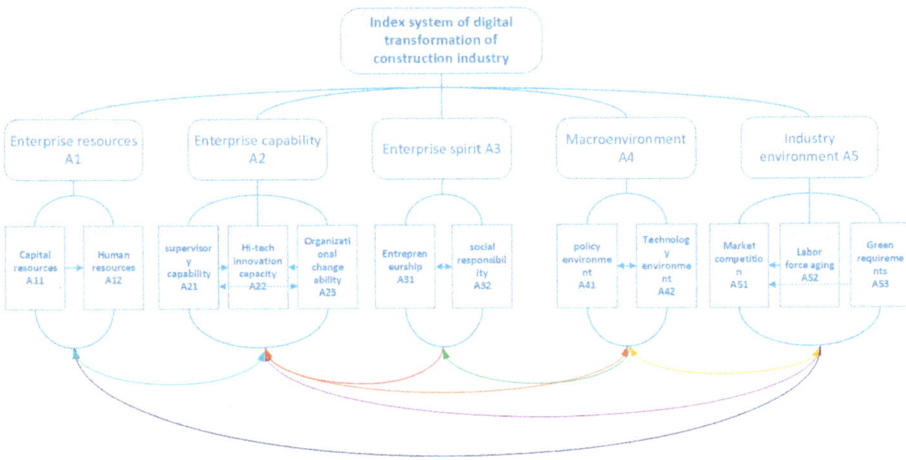

Figure 5. The ANP network structure.

Firstly, based on the comprehensive impact matrix D, the correlation of the first-level and second-level indicators is listed. Due to the limited space, only the correlation of the first-level indicators is shown in Table 4. Then, the 30 experts were invited to fill out the questionnaire by using the nine-level scale method. The scale and meaning are shown in Table 5. The data were averaged and rounded to form a judgment matrix, which was input into SD. After passing the consistency test, the unweighted hypermatrix, weighted hypermatrix and extreme hypermatrix were calculated by SD. See Tables 6–8 for the comprehensive weight of each secondary index. Finally, comprehensive weights of the secondary indicators were sorted, followed by calculation of the weights of the primary indicators, and formation of the index weight table of the comprehensive evaluation system for digital transformation of the construction industry, as shown in Table 9.

Table 4. The correlation of first-level indicators.

Affected Factors	Enterprise Resource A1	Enterprise Capability A2	Enterprise Spirit A3	Macro Environment A4	Industry Environment A5
Enterprise resource A1	1	6	0	0	1
Enterprise capability A2	1	4	0	2	3
Enterprise spirit A3	0	5	1	0	1
Macro environment A4	2	2	1	2	4
Industry environment	2	1	0	1	2

Table 5. The scale degree and meaning of the nine-grade scale method.

Scale a_{ij}	1	3	5	7	9	2, 4, 6, 8
Definition	Two elements, A and B, have the same influence when compared with each other.	Comparing the two elements A and B, A has a slightly greater influence than B.	Comparing the two elements A and B, A has greater influence than B.	Comparing the two elements A and B, A is more influential than B.	Comparing two elements, A and B, A is absolutely more influential than B.	mean value

Table 6. The unweighted hypermatrix.

	A11	A12	A21	A22	A23	A31	A32	A41	A42	A51	A52	A53
A11	0	0	0	0	0	0	0	0.88889	0	0.90000	0	1
A12	1	0	1	0	0	0	0	0.11111	0	0	0	0
A21	0.61538	0.67080	0	0	0.66667	0.65863	0	0	0	0	0	0
A22	0.30769	0.25596	0.75000	0	0.33333	0.26275	0	0	0.80000	1	0	0
A23	0.07692	0.07325	0.25000	0	0	0.07862	0	0	0.20000	0	0	0
A31	0	0	0	0	0	0	1	0	1	0	0	0
A32	0	0	0	0	0	1	0	0	0	0	0	0
A41	0	0	0	0.83333	0	0	0	0	1	0	1	0
A42	0	0	0	0.16667	0	0	0	1	0	0	0	0
A51	0	1	1	1	1	0	0	0.66667	0.66667	0	10	1
A52	0	0	0	0	0	0	0	0	0	0	0	0
A53	0	0	0	0	0	0	0	0.33333	0.33333	0	0	0

Table 7. The weighted hypermatrix.

	A11	A12	A21	A22	A23	A31	A32	A41	A42	A51	A52	A53
A11	0	0	0	0	0	0	0	0.58551	0	0.30499	0	0.83564
A12	0.34188	0	0.31700	0	0	0	0	0.07319	0	0.03389	0	0
A21	0.40543	0.60581	0	0	0.60561	0.52690	0	0	0	0	0	0
A22	0.20271	0.23166	0.46486	0	0.30281	0.21020	0	0	0.51722	0.66112	0	0
A23	0.05068	0.06615	0.15495	0	0	0.06289	0	0	0.12943	0	0	0
A31	0	0	0	0	0	0	1	0	0.18194	0	0	0
A32	0	0	0	0	0	0.20000	0	0	0	0	0	0
A41	0	0	0	0.53335	0	0	0	0	0.10808	0	0.64002	0
A42	0	0	0	0.10667	0	0	0	0.21583	0	0	0	0
A51	0	0.09688	0.06249	0.35998	0.09158	0	0	0.08365	0.04189	0	0.35998	0.16436
A52	0	0	0	0	0	0	0	0	0	0	0	0
A53	0	0	0	0	0	0	0	0.04183	0.02094	0	0	0

Table 8. The weighted limit hypermatrix.

	A11	A12	A21	A22	A23	A31	A32	A41	A42	A51	A52	A53
A11	0.12395	0.12395	0.12395	0.12395	0.12395	0.12395	0.12395	0.12395	0.12395	0.12395	0.12395	0.12395
A12	0.10278	0.10278	0.10278	0.10278	0.10278	0.10278	0.10278	0.10278	0.10278	0.10278	0.10278	0.10278
A21	0.14588	0.14588	0.14588	0.14588	0.14588	0.14588	0.14588	0.14588	0.14588	0.14588	0.14588	0.14588
A22	0.24437	0.24437	0.24437	0.24437	0.24437	0.24437	0.24437	0.24437	0.24437	0.24437	0.24437	0.24437
A23	0.04362	0.04362	0.04362	0.04362	0.04362	0.04362	0.04362	0.04362	0.04362	0.04362	0.04362	0.04362
A31	0.01262	0.01262	0.01262	0.01262	0.01262	0.01262	0.01262	0.01262	0.01262	0.01262	0.01262	0.01262
A32	0.00252	0.00252	0.00252	0.00252	0.00252	0.00252	0.00252	0.00252	0.00252	0.00252	0.00252	0.00252
A41	0.13633	0.13633	0.13633	0.13633	0.13633	0.13633	0.13633	0.13633	0.13633	0.13633	0.13633	0.13633
A42	0.05549	0.05549	0.05549	0.05549	0.05549	0.05549	0.05549	0.05549	0.05549	0.05549	0.05549	0.05549
A51	0.12589	0.12589	0.12589	0.12589	0.12589	0.12589	0.12589	0.12589	0.12589	0.12589	0.12589	0.12589
A52	0	0	0	0	0	0	0	0	0	0	0	0
A53	0.00686	0.00686	0.00686	0.00686	0.00686	0.00686	0.00686	0.00686	0.00686	0.00686	0.00686	0.00686

Table 9. Index weights of the comprehensive evaluation system for digital transformation of the construction industry.

Primary Index	Weight w	Secondary Index	Comprehensive Weight w	Sort
Enterprise resource A1	0.22673	Resources A11	0.12395	5
		Talent resources A12	0.10278	6
Enterprise capability A2	0.43387	Management ability A21	0.14588	2
		Technological innovation capability A22	0.24437	1
		Organizational change capability A23	0.04362	8
Enterprise spirit A3	0.01514	Entrepreneurship A31	0.01262	9
		Social responsibility A32	0.00252	11
Macro environment A4	0.19182	Policy environment A41	0.13633	3
		Technical environment A42	0.05549	7
Industry Environment A5	0.13273	Market competition A51	0.12587	4
		Aging workforce A52	0	12
		Green A53	0.00686	10

5. Conclusions and Discussion

5.1. Main Conclusions of the Study

Based on the LDA-ANP model, this study determines the key factors influencing for the digital transformation of the construction industry. The study systematically explains the key factors and the theoretical logic of digital transformation of the construction industry and constructs a comprehensive quantitative evaluation system. The specific conclusions arising from the empirical research study are as follows:

Firstly, the study identifies the key factors for the digital transformation of the construction industry. Among them, enterprise resources (capital resources, human resources), enterprise capabilities (management capabilities, technological innovation capabilities, enterprise transformation capabilities) and enterprise spirit (entrepreneurship, social responsibility) are internal influencing factors, whereas macro-environment (policy environment, technical environment) and industry environment (market competition, aging labor force, green requirements) are external influencing factors.

Secondly, the study establishes a comprehensive evaluation system for digital transformation of the construction industry through ANP and reveals that the five elements have different levels of influence on digital transformation in the construction industry. Among them, enterprise capability has the most significant impact on the digital transformation of the construction industry. In the era of the digital economy, organizational boundaries have been broken. Therefore, only by actively integrating digital technology with building related entities and making adaptive adjustments to management methods and internal functions can construction enterprises effectively enhance their competitiveness and occupy a competitive advantage in the era of the digital economy.

5.2. Theoretical Contribution

This study gives rise to a number of theoretical contributions to the body of literature, which are summarized as follows.

Firstly, the LDA theme model is applied to the construction industry, and based on Wos and Cnki databases, the frontier hotspots of digital transformation in the construction industry are deeply excavated, which to some extent avoids the limitations caused by insufficient samples. For example, Yiyue Wang et al. studied the influence of human capital, management level and technical ability on the transformation of construction industry [73], but they lacked the factors influencing the transformation of construction industry from the perspective of entrepreneurs' own characteristics and international competition. This study discusses the key influencing factors of the digital transformation of the construction industry as a whole, and clarifies the key factors affecting the digital transformation of the construction industry. However, the existing research results mostly explore certain influencing factors from a limited perspective, lacking the overall perspective achieved in this study.

Secondly, the comprehensive evaluation system of key influencing factors for digital transformation in the construction industry is constructed by ANP, and each key influencing factor is systematically analyzed and quantitatively evaluated. This approach provides theoretical support for digital transformation of construction industry, and also provides a reference for management decisions in construction enterprises to adapt to the digital economy era.

Thirdly, based on the fusion method of machine learning and network analytic hierarchy process, the key influencing factors of digital transformation in the construction industry are excavated and analyzed, and the comprehensive evaluation system of digital transformation of the construction industry is obtained, which represents an innovation application of the research method.

5.3. Management Implications

With the rapid development of the global digital economy, the need to understand how to carry out digital transformation is a challenge that every enterprise in the construction sector needs to face. According to the current situation in the construction industry, addressing this question has become an urgent problem to be solved. Therefore, this study provides guidance for the construction industry and corresponding management functions to devise effective digital transformation strategies according to the following directives:

Firstly, there is a need to establish an improved awareness of digital transformation. Traditionally, construction enterprises mainly focus on the quantity, quality and the price of products and services. Whereas in the digital economy, the management of construction enterprises will be integrated with users, and the corresponding models of products and services are centered on the creation and supply of value. Digital transformation is not only the use of digital technology to improve efficiency, but also represents a potential change of competition mode, as well as the change of enterprise management thinking and internal organizational structure. Therefore, if an enterprise wants to secure a competitive advantage in the new era of the digital economy, such an enterprise must establish a sense of transformation as soon as possible, fundamentally change the traditional management concept, and accelerate the deep integration of digital technology and architectural entities.

Secondly, there is a need to formulate a digital transformation strategy suitable for the context of the construction enterprise. In regard to the general trend of digital transformation, construction enterprises urgently need to formulate digital transformation strategies. According to the internal strategic management elements and external environment of the enterprise, there is a need for enterprises to analyze their strengths and weaknesses and devise a feasible digital strategy and corresponding implementation plan. Although there is no universal digital strategy, "successful enterprises are the same, and failed enterprises have their own misfortunes" [49], which requires enterprises to meet the key requirements of digital transformation in advance according to the concept of "system determines success or failure", and at the same time, to "make up for shortcomings" in time according to the characteristics of "taking one lead and moving the whole body". Although in reality, the process of digital strategic transformation is a complex system engineering, which requires not only the focus of construction enterprises at the technical level, but also adequate consideration by management functions, system implementation as well as careful consideration and effective engagement of various interest groups.

5.4. Shortcomings and Prospects

Although the study seeks to develop a reliable model through rigorous verification, there are some research limitations: (1) The number of samples selected in this study is large (articles), which has certain practical significance, but only papers and meeting are taken into account, and patents, corporate reports and other contents are not taken into account. In the future, we can consider expanding the sample size to fully understand the development state of the digital transformation field of the construction industry; (2) The labels of each theme are summarized by the author according to the key words and their own subjective judgment, which has a certain level of subjectivity. Therefore, future studies are proposed that would adopt methods to reduce the impact of such subjectivity.

Author Contributions: Conceptualization, H.L. and J.Z.; formal analysis, Z.H., D.L., S.P.P. and Y.K.; methodology, Z.H.; project administration, H.L. and S.P.P.; resources, Z.H. and S.P.P.; software, Z.H., D.L. and Y.K.; supervision, J.Z., S.P.P. and Y.K.; validation, H.L., J.Z., S.P.P. and Y.K.; visualization, J.Z. and D.L.; writing—original draft, H.L. and Z.H.; writing—review & editing, H.L., Y.K. and S.P.P. All authors have read and agreed to the published version of the manuscript.

Funding: This research was funded by [National Social Science Fund projects] grant number [No. 20BJY010], [National Social Science Fund Post-financing projects] grant number [No. 19FJYB017], [Sichuan-tibet Railway Major Fundamental Science Problems Special Fund] grant number [No.71942006], [Qinghai Natural Science Foundation] grant number [No. 2020-ZJ-736], [List of Key Science and Technology Projects in China's Transportation Industry in 2018-International Science and Technology Cooperation Project] grant number [No. 2018-GH-006 and No. 2019-MS5-100], [Shaanxi Social Science Fund] grant number [No. 2017S004], [Xi'an Construction Science and Technology Planning Project] grant number [No. SZJJ2019-15 and No. SZJJ2019-16] and [Fundamental Research for Funds for the Central Universities (Humanities and Social Sciences)] grant number No. 300102282601].

Institutional Review Board Statement: Not applicable.

Informed Consent Statement: Not applicable.

Data Availability Statement: Not applicable.

Acknowledgments: The National Social Science Fund projects, National Social Science Fund Post-financing projects, Sichuan-tibet Railway Major Fundamental Science Problems Special Fund, Qinghai Natural Science Foundation, List of Key Science and Technology Projects in China's Transportation Industry in 2018-International Science and Technology Cooperation Project, Shaanxi Social Science Fund, Xi'an Construction Science and Technology Planning Project and Fundamental Research for Funds for the Central Universities enabled this research, for which the authors are most grateful.

Conflicts of Interest: The authors declare no conflict of interest.

References

1. Reis, J.; Amorim, M.; Melao, N.; Matos, P. *Digital Transformation: A Literature Review and Guidelines for Future Research*; Springer: Cham, Switzerland, 2018; pp. 411–421.
2. Frank, A.G.; Mendes, G.H.S.; Ayala, N.F.; Ghezzi, A. Servitization and Industry 4.0 convergence in the digital transformation of product firms: A business model innovation perspective. *Technol. Forecast. Soc. Chang.* **2019**, *141*, 341–351. [CrossRef]
3. Ghobakhloo, M. Industry 4.0, digitization, and opportunities for sustainability. *J. Clean. Prod.* **2020**, *252*, 119869. [CrossRef]
4. Verhoef, P.C.; Broekhuizen, T.; Bart, Y.; Bhattacharya, A.; Dong, J.Q.; Fabian, N.; Haenlein, M. Digital transformation: A multidisciplinary reflection and research agenda. *J. Bus. Res.* **2021**, *122*, 889–901.
5. Dengler, K.; Matthes, B. The impacts of digital transformation on the labour market: Substitution potentials of occupations in Germany. *Technol. Forecast. Soc. Chang.* **2018**, *137*, 304–316. [CrossRef]
6. Kornelakis, A.; Kirov, V.; Thill, P. The digitalisation of service work: A comparative study of restructuring of the banking sector in the United Kingdom and Luxembourg. *Eur. J. Ind. Relat.* **2022**, *28*, 253–272. [CrossRef]
7. Mićić, L. Digital transformation and its influence on GDP. *Econ.-Innov. Res. J.* **2017**, *5*, 135–147. [CrossRef]
8. Llopis-Albert, C.; Rubio, F.; Valero, F. Impact of digital transformation on the automotive industry. *Technol. Forecast. Soc. Chang.* **2021**, *162*, 120343. [CrossRef]
9. Savastano, M.; Amendola, C.; Bellini, F.; D'Ascenzo, F. Contextual impacts on industrial processes brought by the digital transformation of manufacturing: A systematic review. *Sustainability* **2019**, *11*, 891. [CrossRef]
10. Finelli, L.A.; Narasimhan, V. Leading a digital transformation in the pharmaceutical industry: Reimagining the way we work in global drug development. *Clin. Pharmacol. Ther.* **2020**, *108*, 756–761. [CrossRef]
11. Parker, L.E. Creation of the National Artificial Intelligence Research and Development Strategic Plan. *AI Mag.* **2018**, *39*, 25–31. [CrossRef]
12. Rübmann, M. Future of Productivity and Growth in Manufacturing. *Bost. Consult.* **2015**, *9*, 54–89.
13. Liu, Q.; Ma, Y.; Xu, S. Has the development of digital economy improved the efficiency of China's green economy? *China Popul. Resour. Environ.* **2022**, *32*, 72–85.
14. Ngowi, A.; Pienaar, E.; Talukhaba, A.; Mbachu, J. The globalisation of the construction industry—A review. *Build. Environ.* **2005**, *40*, 135–141. [CrossRef]
15. Adekunle, S.A.; Aigbavboa, C.O.; Ejohwomu, O.; Adekunle, E.A.; Thwala, W.D. Digital transformation in the construction industry: A bibliometric review. *J. Eng. Des. Technol.* **2021**. [CrossRef]
16. Pfnür, A.; Wagner, B. Transformation of the real estate and construction industry: Empirical findings from Germany. *J. Bus. Econ.* **2020**, *90*, 975–1019. [CrossRef]
17. Silverio-Fernández, M.A.; Renukappa, S.; Suresh, S. Strategic framework for implementing smart devices in the construction industry. *Constr. Innov.* **2021**, *21*, 218–243. [CrossRef]
18. Sun, J.; Gong, X.; Zhang, H.; Su, X. Strategic Path for High-Quality Development of Construction Industry Driven by Digitalization. *Eng. Sci.* **2021**, *23*, 56–63. [CrossRef]
19. Panenkov, A.; Lukmanova, I.; Kuzovleva, I.; Bredikhin, V. Methodology of the theory of change management in the implementation of digital transformation of construction: Problems and prospects. In Proceedings of the E3S Web of Conferences, Voronezh, Russia, 8–10 December 2020.
20. Wang, Y.; Wang, L. Research on the Type Identification and Transformation Mode Selection of Traditional Enterprises' Digital Transformation Strategy. *J. Manag. Rev.* **2021**, *33*, 84–93. [CrossRef]
21. Xu, Z.; Song, T.; Guo, S.; Peng, J.; Zeng, L.; Zhu, M. Robotics technologies aided for 3D printing in construction: A review. *Int. J. Adv. Manuf. Technol.* **2022**, *118*, 3559–3574. [CrossRef]
22. Liu, H.; Song, J.; Wang, G. A Scientometric Review of Smart Construction Site in Construction Engineering and Management: Analysis and Visualization. *Sustainability* **2021**, *13*, 8860. [CrossRef]
23. Baek, S.; Jung, W.; Han, S.H. A critical review of text-based research in construction: Data source, analysis method, and implications. *Autom. Constr.* **2021**, *132*, 103915. [CrossRef]
24. Han, Q.; Wang, S.; Yang, X.; Li, J. The development mechanism, influencing factors and countermeasures of industrial intelligence-a qualitative study based on interviews with industry experts. *China Sci. Technol. Forum* **2021**, *8*, 59–69.
25. Chen, K.; Du, P.; Fang, W.; Gao, F. Digital transformation of China's construction industry: Connotation, participants and policy tools. *J. Civ. Eng. Manag.* **2021**, *38*, 23–29.
26. Xie, X.; Deng, L.; Xiao, M.; Luo, H.; Fang, R. New-Generation Quality and Safety Management of the Construction Industry. *Eng. Sci.* **2021**, *23*, 71–78. [CrossRef]
27. Yang, Q.; Liang, S.; Yang, D. A Comparative Study of Digital Transformation Policies of Foreign Governments. *J. Inf.* **2021**, *40*, 128–138.
28. Westerman, G.; Bonnet, D.; McAfee, A. The nine elements of digital transformation. *MIT Sloan Manag. Rev.* **2014**, *55*, 1–6.
29. Karagiannaki, A.; Vergados, G.; Fouskas, K. The impact of digital transformation in the financial services industry: Insights from an open innovation initiative in fintech in Greece. In Proceedings of the Mediterranean Conference on Information Systems (MCIS), Genova, Italy, 4–5 September 2017.
30. Fitzgerald, M.; Kruschwitz, N.; Bonnet, D.; Welch, M. Embracing digital technology: A new strategic imperative. *MIT Sloan Manag. Rev.* **2014**, *55*, 1.

31. Liere-Netheler, K.; Packmohr, S.; Vogelsang, K. Drivers of Digital Transformation in Manufacturing. In Proceedings of the 51st Annual Hawaii International Conference on System Sciences (HICSS), Hilton Waikoloa Village, HI, USA, 2–6 January 2018.
32. Demirkan, H.; Spohrer, J.C.; Welser, J.J. Digital innovation and strategic transformation. *It Prof.* **2016**, *18*, 14–18. [CrossRef]
33. Haffke, I.; Kalgovas, B.; Benlian, A. The Transformative Role of Bimodal IT in an Era of Digital Business. In Proceedings of the 50st Annual Hawaii International Conference on System Sciences (HICSS), Hilton Waikoloa Village, HI, USA, 3–7 January 2017.
34. Vial, G. Understanding digital transformation: A review and a research agenda. *J. Strateg. Inf. Syst.* **2019**, *28*, 118–144. [CrossRef]
35. Boyes, H.; Hallaq, B.; Cunningham, J.; Watson, T. The industrial internet of things (IIoT): An analysis framework. *Comput. Ind.* **2018**, *101*, 1–12. [CrossRef]
36. Smith, S.D. Construction 4.0: An Innovation Platform for the Built Environment. *Proc. Inst. Civ. Eng.-Manag. Procure. Law* **2020**, *173*, 190. [CrossRef]
37. Maskuriy, R.; Selamat, A.; Maresova, P.; Krejcar, O.; David, O.O. Industry 4.0 for the construction industry: Review of management perspective. *Economies* **2019**, *7*, 68. [CrossRef]
38. Chao, N.; Han, S.; Wu, X. Research Theme Discovery and Evolution Analysis of Digital Journalism: A Survey Based on Web of Science Literature. *Press* **2021**, *9*, 4–13. [CrossRef]
39. Dalvi-Esfahani, M.; Niknafs, A.; Kuss, D.J.; Nilashi, M.; Afrough, S. Social media addiction: Applying the DEMATEL approach. *Telemat. Inform.* **2019**, *43*, 101250. [CrossRef]
40. Saaty, T.L. Decision making with dependence and feedback. In *The Analytic Network Process*; RWS Publications: Pittsburgh, PA, USA, 1996.
41. Wang, B.; Liu, S.; Ding, K.; Liu, Z.; Xu, J. Identifying technological topics and institution-topic distribution probability for patent competitive intelligence analysis: A case study in LTE technology. *Scientometrics* **2014**, *101*, 685–704. [CrossRef]
42. Sepasgozar, S.M.E. Differentiating Digital Twin from Digital Shadow: Elucidating a Paradigm Shift to Expedite a Smart, Sustainable Built Environment. *Buildings* **2021**, *11*, 151. [CrossRef]
43. Boje, C.; Guerriero, A.; Kubicki, S.; Rezgui, Y. Towards a semantic Construction Digital Twin: Directions for future research. *Autom. Constr.* **2020**, *114*, 103179. [CrossRef]
44. Hinings, B.; Gegenhuber, T.; Greenwood, R. Digital innovation and transformation: An institutional perspective. *Inf. Organ.* **2018**, *28*, 52–61. [CrossRef]
45. Yang, W.; Liu, J.; Li, L.F.; Zhou, Q.; Ji, L.X. How Could Policies Facilitate Digital Transformation of Innovation Ecosystem: A Multiagent Model. *Complexity* **2021**, *2021*, 8835067. [CrossRef]
46. Wang, C.; Yang, Y. Five bottlenecks in the high-quality development of digital economy and countermeasures. *Macroecon. Res.* **2022**, *2*, 107–114. [CrossRef]
47. Bai, C.G.; Quayson, M.; Sarkis, J. COVID-19 pandemic digitization lessons for sustainable development of micro-and small-enterprises. *Sustain. Prod. Consum.* **2021**, *27*, 1989–2001. [CrossRef] [PubMed]
48. Li, H.; Wu, Y.; Cao, D.; Wang, Y. Organizational mindfulness towards digital transformation as a prerequisite of information processing capability to achieve market agility. *J. Bus. Res.* **2021**, *122*, 700–712. [CrossRef]
49. Qi, Y.; Xiao, X. Enterprise Management Reform in the Digital Economy Era. *Manag. World* **2020**, *36*, 135–152+250. [CrossRef]
50. Sun, Z. The impact of digital transformation on production mode and international economic structure and its response. *Qual. Mark.* **2021**, *18*, 150–152.
51. Mubarak, M.F.; Petraite, M. Industry 4.0 technologies, digital trust and technological orientation: What matters in open innovation? *Technol. Forecast. Soc. Chang.* **2020**, *161*, 120332. [CrossRef]
52. Scuotto, V.; Nicotra, M.; Del Giudice, M.; Krueger, N.; Gregori, G.L. A microfoundational perspective on SMEs' growth in the digital transformation era. *J. Bus. Res.* **2021**, *129*, 382–392. [CrossRef]
53. Lu, C.; Wu, J.; Wang, M.; Li, X.; Liu, B.; Liang, C.; Hu, X. Improvement of Core Competitiveness of Chinese Construction Enterprises Against the Background of High-Quality Development. *Eng. Sci.* **2021**, *23*, 79–86. [CrossRef]
54. Chua, T.; Zhang, Q.; Su, X. Cross-regional analysis of the aging problem of China's construction labor force and its countermeasures. *J. Civ. Eng. Manag.* **2021**, *38*, 63–69.
55. Hildebrandt, J.; Kluge, J.; Ziefle, M. Barriers and Concerns of Elderly Workers towards the Digital Transformation of Work. In *Human Aspects of IT for the Aged Population. Design for the Elderly and Technology Acceptance, Proceedings of the 5th International Conference, ITAP 2019 Held as Part of the 21st HCI International Conference, HCII 2019, Orlando, FL, USA, 26–31 July 2019*; Lecture Notes in Computer Science (LNCS 11592); Springer: Berlin/Heidelberg, Germany, 2019; pp. 158–169. [CrossRef]
56. Li, Q.; Liu, G.; Shao, J. Digital Transformation, Supply Chain Integration and Enterprise Performance-Moderating Effect of Entrepreneurship. *Econ. Manag.* **2021**, *43*, 5–23. [CrossRef]
57. Schiuma, G.; Schettini, E.; Santarsiero, F.; Carlucci, D. The transformative leadership compass: Six competencies for digital transformation entrepreneurship. *Int. J. Entrep. Behav. Res.* **2021**, *28*, 1273–1291. [CrossRef]
58. Chen, M.; Zhang, W. Research on the Mechanism of Digital Economy to Economic Growth. *Soc. Sci.* **2021**, *1*, 44–53. [CrossRef]
59. Zhang, J.; Li, M. Research on the Relationship between Corporate Social Responsibility, Internal Control and Financial Performance: Based on the Perspective of Technological Innovation. *Forecast* **2021**, *40*, 81–87.
60. Agafonova, A.N.; Yakhneeva, I.V.; Mukhametshina, G.R. Human-Centric Marketing in the Digital Era. In *Innovative Economic Symposium*; Springer: Cham, Switzerland, 2021; pp. 10–17. [CrossRef]

61. Ren, L. Internal mechanism and external perception in digital transformation of physical retail enterprises. *Bus. Econ. Res.* **2019**, *9*, 115–118.
62. Zhang, S.; Wang, K.; Yang, X.; Xu, W. Research on Emission Goal of Carbon Peak and Carbon Neutral in Building Sector. *Build. Sci.* **2021**, *37*, 189–198.
63. Chen, X.; Hu, D.; Cao, W.; Liang, W.; Xu, X.; Tang, X.; Wang, Y. Path of Digital Technology Promoting Realization of Carbon Neutrality Goal in Chinas Energy Industry. *Bull. Chin. Acad. Sci.* **2021**, *36*, 1019–1029.
64. Li, J.; Zhang, R.; Li, B. Research on the Influence Mechanism of Big Data Capability on the Green Competitiveness of Manufacturing Industry from the Perspective of Environmental Dynamics—Based on SBM-GML Index Model. *J.Sci. Technol. Prog. Countermeas.* **2021**, *38*, 67–75.
65. Ma, L.; Wang, S.; Zhang, J. Research on the Path of Digital Economy Driving Enterprise Business Model Innovation. *Tech. Econ. Manag. Res.* **2021**, *10*, 37–42.
66. Ziyadin, S.; Suieubayeva, S.; Utegenova, A. Digital transformation in business. In *Digital Age: Chances, Challenges and Future. Lecture Notes in Networks and Systems (LNNS 84)*; Springer: Cham, Switzerland, 2020; pp. 408–415. [CrossRef]
67. Yi, J.; Zhang, Z.; Yang, X.; Wang, Y. Organizational inertia, digital capability and business model innovation of Internet enterprises: The moderating effect of enterprise types. *Nankai Bus. Rev.* **2021**.
68. Cai, C.; Liu, W.; Jiang, H. The Impact of Business Model Scenarios on Value Creation—A Longitudinal Case Study of Digital Transformation of over the rainbow Stock Company from 2007 to 2018. *Nankai Bus. Rev.* **2020**, *23*, 98–108.
69. Svarc, J.; Laznjak, J.; Dabic, M. The role of national intellectual capital in the digital transformation of EU countries. Another digital divide? *J. Intellect. Cap.* **2021**, *22*, 768–791. [CrossRef]
70. Seliverstov, Y.I.; Moiseev, V.V.; Komarova, O.A. Digital transformation of russian enterprise and human capital development: Challenges and opportunities. In Proceedings of the 2nd International Scientific and Practical Conference on Modern Management Trends and the Digital Economy-from Regional Development to Global Economic Growth (MTDE), Yekaterinburg, Indonesia, 16–17 April 2020; pp. 133–138.
71. Gabriel, J.; Mayzira, A.; Aditya, J.; Itsari, M.; Satrio, S.; Ruldeviyani, Y. Critical success factors of data integration on digital human capital information system to support digital transformation—A case study at PTXYZ. In Proceedings of the 2020 8th International Conference on Cyber and IT Service Management (CITSM), Pangkal, Indonesia, 23–24 October 2020; pp. 1–7. [CrossRef]
72. Xu, R.; Li, C.; Cao, C.; Fang, M. Does science–industry cooperation policy enhance corporate innovation: Evidence from Chinese listed firms. *Account. Financ.* **2021**, *61*, 3823–3853. [CrossRef]
73. Wang, Y.; Li, Z.; Shi, F. Factors Influencing Mechanism of Construction Development Transformation in China Based on SEM. *Discret. Dyn. Nat. Soc.* **2015**, *2015*, 219865. [CrossRef]

buildings

MDPI

Article

The Creation of Construction Schedules in 4D BIM: A Comparison of Conventional and Automated Approaches

Omar Doukari [1,*], Boubacar Seck [2] and David Greenwood [1]

[1] Digital Built Environment Research Group, Northumbria University, Newcastle-upon-Tyne NE7 7YT, UK; david.greenwood@northumbria.ac.uk

[2] Centre de Paris-Nanterre, CESI Ecole d'Ingénieurs, 92000 Nanterre, France; bseck@cesi.fr

* Correspondence: omar.doukari@northumbria.ac.uk; Tel.: +44-(0)191-227-3006

Abstract: Building Information Modelling (BIM) is now a globally recognised phenomenon, though its adoption remains inconsistent and variable between and within the construction sectors of different countries. BIM technology has enabled a wide range of functional applications, one of which, '4D BIM', involves linking the tasks in a project's construction schedule to its object-orientated 3D model to improve the logistical decision making and delivery of the project. Ideally, this can be automatically generated but in reality, this is not currently possible, and the process requires considerable manual effort. The level of maturity and expertise in the use of BIM amongst the project participants still varies considerably; adding further obstacles to the ability to derive full benefits from BIM. Reflecting these challenges, two case studies are presented in this paper. The first describes a predominantly manual approach that was used to ameliorate the implementation of 4D BIM on a project in Paris. In fact, there is scope for automating the process: a combination of BIM and Artificial Intelligence (AI) could exploit newly-available data that are increasingly obtainable from smart devices or IoT sensors. A prerequisite for doing so is the development of dedicated ontologies that enable the formalisation of the domain knowledge that is relevant to a particular project typology. Perhaps the most challenging example of this is the case of renovation projects. In the second case study, part of a large European research project, the authors propose such an ontology and demonstrate its application by developing a digital tool for application within the context of deep renovation projects.

Keywords: 4D BIM; construction scheduling; project planning; building renovation; automation; ontology; artificial intelligence; lean methodology

Citation: Doukari, O.; Seck, B.; Greenwood, D. The Creation of Construction Schedules in 4D BIM: A Comparison of Conventional and Automated Approaches. *Buildings* **2022**, *12*, 1145. https://doi.org/10.3390/buildings12081145

Academic Editors: Yongjian Ke, Jingxiao Zhang and Simon P. Philbin

Received: 8 July 2022
Accepted: 28 July 2022
Published: 1 August 2022

1. Introduction

The construction sector has undergone many technical and regulatory developments over recent decades. Concerned about the impact of their projects on the environment and the safety of end-users and other project stakeholders, project owners have become more and more demanding when it comes to commissioning projects [1]. However, those involved in construction have, according to some commentators, changed their working methods very little [2]. This gap between changes in rules and standards that affect the demand-side, and the relative stagnation of those on the supply-side, has significant impacts on the duration, cost and quality of the works delivered. A project is an environment where stakeholders with different profiles are required to achieve specific objectives. The success or failure of a project depends on the strategy adopted to organize, coordinate, and supervise all the activities and works that need to be scheduled and then delivered while taking into account the various internal and external project constraints [3].

The planning and scheduling of construction projects represents an important part of the management of the construction process. It plays a crucial role in a project's success, since it facilitates the allocation of resources (such as equipment, materials, and labour)

to project activities over time, to ensure the completion of the project on time and within budget [4]. In addition to determining the pace of the work, good scheduling enables project stakeholders to check project feasibility, estimate the preliminary costs, maintain safety, optimise the use of resources, and allow the project team to monitor and control progress and determine if the work is proceeding efficiently, ensuring that the client's objective is achieved [5]. Furthermore, planning and scheduling deficiencies [6] and poor communication among project participants [7] have been identified as major factors that can lead to project delays and cost overruns, and ultimately to claims and disputes [8]. As a major contributor to the global economy (13% of global Gross Domestic Product) and one that is expected to rise by 85% to $15.5 billion globally by the year 2030 [9], the construction sector is still under-achieving and inefficient, since 9 out of 10 global mega projects encounter delay and cost overruns [10]. Scheduling software packages, such as Primavera and Microsoft Project [11], as well as planning techniques such as bar charts, time charts, and network approaches, are used to assist and help project managers in planning construction projects [12]. However, these tools are still limited and insufficient when considering the massive amount of newly-available data (e.g., feedback, images captured from smart devices, IoT sensors, etc.) that can continuously be produced on every project. Effective use of these data could enable valuable insights through a combination of Building Information Modelling (BIM), Artificial Intelligence (AI) and Machine Learning (ML) [13].

The advent of BIM has changed the practice of project management and has assisted project managers in expediting their duties more effectively than they have in the past. BIM can be defined as a set of tools, processes, and technologies that are enabled by a digital representation of the physical and functional characteristics of a built asset [14] expressed in data-enriched 3D models and their relationships. This digital replica constitutes a shared and central source of data about a facility, forming a reliable basis to produce information that supports insightful decision making for planning and managing a construction project throughout its entire life cycle [15]. Such information could include onsite spatial and topographical information, temporal and schedule information, and resources and cost information, among others [16–18]. BIM models are characterised by a level of development (LoD) which varies from 100 to 500 (i.e., from least to most developed) and serves to specify the appropriate amount of information required for specific uses [19]. Such digital representation is multi-dimensional, or 'nD', where each dimension indicates an information-processing capacity for various aspects [20]. The fourth dimension, known as 4D BIM, incorporates time-related information in the 3D information model to simulate and optimise the project construction process [21]. Practically, this consists of linking units of work or elements in the form of objects from the geometric 3D model to the construction scheduling activities using proprietary software, such as Navisworks or Synchro Pro [22,23]. Beyond the fifth dimension, understood as cost-estimating capacity [24], there appears to be lack of consensus [25].

Research has shown that 4D BIM can be a solution to overcoming many deficiencies of current planning practices [26]. The enrichment of a 3D BIM model with scheduling data has increasingly improved the quality of the construction planning process through the development and integration of several use cases, such as dynamic site analysis with temporary components including equipment movement, resource availability, the management of congestion and other operational constraints [27–29], spatiotemporal analysis for health and safety management [30,31], evacuation path planning [32], logistics management [16], augmented vehicle tracking and transportation route planning [33], construction waste management [34], spatial conflict detection and workspace congestion avoidance [16], and the monitoring of construction progress with site layout designs [35,36]. Overall, according to Candelario-Garrido et al. [37], 4D BIM simulation is 40% more efficient than conventional planning procedures. Furthermore, 4D-BIM-based visualisations provide an intuitive comprehension of the construction process which enables more effective communication and therefore better collaboration between all project stakeholders [38,39].

Although the benefits of 4D BIM are clear and much reported in the literature, few studies have considered the actual implementation of such tools and the corresponding processes during the construction phase that involves many actors. The use of 4D BIM is currently only adapted for small projects with few activities, since its use can be very expensive and time- and effort-consuming [23]. Moreover, there is little research addressing how 4D BIM can best be coupled and used with AI tools; e.g., to optimise and develop more effective strategies for construction project management through the automatic generation and simulation of different construction scenarios [40]. To bridge this gap, this study first presents case-based evidence of the use of 4D BIM during the construction phase of a real project to understand how this tool can be practically implemented to support and assist the project participants in their mission. Second, to enable process automation and 4D tools development, this paper presents an ontology—known to be a useful AI tool in formalising specific domain knowledge including concepts, relations, and constraints [41,42]—dedicated to scheduling, planning, and 4D simulation, and demonstrates its application by populating the corresponding database and developing a digital tool for application within the context of deep renovation projects that are part of a large European research project known as the RINNO project.

The remainder of the paper is organized into five parts. After a brief review of planning in construction projects in general, Section 2 reviews the literature on the use of 4D BIM, concluding in the identification of remaining barriers and problems for investigation. In Section 3, the sources of empirical and theoretical data are introduced. These include a literature review of 4D BIM applications and a survey of 4D BIM practice in France. The construction of a new building at the CESI campus in Nanterre-Paris prompted the overall study of a newly proposed approach to the process of 4D BIM implementation, which is detailed in Section 4. The proposed method is then demonstrated on the Nanterre 2 CESI Project in Section 5. The ontology proposed, along with a digital tool to automate the process of 4D BIM simulation specifically dedicated to scheduling of renovation projects, is introduced in Section 6. The CESI 4D BIM methodology was applied under unique circumstances and its more general applicability to the construction industry is discussed in Section 6 alongside the RINNO project case study, and the future work envisaged by the research team. The content is summarised in Figure 1, below.

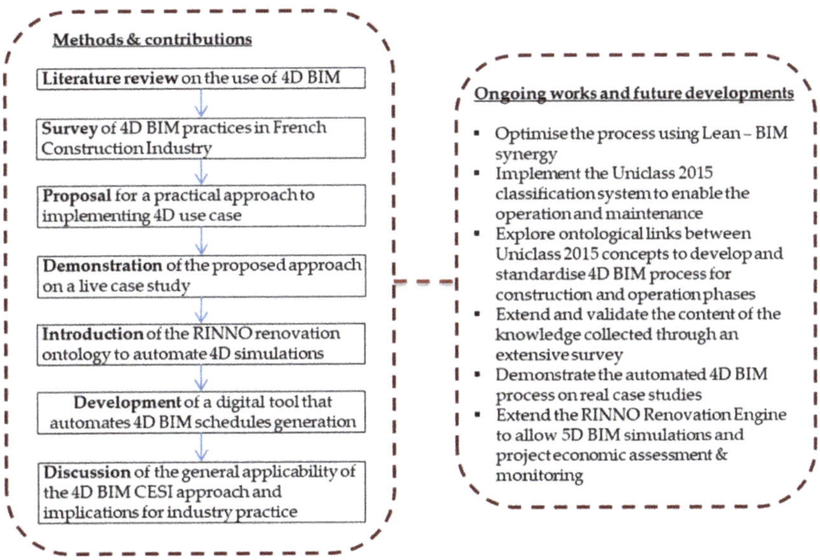

Figure 1. Methods, contributions, ongoing works, and future developments presented in this study.

2. Planning of Construction Projects

The ISO 21500 standard defines a project as "a unique set of processes consisting of coordinated and controlled activities with start and finish dates, undertaken to achieve an objective" [43]. In order to achieve these deliverables, the processes comprise a set of activities to which human, financial, and material resources are allocated. These activities are also limited in time. The total duration of the project is the sum of the durations of those activities that are 'critical' (i.e., those in which a delay would cause a comparable delay in the completion of the project [44]). The definition given by the standard [43] also addresses the importance of the constraints that may be external to the project (regulations, socio-economic situations, environmental issues, etc.) or internal (the availability of resources, the degree of skill of the actors, budget envelopes, etc.). A project is also characterized by a life cycle in which each of the stages requires the implementation of groups of processes to organize and control the work of the stakeholders in order to achieve the initial objectives. This life cycle depends on the nature, size, and field of activity, as well as the constraints of the project. Project management consists of identifying, planning, and controlling the processes required to increase the probability of the project's success at each phase of the life cycle.

2.1. *Construction Project Management*

The construction sector brings together all the activities of the design, construction, operation, and demolition of public works, civil engineering, and building works. The vast majority of construction projects follow a sequential development where each of the stages corresponds to a phase of the project. There are four main phases covering the life cycle of a built asset: Pre-project, Pre-construction, Construction, and Post-construction [45]. At each stage of the project, different actors are needed to plan, execute, and control the activities. The complexity of construction projects lies in their rapid pace [46], their unique environment (regulations, location, type of market, etc.) [47] as well as the multiplicity of stakeholders with different roles and periods of intervention [48].

To successfully manage a project, effective planning in all its phases is essential. The corresponding tools and methods can be counted by the dozen, and they must be adopted and put into practice by all of those involved in the construction industry in order to facilitate coordination and time control [49]. The most commonly used tools and methods are detailed in the next section.

2.2. *Planning Methods for Construction Projects*

2.2.1. Traditional French Job Titles and Planning Process

Planning a construction project involves identifying, for each stage of the project, the activities to be carried out in order to achieve the desired objectives. The activities must then be broken down into elementary tasks so as to be scheduled with respect to the technical, economic, and specific constraints of the project. By doing so, the resulting deliverable makes it possible to clearly explain the detailed progress of the project, while validation milestones ensure the measuring of progress and the quality of the services to be performed.

In a construction project, a schedule is developed by the client, noting the project's feasibility. This preliminary schedule rarely details each of the project tasks because many of them are still unknown or might be deleted and/or modified. However, it enables a quick assessment of the project duration in its early stages, despite a significant degree of uncertainty. To respond to project tenders, bids are expressed using the envelope planning of the project owner (in French the maître d'ouvrage, or MOA) as input. The granularity of tasks is reduced to make improvements and confirm or reject the initial duration expectations. Finally, before the construction phase, companies must also create execution schedules based on the general planning of the project entity in charge of the implementation of the project (the company, firm, department, or person called in French the maître d'oeuvre, or MOE). These schedules take into account their ability to mobilize the necessary resources, delivery times for supplies, and scheduling constraints that link them to other onsite activities [50].

As the activities of all project participants are strongly interconnected in time and space, it is fundamental to collaborate throughout the whole project in order to reduce the impacts of potential causes of delay. This is precisely the role of the scheduling, management, and coordination team (in French the Ordonnancement, Pilotage, Coordination or OPC); to bring together and synchronise all the schedules as well as to be the unique contact for project participants to coordinate activities in case of any conflicts or problems. Depending on the nature and context of the project, the OPC may implement and use different planning methods. It should be noted that, in this paper, although generic role descriptors have been used where possible, occasionally it has been necessary to use French job titles. These, and their English equivalents, are provided in Appendix A, Table A1.

2.2.2. Planning Methods

Several classes of scheduling methods have been identified through the review and analysis of various literature concerning construction planning [51]. The most known and used are (a) the Critical Path method, (b) the Line-of-Balance method, (c) the Program Evaluation and Review Technique (PERT), (d) simulation methods, (e) AI-based methods, (f) visualisation methods, (g) Critical Chain scheduling, (h) location-based scheduling, and (i) Lean methods. These are detailed in the following paragraphs.

(a) The Critical Path method was first used in the 1950s in the United States [52]. It is most used in construction projects since it is easily applicable to sequential projects. However, it has many drawbacks, the most important of which is the use of ratios derived from feedback. Therefore, whether the duration of the tasks is under- or overestimated, this method does not leave the freedom to the actors to easily adjust their schedule to the pace of the actual production.
(b) The Line-of-Balance method was first created by the Goodyear Company in the early 1940s [53], before being adopted and developed by the U.S. Naval Forces in the early 1950s [54]. It is a powerful tool used for scheduling and controlling construction projects that contain repetitive blocks of activities, such as high-rise buildings, roads, tunnels, railways, and pipelines. Its focus is on resource allocation for each construction phase so that activities are achieved in time without any interference from the following phases [53].
(c) The PERT method was also developed by the U.S. Naval Forces. It has the advantage of making the scheduling of tasks more flexible [55]. The PERT method is based on the calculation of total margins and free margins that will enable the OPC to easily evaluate their degree of freedom to modify activity dates in the case of any hazards or to optimise space and resource allocation. Although it allows some flexibility compared to the Critical Path method, the PERT method inherits from it several disadvantages, in particular the absence of indications on the occupation of space or the allocation of resources.
(d) Simulation methods [56] are powerful tools [57,58] that have been developed and used during the last 60 years in construction to enable project managers to optimise project resources and improve productivity [56]. However, only repetitive or cyclical construction activities could be simulated [56].
(e) AI-based scheduling techniques have been used in construction since the 1980s [59]. They allow project managers to automatically generate optimised schedules based on different scenarios and AI algorithms [60–63]. However, representing and integrating uncertain knowledge while generating onsite construction schedules is still the main weakness of these methods [51].
(f) Since the 1990s, several visualisation methods [64] and technologies (BIM [65], VR [66], AR [67], MR [68]) have been developed in construction to improve communication between project participants and enhance the visualisation and validation of the construction plans.

(g) The Critical Chain method [69] complements the impact of variation within a construction project by introducing the concept of buffer. It aims to reduce contingencies and allows project managers to take into account the limitations of resources when developing project schedules through the insertion of aggregated buffers.

(h) The location-based scheduling method is used mainly for projects where the activities have to be repeated many times (high-rise buildings, collective housing, linear infrastructure, etc.). It is a spatiotemporal representation of a project that focuses on the rate of production capacity of each team. A grid is used to describe the duration and location of each task. The space is represented on the ordinate by means of different points of the project; the starting point, the ending point, as well as specific points between (floor numbers, kilometre posts, intersections, structural work, etc.) [70]. Although it enables the checking of the production rate while taking into account the use of site space, the implementation of this method requires time and advanced planning skills.

(i) Lean methodology is based on a production philosophy which originated in the automobile industry [71]. The aim of Lean-based planning methods [72] is to continuously improve the added value of the construction tasks (performance, cost, quality, and duration), with the goal of enhancing the overall profitability of the project. In the construction domain, nine Lean planning and control techniques have been identified [73]. Based on several planning levels (Master, Phase, Look-ahead, Commitment), the Last Planner method [74] is one of the most used and implemented Lean techniques in construction [75–78]. This method focuses on short-term planning at crew level. Generally, Lean methods are implemented using a support (standard or digital board) on which the actors indicate the nature, location, and type of material to be mobilized, as well as the number of workers required to execute the task. This is repeated weekly in order to study the interdependencies of the tasks, prevent possible conflicts, and collectively propose optimisations. The main advantage of this method is that it takes into account the needs and constraints of each project actor by promoting and encouraging collaboration before task executions. Stakeholders combine their knowledge and effort for the general interest of the project and not just to achieve their own activities and objectives. As for any collaborative work, this method requires the real involvement of all the project stakeholders, along with rigorous and constant participation.

2.3. *Problems for Investigation*

Clearly, it is up to the project team to choose the most suitable planning method with respect to the project context and the level of expertise of the project stakeholders. However, despite the fact that these methods have been used for decades, it should be noted that few construction projects have succeeded in achieving their initial objectives in terms of cost and duration [79]. Furthermore, traditional planning methods are very limited and unsuitable when considering the massive amount of data that the construction industry is producing on a daily basis over many construction projects. This knowledge base could be used to enable stakeholders to gain useful insights by using BIM and AI tools, as outlined by Gondia et al. [13] in their survey. Consequently, construction actors are increasingly adopting IT-based tools to develop more comprehensive, flexible, and efficient construction and planning methods, such as those based on BIM [80].

3. Construction Planning and the Advent of 4D BIM

BIM makes it possible to design, build and operate a structure over its entire life cycle [14]. It involves a collaborative process within the company and between external partners around a digital model. The latter can be described as a 3D representation of the physical and functional characteristics of a built asset [81]. It is a technical database, made up of objects defined by their characteristics and the relationships between them. The whole forms a structured set of information about a built asset. The 3D model is useful for visualizing the planned project. However, on its own, this virtual structure remains static and does not allow a clear appreciation of the implementation of the construction process and the dynamic that characterises the sequence works. To make it dynamic, it is necessary to integrate a fourth dimension: time. This is the principle of 4D planning. In practice, this corresponds to linking the 3D elements of the BIM model with the project schedule activities. The next two subsections review the different applications and use cases in which 4D BIM has been used so far, and present a survey about the most-used 4D BIM tools, practices, and methods in France.

3.1. *Review of 4D BIM Applications*

Research has shown that 4D BIM can be a solution to many of the deficiencies of current planning practices [26]. The enrichment of a 3D BIM model with scheduling data has increasingly improved the quality of the construction planning process through the development and integration of several use cases. For example, simulating the progress of work over time is an efficient communication tool for explaining to a client the progress of a project and the construction methods used. Indeed, 4D-BIM-based visualisations provide an intuitive comprehension of the construction process, which enables more effective communication and thereby better collaboration between all project stakeholders [38,39]. The use of 4D BIM has made several applications possible, including the following:

(i) **Site analysis including temporary components** such as equipment movement, resource availability, the management of congestion, and other operational constraints. Construction project productivity can be decreased by about 65% due to space congestion [82]. For instance, Huang et al. [29] proposed the integration of construction virtual prototyping systems with 4D models to provide realistic graphical simulations that incorporate both the layout and dynamic analysis of the construction site.

(ii) **Spatial conflict detection and workspace congestion avoidance.** Chavada et al. [83] analysed conflict detection between workspaces using a 4D BIM visualization of construction schedules. Moon et al. [17] developed an active simulation system based on 4D models and an optimisation process to minimise the simultaneous interference level of the schedule workspace. Trebbe et al. [84] explored the use of 4D models to coordinate different construction engineering designs, schedules, and the operations of contractors on a real railway station renovation project.

(iii) **The monitoring of onsite construction work progress with site layout designs.** Tran et al. [85] explored the use of 4D BIM and a visual programming language to develop a conceptual framework of camera planning that enables the monitoring of construction site progress. To detect deviations in the construction process between the actual state of a construction and its planned state, Braun et al. [86] proposed an automated framework based on photogrammetric surveys and 4D BIM models.

(iv) **The production of short-term work plans.** Sriprasert and Dawood [87] implemented a visual multi-constraints planning framework based on the use of 4D BIM and Lean methodology principles. The LEWIS system enables the integration of construction-related information and constraints with 4D planning to ensure the generation of constraint-free commitment work plans.

(v) **Spatial–temporal analysis for health and safety management.** To prevent construction accidents, Tran et al. [30] proposed a hazard identification approach based on spatial–temporal conflicts that may lead to accidents using 4D BIM models. Han et al. [88] proposed the 3D-CES system to design, verify and simulate the 3D visualisation of mobile crane operation. To assist in elaborating the crane lift schedule, this system enables the identification of safety and productivity aspects while selecting the most efficient crane operation. Tan et al. [89] investigated the use of BIM and 4D acoustics simulation tools to mitigate the noise impact on maintenance workers of offshore platforms. In this framework, the BIM model was used as a source of information, whereas the 4D acoustics approach was used to provide the spatial–temporal sound pressure level distribution of the generated noise;

(vi) **Evacuation path planning.** Kim et al. [32] proposed a 4D-BIM-based framework that automatically analyses, generates, and visualises the evacuation paths of workers. The prototype developed takes into account construction activities and site constraints to enable the identification of accessible evacuation paths considering customised parameters, such as workspaces, temporary structures, and storage areas.

(vii) **Logistics management.** For example, to facilitate the visualisation and analysis of construction progress and workspace logistics, Golparvar-Fard et al. [90] proposed the use of 4D photographs. Said and El-Rayes [91] developed an automated multi-objective construction logistics optimisation system to integrate and optimise material supply and site layout decisions. Chin et al. [92] proposed a framework to support the logistics and progress management of structural steel works by using 4D with radio frequency identification technology.

(viii) **Augmented vehicle tracking and transportation route planning.** Chen and Nguyen [33] investigated the use of BIM with a web map service for selecting between construction material sources. The BIM–WMS decision-making system developed was based on the evaluation of the final cost, material delivery time, and location credits to help designers and project managers in the selection of materials, and cost and schedule planning.

(ix) **Construction waste management**. Hewage and Porwal [93] proposed a 4D-BIM-based framework in order to predict material waste and propose recycling strategies in construction. Won and Cheng [94] developed a 4D BIM construction waste estimation system which relies on construction waste factors for the estimation of waste generation. Bakchan et al. [95] presented a theoretical 4D BIM solution for construction waste disposal scheduling. However, according to Guera et al. [34], existing 4D BIM applications for construction waste management present some deficiencies, such as only focusing on waste generated by rework activities, a lack of estimation for off-site recycling, and/or the use of fixed factors. To overcome these issues, they proposed the integration of temporal-based algorithms with 4D BIM and demonstrated the proposed solution on two different case studies located in North America for the case of concrete and drywall waste reuse and recycling planning [34].

Furthermore, Candelario-Garrido et al. [37] have investigated the general benefits of using 4D BIM simulation for the construction industry and have estimated this tool and corresponding approaches to be 40% more efficient than conventional planning procedures. Linking the digital model to the construction schedule allows project managers to identify planning errors [39] given that 70% of traditional schedules produced are wrong and non-optimised [96]. Non-compliance with a schedule has a direct effect on the duration and costs of the work as well as possible indirect effects on its quality [97]. Extending incorrect schedules usually disrupt the smooth execution and coordination of works, and companies must use additional resources and accommodate the new activities within a very short time.

Despite the clear benefits of 4D planning methods, examples of their adoption, use, and application on real projects are rarely documented. Several research works propose 4D planning methods based on the development of specific and bespoke workflows requiring an advanced level of expertise in using many computerisation tools in different specialisation domains [98]. In addition, the use of 4D BIM is currently only adapted for small projects with few activities, rather than very expensive, time- and effort-consuming projects [23,31]. This is mainly due to the complex nature of 4D planning methods which are usually defined and based on interfacing BIM authoring tools (e.g., Revit, ArchiCAD, Bentley), BIM management tools (e.g., Navisworks, Syncro Pro), and legacy scheduling tools (e.g., Primavera, MS Project). There exists a lack of interoperability between the three groups of tools, since BIM authoring platforms are usually more suited to the design stage [23]. In particular, because of this lack of interoperability between BIM tools and BIM management platforms, the modification of any activity or 3D element generates a lot of manipulations, adjustments, and repetitions of the 4D planning process. Moreover, there is a dearth of studies that investigate how 4D BIM can be integrated with AI tools which will enable the development of more effective methods for construction project management by considering the massive amount of data produced on every project and thus enable the gaining of valuable insights [40].

To capture the industry perspective on BIM-based 4D solutions and understand existing practices, a survey of French construction companies was performed. The results of the survey are described in the next subsection.

3.2. 4D BIM Tools and Methods Used in France

A questionnaire survey was conducted in order to understand 4D BIM practices in France. The questionnaire was first designed then shared online on professional platforms, such as LinkedIn, to allow construction companies to complete it and share their experience and practices. It consisted of three main items: (i) the 4D planning tools used by the company and why (advantages and disadvantages), (ii) the strategy and methods used to implement these tools, and (iii) the different use cases and applications for which these tools were used. The survey was conducted remotely from February to April 2020. Fifty-one participants, including BIM managers and experts from different French construction companies with different backgrounds and specialisations, such as architecture, civil engineering, construction informatics, and project management, received and completed the survey.

As shown in Figure 2, Navisworks and Synchro Pro are the most widely-used BIM-based 4D planning softwares in France with 55% and 25%, respectively. Other 4D planning tools, such as Vico Control (6%), XD Builder (4%) and others (e.g., ITwo 4.0) are used by some French construction companies as well. However, not all the respondents had experienced the use of 4D BIM planning (6%), which demonstrates that the 4D BIM tool is not yet fully implemented and adopted within the French construction ecosystem.

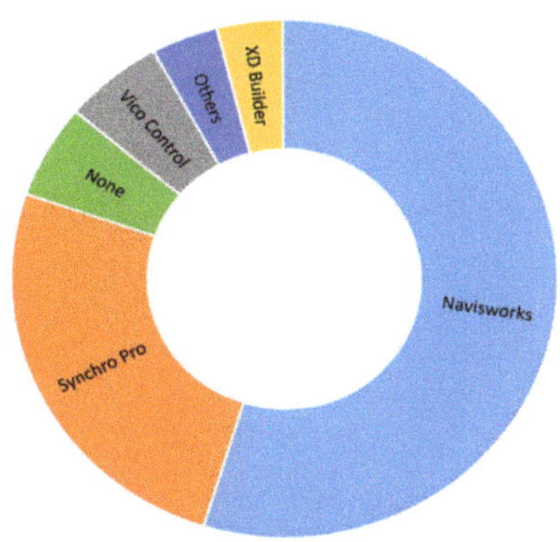

Figure 2. Most used 4D BIM tools in France.

The most used 4D tool in France is Navisworks. It is easy to use and enables the linking of activities on a bar chart to corresponding 3D elements in the BIM model. Navisworks can be used for operation control, conflict analysis, construction sequences, and coordination between disciplines. However, this tool requires many manipulations of the BIM model and planning to integrate changes. In addition, to ensure consistency between these two entities (i.e., BIM and planning), it necessitates the adoption of a relevant LoD. The second most popular 4D tool in France, according to the survey, is Synchro Pro. It integrates some planning functionalities and so enables an integrated planning process performed in the same interface as the 4D simulation. However, this tool is still difficult to learn and therefore to implement on construction projects. Compared to Navisworks, it requires more time and practice before project managers can handle it correctly. Indeed, the planning method used by Synchro Pro seems to make the study of spatial constraints more difficult.

The activities are defined with respect to their locations, which generates a huge number to manipulate and deal with, in addition to their logical and resource links.

The survey showed that two strategies exist for setting up the 4D planning use case. The first is a real implementation which consists of identifying the needs in terms of 4D planning, selecting the most suitable software by acquiring a license, and finally, developing a training plan to master the corresponding tools and methods. The second strategy to implement 4D planning on a construction project is to outsource this use case by subcontracting it to another engineering company that will use its own solution and methods. This article takes part in the first strategy by proposing a method that allows construction companies to initiate and internally implement 4D planning works for any projects.

However, the results revealed that the use of 4D planning is still limited and mainly restricted to creating and simulating 4D project videos to communicate internally within the same company and/or with the client. These video animations are usually non-interactive and disconnected from the progress of actual onsite works. However, some big contractors, such as Vinci Construction France or Bouygues Construction, have good expertise in 4D planning, and they use it for advanced applications, such as project progress monitoring and control, and employ digital tools that allow site workers to provide daily information on the quantity of tasks performed. The use of these data enables project managers to evaluate onsite production, monitor and control progress, and make the appropriate adjustments to the 4D planning if necessary [16].

Furthermore, the survey highlighted two methods used by the French companies to implement 4D BIM. The first method is based on the 'Hardin and McCool method'. This method [96] is the most commonly used because it applies the basic 4D principle of linking a schedule to a digital model. Indeed, it proposes a solution based around the Navisworks software and its TimeLiner module. By importing a Gantt schedule and a digital model of a project, it is possible to link the corresponding digital objects to each of the project tasks. This link can be performed manually or automatically if the 4D parameters of the digital model have been correctly created and entered. The second method is developed by the iBIM teams at the Vinci Construction France company. This method is based on the use of Synchro Pro software. It consists of remodelling and restructuring the BIM model into elementary objects, creating a task for each item in the schedule, and then linking each elementary object to its corresponding task. This results in a 4D schedule with a very high granularity, since the BIM objects in the original BIM model will be split into many new elementary BIM objects according to their actual onsite construction schedule. For example, an existing slab object (or any other element, such as a floor or wall) within the original BIM model could be split into more than one elementary slab in the new restructured BIM model to be suitable for the 4D planning method. Although this solution allows for the precise planning of the schedule, its main drawbacks come from its highly detailed breakdown. The implementation of this method represents a manual, laborious, and time-consuming process of restructuring and reorganising all the BIM model elements of the built asset. For instance, according to an iBIM's BIM manager, a structural BIM model for a medium-sized project (~10M €) includes more than a thousand elementary building components. Therefore, even if the laborious data restructuring work of the BIM model has been carried out in advance, it is relatively complicated for onsite teams to quickly identify the work in progress and indicate the level of progress.

3.3. *Summary of the Identified Problems*

The difficulties encountered by survey respondents corresponded closely with the observations noted in the literature, summarised in Section 2.3, above. This suggests that current methods of creating 4D tools from 3D BIM models are either (i) relatively easy but not suitable for the full range of construction site planning functions (as in the case of the Hardin and McCool approach); or (ii) reasonably suitable for such purposes in terms of their granularity but involve a laborious and time-consuming manual conversion process.

The requirement for the CESI BIM team to undertake the implementation of 4D BIM planning on the Nanterre 2 project presented an opportunity to develop a workflow that addressed these concerns. The aim was to produce an approach to 4D BIM implementation that could produce a construction schedule of maximum functionality with minimum extra effort expended in data exchange. The results, which are presented in the following section, could help project participants more fully exploit the potential of 4D BIM.

Furthermore, to address the problem of how 4D BIM can best be coupled and used with AI technologies to exploit the massive and increasing amount of construction data and enable 4D BIM automation, Section 6 proposes an ontology and demonstrates its application by developing an automated 4D BIM process for application within the context of the RINNO research project.

4. A New Proposed Methodology for 4D BIM Planning: The CESI Process

The previous sections and literature reviews on construction project planning generally and 4D BIM methods in particular were used as a starting point for understanding the construction industry's needs and then developing the CESI planning process. The purpose of this method is to identify and develop all the steps to be followed to guarantee the achievement of the MOA objectives concerning 4D planning. Following a four-phase process (Figure 3), the CESI method details how this BIM use case should be required and specified by the client until the production and delivery of related deliverables. This solution is intended to be simple and pragmatic. The following subsections detail the different phases of the CESI method.

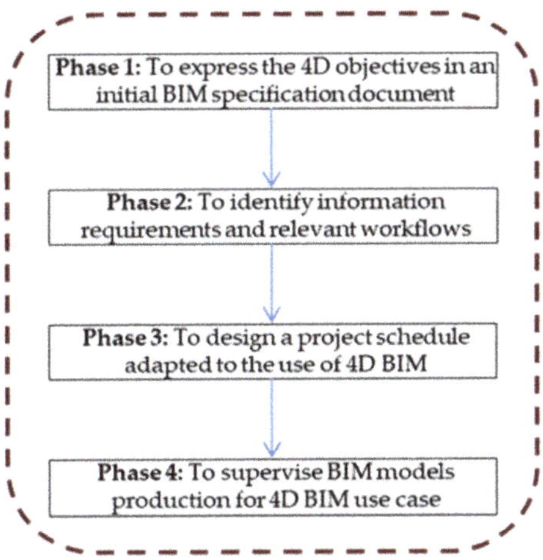

Figure 3. The CESI 4DBIM approach.

4.1. Phase 1: To Express the 4D Objectives in an Initial BIM Specification Document

The first actor in the project that allows the implementation of BIM is the client. Their role is to express needs which can be explicit, implicit, precise, or vague, depending on their familiarity with construction techniques and rules. However, the client can be supported by an assistant project manager (in French the assistant à maîtrise d'ouvrage, or AMO) who will help to precisely define the client's needs and achieve their strategic objectives. The method proposed recommends the use of SMART objectives (Specific, Measurable, Achievable, Realistic and Time-bound) for each of the BIM use cases desired by the client. For the 4D BIM use case, several SMART objectives can be expressed by the client. For the 4D BIM use case, examples of such SMART objectives are:

- The contractor's response to the invitation to tender must include 4D planning of construction methods for the works in a visual form that demonstrates its feasibility.

- The appointed contractor's BIM manager is responsible for the provision of a 4D schedule for the construction phase and its submission to the client. This must allow for progress monitoring of the works at any time during their execution and include the work of all the project delivery team.
- The contractor's 4D planning solution must enable the identification of all co-contracting zones during the progress of the works with the objective of ensuring and optimising the safety of the site staff and operatives of the companies involved in the works.

This formulation of 4D planning objectives makes it possible to explicitly identify the client's expectations. In this case, the BIM manager will be able to propose detailed solutions to organise the production of BIM models that enable this use case implementation.

4.2. Phase 2: To Identify Information Requirements and Relevant Workflows

The BIM manager precisely determines all the parameters to be taken into account and integrated into the 4D planning. This preliminary work is important because it helps explain to project contributors how to adapt their traditional deliverables (schedules and models) in order to be usable within the framework of 4D planning. As such, the BIM manager can proceed in three steps as follows:

Step 1: To convert the client's objectives into SMART objectives if this has not been carried out in the specification document. Indeed, non-SMART goals are often open to interpretation, unfeasible, or incomplete.

Step 2: To determine what information must be produced to perform relevant 4D planning. When the 4D objectives require expertise that is beyond those of the BIM manager (e.g., environmental risk management), project actors, such as QSE (Quality Security Environment), CSPS (Construction Safety and Health Protection Coordination), and specialised BET (an engineering company), should be appointed by the client at the request of the BIM manager.

Step 3: To draw up a functional and pragmatic strategy and formalise it in the BIM Execution Plan (in French the convention BIM). Indeed, to share and use the identified information, a workflow must be put in place to specify: what digital tools are capable of exploiting the information; in which formats the information should be delivered; how the information should be produced, controlled and shared; and finally, what deliverables are expected.

4.3. Phase 3: To Design a Project Schedule Adapted to the Use of 4D BIM

To ensure quality, optimal coherence, and consistency between the BIM model(s) and the construction schedule, a working method is formalised, as follows:

Step 1: The provisional or final construction schedule (depending on the stage of the project) is obtained. This schedule will be the input to the design process described below.

Step 2: The MOE, OPC, and CSPS meet to study the general planning of the project and identify the phases that could represent health and safety risks for the project (e.g., complex construction methods, areas of high co-contracting, significant environmental risks, heavy traffic flow).

Step 3: Meetings are held as necessary between the OPC, contractors and project actors to study in detail the scheduling of tasks. The synthesis of all this information will make it possible to propose a preliminary solution that will then be integrated into the 4D BIM use case and using visualisation, will then be validated or amended.

Step 4: The OPC in charge of monitoring and coordinating the work must master traditional planning and 4D BIM based planning tools, because they are in charge of linking the planning to the BIM model. The BIM manager is also asked to verify that the BIM models have been modified according to the new granularity of the construction schedule.

This process should be formalised in a contractual document, such as the PGC (site management plan, in French Plan de Gestion de Chantier), under the responsibility of the project manager.

4.4. Phase 4: To Supervise BIM Models Production for 4D BIM Use Case

As noted earlier, it is common for French project stakeholders to partially achieve only certain BIM use cases initially required by the client. This is partly due to the BIM maturity and expertise of some project stakeholders and partly due to the disparity between the required and available LoDs. A further important factor is the way the BEP is elaborated. For the use of 4D BIM, it often consists of a brief and summary description contained in a section that is less than one page in length. The proposals below aim to optimise the content of the BEP to better elaborate and specify the implementation of 4D BIM.

Step 1: To explicitly mention the obligation to achieve 4D use and above all what its purposes are. Thanks to the work carried out in Phase 1, 4D applications are easily understood using the SMART objectives identified.

Step 2: To detail which actor must perform or contribute to which task(s) in order to achieve the 4D BIM use objectives. To do so, the RACI matrix (Responsible, Accountable, Consulted, Informed) was proposed. This project management tool [99] is usually used for classifying project stakeholders through a 'responsibility assignment matrix', thus developing a grid specifying who does what in the project. Here it was adapted by the authors for specifying the scheduling process (Figure 4) and the BIM model production (Figure 5) to facilitate the use of 4D BIM.

Step 3: To indicate in the general process describing the BIM model's production when the tasks related to the 4D BIM use case must be performed. This must be undertaken before the BIM models are checked and coordinated so that the BIM manager can control deliverables produced for executing the 4D BIM use case and verify the presence of the parameters that will serve to link the BIM models to the schedule.

Step 4: The process of 4D BIM implementation needs to be very detailed to indicate to the project actors the method adopted. It should begin with a reminder of the SMART objectives envisioned; the modelling rules to comply with the rationale for model structuring and breakdown; the parameters to be created and information on who provides them and when; information on when and how the work will be controlled; the criteria for choosing 4D BIM software; and information on who links the model to the schedule to produce the expected deliverables, and when this will occur.

Clarifying and detailing the process and procedure for implementing a BIM use case, whether it is 4D-BIM-related or not, allows project stakeholders to have a good understanding and clear idea of the tasks to be performed. The more the production and control process is explained, the more the probability of reaching the project objectives increases.

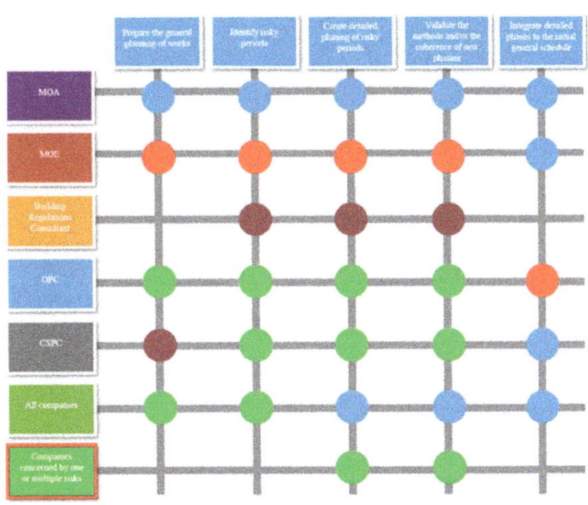

Figure 4. RACI matrix for implementing the scheduling process when using 4D BIM.

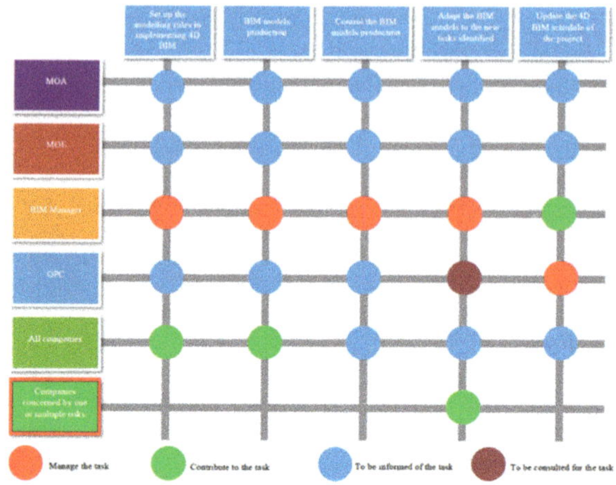

Figure 5. RACI matrix for BIM model production when using 4D BIM.

5. Case Study of the Nanterre 2 CESI Project

This first case study extends the 4D BIM literature by presenting empirical evidence on the implementation and use of 4D BIM tools during the construction phase of an educational building. The Nanterre 2 CESI project (Figure 6) is a four-storey building intended to increase the teaching capacity of the CESI campus in Nanterre. There were two reasons for adopting BIM for the design and delivery of the project. Firstly, so that the deliverables produced by the project participants could be saved at each phase in order to make them available to CESI teaching teams and serve as real case studies. Secondly, so that the information contained in the deliverables allowed CESI to manage the operation and maintenance of the building using BIM-based workflows.

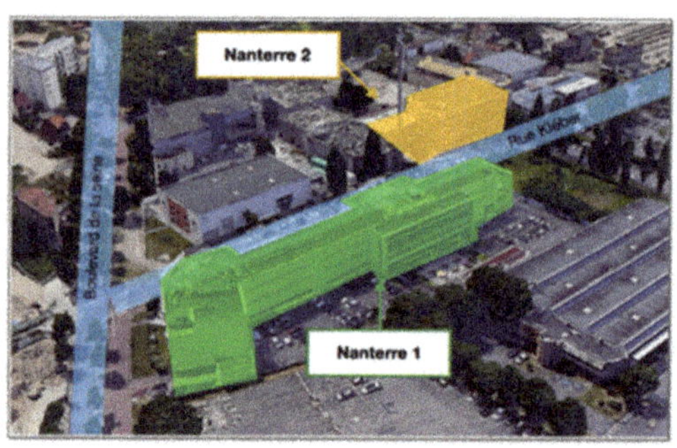

Figure 6. Location of Nanterre 2 (new) and Nanterre 1 (existing) buildings on Rue Kléber.

5.1. Nanterre 2 CESI Project

The Nanterre 2 CESI project is composed of: (i) a ground floor of 800 m^2: coworking and catering space; (ii) the first floor to the third floor (800 m^2 per floor): teaching spaces; (iii) the fourth floor (600 m^2): Living Lab. (i.e., BIM modelling, RA/RV, and digital simulation spaces); (iv) parking for employees, professors and/or researchers, and green spaces. Construction works were planned for a 12-month period (between 5th May 2018, and 31st May 2019). CESI, as client, specified the use of BIM for the design and construction of the Nanterre 2 building. Apart from its intrinsic commitment to the use of the latest available technologies, CESI also wanted to use the project as a case study that provided data for teaching and research programmes in digital construction. Therefore, in commissioning the Nanterre 2 building, the client, CESI, followed a similar process to that described in (the later published) ISO 19650: 2 [81]. This involves the 'appointing party' specifying, as part of its tender documents, a set of 'exchange information requirements' (EIRs) which prospective appointees then reflect in their tenders, and which form the basis (subject to pre-contract negotiation) of the appointed delivery team's BEP. In this case, a BIM specification document (equivalent to a set of EIRs [81]) was produced that included 10 BIM use cases. These were:

- Site modelling;
- Project communication;
- Project review;
- Analytical studies *;
- 4D BIM*;
- 5D BIM* (an application integrating the model with cost management activities);
- Management of design conflicts and clashes;
- Logistics support *;
- Regulatory compliance checking;
- Management of construction works and equipment *

Ultimately, however, for the works to proceed, it was agreed that the appointed delivery team could omit several (marked as *, above) of the BIM use cases. The main reasons for this were the deficiency of certain aspects of the requirement specifications as well as the diversity in the level of BIM expertise and maturity within the delivery team. One such omission was the requirement to provide 4D BIM. For this use case in particular, the BIM specification document only stated that a 4D software must be mastered, indicating Navisworks as an example. The mistake made was to only cite the type of software without specifying its capabilities. In this case, Navisworks was not capable of managing the construction site and monitoring, in real time, construction work progress. These were precisely the two uses of 4D BIM expected by CESI. The second shortcoming was due to the procedure described. It was simply indicated that a 3D model and a schedule must be linked to enable the extraction of a video animation and the phasing logbook (in French the carnet de phasage). This procedure was based on the Hardin and McCool method [96]. It was not indicated which 4D parameters should be produced, by whom, and how to ensure overall planning consistency. This being the case, and believing that it had sufficient in-house expertise, CESI decided to implement the 4D BIM planning of Nanterre 2 internally with the help of academic research teams that were familiar with digital construction. Their experience in doing so is the basis of the developmental case study reported here.

5.2. Application and Illustration

The application of the proposed 4D planning method is structured in four phases as follows:

5.2.1. Phase 1: To Express the 4D Objectives in an Initial BIM Specification Document

Table 1 proposes a reformulation of the 4D requirements so that they can be understood without ambiguity and therefore allow the project teams to propose relevant solutions to answer them.

Table 1. Phase 1: Expression of 4D objectives in the BIM specification document.

N°	Initial Formulation	SMART Formulation
1	The BIM model must include data related to equipment and temporary structures required for onsite construction works.	The use of 4D BIM must enable visualisation and analysis of space occupation by equipment and temporary structures when two companies work within the same area.
2	The update of the schedule must be conducted in real time and correspond to the exact progress of onsite construction works.	At any time during the construction phase, 4D BIM planning must enable the measuring of construction work progress within +/−3 days.

5.2.2. Phase 2: To Identify Information Requirements and Relevant Workflows

Step 2 (*here Step 1 is not necessary*): To determine the information to be produced for each 4D objective (Table 2), the BIM manager must understand how 4D BIM tools work. The more this BIM actor is aware of the latest technological advances, the more technically and economically viable its 4D planning strategy is. Thus, a review of existing 4D BIM tools may be necessary if this use case is not usually implemented by the BIM manager.

Table 2. 4D information requirements related to the 4D BIM use case.

4D BIM Objectives	Information to Be Produced
The use of 4D BIM must enable visualisation and analysis of space occupations by equipment and temporary structures when two companies work within the same area.	- Dates and durations of the co-contracting periods - Names of the companies concerned - Nature of the work - Locations of occupied areas - Dimensions and footprints of equipment/temporary structures - Equipment circulation areas - Equipment type
At any time during the construction phase, 4D BIM planning must enable the measuring of construction work progress within +/- 3 days.	- 4D parameters linked to the schedule - Project task flowchart level - Nature of the links between the tasks

Step 3: BIM modelling and collaboration platforms were imposed. A workflow was established based on these tool constraints. However, no 4D planning tool nor project participant with expertise in using this tool was involved in the project. Therefore, the following two assumptions were made:

Hypothesis 1. 'The BIM manager masters the 4D planning tool. If necessary, he can be called upon to resolve the technical problems encountered.'

Hypothesis 2. 'As soon as the phase is launched, under the responsibility of the BIM manager, the project OPC is introduced to BIM and sufficiently trained to master the 4D planning tool.'

As for the Nanterre 2 project, Hypothesis 2 was relevant, since the BIM manager had had expertise with regard to using Synchro Pro in previous projects. Therefore, the workflow proposed and implemented is illustrated in Figure 7.

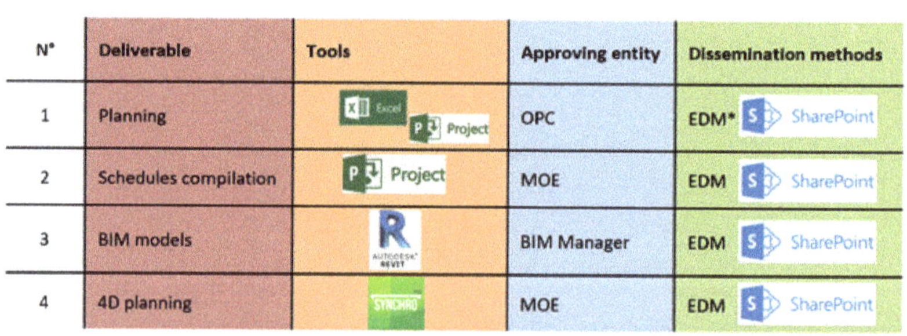

N°	Deliverable	Tools	Approving entity	Dissemination methods
1	Planning	Excel, Project	OPC	EDM* SharePoint
2	Schedules compilation	Project	MOE	EDM SharePoint
3	BIM models	Autodesk Revit	BIM Manager	EDM SharePoint
4	4D planning	Synchro	MOE	EDM SharePoint

*Electronic Document Management

Figure 7. Tools and methods for producing 4D BIM deliverables.

5.2.3. Phase 3: To Design a Project Schedule Adapted to the Use of 4D BIM

The process represented in Figure 8 can be attached to the project's BEP in order to guide stakeholders in designing and developing schedules that are suitable for the 4D BIM use case. The following steps should enable the BIM manager to transform the traditional project planning process by taking into account the 4D BIM objectives.

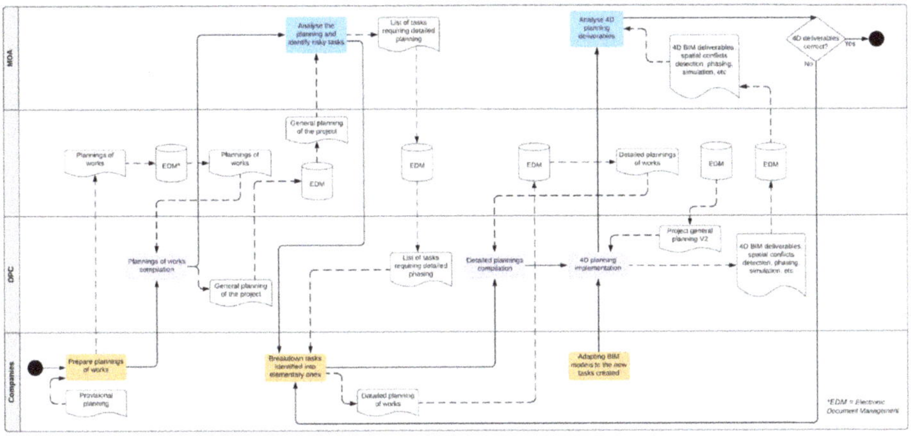

Figure 8. Summary of Phase 2's process - Design of the project schedule adapted for the use of 4D BIM.

Step 1: The traditional schedule is used as an input. This document is produced by the project OPC through combining and including all business planning of the project. The project tasks flowchart is of level 3 (Figure 9).

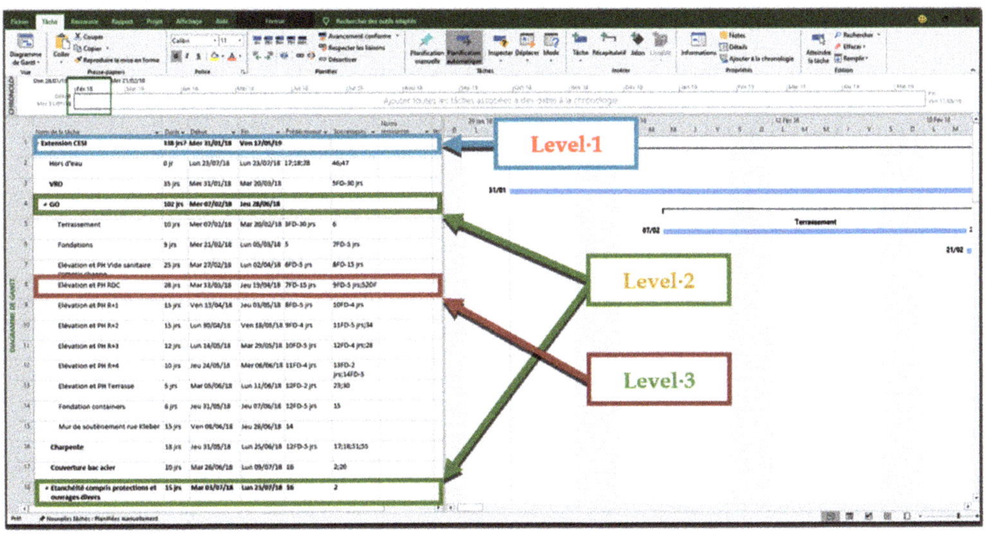

Figure 9. Screenshot of Nanterre 2 CESI project schedule.

Step 2: The first objective that must be studied is the identification of periods that could represent a certain risk and for which 4D planning can provide an added value in terms of safety and space optimisation. The previously identified actors must analyse the sequencing of the tasks of the initial planning and list each of its periods in order to study them in detail.

Step 3: The second objective is to produce the information listed by the BIM manager in Phase 2. Collaboration between project stakeholders should enable the production of precise information and allow for the performance of a first optimisation for the initial planning. Thereafter, 4D planning will enable the optimisation and the validation of the planning generated. At this stage, all of the 4D parameters related to the analysis of occupied spaces (4D BIM Objective N° #1) are known. As for the 4D objective linked to monitoring construction work progress, it is up to the OPC, possibly with the support of the BIM manager, to choose their own method. For the Nanterre 2 project, the initial schedule already included the links between the tasks. It therefore remains to determine the 4D parameter to be created so as to link the schedule to the BIM models, and to increase the level of the project tasks flowchart in order to optimise the precision of the progress monitoring process. Since the BIM model was created using Autodesk Revit software, a simple and straightforward solution is to create a text parameter and name it '*BIM_WBS*'. This parameter is created and assigned to all the BIM model objects. Then, the exact name of the planning task to which it must be linked needs to be entered. By doing so, it was possible to automatically assign all the objects of the BIM model to the project schedule. Table 3 summarises the information produced.

Table 3. Summary of the information produced at Step 3 of Phase 3.

4D BIM Objectives	Information to be Produced	Information Identification or Its Location
The use of 4D BIM must enable visualisation and analysis of space occupations by equipment and temporary structures when two companies work within the same area	Dates and duration of co-contracting periods	Planning
	Name of the companies concerned	Planning
	Nature of the work	Planning
	Location of occupied areas	Planning
	Dimensions and dimensions of equipment/temporary structures	Planning
	Equipment circulation areas	Planning
	Equipment type	Planning
At any time during the construction phase, 4D BIM planning must enable the measuring of construction work progress within +/- 3 days.	4D parameter linked to the schedule	'BIM_WBS': Planning
	Project task flowchart level	Planning
	Nature of the links between the tasks	Planning

Step 4: Phase 3 ends when all the information produced is gathered and combined. Using the initial schedule, the OPC transforms all tasks by integrating 4D information, and then shares the resulting schedule with the MOE for approval (Figure 10).

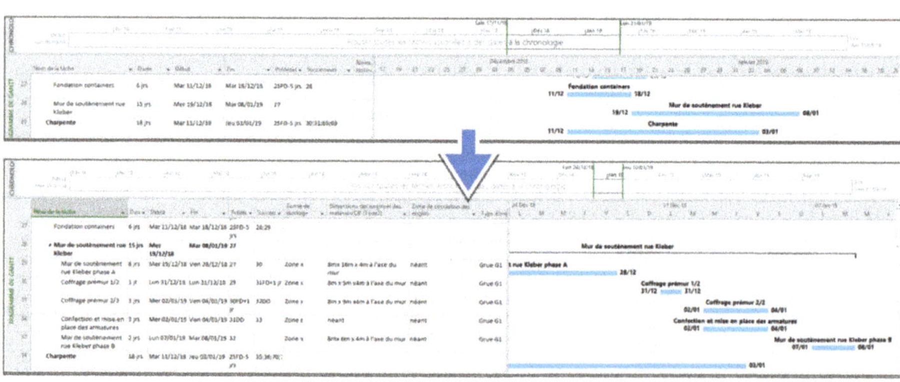

Figure 10. Evolution from initial planning to final planning (screenshot from Microsoft Project).

5.2.4. Phase 4: To Supervise BIM Models Production for 4D BIM Use Case

This phase allows the project contractors to integrate 4D BIM information into their BIM models to achieve the 4D BIM objectives of the project. By relying on the project BEP, construction companies must be able to use the information provided in the schedule resulting from phase 3.

Step 1: The BIM manager must indicate in the BEP the main BIM objectives of the project. At this stage, the project BIM contributors should be informed that a 4D planning process will be implemented. The BEP section following the general objectives must indicate in detail and in a SMART way, what the objectives of each BIM use case are, and in particular, the objectives related to 4D BIM.

Step 2: This step consists of precisely determining the degree of responsibility and involvement of each project participant towards the achievement of the 4D BIM objectives. Therefore, it was necessary to use the RACI matrix templates proposed in Figures 4 and 5 and substitute the actor names by the actual companies that will execute the project.

Step 3: The two production processes given in Figures 8 and 9 were integrated into the project BEP to explain to the BIM contributors how the 4D BIM use case should be implemented.

Step 4: Once the objectives have been identified and the production methods thoroughly detailed through Steps 1, 2 and 3 of the current Phase 4, the workflow to be implemented was clearly explained to enable efficient project communication and information exchange.

5.3. *4D BIM Planning Implementation*

In this section, the 4D planning of structural works of the Nanterre 2 CESI project is performed. The initial planning is first linked to the digital model without taking into account the 4D BIM objectives of the project. Then, the proposed method is applied to integrate the information produced in the previous subsections.

5.3.1. Primary 4D BIM Planning

First, the 4D parameter 'BIM_WBS' is created and integrated into the BIM model of structural works. Then, the BIM model is exported from the authoring platform using the Synchro Plug-in. The project schedule is also imported and the links between the tasks are preserved. Finally, the objects of the BIM model are assigned to the corresponding tasks. This first summary 4D BIM schedule, which can be exported in video format (.AVI) or in phasing form (.JPEG), allows project teams to check that the project workflow is correct. It is now possible to work directly on the 4D planning software to make changes to the project schedule. However, the BIM objectives of the project have not been achieved yet.

5.3.2. 4D BIM Planning According to the CESI Method

To test and validate the proposed method, it should be confronted with realistic co-contracting issues. To do so, the primary 4D planning performed in the previous subsection was used along with a detailed scenario including periods where four tasks overlap, and three companies 'coexist' on the construction site. The 4D parameters to be produced to meet the 4D BIM requirements of the Nanterre 2 project have been identified in Section 4.2. The OPC integrates this information into their planning (see process in Figure 8) while companies make the necessary changes to fill in the information in the BIM models (see process in Figure 11). The compilation of these two deliverables makes it possible to create a more detailed 4D plan which was used to validate and/or optimise constructive choices. Figure 12 illustrates the result of 4D BIM planning. By creating the spaces linked to each activity, an analysis of the workflow can be performed to detect hazard sources and reduce their impact. In this planning scenario, it has been detected that the work areas on the ground floor overlap during HVAC and ELEC interventions. Preventive actions can therefore be taken to manage the space where these two activities are located so that no hazards due to this co-contracting slow down the work progress.

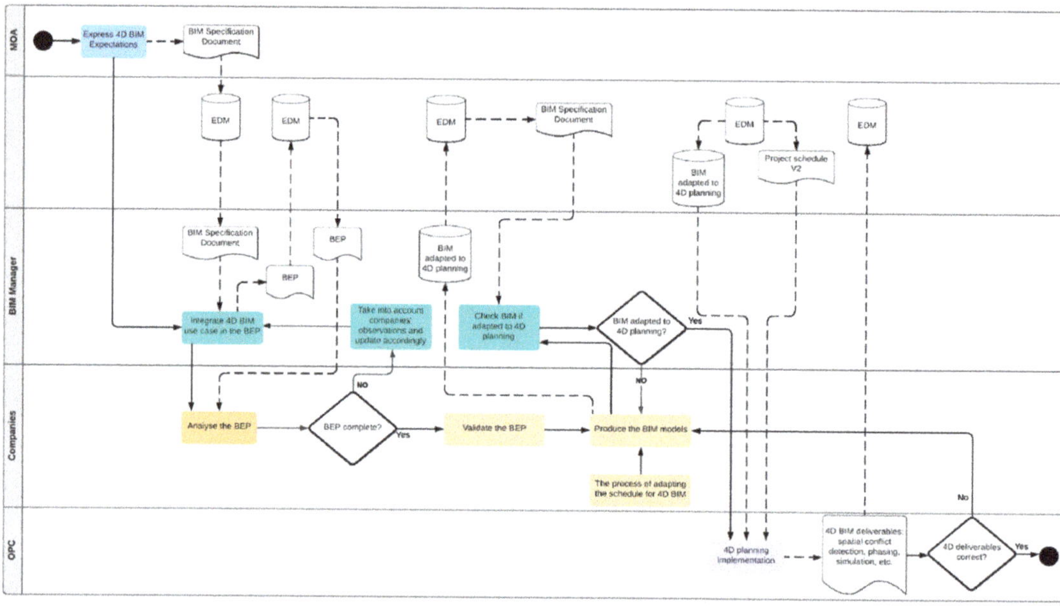

Figure 11. Summary of the Phase 3 process - Supervision of BIM model production for 4D BIM use case.

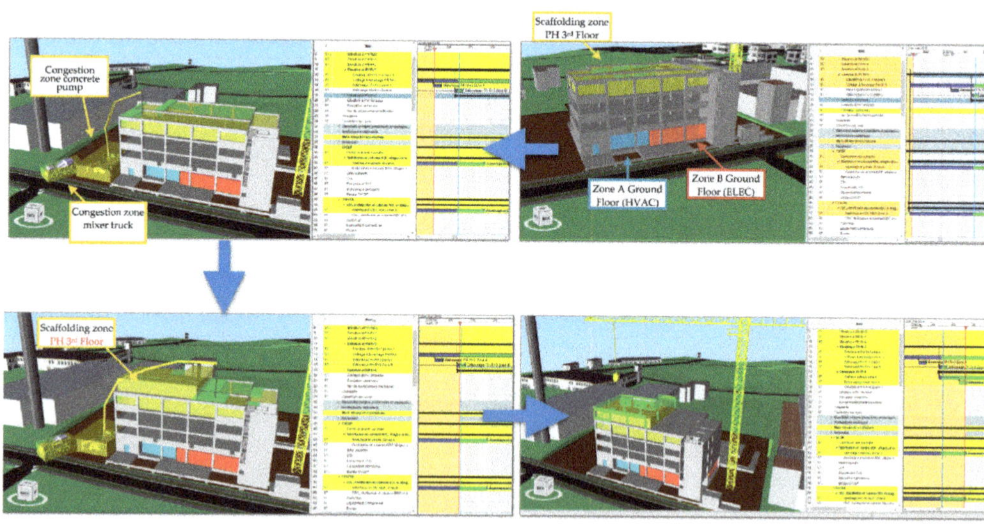

Figure 12. 4D BIM planning of the Nanterre 2 CESI project using the CESI method.

The 4D BIM planning performed using the proposed method enables the achievement of the client's objectives better than the primary 4D planning. Indeed, the use of 4D information, identified from the second phase of this method, allowed the actors to focus effectively on the expectations of the MOA (namely, the evaluation of the result of 4D BIM planning with regard to the 4D objectives) as shown in Table 4.

Table 4. Proposed method evaluation regarding the client's 4D BIM objectives achievement.

4D BIM Objectives	Evaluation
The use of 4D BIM must enable visualisation and analysis of space occupations by equipment and temporary structures when two companies work within the same area	- The zones have been identified and taken into account for the planning of the works. - An analysis of the zone interfaces was performed, which made it possible to detect an overuse of certain spaces. - The equipment circulation zone has been taken into account by means of '3D paths' linked to the tasks.
At any time during the construction phase, 4D BIM planning must enable the measuring of construction work progress within +/−3 days	As shown in Figure 12, the new 4D planning required the creation of new sub-tasks and therefore the increasing of the level of the project tasks flowchart. This makes it possible to specify the exact nature of the work to be carried out and therefore to assess more precisely its progress.

6. Automating 4D BIM Planning: The RINNO Case Study

Research leveraging the use of 4D BIM data using AI tools is lacking [13] although such an integration is crucial to optimise and develop more effective strategies for construction project planning through the development of automated tools to, for example, automatically generate and simulate different scenarios [40]. Furthermore, there is a dearth of methods and tools dedicated to renovation project management. New construction projects have usually been prioritised in terms of design, planning and management tool development. When possible, these tools are adapted to the context of renovation projects [100] which represents one of the main factors that make their performance typically lower than that for new constructions [101,102]. The research study reported in this section is part of the RINNO research project [103,104] which aims to develop a holistic multi-disciplinary platform that will ensure the acceleration of the rate of deep renovation in EU residential buildings. Here, an ontology is introduced, and its application is illustrated within the context of building renovation. Ontologies are useful AI tools in formalising specific domain knowledge, including their concepts, relations, and constraints [41]. They enable process automation and tool development as they provide a machine-readable representation of knowledge.

6.1. *Renovation 4D Planning Ontology*

Figure 13 presents the ontology developed within the RINNO project [103,104] dedicated to generating 4D BIM schedules for renovation projects. The overview given in Figure 13 is a UML (Unified Modelling Language) class diagram illustrating the ontological concepts, their relations and constraints, as well as the attributes or properties that define each concept (here class or entity) to facilitate its implementation as a renovation knowledge base in the case of this study.

As presented in Figure 13, a 'Built Asset' is composed of several 'Building Elements' (e.g., windows, walls) and WBSs (Work Breakdown Structures, e.g., floors, external façades) where each 'Building Element' belongs to a WBS. A 'Building Element' requires one or many 'Renovation Activity' components to be renovated, and this may be carried out by installing one or many 'Innovative Products' (e.g., photovoltaic panels, solar collectors). Each 'Renovation Activity' is temporally constrained by activities that should start and finish before it, whereas some others will be triggered and executed after its completion. It also requires a set of 'Material', 'Workforce', and 'Equipment' so it can be executed and may cause 'Disruption' to the building occupants. The 'Disruption' concept is beyond the scope of this paper and will be detailed in future research, since its estimation and simulation is one of the RINNO project targets.

The ontology proposed was populated into a database using the SQL Server. This database included several tables, and each one enabled the implementation of a concept from the ontology. The tables describe in detail data related to renovation activities, their sequencing rules, constraints, duration, cost, equipment, etc., all gathered, structured, and verified with the help of the RINNO project partners.

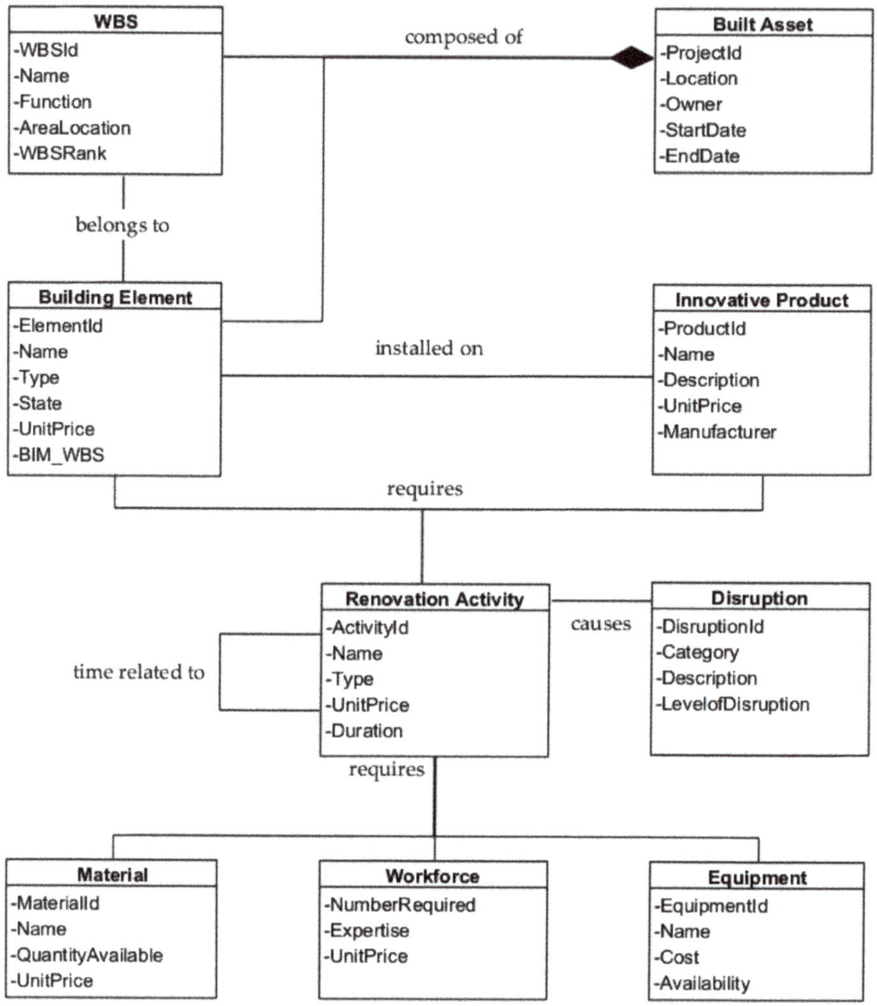

Figure 13. Renovation 4D planning ontology—UML class diagram.

6.2. Automated 4D BIM Planning and Simulation Process

Figure 14 presents the system architecture of the digital tool developed and Figure 15 the automated 4D BIM planning process implemented; both are based on the ontology introduced in the previous subsection. The automated process, represented here as a UML scenario diagram, enables the BIM manager to leverage the BIM data and automatically generate and simulate the 4D BIM planning based on a renovation scenario identified at the beginning of the process. Indeed, after checking the BIM model and selecting a renovation scenario via two main GUIs (Graphical User Interfaces) (Figure 16C,D), the 'RINNO Renovation Engine' component coordinates the 4D BIM planning process by: (i) updating the BIM model if necessary to integrate project-related information (Figure 16A); (ii) generating traditional (Gantt) planning using (a) the scenario selected by the BIM manager in order to identify WBS, renovation activities to be performed, and equipment used, (b) the BIM model to extract BIM elements considered by the scenario simulated and corresponding quantity take-off (QTO), and (c) the ontology to map activities to WBS and BIM elements and assign relevant equipment to activities and estimate their durations; (iii) updating the BIM model using the traditional planning generated, particularly by integrating the activities data and initialising the 'WBS_BIM' parameter; and (iv) simulating the 4D BIM planning created through the BIM management tool.

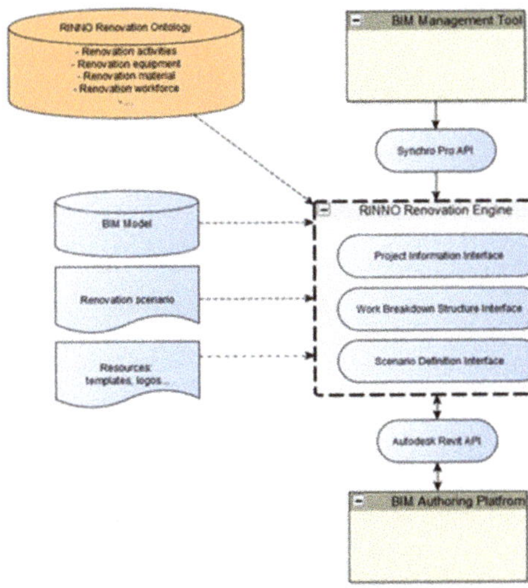

Figure 14. System architecture of the 4D BIM digital tool.

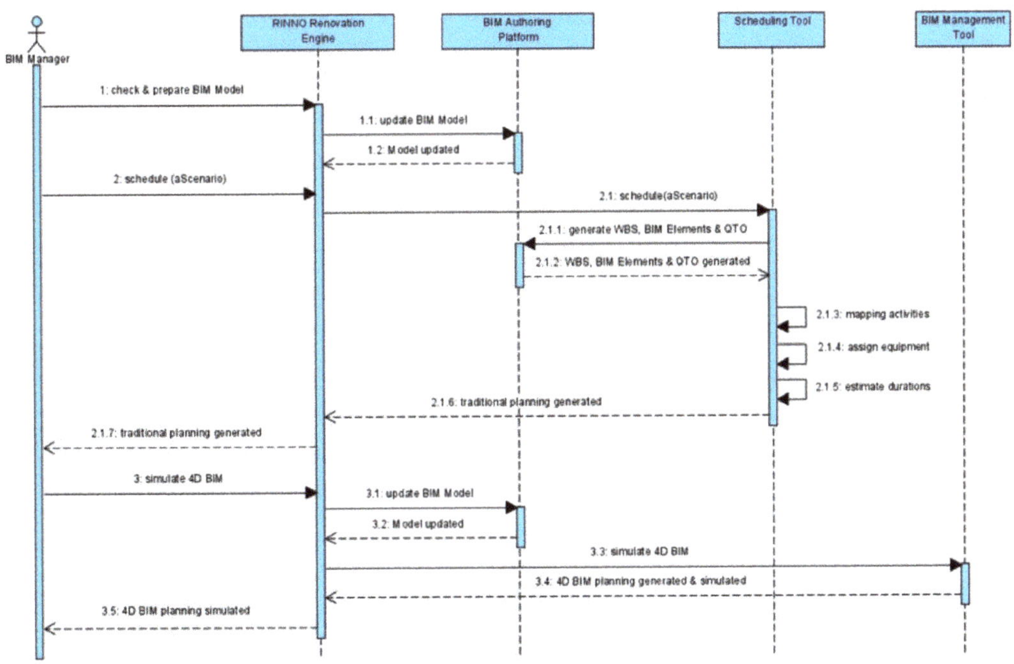

Figure 15. Automated 4D BIM planning and simulation.

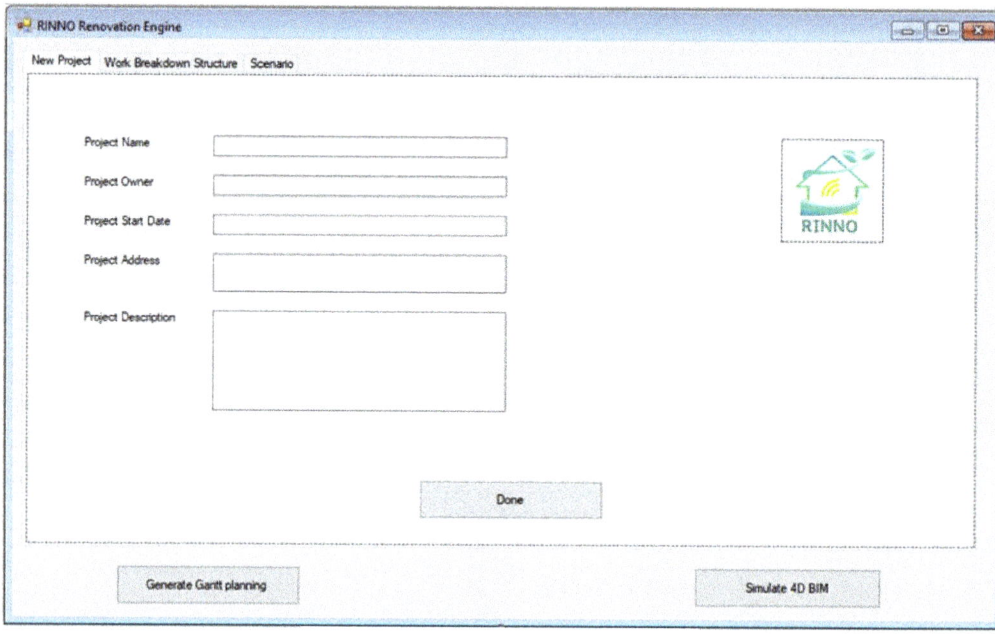

(A)

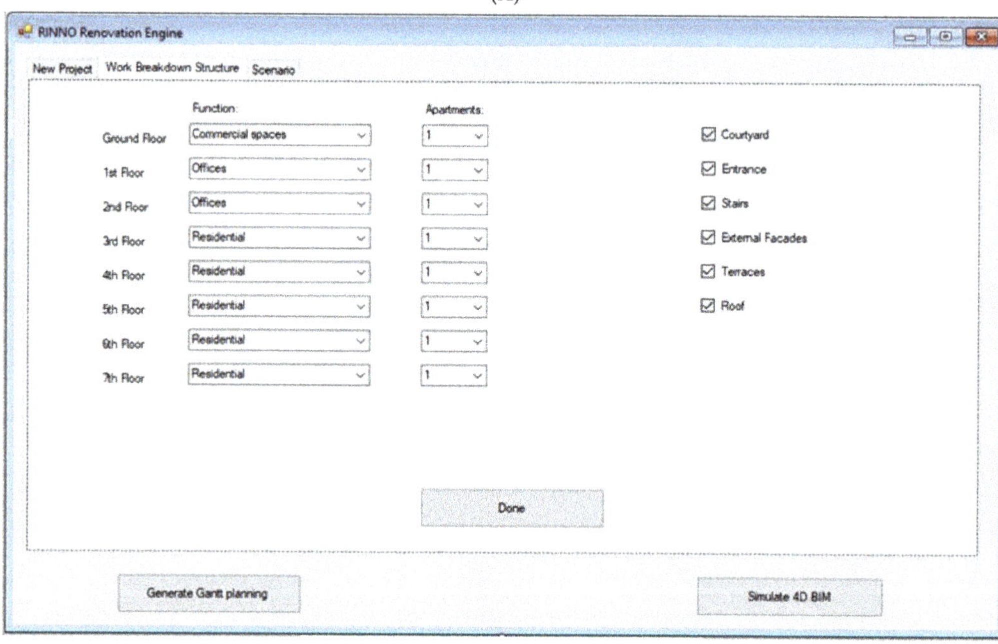

(B)

Figure 16. *Cont.*

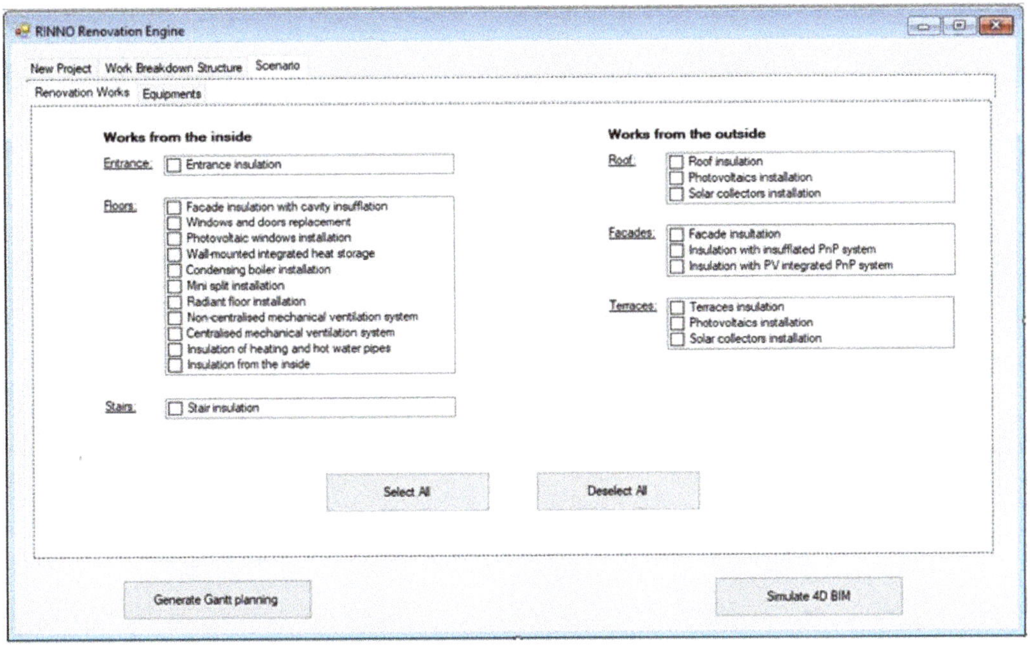

(C)

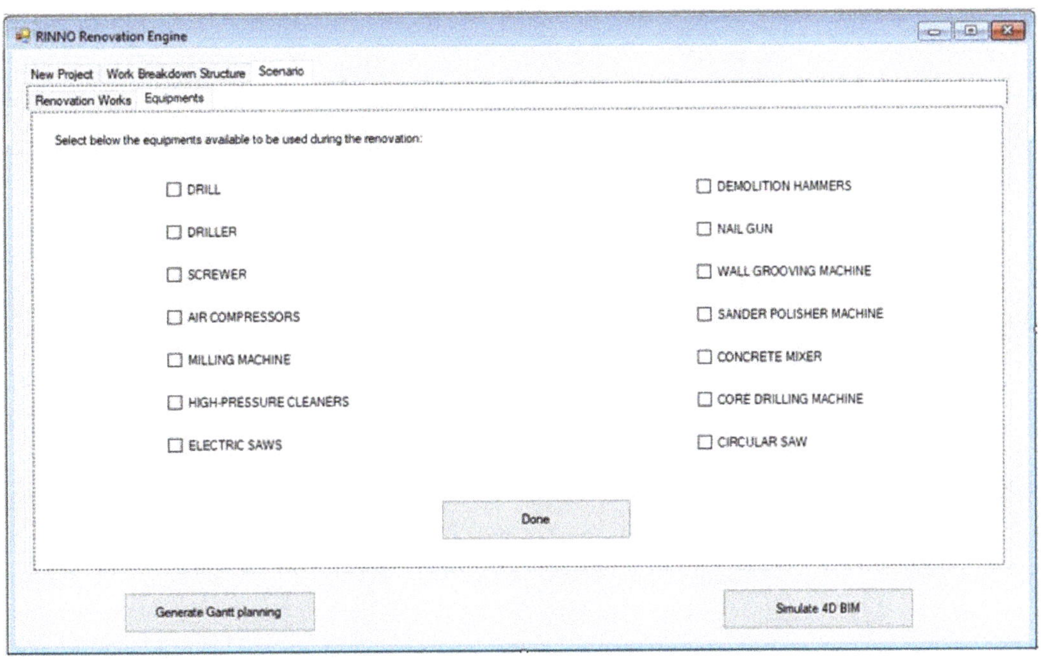

(D)

Figure 16. RINNO Renovation Engine GUI. (**A**) Project details interface. (**B**) WBS interface. (**C**) Scenario interface 1 for renovation activities selection. (**D**) Scenario interface 2 for renovation equipment selection.

Figure 16 illustrates the GUI of the RINNO Renovation Engine developed to: (i) facilitate interaction with the tool; (ii) automate and assist users for renovation scenario definition and generation; and (iii) streamline the whole automated 4D planning and simulation process.

7. Discussion, Conclusions, and Perspectives

This work was an opportunity to study project management in general, and then to look at the new methods that have emerged with recent technological advances. Traditional planning methods are often ineffective in taking into account all the uncertainties and hazards that may occur during a construction project. As a result, time and cost targets are rarely met. BIM represents an evolution of these practices and can be a useful way to optimise planning methods. Numerous studies show that 4D BIM helps to reduce planning errors, promotes collaboration between stakeholders, and helps project teams make effective decisions.

7.1. The Nanterre 2 CESI Project Case Study

The first case study showed that not all actors can afford to innovate and adopt BIM for project management. Although the use of 4D BIM was specified and required by the client, it was not implemented in the Nanterre 2 CESI project, mainly due to a lack of practical and simple methods of implementation, as well as the low and heterogeneous level of BIM training and expertise of the project actors.

To contribute to the democratisation of the 4D BIM use case, a structured and practical method was proposed. This method was based on a four-step process to enable project participants to (i) clearly and in a SMART way, express the 4D BIM objectives from the beginning of the project in the BIM specification document, (ii) identify information requirements and relevant workflows to achieve these objectives, (iii) design an adapted 4D BIM plan, and (iv) supervise the BIM model's production to enable the implementation of the 4D BIM use case. The proposed method promotes collaboration between project actors and guides them towards information production and management. Pragmatic and detailed processes and methods were introduced and applied on the Nanterre 2 CESI project to implement the 4D BIM use case and related objectives required by the client. This implementation enabled the illustration of some of the advantages 4D BIM can bring to a construction project. The effective adoption and development of 4D BIM can provide efficient tools for monitoring work progress and so represents a significant added value to ensure that time and cost constraints, when initially planned, are constantly satisfied. Sacks et al. [105] proposed a matrix that contains 56 interactions between Lean methodology principles and BIM functionalities to clarify possible synergies between the two methodologies. Using this matrix as a starting point to develop the proposed method would have enabled more pragmatism and better streamlined and optimised processes alongside the workflows proposed in this paper.

Furthermore, it should be noted that objective N° #2 (i.e., 'At any time during the construction phase, 4D BIM planning must enable measuring construction works progress within +/- 3 days') was difficult to achieve. This objective should be studied in detail so as to clearly understand what the actual intention of the client was. Objectively, 4D BIM provides very little added value for monitoring the work progress when data are not collected in time and the BIM models are not updated regularly. Indeed, the accuracy of the work progress depends on the frequency with which the OPC carries out their onsite supervision and monitoring missions. In general, clients have little knowledge of the possibilities and limits of digital tools. It would be very difficult to visualise the 'exact' onsite work progress as it would require a BIM model containing each element, equipment, or work structure installed and used, temporary or permanently, during each construction phase. Therefore, the assistance of an AMO BIM to help the client express and formalise their requirements in a SMART manner is crucial. Furthermore, IoT and autonomous systems [106] can play a fundamental role in collecting onsite data efficiently and regularly so to enable the automatic and regular updating of BIM data.

Moreover, the proposed method, if it were to be implemented in more typical circumstances, would require certain levels of collaboration between key project actors. The two RACI matrices (Figures 4 and 5) show the required interactions between key project actors for the effective integration of the two processes (construction schedule and BIM model creation). The collaboration of these key project actors is crucial. The issue of collaboration is frequently raised when the realisation of the benefits of BIM are considered. Increased collaboration is not only a prospective beneficial outcome of BIM adoption [107]; it is a prerequisite for the full attainment of these potential benefits [108]. Thus, studies have suggested that the realisation of BIM's key benefits relies upon the degree of collaboration achieved and that this is not achievable with traditional project procurement approaches [109–111]. Objectively, the adoption of 4D BIM provides very little added value without the necessary and timely integration of construction scheduling and BIM model creation; a process that is difficult to achieve in traditional project frameworks.

The Nanterre 2 CESI building has been open to students since September 2019. The operation and maintenance management of this building is expected to be implemented using BIM technology. To undertake this, it is essential to implement and use a classification system such as Uniclass 2015. Based on a set of consistent and hierarchically organised tables, Uniclass 2015 allows the classification of all types of elements that could be considered in the context of a construction project, from the most complex items such as industrial or residential complexes, to the most detailed items such as door locks or the covering and finishing of products. This system also allows for the classification of physical objects using the 'Entities', 'Elements', 'Systems' and 'Products' tables, as well as construction processes and activities through the 'Activities' table. Consequently, it allows for the simultaneous integration of PBS (Product Breakdown Structure) and WBS [51] and therefore the implementation of BIM for the operation phase in which the building objects are linked to their respective operation and maintenance activities. This ontological link between objects and activities should also enable the standardisation and automation of 4D BIM planning and simulation for both construction and operation phases.

7.2. The RINNO Project Case Study

This paper also introduced an ontology that is dedicated to 4D BIM planning within the context of renovation projects and demonstrated its applicability and value by designing and developing the RINNO Renovation Engine to enable automatic 4D planning generation and simulation. Ontologies are known to be powerful AI tools that allow the formalization of specific domain knowledge by providing it in a machine-readable format. To allow users to test and simulate different 4D planning strategies, the automated 4D planning process proposed enables 4D scenario identification and generation in a simple, interactive, and automatic way through a set of user-friendly interfaces. In the literature, ontologies are usually implemented using the Protégé platform [112]. Based on expert's input and knowledge from renovation engineering documents, Amorocho et al. [99] developed an ontology restricted to the installation of common renovation products, such as windows, HVAC components, and external thermal insulation panels. The 4D BIM ontology proposed in this paper considers the general case of renovation projects and so includes all related activities: innovative products installation as well as general activities, such as 'façade insulation with cavity insufflation'. Furthermore, to improve interoperability and interfacing between the BIM authoring tool (Autodesk Revit), the BIM management tool (Synchro Pro), and the scheduling tool (MS Excel), the RINNO Renovation Engine was developed as a plugin using the C# programming language, and the ontology was implemented as a knowledge base and populated using both the SQL Server tool and the RINNO partners' expertise and knowledge. However, further research is needed in order to extend and validate the content of the domain expert knowledge in terms of renovation activities, equipment, innovative products, sequencing rules and constraints between activities, costs, etc. For that, a survey has already been launched across many EU countries, and the results will be soon reported. Since costs data will be validated by the survey, it can be then integrated to implement a 5D BIM use case, thus enabling the economic assessment and monitoring of renovation projects.

Disruption was one of the concepts represented through the UML class diagram of the 4D BIM planning ontology; however, it was not further developed nor detailed. This concept usually refers to a disturbance which interrupts, diminishes, or alters the usual functioning of an activity, service, or system, and its impacts on the overall efficiency and productivity of the project [113]. Disruptions and delays are two interrelated concepts. While disruptions can cause delays in the progress of construction works, delays similarly can generate disruptions and a loss of productivity. Although renovation may be disruptive per definition, project participants should ensure the managing and minimisation of the impact on occupants. To build trust and minimise this impact, communicating with residents and explaining the renovation process in advance can play a crucial role [114]. Certainly, BIM-based disruption simulations, along with detailed 4D BIM planning for retrofitting works, can help address this need, which is one of the RINNO project targets that will be presented in future.

Ultimately, to validate and evaluate its impact, the automated 4D BIM planning tool (i.e., the RINNO Renovation Engine) should be tested and applied using real case studies. For this, the RINNO project and its partners offer four demonstration sites in France, Denmark, Greece, and Poland, respectively (Figure 17) with a total floor area of 3 386 m2 including both single-family and multi-family residential dwellings. These buildings reside in different climatic regions, and comply with diverse building codes and regulations, having been built using very different construction components and tools, and they are equipped with different construction systems and other building amenities. Previous research [115] recommended simplifying the 4D BIM model to ensure it does not contain too many unnecessary details and that it enables the clear visualisation of retrofitting works. For the design and planning phase, other research [116,117] specified that LoD 200 should be adopted. These hypotheses, along with the BIM model with granularity relevant to 4D simulations, will be thoroughly studied while demonstrating the 4D automated process using the RINNO's demo sites.

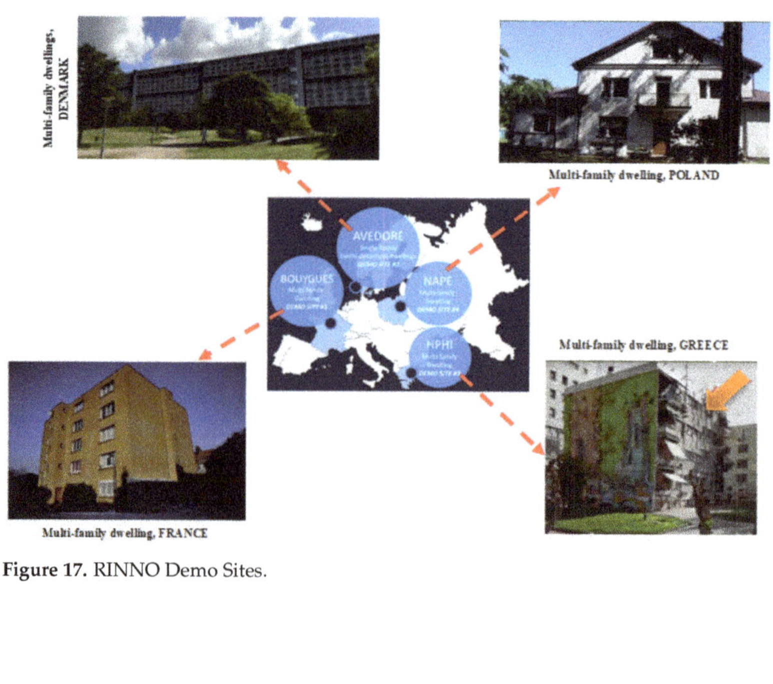

Figure 17. RINNO Demo Sites.

Author Contributions: Conceptualization, O.D.; methodology, O.D.; software, O.D.; validation, O.D., B.S., D.G.; formal analysis, O.D.; investigation, O.D.; resources, O.D.; data curation, O.D.; writing—original draft preparation, O.D., B.S. and D.G.; writing—review and editing, O.D., B.S. and D.G.; visualization, O.D., B.S. and D.G.; supervision, O.D.; project administration, O.D.; funding acquisition, O.D., B.S. and D.G. All authors have read and agreed to the published version of the manuscript.

Funding: This work was funded by the European Union's Horizon 2020 Research and Innovation Programme through the RINNO project (https://RINNO-h2020.eu/ accessed on 1 August 2022 [103,104]) under Grant Agreement Number 892071.

Institutional Review Board Statement: Not applicable.

Informed Consent Statement: Not applicable.

Data Availability Statement: All data, models and code generated or used during the study appear in the submitted article.

Acknowledgments: The authors would like to gratefully acknowledge the useful assistance, help and support received from Ridha Bensahaila, Célia Guilloteau, Jean-Daniel Penot, and David Failly during the development of this work.

Conflicts of Interest: The authors declare no conflict of interest.

Appendix A

Table A1. French terminology for project actors with English equivalents.

French Role	Abbreviation	English Equivalent
Maitre d'ouvrage	MOA	Client, Employer, Owner
Maitre d'œuvre	MOE	[varies] Client's Representative, Project Manager, Architect, Engineer
Assistant à la maitrise d'ouvrage	AMO	On-site assistant to the above
Ordonnancement, planning, coordination	OPC	Consultant planning specialist [unusual in UK context]
Coordinateur santé et protection de la sécurité	CSPS	Health and Safety consultant. Planning supervisor.
Assistant à la maitrise d'ouvrage BIM	AMO (BIM)	BIM Manager

References

1. Cavka, H.B.; Staub-French, S.; Poirier, E.A. Developing Owner Information Requirements for BIM-Enabled Project Delivery and Asset Management. *Autom. Constr.* **2017**, *83*, 169–183. [CrossRef]
2. Buisman, A. *How Are Engineering and Construction Companies Adapting Digital to Their Businesses;* Ernst&Young: London, UK, 2018.
3. Camacho, A.; Cañizares, P.C.; Estévez, S.; Núñez, M. A Tool-Supported Framework for Work Planning on Construction Sites Based on Constraint Programming. *Autom. Constr.* **2018**, *86*, 190–198. [CrossRef]
4. Conlin, J.; Retik, A. The Applicability of Project Management Software and Advanced IT Techniques in Construction Delays Mitigation. *Int. J. Proj. Manag.* **1997**, *15*, 107–120. [CrossRef]
5. Importance of Scheduling in Construction Projects. Available online: https://theconstructor.org/construction/const-management/importance-scheduling-construction-projects/1710/ (accessed on 22 April 2022).
6. Egwim, C.N.; Alaka, H.; Toriola-Coker, L.O.; Balogun, H.; Sunmola, F. Applied Artificial Intelligence for Predicting Construction Projects Delay. *Mach. Learn. Appl.* **2021**, *6*, 100166. [CrossRef]
7. Chen, G.-X.; Shan, M.; Chan, A.P.C.; Liu, X.; Zhao, Y.-Q. Investigating the Causes of Delay in Grain Bin Construction Projects: The Case of China. *Int. J. Constr. Manag.* **2019**, *19*, 1–14. [CrossRef]
8. Aravindhan, C.; Santhoshkumar, R.; Bonny, K.; Vidhya, K.; Manishankar, S.; Dhamodharam, P. Delay Analysis in Construction Project Using Primavera & SPSS. *Mater. Today Proc.* **2021**, *7*, 186. [CrossRef]
9. Robinson, T.G. *Global Construction Market to Grow $8 Trillion by 2030: Driven by China, US and India;* Global Construction Perspectives and Oxford Economics: London, UK, 2015; p. 44.
10. Flyvbjerg, B. What You Should Know about Megaprojects and Why: An Overview. *Proj. Manag. J.* **2014**, *45*, 6–19. [CrossRef]
11. Chen, S.-M.; Chen, P.-H.; Chang, L.-M. Simulation and Analytical Techniques for Construction Resource Planning and Scheduling. *Autom. Constr.* **2012**, *21*, 99–113. [CrossRef]
12. Yogesh, G.; Hanumanth Rao, C. A Study on Linear Scheduling Methods in Road Construction Projects. *Mater. Today Proc.* **2021**, *47*, 5475–5478. [CrossRef]

13. Gondia, A.; Siam, A.; El-Dakhakhni, W.; Nassar, A.H. Machine Learning Algorithms for Construction Projects Delay Risk Prediction. *J. Constr. Eng. Manag.* **2020**, *146*, 04019085. [CrossRef]
14. Sacks, R.; Eastman, C.; Lee, G.; Teicholz, P. *BIM Handbook: A Guide to Building Information Modeling for Owners, Designers, Engineers, Contractors, and Facility Managers*, 3rd ed.; Wiley: Hoboken, NJ, USA, 2018; ISBN 978-1-119-28753-7.
15. Dashti, M.S.; RezaZadeh, M.; Khanzadi, M.; Taghaddos, H. Integrated BIM-Based Simulation for Automated Time-Space Conflict Management in Construction Projects. *Autom. Constr.* **2021**, *132*, 103957. [CrossRef]
16. Bortolini, R.; Formoso, C.T.; Viana, D.D. Site Logistics Planning and Control for Engineer-to-Order Prefabricated Building Systems Using BIM 4D Modeling. *Autom. Constr.* **2019**, *98*, 248–264. [CrossRef]
17. Moon, H.; Dawood, N.; Kang, L. Development of Workspace Conflict Visualization System Using 4D Object of Work Schedule. *Adv. Eng. Inform.* **2014**, *28*, 50–65. [CrossRef]
18. Popov, V.; Juocevicius, V.; Migilinskas, D.; Ustinovichius, L.; Mikalauskas, S. The Use of a Virtual Building Design and Construction Model for Developing an Effective Project Concept in 5D Environment. *Autom. Constr.* **2010**, *19*, 357–367. [CrossRef]
19. American Institute of Architects. *G202-2013 Project BIM Protocol*; AIA: Columbia, SC, USA, 2013.
20. Koutamanis, A. Dimensionality in BIM: Why BIM Cannot Have More than Four Dimensions? *Autom. Constr.* **2020**, *114*, 103153. [CrossRef]
21. Ding, L.; Zhou, Y.; Akinci, B. Building Information Modeling (BIM) Application Framework: The Process of Expanding from 3D to Computable ND. *Autom. Constr.* **2014**, *46*, 82–93. [CrossRef]
22. Staub-French, S.; Khanzode, A. 3D and 4D Modeling for Design and Construction Coordination: Issues and Lessons Learned. *Electron. J. Inf. Technol. Constr.* **2007**, *12*, 381–407.
23. Mazars, T.; Francis, A. Chronographical Spatiotemporal Dynamic 4D Planning. *Autom. Constr.* **2020**, *112*, 103076. [CrossRef]
24. Mesároš, P.; Smetanková, J.; Mandičák, T. The Fifth Dimension of BIM—Implementation Survey. *IOP Conf. Ser. Earth Environ. Sci.* **2019**, *222*, 012003. [CrossRef]
25. Charef, R.; Alaka, H.; Emmitt, S. Beyond the Third Dimension of BIM: A Systematic Review of Literature and Assessment of Professional Views. *J. Build. Eng.* **2018**, *19*, 242–257. [CrossRef]
26. Sheikhkhoshkar, M.; Pour Rahimian, F.; Kaveh, M.H.; Hosseini, M.R.; Edwards, D.J. Automated Planning of Concrete Joint Layouts with 4D-BIM. *Autom. Constr.* **2019**, *107*, 102943. [CrossRef]
27. Antwi-Afari, M.F.; Li, H.; Pärn, E.A.; Edwards, D.J. Critical Success Factors for Implementing Building Information Modelling (BIM): A Longitudinal Review. *Autom. Constr.* **2018**, *91*, 100–110. [CrossRef]
28. Zhou, Y.; Ding, L.; Wang, X.; Truijens, M.; Luo, H. Applicability of 4D Modeling for Resource Allocation in Mega Liquefied Natural Gas Plant Construction. *Autom. Constr.* **2015**, *50*, 50–63. [CrossRef]
29. Huang, T.; Kong, C.W.; Guo, H.L.; Baldwin, A.; Li, H. A Virtual Prototyping System for Simulating Construction Processes. *Autom. Constr.* **2007**, *16*, 576–585. [CrossRef]
30. Tran, S.V.-T.; Khan, N.; Lee, D.; Park, C. A Hazard Identification Approach of Integrating 4D BIM and Accident Case Analysis of Spatial–Temporal Exposure. *Sustainability* **2021**, *13*, 2211. [CrossRef]
31. Sloot, R.N.F.; Heutink, A.; Voordijk, J.T. Assessing Usefulness of 4D BIM Tools in Risk Mitigation Strategies. *Autom. Constr.* **2019**, *106*, 102881. [CrossRef]
32. Kim, K.; Lee, Y.-C. Automated Generation of Daily Evacuation Paths in 4D BIM. *Appl. Sci.* **2019**, *9*, 1789. [CrossRef]
33. Chen, P.-H.; Nguyen, T.C. A BIM-WMS Integrated Decision Support Tool for Supply Chain Management in Construction. *Autom. Constr.* **2019**, *98*, 289–301. [CrossRef]
34. Guerra, B.C.; Leite, F.; Faust, K.M. 4D-BIM to Enhance Construction Waste Reuse and Recycle Planning: Case Studies on Concrete and Drywall Waste Streams. *Waste Manag.* **2020**, *116*, 79–90. [CrossRef]
35. Costin, A.; Adibfar, A.; Hu, H.; Chen, S.S. Building Information Modeling (BIM) for Transportation Infrastructure—Literature Review, Applications, Challenges, and Recommendations. *Autom. Constr.* **2018**, *94*, 257–281. [CrossRef]
36. Hosseini, M.R.; Maghrebi, M.; Akbarnezhad, A.; Martek, I.; Arashpour, M. Analysis of Citation Networks in Building Information Modeling Research. *J. Constr. Eng. Manag.* **2018**, *144*, 04018064. [CrossRef]
37. Candelario-Garrido, A.; García-Sanz-Calcedo, J.; Reyes Rodríguez, A.M. A Quantitative Analysis on the Feasibility of 4D Planning Graphic Systems versus Conventional Systems in Building Projects. *Sustain. Cities Soc.* **2017**, *35*, 378–384. [CrossRef]
38. Hartmann, T.; Gao, J.; Fischer, M. Areas of Application for 3D and 4D Models on Construction Projects. *J. Constr. Eng. Manag.* **2008**, *134*, 776–785. [CrossRef]
39. Mahalingam, A.; Kashyap, R.; Mahajan, C. An Evaluation of the Applicability of 4D CAD on Construction Projects. *Autom. Constr.* **2010**, *19*, 148–159. [CrossRef]
40. Blanco, J.L.; Fuchs, S.; Parsons, M.; Ribeirinho, M.J. Artificial Intelligence: Construction Technology's next Frontier/McKinsey. Available online: https://www.mckinsey.com/business-functions/operations/our-insights/artificial-intelligence-construction-technologys-next-frontier (accessed on 23 April 2022).
41. Hartmann, T.; Trappey, A. Advanced Engineering Informatics—Philosophical and Methodological Foundations with Examples from Civil and Construction Engineering. *Dev. Built Environ.* **2020**, *4*, 100020. [CrossRef]
42. Ait-Lamallam, S.; Yaagoubi, R.; Sebari, I.; Doukari, O. Extending the IFC Standard to Enable Road Operation and Maintenance Management through OpenBIM. *ISPRS Int. J. Geo-Inf.* **2021**, *10*, 496. [CrossRef]

43. ISO. ISO 21500: 2021. Available online: https://www.iso.org/cms/render/live/en/sites/isoorg/contents/data/standard/07/57/75704.html (accessed on 16 April 2022).
44. Kelley, J.E. Critical-Path Planning and Scheduling: Mathematical Basis. *Oper. Res.* **1961**, *9*, 296–320. [CrossRef]
45. Boton, C. Conception de Vues Métiers Dans Les Collecticiels Orientés Service. Vers Des Multi-Vues Adaptées Pour La Simulation Collaborative 4D/ND de La Construction. Ph.D. Theses, Université de Lorraine, Nancy, France, 2013.
46. Heaton, J.; Parlikad, A.K.; Schooling, J. Design and Development of BIM Models to Support Operations and Maintenance. *Comput. Ind.* **2019**, *111*, 172–186. [CrossRef]
47. Santana, G. Classification of Construction Projects by Scales of Complexity. *Int. J. Proj. Manag.* **1990**, *8*, 102–104. [CrossRef]
48. Qazi, A.; Quigley, J.; Dickson, A.; Kirytopoulos, K. Project Complexity and Risk Management (ProCRiM): Towards Modelling Project Complexity Driven Risk Paths in Construction Projects. *Int. J. Proj. Manag.* **2016**, *34*, 1183–1198. [CrossRef]
49. Pellerin, R.; Perrier, N. A Review of Methods, Techniques and Tools for Project Planning and Control. *Int. J. Prod. Res.* **2019**, *57*, 2160–2178. [CrossRef]
50. Habibi, F.; Barzinpour, F.; Sadjadi, S.J. Resource-Constrained Project Scheduling Problem: Review of Past and Recent Developments. *J. Proj. Manag.* **2018**, *3*, 55–88. [CrossRef]
51. Sriprasert, E.; Dawood, N. Multi-Constraint Information Management and Visualisation for Collaborative Planning and Control in Construction. *J. Inf. Technol. Constr.* **2003**, *8*, 341–366.
52. Jaafari, A. Criticism of CPM for Project Planning Analysis. *J. Constr. Eng. Manag.* **1984**, *110*, 222–233. [CrossRef]
53. Harris, P.F.; McCaffer, P.R.; Edum-Fotwe, F. *Modern Construction Management*, 7th ed.; Wiley-Blackwell: Chichester, UK, 2013; ISBN 978-0-470-67217-4.
54. Line of Balance (LOB). Available online: https://www.designingbuildings.co.uk/wiki/Line_of_balance_(LOB) (accessed on 18 April 2022).
55. Mawdesley, M.; Askew, W.; Oreilly, M. *Planning & Controlling Construction Projects (Chartered Institute of Building): The Best Laid Plans*, 1st ed.; Prentice Hall: Essex, UK, 1997; ISBN 978-0-582-23409-3.
56. Halpin, D.W.; Martinez, L.-H. Real World Applications of Construction Process Simulation. In Proceedings of the 31st conference on Winter simulation: Simulation—A Bridge to the Future, Association for Computing Machinery, New York, NY, USA, 1 December 1999; Volume 2, pp. 956–962.
57. Planning and Analysis of Construction Operations/Wiley. Available online: https://www.wiley.com/en-us/Planning+and+Analysis+of+Construction+Operations-p-9780471555100 (accessed on 18 April 2022).
58. Martínez, J.C. STROBOSCOPE: State and Resource Based Simulation of Construction Processes. University of Michigan, USA. Available online: https://books.google.co.uk/books?id=xsceAQAAMAAJ (accessed on 18 April 2022).
59. Kartam, N.A.; Levitt, R.E.; Wilkins, D.E. Extending Artificial Intelligence Techniques for Hierarchical Planning. *J. Comput. Civ. Eng.* **1991**, *5*, 464–477. [CrossRef]
60. Soman, R.K.; Molina-Solana, M. Automating Look-Ahead Schedule Generation for Construction Using Linked-Data Based Constraint Checking and Reinforcement Learning. *Autom. Constr.* **2022**, *134*, 104069. [CrossRef]
61. Amer, F.; Golparvar-Fard, M. Modeling Dynamic Construction Work Template from Existing Scheduling Records via Sequential Machine Learning. *Adv. Eng. Inform.* **2021**, *47*, 101198. [CrossRef]
62. Jaklinović, K.; Đurasević, M.; Jakobović, D. Designing Dispatching Rules with Genetic Programming for the Unrelated Machines Environment with Constraints. *Expert Syst. Appl.* **2021**, *172*, 114548. [CrossRef]
63. Amer, F.; Jung, Y.; Golparvar-Fard, M. Transformer Machine Learning Language Model for Auto-Alignment of Long-Term and Short-Term Plans in Construction. *Autom. Constr.* **2021**, *132*, 103929. [CrossRef]
64. McKinney, K.; Fischer, M. Generating, Evaluating and Visualizing Construction Schedules with CAD Tools. *Autom. Constr.* **1998**, *7*, 433–447. [CrossRef]
65. Rolfsen, C.N.; Merschbrock, C. Acceptance of Construction Scheduling Visualizations: Bar-Charts, Flowline-Charts, Or Perhaps BIM? *Procedia Eng.* **2016**, *164*, 558–566. [CrossRef]
66. Retik, A.; Shapira, A. VR-Based Planning of Construction Site Activities. *Autom. Constr.* **1999**, *8*, 671–680. [CrossRef]
67. Sidani, A.; Matoseiro Dinis, F.; Duarte, J.; Sanhudo, L.; Calvetti, D.; Santos Baptista, J.; Poças Martins, J.; Soeiro, A. Recent Tools and Techniques of BIM-Based Augmented Reality: A Systematic Review. *J. Build. Eng.* **2021**, *42*, 102500. [CrossRef]
68. Casini, M. Chapter 9—Advanced Site Management Tools and Methods. In *Construction 4.0*; Casini, M., Ed.; Woodhead Publishing Series in Civil and Structural Engineering; Woodhead Publishing: Sawston, UK, 2022; pp. 471–521; ISBN 978-0-12-821797-9.
69. Goldratt, E.M. *Critical Chain*; North River Press: Great Barrington, MA, USA, 1997; ISBN 978-0-88427-153-6.
70. Jongeling, R.; Olofsson, T. A Method for Planning of Work-Flow by Combined Use of Location-Based Scheduling and 4D CAD. *Autom. Constr.* **2007**, *16*, 189–198. [CrossRef]
71. Womack, J.P.; Jones, D.T.; Roos, D. *The Machine That Changed the World: The Story of Lean Production—Toyota's Secret Weapon in the Global Car Wars That Is Now Revolutionizing World Industry*; Simon and Schuster: New York, NY, USA, 2007; ISBN 978-1-4165-5452-3.
72. Howell, G.A. *What Is Lean Construction?* Lean Construction Institute: Arlington, VA, USA, 1999.
73. Babalola, O.; Ibem, E.O.; Ezema, I.C. Implementation of Lean Practices in the Construction Industry: A Systematic Review. *Build. Environ.* **2019**, *148*, 34–43. [CrossRef]
74. Ballard, H.G. The Last Planner System of Production Control. Ph.D. Thesis, University of Birmingham, Birmingham, UK, 2000.

75. Murguia, D.; Brioso, X.; Pimentel, A. Applying Lean Techniques to Improve Performance in the Finishing Phase of a Residential Building. In Proceedings of the 24th Annual Conference of the International Group for Lean Construction, Boston, MA, USA, 18–24 July 2016; pp. 43–52.
76. Adamu, S.; Howell, G.; Abdulhamid, R. Adapting Lean Construction Tecniques in Nigerian Construction Industry. *Int. J. Sci. Eng. Res.* **2012**, *3*, 1–11.
77. Alsehaimi, A.O.; Tzortzopoulos, P.; Koskela, L.J. Last Planner System: Experiences from Pilot Implementation in the Middle East. In Proceedings of the 17th Annual Conference of the International Group for Lean Construction, Taipei, Taiwan, 15–17 July 2009; pp. 53–65.
78. Alarcón, L.F.; Diethelmand, S.; Rojo, O. Collaborative Implementation of Lean Planning Systems in Chilean Construction Companies. In Proceedings of the Tenth Annual Conference of the International Group for Lean Construction (IGLC-10), Gramado, Brazil, 6–8 August 2002; pp. 1–11.
79. Sanni-Anibire, M.O.; Mohamad Zin, R.; Olatunji, S.O. Causes of Delay in the Global Construction Industry: A Meta Analytical Review. *Int. J. Constr. Manag.* **2020**, *22*, 1395–1407. [CrossRef]
80. Michalski, A.; Głodziński, E.; Böde, K. Lean Construction Management Techniques and BIM Technology—Systematic Literature Review. *Procedia Comput. Sci.* **2022**, *196*, 1036–1043. [CrossRef]
81. ISO. ISO 19650-1: 2018. Available online: https://www.iso.org/cms/render/live/en/sites/isoorg/contents/data/standard/06/80/68078.html (accessed on 16 April 2022).
82. Sanders, S.R.; Thomas, H.R. Factors Affecting Masonry-Labor Productivity. *J. Constr. Eng. Manag.* **1991**, *117*, 626–644. [CrossRef]
83. Chavada, R.; Dawood, N.; Kassem, M. Construction Workspace Management: The Development and Application of a Novel ND Planning Approach and Tool. *Electron. J. Inf. Technol. Constr.* **2012**, *17*, 213–236.
84. Trebbe, M.; Hartmann, T.; Dorée, A. 4D CAD Models to Support the Coordination of Construction Activities between Contractors. *Autom. Constr.* **2015**, *49*, 83–91. [CrossRef]
85. Tran, S.; Ali, A.K.; Khan, N.; Lee, D.; Park, C. A Framework for Camera Planning in Construction Site Using 4D BIM and VPL. *ISARC Proc.* **2020**, *37*, 1404–1408.
86. Braun, A.; Tuttas, S.; Borrmann, A.; Stilla, U. A Concept for Automated Construction Progress Monitoring Using BIM-Based Geometric Constraints and Photogrammetric Point Clouds. *J. Inf. Technol. Constr.* **2015**, *20*, 68–79.
87. Sriprasert, E.; Dawood, N. Next Generation of Construction Planning and Control System: The Lewis Approach. Available online: https://www.semanticscholar.org/paper/NEXT-GENERATION-OF-CONSTRUCTION-PLANNING-AND-SYSTEM-Sriprasert-Dawood/b34eb601bbae61605e89f563449aa3a72f6c019d (accessed on 20 April 2022).
88. Han, S.; Bouferguene, A.; Al-Hussein, M.; Hermann, U. (Rick) 3D-Based Crane Evaluation System for Mobile Crane Operation Selection on Modular-Based Heavy Construction Sites. *J. Constr. Eng. Manag.* **2017**, *143*, 04017060. [CrossRef]
89. Tan, Y.; Fang, Y.; Zhou, T.; Gan, V.J.L.; Cheng, J.C.P. BIM-Supported 4D Acoustics Simulation Approach to Mitigating Noise Impact on Maintenance Workers on Offshore Oil and Gas Platforms. *Autom. Constr.* **2019**, *100*, 1–10. [CrossRef]
90. Golparvar-Fard, M.; Peña-Mora, F.; Arboleda, C.A.; Lee, S. Visualization of Construction Progress Monitoring with 4D Simulation Model Overlaid on Time-Lapsed Photographs. *J. Comput. Civ. Eng.* **2009**, *23*, 391–404. [CrossRef]
91. Said, H.; El-Rayes, K. Automated Multi-Objective Construction Logistics Optimization System. *Autom. Constr.* **2014**, *43*, 110–122. [CrossRef]
92. Chin, S.; Yoon, S.; Choi, C.; Cho, C. RFID+4D CAD for Progress Management of Structural Steel Works in High-Rise Buildings. *J. Comput. Civ. Eng.* **2008**, *22*, 74–89. [CrossRef]
93. Hewage, K. Sustainable Construction: An Information Modelling Approach for Waste Reduction. In Proceedings of the International Conference on Building Resilience, Kandalama, Sri Lanka, 19–21 July 2011.
94. Won, J.; Cheng, J.C.P. Identifying Potential Opportunities of Building Information Modeling for Construction and Demolition Waste Management and Minimization. *Autom. Constr.* **2017**, *79*, 3–18. [CrossRef]
95. Bakchan, A.; Guerra, B.C.; Faust, K.M.; Leite, F. BIM-Based Estimation of Wood Waste Stream: The Case of an Institutional Building Project. In *Computing in Civil Engineering 2019: Visualization, Information Modeling, and Simulation*; American Society of Civil Engineers: Reston, VA, USA, 2019; pp. 185–192. [CrossRef]
96. Hardin, B.; McCool, D. *BIM and Construction Management: Proven Tools, Methods, and Workflows*, 2nd ed.; Wiley: Hoboken, NJ, USA, 2015; ISBN 978-1-118-94276-5.
97. Gebrehiwet, T.; Luo, H. Analysis of Delay Impact on Construction Project Based on RII and Correlation Coefficient: Empirical Study. *Procedia Eng.* **2017**, *196*, 366–374. [CrossRef]
98. Sriprasert, E.; Dawood, N. Genetic Algorithms for Multi-Constraint Scheduling: An Application for the Construction Industry. In Proceedings of the 20th International Conference on Construction IT: Construction IT Bridging the Distance, Auckland, New Zealand, 23–25 April 2003.
99. Susanto, N.; Putranto, T.T. Stakeholder Interactions Model of Groundwater Management in Semarang City/Indonesia. *Int. J. Geomate* **2018**, *15*, 170–177. [CrossRef]
100. Amorocho, J.A.P.; Hartmann, T. Reno-Inst: An Ontology to Support Renovation Projects Planning and Renovation Products Installation. *Adv. Eng. Inform.* **2021**, *50*, 101415. [CrossRef]
101. Singh, Y.; Abdelhamid, T.; Mrozowski, T.; El-Gafy, M.A. Investigation of Contemporary Performance Measurement Systems for Production Management of Renovation Projects. *J. Constr. Eng.* **2014**, *2014*, 1–9. [CrossRef]

102. Aldanondo, M.; Barco-Santa, A.; Vareilles, E.; Falcon, M.; Gaborit, P.; Zhang, L. Towards a BIM Approach for a High Performance Renovation of Apartment Buildings. In *Product Lifecycle Management for a Global Market*; Fukuda, S., Bernard, A., Gurumoorthy, B., Bouras, A., Eds.; Springer: Berlin/Heidelberg, Germany, 2014; pp. 21–30.
103. Doukari, O.; Lynn, T.; Rosati, P.; Egli, A.; Krinidis, S.; Angelakoglou, K.; Sougkakis, V.; Tzovaras, D.; Kassem, M.; Greenwood, D. RINNO: Transforming Deep Renovation through an Open Renovation Platform. In Proceedings of the ICDS The Fifteenth International Conference on Digital Society, Nice, France, 18–22 July 2021.
104. Lynn, T.; Rosati, P.; Egli, A.; Krinidis, S.; Angelakoglou, K.; Sougkakis, V.; Tzovaras, D.; Kassem, M.; Greenwood, D.; Doukari, O. RINNO: Towards an Open Renovation Platform for Integrated Design and Delivery of Deep Renovation Projects. *Sustainability* **2021**, *13*, 6018. [CrossRef]
105. Sacks, R.; Koskela, L.; Dave, B.A.; Owen, R. Interaction of Lean and Building Information Modeling in Construction. *J. Constr. Eng. Manag.* **2010**, *136*, 968–980. [CrossRef]
106. Kanan, R.; Elhassan, O.; Bensalem, R. An IoT-Based Autonomous System for Workers' Safety in Construction Sites with Real-Time Alarming, Monitoring, and Positioning Strategies. *Autom. Constr.* **2018**, *88*, 73–86. [CrossRef]
107. Barlish, K.; Sullivan, K. How to Measure the Benefits of BIM—A Case Study Approach. *Autom. Constr.* **2012**, *24*, 149–159. [CrossRef]
108. Matthews, J.; Love, P.E.D.; Mewburn, J.; Stobaus, C.; Ramanayaka, C. Building Information Modelling in Construction: Insights from Collaboration and Change Management Perspectives. *Prod. Plan. Control* **2018**, *29*, 202–216. [CrossRef]
109. Ilozor, B.; Kelly, D. Building Information Modeling and Integrated Project Delivery in the Commercial Construction Industry: A Conceptual Study. *J. Eng. Proj. Prod. Manag.* **2012**, *2*, 23–26. [CrossRef]
110. Collins, W.; Parrish, K. The Need for Integrated Project Delivery in the Public Sector. In Proceedings of the Construction Research Congress 2014: Construction in a Global Network, Atlanta, GA, USA, 19–21 May 2014; pp. 719–728. [CrossRef]
111. Vass, S.; Gustavsson, T.K. Challenges When Implementing BIM for Industry Change. *Constr. Manag. Econ.* **2017**, *35*, 597–610. [CrossRef]
112. Stanford University Protégé. Available online: https://protege.stanford.edu/ (accessed on 28 April 2022).
113. Designing Buildings Disruption Claims in Construction. Available online: https://www.designingbuildings.co.uk/wiki/Disruption_claims_in_construction (accessed on 17 February 2022).
114. Estate Regeneration—Building out Disruption. Available online: http://thoughtleadership.trowers.com/estate-regeneration/building-out-disruption/ (accessed on 17 February 2022).
115. Chaves, F.J.; Tzortzopoulos, P.; Formoso, C.T.; Biotto, C.N. Building Information Modelling to Cut Disruption in Housing Retrofit. In *Proceedings of the Institution of Civil Engineers-Engineering Sustainability*; Thomas Telford Ltd: London, UK, 2017; Volume 170, pp. 322–333. [CrossRef]
116. Bedrick, J.; Builders, W. Organizing the Development of a Building Information Model. *Am. Inst. Archit.* **2008**, *9*, 4.
117. Leite, F.; Akcamete, A.; Akinci, B.; Atasoy, G.; Kiziltas, S. Analysis of Modeling Effort and Impact of Different Levels of Detail in Building Information Models. *Autom. Constr.* **2011**, *20*, 601–609. [CrossRef]

buildings

MDPI

Article

Impacts of COVID-19 on the Use of Digital Technology in Construction Projects in the UAE

Omar Elrefaey [1], Salma Ahmed [2], Irtishad Ahmad [3] and Sameh El-Sayegh [3,*]

[1] Master of Science in Construction Management (MSCM) Program, American University of Sharjah, Sharjah P.O. Box 26666, United Arab Emirates; b00069102@aus.edu
[2] Engineering System Management—Ph.D. Program, American University of Sharjah, Sharjah P.O. Box 26666, United Arab Emirates; g00043157@alumni.aus.edu
[3] Civil Engineering Department, American University of Sharjah, Sharjah P.O. Box 26666, United Arab Emirates; irahmad@aus.edu
* Correspondence: selsayegh@aus.edu

Abstract: The construction industry has been incorporating digital technology over the last two decades, albeit gradually, as "technology-push" continues to overcome customary and traditional passivity typical in the sector. The objective of the study presented in this paper is to investigate how digital technology is making a headway in the construction industry as a consequence of COVID-19. For the purpose of this paper, digital technology applications are divided into three groups: data acquisition, processing, and communication. The methodology involved conducting a questionnaire survey among the construction professionals in the UAE. The survey included questions on the extent of use and level of investment on the three types of technology in three periods—pre-COVID, during COVID, and post-COVID. The results clearly show the increasing level of usage of digital technology in the construction industry from pre-COVID to during COVID and post-COVID periods. Among the three categories, communication technology indicated higher extent of use as compared to the other two. In addition, a marked difference was observed between the "small" project organizations and the "large" ones. Unsurprisingly, both usage and investment in digital technology, in smaller organizations, indicated higher extent of increase in during and post-pandemic periods when compared to the larger organizations.

Keywords: digital technology; construction industry; UAE; COVID-19

Citation: Elrefaey, O.; Ahmed, S.; Ahmad, I.; El-Sayegh, S. Impacts of COVID-19 on the Use of Digital Technology in Construction Projects in the UAE. *Buildings* **2022**, *12*, 489. https://doi.org/10.3390/buildings12040489

Academic Editors: Yongjian Ke, Jingxiao Zhang and Simon P. Philbin

Received: 12 March 2022
Accepted: 13 April 2022
Published: 14 April 2022

Publisher's Note: MDPI stays neutral with regard to jurisdictional claims in published maps and institutional affiliations.

1. Introduction

Global construction is a $12 trillion (USD) industry, accounting for close to 13.5% of global GDP [1]. However, in terms of sector productivity, construction lags manufacturing and the total world economy. "Globally, construction sector labor productivity growth averaged 1 percent a year over the past two decades, compared with 2.8 percent for the total world economy and 3.6 percent for manufacturing" [2]. There are strong indications from professionals and researchers in the construction industry that advanced digital technology (DT) can play a pivotal role, either directly or as a catalyst, in improving construction productivity. However, it is also a fact that construction is traditionally slow in adopting technology. Despite being sluggish as compared to the service and the manufacturing industries, construction has been incorporating digital technology gradually over the last two decades—as "technology-push" continues to overcome customary and traditional passivity. BIM, or building information modeling, is a prime example of a digital technology that took a long time to gain acceptance in the mainstream construction industry. It is now widely accepted and is proving itself almost essential in the industry. It is more apparent now than ever before in the wake of the COVID-19 crisis. A recent IFS study based on a survey reported that companies concerned with economic disruption were 20% more likely to plan increased spending on digital transformation [3]. It appears that the construction industry is ready to embrace a technological revolution to address these issues and that the global pandemic is perhaps the catalyst that the industry needed. Numerous studies listed contemporary digital technologies suitable for use and application in construction. For instance, PwC [4] lists eight essential technologies that are making large impacts on the business of construction. They are artificial intelligence, virtual reality, augmented reality, blockchain, drones, Internet of Things (IoT), robotics, and 3D printing.

The early months of 2020 witnessed the rise of the COVID-19 pandemic, which had its adverse effects on almost all regions and industry sectors in the globe. However, many claim that the Architecture, Engineering, and Construction (AEC) industry was one of the sectors that was hit the hardest by the pandemic. According to a recent Los Angeles investigation, construction crews reported the highest number of positive cases when compared to workers in other sectors, such as transport and manufacturing [5]. It is with no doubt that the COVID-19 pandemic affected the United Arab Emirates (UAE) AEC industry in many ways. Dadlani [6] explained how precautionary measures were enforced in construction companies to ensure prioritization of health and safety of workers. These measures included frequent disinfection programs, social distancing measures, construction worker body temperature checks, and isolation rooms in the event of workers showing any symptoms of the coronavirus. Additionally, during the first six months of 2020, contract awards in the GCC region fell by 20%; this coincided with an increase in payment delays and disputes [7]. In addition, supply chains and subcontractors in the UAE were pressurized by such marketing challenges, causing some of them to go out of business [8].

The premise of this research is that most construction organizations' (owner, consultants, and contractors) investment in digital technology prior to the pandemic was inadequate to deal with the demands imposed by the crisis. It is likely that several technologies already available and deployed were not utilized at the expected and optimum level. The pandemic abruptly made them more useful than before. As an example, the use of technology necessary for virtual meetings can be cited. This technology has become essential during the lockdown phase, although it most likely remained underutilized before the crisis. Perhaps it was not considered necessary or did not have the necessary bandwidth for smooth operations. Therefore, it is important to assess the level of use and investments in digital technology before, during, and after the pandemic. The main goal of this paper is to investigate the use and investment in digital technology for construction projects amid, and in the aftermath of, the COVID-19 pandemic in the context of the UAE construction industry.

The rest of the paper is organized as follows. A literature review about the benefits of digital technology applications in the AEC industry is discussed in the next section. It is followed by an elaborate explanation of three facets of digital technology applications—data acquisition, processing of information, and communication. The section afterward explains the methodology adopted in this research. The goal was to differentiate between the levels of usage and investment in the three digital technology categories across the three groups, namely, pre-, during, and post-COVID-19. Finally, the results are interpreted and discussed in order to draw significant conclusions that provide valuable insights regarding usage and investments in digital technology in construction in the context of the UAE (United Arab Emirates).

2. Background

Digital technology, although not new and having existed for some time, was not utilized extensively prior to the "lockdown" situation caused by the pandemic. The AEC industry has been incorporating digital technology gradually over the last two decades but remains sluggish as compared to other sectors, such as the service and the manufacturing industries, for instance. Ironically, the COVID-19 pandemic has developed an urgency towards adopting new technologies that will play a key role in reshaping the future of the AEC industry, allowing collaboration, data-driven decision-making, and greater control. The utilization of such technologies shall also facilitate the move towards a sustainable future for contractors and developers [9]. Wallet [10] noted that this pandemic has provided a silver lining, as construction professionals grew more confident in dealing with these technologies, which not only helped keep the industry moving during those drastic times but also provided cost and time saving tools that construction professionals can benefit from in the long run.

Despite their attractive potential capabilities, many advanced digital technologies were viewed as tools that could lead to poor results and inefficiencies in the AEC industry prior to the new realization during the pandemic [11]. For instance, Hinks and Allen [12] discovered that videoconferencing was not in sync with existing processes. Even in the period post-2010, digital technologies remained in an "under-utilization" state. In a study conducted in 2014, only two out of 14 interviewed AEC firms mentioned that video conferencing software was being utilized to communicate information [13]. To sum up, the capacity and value of digital technologies in the business context was poorly recognized by the AEC firms prior to the COVID-19 pandemic.

The disruptive nature of COVID-19 has pressured many AEC companies to immediately remodel their business strategies via the usage of digital technologies [14]. In many cases, AEC firms lacked the essential digital infrastructure to make the move simple. As a result, obtaining necessary software packages and other resources posed significant challenges, leading to severe inefficiencies [15]. For instance, in the early stages of the pandemic, it was reported that several coordination problems arose due to online communication. Considering a design consultant, such coordination issues caused a delay in the design process during work from home. Additionally, difficulties in meeting with clients were evident, as some of the online meetings turned out to be ineffective and unnecessarily prolonged [16]. In many cases, organizations were required to make additional technological investments to improve their ability to work under the changed circumstances. Several businesses, for example, have invested in Virtual Private Networks (VPNs) to acquire remote access to resources and software packages. Only few companies reported that they already had cloud solutions in place to access licensed software and company databases, making the shift smooth [15].

The benefits of using digital technology (DT) in the AEC industry are multifarious and significant. The adoption of technological innovation gives construction companies the opportunity to rebrand themselves as the providers of smart engineering solutions. In fact, adopting such technological innovations is a strategic decision to improve the image of the construction company and its reputation in the market. Not only this but digitalizing the construction industry makes it more appealing to young graduates, as it challenges the traditional notion of construction jobs being dangerous, difficult, and dirty [17] in addition to being labeled as backward. Furthermore, the efficient use of emerging digital technology helps improve communication, collaboration, better project comprehension, improved information retrieval, and increased productivity rates as well as time and cost savings. These benefits contribute to offering a distinct advantage to the market as well as creating a healthier organizational culture where project team members feel more connected to each other through advanced means of communication that raise a sense of belonging and commitment to the construction project [18].

The adoption of these technological innovations helps reduce the risks associated with construction projects by directing the construction industry away from the high-risk and towards the low-risk sector zone [19]. In addition, the nature of today's construction industry demands project teams that are geographically dispersed in different time zones across diverse organizational cultures and boundaries. It is only through using such advanced technological innovations that a collaborative project delivery system can be established where teams can work closely with various disciplines in an effective way [20]. Besides, adopting technological innovations can also help improve the monitoring and inspection of construction sites. For instance, Zhou et al. [21] introduced an AR technique to overlook the inspection process of segment displacement during tunneling construction.

The technique enabled overlaying a quality control baseline model onto the real segment and reported the discrepancies.

Despite the benefits offered by digital technology, its adoption in the construction sector is still very low. In fact, Manyika [22] stated that the level of digitalization index for the construction industry ranked the lowest among 22 other industries. Delgado et al. [23] divided the challenges of creation of virtual environment in the construction industry, in terms of limitations, into three categories: social, technical, and economic limitations. The social limitations included (but were not limited to) lack of trained workforce, repugnance to change, and data privacy issues, whereas technical limitations included large space as well as prohibitive processing requirements, lack of user friendliness, and accuracy issues. Economic limitations included expensive hardware, expensive training needs, lack of client interest, and lack of funds for research and development. It can be concluded that the last one, namely, the economic issues, are mainly responsible for the lack of necessary investment for promoting digital technology in construction. Thus, in this paper the discourse is centered around the issue of investment in digital technology. To that end, digital technology applications are broken down into three facets: data acquisition, data processing, and data communication.

2.1. Data Acquisition

These are technology and tools in which their utilization in the construction industry would assist in a faster and more efficient data gathering. Regardless of whether the information is gathered onsite or offsite, the data are essential for a successful project delivery. Data acquisition tools include RFID (Radio Frequency Identification), drones, and 3D laser scanning. Drones, for instance, are aircrafts designed to fly without a pilot or passengers, controlled remotely using radio waves [24]. These innovations have high mobility and visual data acquisition capabilities. As a matter of fact, drones can play a significant role in conducting quantitative analysis of productivity and safety-related metrics through the reconstruction of three-dimensional (3D) point clouds from video images produced by the drones [25]. Furthermore, a site staff equipped with a drone can do the same job as multiple personnel without a drone. Thus, using drones can be a powerful tool to comply with the COVID-19 imposed social distancing measures, as the number of people on site would be few. While some firms may be facing major financial constraints to invest in drone technology, it is critical to note that this technology helped construction firms reduce wastage in time by 18.4% and the time to survey a site by 98%, translating into money savings in the long term [8]. Lastly, this smart system gives construction professionals access to real time data, which allows organizations to keep track of their inventory plan and better plan their construction site in general, making room for any adjustments needed as soon as possible [26].

Another data acquisition tool that could be well integrated and utilized in the construction industry, as mentioned earlier, is RFID. Radio Frequency Identification (RFID) technology allows for automated tracking of equipment and materials in storage or during delivery and makes information readily available for the personnel who handle material tracking and delivery processes [27]. Furthermore, laser scanning technologies are powerful data acquisition tools that have the capability of capturing complex geometries, angles, and distances [28]. While laser scanning technologies require expert operators to run, these technologies provide a great value, as the output of these technologies can be used as the basis to develop as-built BIM models of a building [28]. Table 1 shows some of the potential applications of RFID, drones, and laser scanning technologies in the construction industry.

Table 1. Examples of applications of data acquisition technologies.

Data Acquisition Technology	Application(s)
Radio Frequency Identification Device (RFID)	• Three-dimensional (3D) location of buried utilities [29] • Equipment tracking to prevent collision accidents with heavy equipment [30]
Drones	• Live site surveillance • Quality and safety checks and monitoring
Laser Scanning	• Dimensional accuracy and structural performance assessment [31] • Characterization of steel reinforcement corrosion [32]

2.2. Data Processing

The second stage that ultimately comes after data collection is data processing. A virtual innovation that facilitates such a task is Building Information Modeling (BIM). BIM is defined by the US National Institute of Building Sciences as a "digital representation of physical and functional characteristics of a facility creating a shared knowledge resource for information about it forming a reliable basis for decisions during its life cycle, from earliest conception to demolition" [33]. Kaner et al. [34] revealed that BIM enables the design and detail processing of complex 3D geometric shapes to a high level of accuracy with error-free drawings. Additionally, through features of clash detection and early error identification, the probability of disputes and claims is greatly minimized due to the utilization of BIM [35]. Besides this, other features of BIM include automated quantity takeoff that enables users to avoid the traditional error-prone method of measuring elements of a project and assigning related costs to each. This feature also allows stakeholders to know costs early in the design phase so that more informed decisions can be made [36,37].

Lastly, there exist tools/software that process schedule-related and cost-related data, such as Primavera P6 and Esti-mate, respectively. Primavera P6, for instance, is a very commonly used scheduling and resource-planning tool in the AEC industry. "P6" has the capacity to allow users to develop complex and large-scale programs, forecast scenarios using what-if analysis, and track resources and costs through the project life cycle [38]. Comparably, "Esti-mate" is an estimation tool that could be utilized by quantity surveyors, project managers, contract managers, and vendors to process and coordinate Bills of Quantities (BoQs), subcontractor and material inquiries, and tender adjudications, allowing for accurate and timely project cost tracking [39]. Table 2 illustrates some applications of data processing tools.

Table 2. Examples of applications of data processing technologies.

Data Processing Technology	Application(s)
BIM	• Allows staff to edit, manage, and document design [40] • Documenting the design and project reviews [40]
Primavera P6	• CPM calculations • Resource allocation, delay analysis, etc.
Esti-mate	• Quantity takeoffs

2.3. Data Communication

As far as communication is concerned, some tools provide features such as discussion boards, work sharing, project websites, and videoconferencing with data sharing as well as real time data manipulation and exchange through virtual teaming [41].

Furthermore, a virtual platform that has a huge potential to transform coordination in the construction industry is cloud computing. According to Mell and Grance [42], clouding computing is a model for "enabling ubiquitous, convenient, on-demand network access to a shared pool of configurable computing resources that can be rapidly provisioned and released with minimal management effort or service provider interaction". In a recent study by Du et al. [43], they proposed a cloud-based multiuser virtual reality headset system, which is an innovation that collects BIM metadata and translates them into a multiuser virtual communication environment. The technology allows remote project stakeholders to interact and connect simultaneously, thus improving collaboration. Cloud computing software (e.g: Dropbox) can also provide the necessary features to complement intelligent contracts, or "icontracts". An icontract is an advanced smart contract that utilizes computer codes to automate the execution of contractual clauses between project parties [44]. For instance, an icontract connected to a BIM model would enable the release of payments to a design consultant upon completion of a pre-specified BIM milestone in the contract. Through Dropbox and other cloud computing technologies, data sources such as models and emails, which may be linked to icontracts, can be stored, accessed, and updated by various project stakeholders [44]. In a similar manner, blockchain-based intelligent contracts are envisaged to upgrade the supply chain financing industry through "smart factoring" [45].

Furthermore, Google Meet and Zoom calls, along with the utilization of Dropbox and e-mail communication, are a few of several communication technologies that are adopted in the AEC industry. Due to the ongoing pandemic, video conferencing platforms have now become embedded into the day-to-day AEC work life in the UAE [46]. Despite their current recognized benefits, video conferencing mediums have the potential to be further optimized to ensure high productivity, which would be especially beneficial for countries that adopt a four- or a four-and-a-half-day work week, such as Iceland and the UAE [47]. Table 3 identifies the applications of these technologies.

Table 3. Examples of applications of data communication technologies.

Data Communication Technology	Application(s)
Google Meet/Zoom	• Virtual employees with candidates • Conducting from-home coordination meetings
Dropbox (cloud computing)	• Sharing of contract documents/other relevant files among project stakeholders [44].
Email	• Contractor–consultant communication (e.g.,: Request for Information-RFI) • Direct file sending/receiving

To explore the levels of use of communication technologies in the Nigerian construction industry, a questionnaire [48] revealed that different positions within an AEC firm utilize electronic communications at different levels of sophistication, based on the position's need. The questionnaire revealed that email systems were the most frequently used tools to achieve effective information management [48]. Oliver [49] stated that current communication platforms provide the necessary features to allow for higher levels of interactivity, intimacy, and immediacy for AEC stakeholders to experience better social and technical communications. Thus, digital communication technologies may ease cost pressures and technical complexities [50]. For instance, there is a high potential for video conferencing ICTs to solve corporate difficulties in an international market through overcoming lack of expertise and lowering travel expenses [11]. It is worth noting that Computer-Mediated Communication (CMC) processes have existed for years. However, such communication technologies may not be well-integrated into existing industries to improve their efficacy [51], meaning that they are underutilized.

3. Materials and Methods

The study used a quantitative approach to differentiate between the level of usage and investment in three digital technology categories (data acquisition, processing, and communication) across three classes (phases)—pre-COVID, during COVID, and post-COVID. In order to select the frequently used digital technology for the purpose of this research study, a thorough literature survey of published materials was conducted. This included journal articles, periodicals, and books that discuss the leading digital technologies currently in use in the construction industry. In the next stage, a questionnaire survey was developed and distributed to construction professionals (owners, consultants, contracts, and project managers) in the UAE using online means. The survey was sent to 100 construction professionals, and 39 responses were collected in the end, representing a 39 percent rate. The survey consisted of 21 questions, and the duration of the survey was approximately 10 min. The questions were developed in a clear manner, where examples from the literature review about the different digital technologies in each category were given to make the questions comprehensible and easy to follow for the respondents. The questionnaire questions are presented in Appendix A. The first section of the survey consisted of general information about the respondents' companies or organizations. These included years of experience, respondents' role, and whether the organizations or companies are local or international. The remaining sections of the survey were designed to elicit the perceptions of the respondents on the frequency of usage and level of investment of the three digital technology categories across the three classes: pre-COVID, during COVID and post-COVID. The questions related to the level of usage of all three digital technology categories pre-COVID and during COVID followed a Likert scale of 1–5, where 5 represented "almost always" and 1 represented "never". The level of investment of all three digital technology categories pre-COVID and during COVID followed a Likert scale of 1–5, where 5 represented "very high" and 1 represented "very little". All questions related to the class post-COVID, as being a future state, followed a Likert scale of 1–5, where 5 represented "strongly agree-will increase greatly" and 1 represented "strongly disagree-will return to pre-pandemic level". We focus on the different digital technology (DT) categories rather than the specific technologies. The paper intentionally states only 3–4 examples within each DT category. The respondents are able to identify the example technologies within each category, as they were included in the survey questions. The survey questions are provided in Appendix A.

The data were then analyzed using both descriptive statistics and inferential statistics. The descriptive statistics mainly consisted of comparing the means of the three class populations in the frequency of usage and level of investment of the digital technologies. Additionally, the means of the small projects were compared against those of the large projects across the three class populations in frequency of usage and level of investment of the three digital technology categories, while for the inferential statistics, multivariate analysis of variance (MANOVA) was conducted, as the data comprised multiple dependent variables, namely, the level of investment in, and frequency of usage of the three digital technology categories, and a single independent qualitative variable designated as classes: pre-COVID, during COVID, and post-COVID. MANOVA is used to determine whether there is a difference between the level of investment and frequency of usage of the three digital technologies across these classes. In particular, the researchers were interested in determining whether, during COVID, a peak in the level of usage and investment in digital technology is evident, as expected in order to meet the demands imposed by the pandemic. The statistical analysis mainly examined whether or not the average of the dependent variables differs between the three categories of the class variable. MANOVA is an extension of the analysis of variance (ANOVA) in order to incorporate more than one dependent variable by combining the multiple dependent variables into a single optimum value to maximize the difference between the classes. Furthermore, MANOVA is also more appropriate than ANOVA, because it provides additional insights into the effects of the independent variables on the dependent variable [52,53].

MANOVA is a very effective statistical tool that has been previously used in the construction context in literature. For instance, Zhao et al. [54] have used MANOVA analysis to test whether or not different factors such as construction type, technology level, climate, and conditioned floor area significantly affect energy use and whether or not the effect changes over time. The authors tested the "between-subject" effect to analyze the factor's effect across all building units and used the "within-subject" effects to test the factor's effect over time. The authors modeled the "between-subject" effect by fitting the sum of the repeated measures to the model effect, while the "within-subject" effect was modeled using a function that fits differences in the repeated measures.

There are three main assumptions when conducting MANOVA analysis. First, the data should be multivariate normally distributed, while the second assumption is concerned with the equality of the covariance matrices for all treatments in the study. The third and last assumption states that all observations must be independent of each other, as this will affect the significance level reported in results later on [55]. Furthermore, Wilk's Lambda test statistic (Λ) is used in MANOVA to test whether there are differences between the means of the three classes on the frequency of usage and on the level of the investment of the three digital technologies or, in other words, to test the two-null hypothesis H_0 shown below.

$$H_0 \text{ for frequency of usage of the three digital technologies} = \begin{bmatrix} \mu_{11} \\ \mu_{12} \\ \mu_{13} \end{bmatrix} = \begin{bmatrix} \mu_{21} \\ \mu_{22} \\ \mu_{23} \end{bmatrix} = \begin{bmatrix} \mu_{31} \\ \mu_{32} \\ \mu_{33} \end{bmatrix} \quad (1)$$

$$H_0 \text{ for level of investment of the three digital technologies} = \begin{bmatrix} \mu_{11} \\ \mu_{12} \\ \mu_{13} \end{bmatrix} = \begin{bmatrix} \mu_{21} \\ \mu_{22} \\ \mu_{23} \end{bmatrix} = \begin{bmatrix} \mu_{31} \\ \mu_{32} \\ \mu_{33} \end{bmatrix} \quad (2)$$

where μ_{ip}, i = (1,2,3) is the number of the three classes' populations (pre-COVID, during COVID, post-COVID), while "p" is the number of dependent variables, which, in this study, is three, since there are three digital technology categories (acquisition, processing, and communication) for the variable frequency of usage and the variable level of investment. Furthermore, to test whether there is no difference between the three population mean vectors, Λ will be used according to Equation (3) [55], as it is the test statistic preferred for MANOVA, where H_0 will be rejected if Λ is small. Lambda is a value that ranges between zero and one; a null hypothesis would be rejected if Lambda was close to zero, but it should be considered in conjunction with a small *p*-value as well, where the *p*-value here represents the probability that measures the consistency between the data and the hypothesis being tested [51]. The alpha level that the *p*-value will be compared against in this study is 0.1, indicating a level of confidence of 90%. Wilk's Lambda test statistic is defined as:

$$\Lambda = \frac{|S_{error}|}{|S_{effect} + S_{error}|} \quad (3)$$

where

S_{error} is the variation of the residual within the matrix of sum of squares cross product and

S_{effect} is the variation of the treatment between the matrices of sum of squares cross product.

Figure 1 below further illustrates the methodology that will be adopted in this research.

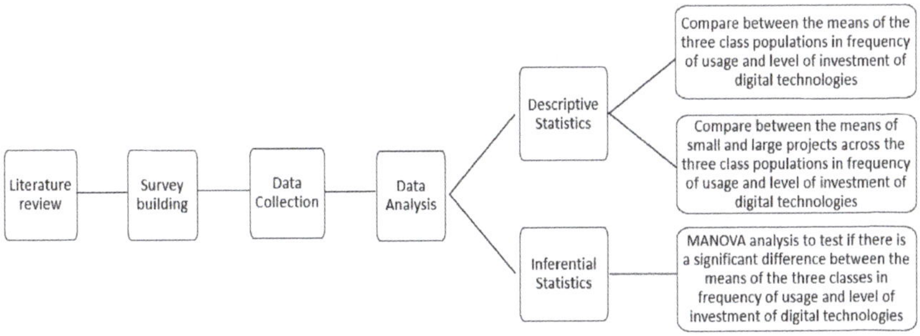

Figure 1. Methodology.

4. Results

A total of 39 responses were received from the online survey. The respondents' profiles showed that 15% of them worked in international offices, while 85% worked in local offices. Moreover, 13% of the respondents had more than 20 years of experience. Thus, a majority of the respondents were local contractors in the building industry with less than 5 years in business. In addition, it was observed that 56.4% of the respondents were involved in projects with an average project size of AED 50 million (considered large) or more, while 43.6% of respondents' average project size was less than AED 50 million (considered small). In this research, these two groups' (large and small) responses were further analyzed to gain important insights.

Descriptive analysis was conducted to compute the mean of the three populations in each variable (Table 4). The results show that the mean of frequency of using the three digital technologies (y1, y3, y5) in class 0 (pre-COVID) population is 3.29, while the mean of the level of investment (y2, y4, y6) in class 0 was 3.30. Similar calculations were performed for class 1 (during COVID) population, and the mean of frequency of usage and level of investment are 4.03 and 3.75, respectively, whereas the mean of frequency of usage in class 2 population (post-COVID) is 4.26, and the mean of the level of investment is 4.23. This indicates that there is in fact an increasing trend from pre-COVID to post-COVID in terms of mean frequency of usage and level of investment in digital technology for data gathering, processing, and communication. The results are summarized in Table 4.

Table 4. Means of the responses of the three classes.

Variable	Questions on Usage and Investment	Class 0 Mean (Pre-COVID)	Class 1 Mean (During COVID)	Class 2 Mean (Post-COVID)
y1	Frequency of using digital technology for data and information gathering	3.10	3.64	4.28
y2	Level of company's investment in digital technology for data and information gathering	3.13	3.59	4.23
y3	Frequency of using digital technology for data and information processing	3.56	3.95	4.23
y4	Level of company's investment in digital technology for communication of data and information	3.33	3.69	4.23
y5	Frequency of using digital technology for communication of data and information	3.21	4.49	4.28
y6	Level of company's investment in digital technology for communication of data and information	3.44	3.97	4.26
	Average of frequency of usage (y1, y3, y5)	3.29	4.03	4.26
	Average of level of investment (y2, y4, y6)	3.30	3.75	4.23

In addition, the questionnaire contained three questions related to "project success". The questionnaire statements and the weighted average of responses, indicating a high degree of agreement, are listed below:

- Use of digital technology for data and information gathering contributes towards project success—4.36 (out of 5).
- Use of digital technology for data and information processing contributes towards project success—4.59 (out of 5).
- Use of digital technology for communication of data and information contributes towards project success—4.28 (out of 5).

Thus, it is substantiated by the above findings that the digital technologies do indeed contribute to overall success of construction projects. Furthermore, the results presented in Table 4 also indicate that the frequency of using digital technology for communication was at its peak (variable y5 with a value of 4.49) during COVID, as the demands imposed by the pandemic were higher than they were before the pandemic, indicated by a value of 3.21, and decreased somewhat from the peak to 4.28 in the post-COVID phase, as in-person communication is expected to return back somewhat but probably not at the pre-COVID level. This is because it is expected that the professionals in the industry will retain some of the communication technologies and conveniences, as they will be increasingly more familiar and comfortable with the technology. It was noticed that there was a slight drop in the use of communication technology from during COVID to post-COVID, apparently as some of the functions and operations are expected to return to "normalcy" after the pandemic. The results also indicate, predictably, that investment in communication technologies (variable y6) had increased during the pandemic, to 3.97 from 3.44 (pre-COVID), and is higher after the pandemic, at 4.26.

In an effort to compare between the responses based on the project size, two groups were created, as noted earlier: small projects with sizes less than 50 million AED and large projects with sizes 50 million AED or more. These two groups were almost equally represented in the survey.

In general, it should be noted that, while in class 0 (pre-COVID), both the average of the frequency of usage and level of investment were both greater in larger organizations than the smaller ones. The difference shrinks in both when compared to class 1 (during COVID) and class 2 (post-COVID), as the data in the last two rows of Table 5 indicate. This is a significant observation, as it substantiated two facts—(1) overall, there has been an increase in both the usage and investment in digital technologies in construction during pandemic, and the trend will continue after the pandemic; and (2) the extent of both usage and investment is significantly greater in smaller organizations than the larger ones at the during and post-COVID stages. This underscores the notion that benefits of digital technologies remained largely unknown or unexperienced by the smaller organizations, and once known they can become increasingly more acceptable in the industry.

Table 5. Means of the three class populations across small vs. large projects.

Variable	Questions on Usage and Investment	Class 0 Mean (Pre-COVID)		Class 1 Mean (During-COVID)		Class 2 Mean (Post-COVID)	
		Small	Large	Small	Large	Small	Large
y1	Frequency of using digital technology for data and information gathering	2.75	3.35	3.69	3.61	4.44	4.17
y2	Level of company's investment in digital technology for data and information gathering	3.00	3.22	3.50	3.65	4.38	4.18
y3	Frequency of using digital technology for data and information processing	3.06	3.91	3.69	4.13	4.25	4.22
y4	Level of company's investment in digital technology for communication of data and information	3.00	3.57	3.56	3.78	4.18	4.26
y5	Frequency of using digital technology for communication of data and information	2.69	3.57	4.63	4.40	4.25	4.30
y6	Level of company's investment in digital technology for communication of data and information	3.25	3.57	3.81	4.10	4.25	4.26
	Average of frequency of usage (y1, y3, y5)	2.83	3.61	4.00	4.05	4.31	4.23
	Average of level of investment (y2, y4, y6)	3.08	3.45	3.62	3.84	4.27	4.23

The results show that in class 2 (post-COVID), as illustrated in Table 5, the average of the frequency of usage and the level of investment were equal to each other in the large projects, but in small projects the average of the frequency of usage was higher than the average of the level of investment. Furthermore, it can be seen that both the averages were greater in small projects than in large projects in class 2, and that could be due to the fact that large projects were ahead of small projects in terms of digital technology usage before the pandemic and, therefore, will not need to invest as much as the small projects to keep up with the demands created by the pandemic. Another observation that can be drawn is the fact that in class 1 the averages of the frequency of digital technology usage were almost the same in both small and large projects—in data gathering (variable y1, 3.69—small to 3.61—large) and communication (variable y5, 4.63—small to 4.40—large) technologies but not so different in processing technologies (variable y3, 3.69—small versus 4.13—large). Again, it is not surprising, as data/information processing technologies are more capital-intensive (high initial investment cost) compared to the other two—acquisition and communication technologies—and organizations dealing with smaller projects would not have investment capital available to them during the pandemic. This observation naturally leads to a closer look at what happens to expected investment responses after the pandemic. Results in Table 5 show that it is expected that organization with smaller projects would increase their investments significantly—in data acquisition (y2) from pre-COVID response of 3.00 to 4.38; in data processing (y4) from 3.00 to 4.18; and, lastly, in data communication (y6) from 3.25 from 4.25. The average of all three variables shows an increase in responses from 3.08 to 4.27. For organizations with larger projects, the extent of this difference in investment level between pre-COVID and post-COVID is much lower—from 3.45 to 4.23—although higher as predicted. These results are shown in Table 5.

To gain further insight, the responses from the survey were then analyzed using SAS statistical software as per the methodology explained in the earlier section. The MANOVA test was conducted using F approximation to test the hypothesis of no significant difference between the means of the three classes' populations (pre-, during, and post-COVID) on the frequency of usage and level of investment of the three digital technologies as shown in Tables 6 and 7, respectively. Table 6 shows that the p-value is less than alpha (0.1) for the null hypothesis of "no overall class difference"; therefore, it can be rejected, and it can be concluded with 90% confidence that there is in fact a significant difference between the means of the three class populations on the frequency of usage of the three digital technologies. Therefore, the analysis was taken a step further to conduct pairwise comparisons between the means of class 0 (pre-COVID) and class 1 (during COVID), class 0 (pre-COVID) and class 2 (post-COVID), and class 1 (during COVID) and 2 (post-COVID). The results have shown that in all three MANOVA tests, the p-values were less than alpha (0.1), which leads to the rejection of the null hypothesis and the conclusion with 90% confidence that there is a significant difference between the means of class 0 and class 1 and between the means of class 0 and 2 as well as between the means of class 1 and 2 on the frequency of usage of the three digital technology categories. Similar conclusions were drawn when MANOVA was conducted to compare between the means of the three class populations on the level of investment of the three digital technology categories, and the results are shown in Table 7.

Table 6. MANOVA Test: Wilk's Lambda for frequency of usage.

Comparison	Overall Class Effect	Difference between Classes 0 and 1	Difference between Classes 0 and 2	Difference between Classes 1 and 2
Wilk's Lambda	0.639	0.728	0.778	0.894
p-Value	0.0001	0.0001	0.0001	0.0055

The first row in the tables above shows the value of Wilk's Lambda (refer to Equation (3)) in each MANOVA test. It measures how well each category of the independent variable (class) contributes to the model. The scale ranges from 0 to 1 as mentioned earlier, where 0 means total discrimination and 1 means no difference. Since all the Lambda values are less than 1 and are associated with a small p-value (significant at the 0.1 level), it can be concluded with 90% confidence that there is in fact a difference between the means of the three classes on both the frequency of usage and level of investment of the three digital technology categories. The *p*-value represents the probability that measures the consistency between the data and the hypothesis being tested. The null hypothesis that there is no overall class effect and no difference between the classes in pairwise comparisons is evaluated with regard to this *p*-value. For a given alpha level, if the *p*-value is less than alpha then this null hypothesis is rejected. The alpha level used in this study is 0.1, and the tables show that all *p*-values are less than 0.1. It should be noted that some of the *p*-values are slightly higher for "no difference between classes 1 and 2" and "no difference between classes 0 and 1" in the level of investment, they are still less than 0.1. Therefore, it is safe to reject all the null hypotheses and conclude with 90% confidence that there is in fact a significant difference between the means of the three classes on both the frequency of usage and level of investment of the three digital technology categories. Future studies can focus on testing a larger sample size in order to achieve a smaller *p*-value and a higher confidence level.

Table 7. MANOVA Test: Wilk's Lambda for level of investment.

Comparison	Overall Class Effect	Difference between Classes 0 and 1	Difference between Classes 0 and 2	Difference between Classes 1 and 2
Wilk's Lambda	0.807	0.935	0.827	0.932
p-Value	0.0005	0.0551	0.0001	0.0486

5. Discussion

The study on COVID-19 and its impact in the construction industry are fairly new. Nevertheless, the following paragraph compares the results of this study with recently published papers. The results of this study are in line with the conclusions of the research done by Cheshmehzang [56], who reported on the impact of COVID-19 on boosting digitalization in the built environment. The author stated that COVID-19 was a driver for the built environment to utilize available technological toolkits even further and formulate new policies on the use of digital technologies. Similarly, Ebekozien and Aigbavboa [57] have highlighted the role of the fourth industrial revolution technologies in curtailing the impacts of COVID-19 in Nigerian construction sites. The authors explained how the use of AI technologies such as RFID have compensated for the absence of workforces and the use of BIM has helped construction professionals to comply with the pandemic rules, such as physical distancing, and still be able to simulate construction site activities.

The study was conducted using a carefully crafted questionnaire survey in the UAE construction industry. The survey instrument (Appendix A) was designed to elicit responses from the professionals. Although it is common knowledge that digital technology contributes towards project success, the findings of this study confirmed the observation across all three categories of digital technology. Importance of data acquisition and processing technologies were rated at a higher level by the respondents than communication, indicating that in-person communication will continue to be used with less dependence on technology as compared with the other two. The study showed that smaller organizations will be using and investing in digital technologies at a greater extent than the larger ones.

One of the limitations of this study was the small size of data, as represented by only 39 respondents. It should be pointed out, however, that the respondents were drawn from a very specific population—the UAE construction professionals. Thus, the value of their responses was considered significant qualitatively for this study despite the small sample size. The second limitation was that the study was conducted in a specific country (UAE), and, thus, the results are difficult to generalize. While this is a shortcoming, the fact is that the UAE is a vibrant economy in the Middle East with a high projected growth rate in the construction sector and is considered a representative example in the global construction industry. Nevertheless, the results and findings derived from this study conducted in the UAE provide valuable future directions and insights for further research on this topic.

The methodology developed and adopted in this study can be replicated in a larger country with a major economy and high volume of construction spending. A similar study can be undertaken with bigger sample sizes derived from different professional groups in the AEC industry. The variations and correlations between the regions and the professional groups will provide valuable insights regarding the use of digital technologies. Moreover, specific digital technologies can be investigated further, in addition to the three broad categories of digital technologies as conducted in this study. Nevertheless, this study provided a framework for future investigations on the use and investment in digital technologies in construction in the wake of catastrophic disasters, such as COVID-19.

The findings of this study can be useful from a practical standpoint by several measures. First, the study highlights the importance of digital technology in improving construction productivity. Second, this study differentiates between data acquisition, information processing, and communication technologies; this differentiation is helpful to identify and prioritize technologies for investing. Finally, and most importantly, this study underscores the importance of 'virtualization' in construction enabled by digital technology in the wake of catastrophes such as COVID-19. While the findings are general in nature in the context of the construction sector, the implications at the project level are imperative and indicative.

6. Conclusions

It is common knowledge that overall productivity increases in the construction sector, both regional and global, are minimal and significantly lower than the manufacturing and the service sectors. One reason frequently cited by the researchers and the practitioners alike is the slow adoption of technology, digital in particular, in construction. This study takes a deep look into this issue of adoption (usage and investment) of digital technology in construction in the context of the prevalent pandemic caused by COVID-19 since early 2020. COVID-19, ironically, presented an opportunity to investigate the status of use of digital technology adoption in the construction industry. Thus, this study is undertaken to determine the status in three levels, pre-COVID, during COVID, and post-COVID. The premise of the study is that digital technologies in three major categories—data acquisition, processing, and communication of data and information—are all impacted by the pandemic. There was a noticeable increase in the use of, and investment in, these digital technologies in the industry since early 2020 as a reaction to the restrictions put in place to reduce the spread of the COVID-19 virus. This crisis caused by the COVID-19 pandemic imposed a de facto mandate by instilling a sense of urgency among the construction professionals for digitalization of many processes and operations in construction and to perform them virtually. Digital technology, although not new and having existed for some time, was not utilized extensively prior to the "lockdown" situation caused by the pandemic. Oddly, COVID-19 provided the necessary impetus for digitalization in the construction industry. This crisis, as unexpected and undesirable as it is, offers a window of opportunity to improve and make the industry better positioned for the future.

This research study, despite the limitations, identified several significant facts and provided important insights. The most significant among them are that COVID-19 revealed that the use of digital technology, although remaining underutilized, is gaining wider acceptance in the industry. It also showed that the benefits of digital technology, once realized, will continue to be used. As a consequence, investment in digital technology in the construction industry will continue to increase. This will have long-term transformational and beneficial impacts on productivity in the construction sector. The methodology developed and employed in this study can be utilized to investigate certain selected technologies for use and investment decisions. A decision-making model for use in the industry can be developed using the categories as outlined in this study.

Author Contributions: Conceptualization, I.A. and S.E.-S.; methodology, O.E., S.A., I.A. and S.E.-S.; software, S.A.; validation, O.E., S.A., I.A. and S.E.-S.; formal analysis, O.E., S.A., I.A. and S.E.-S.; investigation, O.E., S.A., I.A. and S.E.-S.; resources, O.E., S.A., I.A. and S.E.-S.; data curation, O.E. and S.A.; writing—original draft preparation, O.E. and S.A.; writing—review and editing, I.A. and S.E.-S.; visualization, O.E. and S.A.; supervision, I.A. and S.E.-S.; project administration, I.A. and S.E.-S.; funding acquisition, I.A. and S.E.-S. All authors have read and agreed to the published version of the manuscript.

Funding: This research was funded by The American University of Sharjah, grant number FRG21-M-E71. The work in this paper was supported, in part, by the Open Access Program from the American University of Sharjah, and the APC was funded through Grant number: OAPCEN-1410-E00060.

Institutional Review Board Statement: The study was approved by the Institutional Review Board of the AMERICAN UNIVERSITY OF SHARJAH (protocol # 20-072 on 5 May 2021).

Informed Consent Statement: Written informed consent was waived for this study.

Data Availability Statement: Data are available upon request.

Conflicts of Interest: The authors declare no conflict of interest. The funders had no role in the design of the study; in the collection, analyses, or interpretation of data; in the writing of the manuscript; or in the decision to publish the results. This paper represents the opinions of the author(s) and does not mean to represent the position or opinions of the American University of Sharjah.

Appendix A. Survey Instrument

Appendix A.1. General Information

In this section, you will be asked about your general information.

Your company?

- ☐ Local (Main Office in UAE)
- ☐ International (Main Office outside UAE)

Years of experience in UAE

- ☐ <5
- ☐ 5–<10
- ☐ 10–20
- ☐ >20

Project Type/expertise

- ☐ Buildings
- ☐ Infrastructure
- ☐ Other

Current role

- ☐ Owner
- ☐ Consultant
- ☐ Contractor
- ☐ Construction/project management firm

Average size of projects

- ☐ <50 M AED
- ☐ 50 to 200 M AED
- ☐ >200–500 M AED
- ☐ >500 M AED

Appendix A.2. Data and Information Gathering Technologies

In this section, you will be asked to rate the level of usage, investment decisions, and the success potential for data and information gathering technologies (e.g., RFID, Barcode, Drones, GPS, laser scanning).

1. Do you believe that the use of digital technology (such as drones) for data and information gathering contributes towards project success?
   - ☐ Strongly Agree
   - ☐ Agree
   - ☐ Neutral
   - ☐ Disagree
   - ☐ Strongly Disagree
2. Prior to the COVID-19 Pandemic, how frequently your company have used digital technology for data and information gathering?
   - ☐ Almost Always
   - ☐ Often
   - ☐ Sometimes
   - ☐ Rarely
   - ☐ Never
3. During the COVID-19 Pandemic (after March 2020), how frequently your company have used digital technology for data and information gathering?
   - ☐ Almost Always
   - ☐ Often
   - ☐ Sometimes
   - ☐ Rarely
   - ☐ Never
4. Please indicate your choice related to the statement: After COVID-19 Pandemic is essentially over, the usage of digital technology *will increase* beyond the level reached during COVID-19 Pandemic for data and information gathering.
   - ☐ Strongly Agree—Will greatly increase
   - ☐ Agree—Will slightly increase
   - ☐ Neutral—remain the same as already reached
   - ☐ Disagree—Will slightly decrease
   - ☐ Strongly Disagree—Will return to the pre-Pandemic level
5. Prior to the COVID-19 Pandemic, what is the level of your company's investment in digital technology for data and information gathering?
   - ☐ Very high
   - ☐ High
   - ☐ Normal
   - ☐ Adequate
   - ☐ Very little
6. During the COVID-19 Pandemic (after March 2020), what is the level of your company's investment in digital technology for data and information gathering?
   - ☐ Very high
   - ☐ High
   - ☐ Normal
   - ☐ Adequate
   - ☐ Very little
7. Please indicate your choice related to the statement: After COVID-19 Pandemic is essentially over, the investment in digital technology will increase beyond the level reached during COVID-19 Pandemic for data and information gathering.
   - ☐ Strongly Agree—Will greatly increase
   - ☐ Agree—Will slightly increase
   - ☐ Neutral—remain the same as already reached
   - ☐ Disagree—Will slightly decrease
   - ☐ Strongly Disagree—Will return to the pre-Pandemic level

Appendix A.3. Data and Information Processing Technologies

In this section, you will be asked to rate the level of usage, investment decisions, and the success potential for data and information processing technologies (e.g., Building Information Modeling or BIM, Scheduling program, such as Primavera, Power Project, Procore, Job Master, CMIC, Estimating software such as Manifest, Esti-mate, FBS-Estimator).

1. Do you believe that the use of digital technology (such as BIM) for data and information processing contributes towards project success?
   - ☐ Strongly Agree
   - ☐ Agree
   - ☐ Neutral
   - ☐ Disagree
   - ☐ Strongly Disagree
2. Prior to the COVID-19 Pandemic, how frequently your company have used digital technology for data and information processing?
   - ☐ Almost Always
   - ☐ Often
   - ☐ Sometimes
   - ☐ Rarely
   - ☐ Never
3. During the COVID-19 Pandemic (after March 2020), how frequently your company have used digital technology for data and information processing?
   - ☐ Almost Always
   - ☐ Often
   - ☐ Sometimes
   - ☐ Rarely
   - ☐ Never
4. Please indicate your choice related to the statement: After COVID-19 Pandemic is essentially over, the usage of digital technology will increase beyond the level reached during COVID-19 Pandemic for data and information processing.
   - ☐ Strongly Agree—Will greatly increase
   - ☐ Agree—Will slightly increase
   - ☐ Neutral—remain the same as already reached
   - ☐ Disagree—Will slightly decrease
   - ☐ Strongly Disagree—Will return to the pre-Pandemic level
5. Prior to the COVID-19 Pandemic, what is the level of your company's investment in digital technology for data and information processing?
   - ☐ Very high
   - ☐ High
   - ☐ Normal
   - ☐ Adequate
   - ☐ Very little
6. During the COVID-19 Pandemic (after March 2020), what is the level of your company's investment in digital technology for data and information processing?
   - ☐ Very high
   - ☐ High
   - ☐ Normal
   - ☐ Adequate
   - ☐ Very little
7. Please indicate your choice related to the statement: After COVID-19 Pandemic is essentially over, the investment in digital technology will increase beyond the level reached during COVID-19 Pandemic for data and information processing.

- ☐ Strongly Agree—Will greatly increase
- ☐ Agree—Will slightly increase
- ☐ Neutral—remain the same as already reached
- ☐ Disagree—Will slightly decrease
- ☐ Strongly Disagree—Will return to the pre-Pandemic level

Appendix A.4. Information and Communications Technologies

In this section, you will be asked to rate the level of usage, investment decisions, and the success potential for information and communication technologies (e.g., Video Conferencing such as google meet and zoom, Dropbox, mail, web-based systems).

1. Do you believe that the use of digital technology (such as video conferencing and file sharing) for communications of data and information (including virtual meetings) contributes towards project success?
    - ☐ Strongly Agree
    - ☐ Agree
    - ☐ Neutral
    - ☐ Disagree
    - ☐ Strongly Disagree
2. Prior to the COVID-19 Pandemic, how frequently your company have used digital technology for communication of data and information (including virtual meetings)?
    - ☐ Almost Always
    - ☐ Often
    - ☐ Sometimes
    - ☐ Rarely
    - ☐ Never
3. During the COVID-19 Pandemic (after March 2020), how frequently your company have used digital technology for communication of data and information (including virtual meetings)?
    - ☐ Almost Always
    - ☐ Often
    - ☐ Sometimes
    - ☐ Rarely
    - ☐ Never
4. Please indicate your choice related to the statement: After COVID-19 Pandemic is essentially over, the usage of digital technology will increase beyond the level reached during COVID-19 Pandemic for communication of data and information (including virtual meetings).
    - ☐ Strongly Agree—Will greatly increase
    - ☐ Agree—Will slightly increase
    - ☐ Neutral—remain the same as already reached
    - ☐ Disagree—Will slightly decrease
    - ☐ Strongly Disagree—Will return to the pre-Pandemic level
5. Prior to the COVID-19 Pandemic, what is the level of your company's investment in digital technology for communication of data and information (including virtual meetings)?
    - ☐ Very high
    - ☐ High
    - ☐ Normal
    - ☐ Adequate
    - ☐ Very little

6. During the COVID-19 Pandemic (after March 2020), what is the level of your company's investment in digital technology for communications of data and information (including virtual meetings)?
   - ☐ Very high
   - ☐ High
   - ☐ Normal
   - ☐ Adequate
   - ☐ Very little
7. Please indicate your choice related to the statement: After COVID-19 Pandemic is essentially over, the investment in digital technology will increase beyond the level reached during COVID-19 Pandemic for communication of communications of data and information (including virtual meetings).
   - ☐ Strongly Agree—Will greatly increase
   - ☐ Agree—Will slightly increase
   - ☐ Neutral—remain the same as already reached
   - ☐ Disagree—Will slightly decrease
   - ☐ Strongly Disagree—Will return to the pre-Pandemic level

References

1. Global Construction Market to Grow $8 Trillion by 2030: Driven by China, US and India. Institution of Civil Engineers (ICE). 2015. Available online: https://myice.ice.org.uk/ICEDevelopmentWebPortal/media/documents/news/ice%20news/global-construction-press-release.pdf (accessed on 29 January 2022).
2. Reinventing Construction through a Productivity Revolution | McKinsey. McKinsey Global Institute. 2017. Available online: https://www.mckinsey.com/business-functions/operations/our-insights/reinventing-construction-through-a-productivity-revolution (accessed on 29 January 2022).
3. IFS Finds Concerns about Economic Disruption Driving Digital Transformation Spending—Bizness Transform. Business Transformation (BT). 2020. Available online: https://www.biznesstransform.com/ifs-finds-concerns-about-economic-disruption-driving-digital-transformation-spending/ (accessed on 29 January 2022).
4. The Essential Eight Technologies. PricewaterhouseCoopers (PwC). 2022. Available online: https://www.pwc.com/us/en/tech-effect/emerging-tech/essential-eight-technologies.html (accessed on 29 January 2022).
5. Allan-Blitz, L.-T.; Turner, I.; Hertlein, F.; Klausner, J.D. High Frequency and Prevalence of Community-Based Asymptomatic SARS-CoV-2 Infection. *Emerg. Infect. Dis.* **2020**. [CrossRef]
6. Dadlani, D. COVID-19 Guidelines for Construction Workers. Construction Week Online. 9 May 2020. Available online: https://www.constructionweekonline.com/business/264880-covid-19-guidelines-for-construction-workers (accessed on 8 November 2021).
7. Report: UAE Construction after COVID-19, Middle East Business Intelligence (MEED). 10 August 2020. Available online: https://www.meed.com/report-uae-construction-after-covid-19 (accessed on 22 January 2022).
8. The Rise of Drones in UAE Construction. Middle East Business Intelligence (MEED). 15 December 2020. Available online: https://www.meed.com/the-rise-of-drones-in-uae-construction (accessed on 22 January 2022).
9. Here's Why UAE's Construction Industry Needs to Pick up Pace of Change, Gulf News. 8 January 2021. Available online: https://gulfnews.com/business/property/heres-why-uaes-construction-industry-needs-to-pick-up-pace-of-change-1.1610097889474 (accessed on 22 January 2022).
10. Wallett, P. Going Digital in COVID-19: Welcoming Next-Gen Workers on Site. Construction Week Online. 28 June 2020. Available online: https://www.constructionweekonline.com/products-services/266085-going-digital-in-covid-19-welcoming-next-gen-workers-on-site (accessed on 8 November 2021).
11. Bower, D.J.; Hinks, J.; Wright, H.; Hardcastle, C.; Cuckow, H. ICTs, videoconferencing and the construction industry: Opportunity or threat? *Constr. Innov.* **2002**, *1*, 129–144. [CrossRef]
12. Allen, S.; Hinks, A.J. *Using Action Research to Compare the Theory and Practice of Managing Housing Information: International Symposium on the Organisation and Management of Construction (W65)*; International Symposium on the Organization and Management of Construction: Glasgow, UK, 1996; pp. 276–286.
13. Bigliardi, B.; Galati, F.; Petroni, A. How to effectively manage knowledge in the construction industry. *Meas. Bus. Excell.* **2014**, *18*, 57–72. [CrossRef]
14. Kamal, M. The triple-edged sword of COVID-19: Understanding the use of digital technologies and the impact of productive, disruptive, and destructive nature of the pandemic. *Inf. Syst. Manag.* **2020**, *37*, 310–317. [CrossRef]
15. Alsharef, A.; Banerjee, S.; Uddin, S.M.J.; Albert, A.; Jaselskis, E. Early Impacts of the COVID-19 Pandemic on the United States Construction Industry. *Int. J. Environ. Res. Public Health* **2021**, *18*, 1559. [CrossRef]

16. Oey, E.; Lim, J. Challenges and action plans in construction sector owing to COVID-19 pandemic—A case in Indonesia real estates. *Int. J. Lean Six Sigma* **2021**, *12*, 835–858. [CrossRef]
17. The Construction Labor Shortage: Will Developers Deploy Robotics? *Forbes*. 2019. Available online: https://www.forbes.com/sites/columbiabusinessschool/2019/07/31/the-construction-labor-shortage-will-developers-deploy-robotics/ (accessed on 8 November 2021).
18. Buchanan, J.; Kelley, B.; Hatch, A. Digital Workplace and Culture How Digital Technologies Are Changing the Workforce and How Enterprises Can Adapt and Evolve. Deloitte. 2016. Available online: https://www2.deloitte.com/content/dam/Deloitte/us/Documents/human-capital/us-cons-digital-workplace-and-culture (accessed on 29 January 2022).
19. Delgado, J.M.D.; Oyedele, L.O.; Ajayi, A.; Àkànbí, L.; Akinadé, O.O.; Bilal, M.; Owolabi, H. Robotics and automated systems in construction: Understanding industry-specific challenges for adoption. *J. Build. Eng.* **2019**, *26*, 100868. [CrossRef]
20. Kent, D.C.; Becerik-Gerber, B. Understanding Construction Industry Experience and Attitudes toward Integrated Project Delivery. *J. Constr. Eng. Manag.* **2010**, *136*, 815–825. [CrossRef]
21. Zhou, Y.; Luo, H.; Yang, Y. Implementation of augmented reality for segment displacement inspection during tunneling construction. *Autom. Constr.* **2017**, *82*, 112–121. [CrossRef]
22. Digital America: A Tale of the Haves and Have-Mores | McKinsey. Available online: https://www.mckinsey.com/industries/technology-media-and-telecommunications/our-insights/digital-america-a-tale-of-the-haves-and-have-mores (accessed on 29 January 2022).
23. Delgado, M.D.; Oyedele, L.; Demian, P.; Beach, T. A research agenda for augmented and virtual reality in architecture, engineering and construction. *Adv. Eng. Inform.* **2020**, *45*, 101122. [CrossRef]
24. Kardasz, P.; Doskocz, J. Drones and Possibilities of Their Using. *J. Civ. Environ. Eng.* **2016**, *6*, 1–7. [CrossRef]
25. Xiao, Y.; Kamat, V.R.; Lee, S. Monitoring Excavation Slope Stability Using Drones. In Proceedings of the Construction Research Congress 2018, New Orleans, LA, USA, 2–4 April 2018; pp. 169–179. [CrossRef]
26. Liu, P.; Chen, A.; Huang, Y.-N.; Han, J.-Y.; Lai, J.-S.; Kang, S.-C.; Wu, T.-H.; Wen, M.-C.; Tsai, M.-H. A review of rotorcraft Unmanned Aerial Vehicle (UAV) developments and applications in civil engineering. *Smart Struct. Syst.* **2014**, *13*, 1065–1094. [CrossRef]
27. Ergen, E.; Akinci, B. An Overview of Approaches for Utilizing RFID in Construction Industry. In Proceedings of the 2007 1st Annual RFID Eurasia, Istanbul, Turkey, 5–6 September 2007; pp. 1–5. [CrossRef]
28. Dinis, F.M.; Sanhudo, L.; Martins, J.P.; Ramos, N.M.M. Improving project communication in the architecture, engineering and construction industry: Coupling virtual reality and laser scanning. *J. Build. Eng.* **2020**, *30*, 101287. [CrossRef]
29. Dziadak, K.; Kumar, B.; Sommerville, J. Model for the 3D Location of Buried Assets Based on RFID Technology. *J. Comput. Civ. Eng.* **2009**, *23*, 148–159. [CrossRef]
30. Chae, S.; Yoshida, T. Application of RFID technology to prevention of collision accident with heavy equipment. *Autom. Constr.* **2010**, *19*, 368–374. [CrossRef]
31. Liu, J.; Zhang, Q.; Wu, J.; Zhao, Y. Dimensional accuracy and structural performance assessment of spatial structure components using 3D laser scanning. *Autom. Constr.* **2018**, *96*, 324–336. [CrossRef]
32. Zhang, J.; Huang, J.; Fu, C.Q.; Huang, L.; Ye, H. Characterization of steel reinforcement corrosion in concrete using 3D laser scanning techniques. *Constr. Build. Mater.* **2020**, *270*, 121402. [CrossRef]
33. National Institute of Building Sciences. Frequently Asked Questions About the National BIM Standard-United StatesTM | National BIM Standard—United States. Available online: https://www.nationalbimstandard.org/faqs (accessed on 8 November 2021).
34. Kaner, I.; Sacks, R.; Kassian, W.; Quitt, T. Case studies of BIM adoption for precast concrete design by mid-sized structural engineering firms. *Electron. J. Inf. Technol. Constr.* **2008**, *13*, 303–323.
35. Marzouk, M.; Othman, A.; Enaba, M.; Zaher, M. Using BIM to Identify Claims Early in the Construction Industry: Case Study. *J. Leg. Aff. Disput. Resolut. Eng. Constr.* **2018**, *10*, 5018001. [CrossRef]
36. Forgues, D.; Iordanova, I.; Valdivesio, F.; Staub-French, S. Rethinking the Cost Estimating Process through 5D BIM: A Case Study. In Proceedings of the Construction Research Congress 2012: Construction Challenges in a Flat World, West Lafayette, Indiana, 21–23 May 2012; pp. 778–786. [CrossRef]
37. Monteiro, A.; Martins, J. A survey on modeling guidelines for quantity takeoff-oriented BIM-based design. *Autom. Constr.* **2013**, *35*, 238–253. [CrossRef]
38. Oracle. Primavera P6 Enterprise Project Portfolio Management | Oracle United Arab Emirates. Available online: https://www.oracle.com/ae/industries/construction-engineering/primavera-p6/ (accessed on 30 January 2022).
39. Esti-Mate. Describe, Measure & Price Jobs | Esti-Mate. Available online: https://estimate.co.uk/ (accessed on 30 January 2022).
40. Cadalyst. Cadalyst Labs Review: Autodesk Revit Structure 2-Digital Models for Structural Design | Cadalyst. 2005. Available online: https://www.cadalyst.com/aec/cadalyst-labs-review-autodesk-revit-structure-2-digital-models-structural-design-5397 (accessed on 22 January 2022).
41. Chinowsky, P.S.R. Virtual teams: Guide to successful implementation. *J. Manag. Eng.* **2003**, *19*, 98–106. [CrossRef]
42. Mell, P.; Grance, T. The NIST Definition of Cloud Computing. *Natl. Inst. Stand. Technol.* **2011**, *800*, 145. [CrossRef]
43. Du, J.; Shi, Y.; Zou, Z.; Zhao, D. CoVR: Cloud-Based Multiuser Virtual Reality Headset System for Project Communication of Remote Users. *J. Constr. Eng. Manag.* **2018**, *144*, 4017109. [CrossRef]

44. McNamara, A.J.; Sepasgozar, S.M.E. Intelligent contract adoption in the construction industry: Concept development. *Autom. Constr.* **2021**, *122*, 103452. [CrossRef]
45. Zheng, K.; Zhang, Z.; Gauthier, J. Blockchain-based intelligent contract for factoring business in supply chains. *Ann. Oper. Res.* **2022**, *308*, 777–797. [CrossRef]
46. Godinho, V. How Digital Transformation Is Taking Root within the Construction Industry. Gulf Business. 2021. Available online: https://gulfbusiness.com/how-digital-transformation-is-taking-root-within-the-construction-industry/ (accessed on 30 January 2022).
47. UAE's New Workweek: A Bit of Tech and More Creative Thinking Can Get the Job Done. Gulf News. 2022. Available online: https://gulfnews.com/business/analysis/uaes-new-workweek-a-bit-of-tech-and-more-creative-thinking-can-get-the-job-done-1.1641816313034 (accessed on 30 January 2022).
48. Ikediashi, D.I.; Ogwueleka, A.C. Assessing the use of ICT systems and their impact on construction project performance in the Nigerian construction industry. *J. Eng. Des. Technol.* **2016**, *14*, 252–276. [CrossRef]
49. Oliver, S. Communication and trust: Rethinking the way construction industry professionals and software vendors utilise computer communication mediums. *Vis. Eng.* **2019**, *7*, 1. [CrossRef]
50. Wong, C.H.; Sloan, B. Use of ICT for E-Procurement in The UK Construction Industry: A Survey of SMEs Readiness. In Proceedings of the ARCOM Proceedings Twentieth Annual Conference, Edinburgh, UK, 1–3 September 2005; pp. 1–3. Available online: https://www.semanticscholar.org/paper/USE-OF-ICT-FOR-E-PROCUREMENT-IN-THE-UK-CONSTRUCTION-Wong-Sloan/9fc85aaac97515ee4115f22a89b1c686eee4e742 (accessed on 8 November 2021).
51. Giel, B.; Issa, R.R.A. Quality and Maturity of BIM Implementation in the AECO Industry. *Appl. Mech. Mater.* **2013**, *438–439*, 1621. [CrossRef]
52. Kleinbaum, D.G.; Kupper, L.L.; Nizam, A.; Rosenberg, S.E. *Applied Regression Analysis and Other Multivariable Methods*, 5th ed.; Cengage Learning: Boston, MA, USA, 2013.
53. Hair, J.F., Jr.; Black, W.C.; Babin, B.J.; Anderson, R.E. *Multivariate Data Analysis*, 7th ed.; Pearson: Upper Saddle River, NJ, USA, 2009.
54. Zhao, D.; Mccoy, A.; Agee, P.; Mo, Y.; Reichard, G.; Paige, F. Time effects of green buildings on energy use for low-income households: A longitudinal study in the United States. *Sustain. Cities Soc.* **2018**, *40*, 559–568. [CrossRef]
55. Huberty, C.J.; Olejnik, S. *Applied MANOVA and Discriminant Analysis*, 2nd ed.; Wiley-Interscience: Hoboken, NJ, USA, 2006.
56. Cheshmehzangi, A. From transitions to transformation: A brief review of the potential impacts of COVID-19 on boosting digitization, digitalization, and systems thinking in the built environment. *J. Build. Constr. Plan. Res.* **2021**, *9*, 26. [CrossRef]
57. Ebekozien, A.; Aigbavboa, C. COVID-19 recovery for the Nigerian construction sites: The role of the fourth industrial revolution technologies. *Sustain. Cities Soc.* **2021**, *69*, 102803. [CrossRef]

buildings

MDPI

Article

Spatiotemporal Differentiation and Influencing Factors of Green Technology Innovation Efficiency in the Construction Industry: A Case Study of Chengdu–Chongqing Urban Agglomeration

Bo Wang [1], Hongxi Chen [1,*], Yibin Ao [2,*] and Fangwei Liao [3]

[1] School of Civil Engineering and Architecture, Southwest University of Science and Technology, Mianyang 621010, China
[2] College of Environmental and Civil Engineering, Chengdu University of Technology, Chengdu 610059, China
[3] School of Economics and Management, Southwest University of Science and Technology, Mianyang 621010, China
* Correspondence: chenhongxi@mails.swust.edu.cn (H.C.); aoyibin10@mail.cdut.edu.cn (Y.A.)

Abstract: In order to support the green and low-carbon transformation of China's construction industry and accomplish the dual carbon objective, it is vital to accelerate green technology innovation. Therefore, this paper takes the Chengdu–Chongqing urban agglomeration of China as the study area, using the super-efficiency slacks-based measure (SBM)model and the gravity model to assess the efficiency of green technology innovation in the construction industry, utilizing geographical detectors to investigate the drivers of green technology innovation in the construction industry further. Additionally, we consider each influencing factor's level of impact on the efficiency of green technology innovation in the construction sector both under the single factor and double factor scenarios. The findings indicate that there is a considerable difference in the efficiency of green technology innovation in the Chengdu–Chongqing metropolitan agglomeration's construction industry, and the trend is upward. In addition, the research area exhibited spatially heterogeneous characteristics in terms of the efficiency of green technology innovation in the construction industry, and the spatial spillover effect was significantly limited by distance. Further research revealed that environmental legislation, economic development, public environmental concern, urbanization level, and foreign direct investment were the primary driving factors of green technology innovation efficiency in the construction sector, and industrial size was the potential driving factor. The spatial and temporal differentiation of the green technology innovation efficiency in the construction industry was also more affected by the interaction between the dominating factor and the prospective factor than by either factor acting alone. The research's findings are useful in advancing the green and low-carbon transformation of the construction sector in the Chengdu–Chongqing metropolitan agglomeration by offering theoretical support and decision-making reference.

Keywords: construction industry; efficiency of green technology innovation; super-efficiency SBM model; spatiotemporal differentiation; influencing factors

Citation: Wang, B.; Chen, H.; Ao, Y.; Liao, F. Spatiotemporal Differentiation and Influencing Factors of Green Technology Innovation Efficiency in the Construction Industry: A Case Study of Chengdu–Chongqing Urban Agglomeration. *Buildings* **2023**, *13*, 73. https://doi.org/10.3390/buildings13010073

Academic Editor: Saeed Banihashemi

Received: 10 November 2022
Revised: 11 December 2022
Accepted: 23 December 2022
Published: 28 December 2022

1. Introduction

Reducing carbon emissions has become a top priority for many nations due to the state of the environment, and the development of "carbon neutral" awareness is accelerating globally. The accomplishment of China's "dual carbon" objective is significant on a worldwide scale and presents new opportunities and challenges for the growth of China's economic structure. China's environmental quality ranks 120th out of 180 countries and regions in the world with major ecological and environmental issues, according to the 2020 Global Environmental Performance Index study jointly released by Yale University and other institutions [1]. At the same time, figures show that, in 2019, China was responsible for 27.92% of the world's carbon emissions. The problem of carbon emissions is very prominent, and excessive carbon emissions will cause serious harm to the environment [2]. The development of green technology innovation is an important factor to promote the transformation of an economic development mode [3] and also an important means to achieve sustainable development [4]. To achieve the "double carbon" objective in the face of difficult environmental issues, it is imperative to speed up the promotion of green transformation in the new era.

According to data released by the China Building Energy Efficiency Association, the total carbon emission from the whole process of construction is more than half of the total carbon emission in China [5], which seriously hinders the green development of the economy. The construction industry, a pillar of the national economy, is crucial to raising the level of the country's economy. However, with the rapid development of the construction industry, problems such as large resource consumption and extensive construction methods in the construction industry have become increasingly prominent [6], which will also cause irreversible damage to resources and the environment such as carbon dioxide emission [7]. Construction companies urgently need to follow the path of green and low-carbon development due to serious environmental issues, the need to reduce carbon emissions, and the need to actively explore the new model of low-carbon development, which is of great significance for controlling total carbon emissions, realizing green development, and guiding and promoting the conclusion of the Paris Agreement [8]. To achieve the decoupling of green economic development from resource consumption and environmental pollution, green technology innovation seeks to maximize economic, ecological, and social benefits with minimal cost and pollution [9]. When conducting innovation activities, consideration should be given to the efficiency of green technology innovation in order to realize the sustainable and high-quality development of the construction sector [10].

This study examined the spatial and temporal differential characteristics and key influencing elements of green technology innovation efficiency in the construction industry, using the construction industry as the research topic. In order to further share knowledge and make policy recommendations for enhancing the efficiency of green technology innovation in the construction sector, this paper used the following framework: Section 2 outlines the study's theoretical basis; Section 3 introduces the research area, research methods, and selection and measuring of variables; Section 4 summarizes the research findings; Section 5 discusses the research results and draws conclusions.

2. Theoretical Review

Joseph A. Schumpeter (1991) was the first to put forward the theory of technological innovation. Technological innovation can boost productivity, deliver competitive advantages and good economic benefits to society, an industry, or an organization, and alleviate numerous difficulties that humans encounter. The traditional model of technological innovation, however, is overly simplistic, and rapid economic growth results in a shortage of natural resources and environmental pollution, severely impeding the path to sustainable development and impeding the transition to a greener form of economic development. Green and sustainable development is a crucial tool for advancing ecological civilization and superior economic growth [11–14]. Green technology innovation was proposed in 1960 to efficiently address the difficult environmental pollution issues that Western nations were facing and to offer technical help [15]. Green technology was formally defined by Braun and Wield [16] in 1994, who argued that it was important for improving environmental quality [17]. Most academics agree that green technology innovation is one of the key steps to take into account both the ecological environment and a low-carbon economy in order to address the internal conflict between economic expansion and environmental pollution [18–20]. Additionally, the advancement of green technology has emerged as a crucial element in advancing sustainable development [21,22]. Under the background of sustainable development, the relevant policies for green technology innovation have been continuously introduced and implemented, promoting the development of green technology innovation. Green technology innovation is distinct to traditional technology innovation, as it is theoretically based on ecology, information science, sociology, current management, etc. [23]. It pays attention to saving circulation, efficient utilization, and reducing pollution, and plays an important role in realizing sustainable development [24]. Research on green technological innovation is currently focused primarily on connotation, assessment, and influencing variables both domestically and internationally.

Various scholars have different meanings for the meaning of "green technology innovation efficiency" in their research. Pedro (2004) [25], Werf (2003) [26], and others consider "green technology innovation" as technological innovation that takes full account of environmental factors on top of traditional innovation. Hellstm (2007) [27] believed that "green technology innovation" should consider reducing the impact on the environment while meeting product innovation. Manral (2018) [28] and Yudietal et al. (2019) [29] defined the term "green technology innovation" as a process that starts with the goals of preserving the environment, conserving energy, and reducing emissions as the premise and achieves financial gain. Though there is not yet a common definition of "green technology innovation" in academia, it typically refers to novel technologies that can enhance environmental performance [30]. Acemoglu (2012) [21] and Wang et al. (2021) [31] believed that "green technology innovation" is a new technology that can inhibit energy consumption while reducing pollutant emissions, improving environmental quality, and promoting green economic development [32] and is widely recognized by the general public.

The evaluation of green technology innovation focuses on its efficiency, primarily through the development of indicators, stochastic frontier analysis (SFA), and data envelopment analysis (DEA). Sun et al. (2017) [33] applied the entropy weighting method in order to assess green technology innovation from the standpoint of its ecological and economic performance. Li et al. (2022) [34] created a system health evaluation index system to assess the state of green technology innovation. SFA is not suited to complicated systems, with many inputs and multiple outputs since it is a typical parametric analytic technique with little room for error in function selection and parameter configuration [35]. The data envelopment analysis method can overcome the defects caused by the ratio method and index system calculation method. Early data envelopment analysis techniques, such as BCC (Banker-Charnes-Cooper) and CCR (Charnes-Cooper-Rhodes), were used to gauge how well green technology was able to innovate. Lin et al. (2018) [36] used the ideal window width DEA window analysis method to assess the efficiency of green technology innovation. When the results were compared, it was thought that the obtained results were more realistic than the calculation results of the conventional DEA model. The measurement results are not accurate, because the conventional model overlooks the relaxation of variables [37]. In 2001, Tone constructed a DEA-SBM model considering the relaxation of output and input factors [38]. The DEA-SBM model was employed by Feng et al. (2013) [39] to assess the effectiveness of green technology innovation. As research has continued to advance, many academics have taken unwanted output into account when calculating the efficiency of green technology innovation. They have also developed the super-efficiency SBM model [40,41] that incorporates unwanted output in order to increase the precision of efficiency assessment.

The government, the market, the general public, and the industry itself are the primary players in the study of the factors that affect the development of green technologies. The research perspectives of green technology innovation range from macro to micro, mainly focusing on regional, industrial, and enterprise levels [42]. With different perspectives and methods, abundant research results have been obtained on the influencing factors of green technology innovation [43]. Behera et al. (2022) [44] used the mixed mean group, random effect, generalized mixed models, and gaussian mixture model models to analyze Organization for Economic Co-operation and Development (OECD) countries. They felt that effective environmental regulation and foreign direct investment inflow might stimulate the development of green technology. Zhang (2022) [37] built a spatial econometric model to examine the influencing variables and came to the conclusion that environmental regulation, government support, educational attainment, and industry scale all played an important role in fostering the efficiency of green technology innovation. Li (2022) [45] used microfirms as the research subject and employed evolutionary game theory to support the contention that manufacturing companies can be encouraged to promote green technology innovation through subsidies and fair environmental legislation. Green technology innovation efficiency is a typical indicator that has both economic and ecological qualities [37], and it is influenced by a wide range of variables such as environmental regulation [46–48], government subsidies [49], economic development level [50], foreign direct investment [43], education level [51], industrial scale [52], and other aspects factors.

As mentioned above, first of all, the design of an assessment system that takes into account input, output, and undesirable output is the main method of the current measurement of the efficiency of green technology innovation. Secondly, when choosing research methodologies for green technology innovation, multiple linear regression, spatial econometric models, evolutionary games, and other techniques are used. However, only linear influence is taken into account when studying the influencing factors of green technology innovation. Finally, existing research perspectives mostly focus on manufacturing, high-tech industries, and industry. Green technology innovation in the construction business is very important, because it has a reputation for being high-consumption, high-pollution, and one of the most carbon-generating industries in China. As a result, this study examines the efficiency of green technology innovation and the factors that drive it, using the construction sector as its focus. The super-efficiency model is used in this study to quantify the efficiency value by combining the methods of econometrics, geography, and physics. Furthermore, tools such as the gravitational model and geographic detector are brought into the field of green technology innovation in the construction industry to evaluate and affect elements of green technology innovation efficiency. This paper delves into green technology innovation in the construction industry, expanding and enriching the relevant content and helping to foster the coordinated development of the regional economy and regional environment. It also provides a theoretical basis for the development of targeted and regionally differentiated countermeasures for the efficiency of green technology innovation in the construction industry.

3. Materials and Methods

3.1. Study Area

The urban agglomeration of Chengdu–Chongqing is situated in the junction zone of China's "two horizontal and three vertical" urbanization strategic pattern, which benefits from its advantageous location by linking the east and west with the north and south. It is one of the western regions with the best economic foundation and greatest economic power, and it contributes to the promotion of the west's development. To achieve the green, coordinated, and sustainable growth of China, it is crucial to accelerate the development of the construction sector in the Chengdu–Chongqing urban agglomeration. The urban agglomeration of Chengdu–Chongqing was chosen as the research object in this study. According to the Development Planning of the Chengdu–Chongqing Urban Agglomeration, this urban agglomeration specifically includes 16 cities, including Chengdu, Zigong, Luzhou, Deyang, Mianyang, Meishan, Yibin, Neijiang, Leshan, Suining, Nanchong, Ya'an, Dazhou, Guang'an, Ziyang, and Chongqing, as shown in Figure 1.

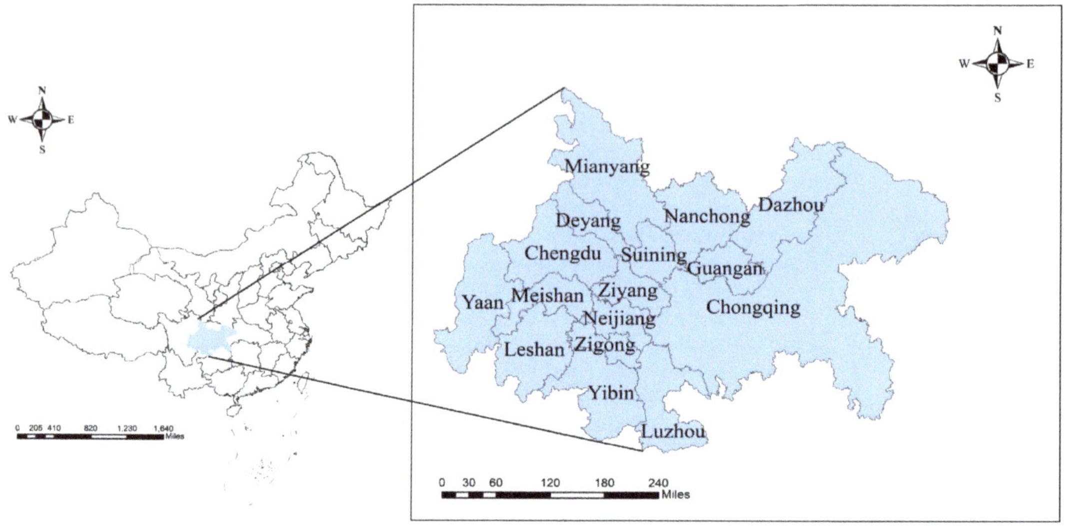

Figure 1. Study area.

3.2. *Methodology*

3.2.1. Super-Efficiency SBM Model

The super-efficiency SBM model is one of the most commonly employed models for efficiency measurement. It combines the benefits of the super-efficiency DEA and SBM models and takes the undesirable output into account to create a super-efficiency model. This addresses the shortcoming of the static DEA model, which cannot measure panel data. Assuming there exist n decision-making units (DMU), each consisting of m input factors, q_1 desired outputs, and q_2 undesired outputs, ρ^* is the efficiency value. The specific model of the SBM model is as follows.

$$\min\rho^* = \frac{1 - \frac{1}{m}\sum\limits_{i=1}^{m}\frac{s_i^-}{x_{ik}}}{1 + \frac{1}{q_1+q_2}\left(\sum\limits_{r=1}^{q_1}\frac{s_r^+}{y_{rk}} + \sum\limits_{t=1}^{q_2}\frac{s_t^{b-}}{b_{tk}}\right)} \tag{1}$$

$$s.t.\begin{cases} \sum\limits_{j=1,j\neq k}^{n} x_{ij}\lambda_j + s_i^- = x_{ik} \\ \sum\limits_{j=1,j\neq k}^{n} y_{rj}\lambda_j - s_r^+ = y_{rk} \\ \sum\limits_{j=1,j\neq k}^{n} b_{tj}\lambda_j + s_t^{b-} = b_{tk} \\ \lambda_j, s_i^-, s_r^+, s_t^{b-} \geq 0; i = 1,2,\ldots,m; r = 1,2,\ldots,q_1 \\ t = 1,2,\ldots,q_2; j = 1,2,\ldots,n(j \neq k) \end{cases} \tag{2}$$

A distinction between decision units that have an efficiency value of 1 as determined by the SBM model is necessary in order to more effectively compare efficiency values. Therefore, this paper further selects the super-efficient SBM model to calculate the efficiency of green technology innovation in the construction industry. The specific model of the super-efficient SBM model is as follows.

$$\min\rho = \frac{1 + \frac{1}{m}\sum_{i=1}^{m}\frac{s_i^-}{x_{ik}}}{1 - \frac{1}{q_1+q_2}\left(\sum_{r=1}^{q_1}\frac{s_r^+}{y_{rk}} + \sum_{t=1}^{q_2}\frac{s_t^{b-}}{b_{tk}}\right)} \tag{3}$$

$$s.t.\begin{cases} \sum_{j=1,j\neq k}^{n} x_{ij}\lambda_j - s_i^- \leq x_{ik} \\ \sum_{j=1,j\neq k}^{n} y_{rj}\lambda_j + s_r^+ \geq y_{rk} \\ \sum_{j=1,j\neq k}^{n} b_{tj}\lambda_j - s_t^{b-} \leq b_{tk} \\ 1 - \frac{1}{q_1+q_2}\left(\frac{\sum_{r=1}^{q_1} s_r^+}{y_{rk}} + \frac{\sum_{t=1}^{q_2} s_t^{b-}}{b_{tk}}\right) > 0 \\ \lambda_j, s_i^-, s_r^+, s_t^{b-} \geq 0; i = 1,2,\ldots,m; r = 1,2,\ldots,q_1 \\ t = 1,2,\ldots,q_2; j = 1,2,\ldots,n(j \neq k) \end{cases} \tag{4}$$

The ρ in Formula (3) is the green technology innovation efficiency of the construction industry. x_{ij} denotes the i-th input variable of the k-th DMU, and y_{rk} denotes the r-th expected output variable of the k-th DMU, b_{tk} denotes the t-th unexpected output variable of the k-th DMU. Where x_{ij} is the total input of the j-th DMU of the i-th type input, y_{rj} is the total expected output of the j-th DMU of the r-th type expected output, and b_{tj} is the total unexpected output of the j-th DMU of the t-th type unexpected output. s_i^-, s_r^+, and s_t^{b-} are the slack variables of input, expected output, and undesired output indicators, respectively, and λ_j is the weight variable.

3.2.2. The Gravity Model

As the fundamental theory of physics, Newton's universal gravity formula is employed extensively in the gravity model to examine and analyze the strength of spatial interaction. Subsequently, the gravity model has been gradually applied to economics [53], economic geography [54], urban and regional planning [55], and other fields. The traditional gravity model's coefficient is adjusted after taking into account how geographic distance between cities affects the spatial spillover impact of green technology innovation efficiency. The gravity intensity of the city is proportional to the product of G_z and G_t, and inversely proportional to the distance [56]. The specific modified gravity model formula is as follows:

$$V_{zt} = \frac{G_z \times G_t}{D_{zt}^2} \tag{5}$$

G_z and G_t are the green technology innovation efficiencies of city z and city t, D_{zt} is the geographical distance between city z and city t, and V_{zt} is the spatial spillover intensity of the green technology innovation efficiencies of city z to that of city t.

3.2.3. Geographic Detector

The geographic detector is a statistical technique for analyzing geographic spatial differentiation that can detect spatial differences and reveal the driving factors behind them [57]. Because the influencing elements in the spatiotemporal development characteristics of green technology innovation efficiency cover a wide range of areas, their driving force and influencing mechanisms can be scientifically and logically recognized by geographic detectors [58]. This paper primarily employs the factor detection and interaction detection methods of the geographical detector model to investigate the degree to which each influencing element explains the efficiency of green technology innovation in the construction sector as well as the interactions between various influencing variables.

Factor detection is to calculate the q-value of each influence factor and detect the degree of explanation of the spatial variation of each influence factor on the efficiency of green technology innovation in the construction industry. The specific formula is as follows:

$$q = 1 - (\sum_{h=1}^{L} N_h - \sigma_h^2)/N\sigma^2 \quad (6)$$

where h = 1, 2, 3 ... , L is the stratification or zoning of influencing factor X and green technology innovation efficiency Y of the construction industry; N_h and N are the number of units in layer h and the whole area, respectively; σ_h^2 and σ^2 are the variance of Y value of layer h and the whole area, respectively; the q value represents the explanatory power of the influencing factors, and its range is 0 to 1. The larger the q value is, the stronger the explanatory power of the influencing factor X on the green technology innovation efficiency Y of the construction industry is, and the weaker it is otherwise.

Interaction detection is to identify the interaction between different influencing factors, and then consider if this interaction has strengthened or weakened the case for the efficiency of green technology innovation in the construction sector or whether each influencing factor X has a separate impact on the construction industry's efficiency Y to innovate and use green technology. The relationship between the influencing factors can be divided into five categories by comparing the result sizes of $q(X1)$, $q(X2)$, and $q(X1 \cap X2)$, and then the relationship between the influencing factors is examined. These categories are nonlinear weakening, single factor nonlinear weakening, double factor enhancement, independent enhancement, and nonlinear enhancement. The specific judgment basis is shown in Table 1.

Table 1. Types of interaction between two covariates.

Basis of Judgment	Interaction
$q(X1 \cap X2) < \text{Min}(q(X1), q(X2))$	Nonlinear attenuation
$\text{Min}(q(X1), q(X2)) < q(X1 \cap X2) < \text{Max}(q(X1), q(X2))$	Single factor nonlinear attenuation
$q(X1 \cap X2) > \text{Max}(q(X1), q(X2))$	Double factors enhancement
$q(X1 \cap X2) = q(X1) + q(X2)$	Independent enhancement
$q(X1 \cap X2) > q(X1) + q(X2)$	Nonlinear enhancement

The q value represents the explanatory power of the influencing factors on the efficiency of green technology innovation in the construction industry and the same below.

3.3. *Index System*

The indicator system is separated into input indicators and output indicators, and output indicators are further divided into desired and non-desired outputs for the selection of indicators for green technology innovation. The construction industry's green technology innovation input–output index system is further formed (as shown in Table 2), and the SBM model with unexpected output super-efficiency is used to gauge the efficiency of green technology in the construction industry.

3.3.1. Input Indicators

(1) Research and development personnel. Since the number of industrial research and development (R&D) personnel was only systematically counted in 2009, data were seriously missing. Therefore, the construction industry was employed to replace the R&D personnel. (2) Research and development capital. R&D capital is usually measured by R&D capital stock or R&D expenditure. Due to the fact that the data on R&D expenditures of different industries were only counted in the second resource inventory of Sichuan Province in 2009, there were not enough pertinent statistics from prior years. In 2009, the R&D expenditure of the construction industry accounted for 1.4% of the internal R&D expenditure of Sichuan Province. The R&D expenditure of each prefecture level city in Sichuan Province in each

year was therefore calculated in this study by multiplying this proportion by the internal R&D expenditure of each prefecture level city in Sichuan Province in each year [59]. The expenditure data of Chongqing's subindustries were only counted in the second national resource inventory in 2009, lacking relevant statistical data from other years. The R&D expenditure of the Chongqing construction sector in 2009 was 0.17% of the city's internal R&D expenditure. To determine the R&D spending of Chongqing in the construction sector each year, the product of this ratio and the internal expenditure of R&D funds in Chongqing were employed. (3) Resource investment. In this paper, the electricity consumption of the whole society each year was selected as the input index.

3.3.2. Output Indicators

(1) Expected output. The building's finished area and the industry's overall output value are chosen as the expected output indicators based on the data's availability. (2) Undesired output. Carbon dioxide emissions were selected as an indicator of undesirable output from the standpoint of environmental contamination. The United Nations' Intergovernmental Panel on Climate Change carbon dioxide emission accounting method was used as a result of the lack of data on carbon dioxide emissions. According to the various energy consumptions of the construction industry, as well as their respective carbon emission coefficients, carbon oxidation factor for the provincial construction industry carbon emissions accounting. The corresponding indicators of the construction industry at the municipal level were used for the top-down conversion, and the carbon emissions of the construction industry at the provincial level were converted into the carbon emissions of the construction industry at the municipal level [60].

Table 2. Green technology innovation efficiency index system.

Indicator Category	Index	Describe	Unit
Input index	Personnel input [61]	Full time equivalent of R&D personnel in construction industry	10^4 persons
	Capital input [59]	R&D expenditure of construction industry	10^8 CNY
	Resource input [62]	Total electricity consumption	10^4 kWh
Expected output index	Output value [63]	Total output value of construction industry	10^4 CNY
	Area output [63]	Completed area of housing construction	10^4 m^2
Unexpected output index	Pollution emission output [64]	Carbon dioxide emission from construction industry	10^4 t

3.4. *Influence Factors*

Based on the previous research, considering the current situation and characteristics of the development of the construction industry, the influencing factors can be summarized into three aspects: construction industry resource endowment, social economy, and environmental awareness. The resource endowment of the construction industry mainly includes the scale of the industry, the rate of technical equipment, and the degree of industrial agglomeration. The socioeconomic factors mainly include the level of economic development, scientific and technological innovation, urbanization, and foreign direct investment. Environmental awareness mainly includes environmental regulation, education level, and public environmental concern. Specific variables are explained as shown in Table 3 below.

3.4.1. Industrial Scale

The industrial scale reflects the economic operation ability of the industry. The size of the industrial scale determines the capital strength of the industry, the number of scientific researchers, and the output of scientific research innovation. It will significantly affect the effective preservation of the ecological environment and the effective use of resources, and it is a crucial factor in advancing the development of green technology. The industrial scale is chosen to be the proportion of the entire production value of the construction industry above the scale toward the majority of construction firms.

3.4.2. Technical Equipment Rate

The ratio between the net value of machinery and equipment and related employees in the construction industry is known as the rate of technical equipment, and it somewhat indicates the degree of technical input in the sector. It has an impact on the advancement of technological innovation in the construction sector and serves as a crucial foundation for green technological innovation. The rate of technical equipment in the construction industry is measured by the coefficient of technical equipment or the degree of technical equipment.

3.4.3. Industrial Agglomeration

On the one hand, it will bring positive external benefits, such as resource sharing and technology spillover, in the process of industrial agglomeration, which contributes to the innovation of green technology in the region. On the other hand, when industrial agglomeration grows, pollutants are concentrated, increasing environmental pressure and the risk of a crowding-out effect, which are detrimental to the advancement of new green technologies. The location entropy is used to gauge the degree of concentration in the building business. It reflects the occupancy of an industry and can accurately and reasonably reflect the agglomeration level of enterprises. The proportion of the overall output value of the construction industry to the regional gross domestic product (GDP) of each city is divided by the proportion of the total output value of the construction industry to the regional GDP of the Chengdu–Chongqing urban agglomeration to calculate the location entropy of the construction industry.

3.4.4. Economic Development Level

While the funding of green technology research and development is highly tied to the level of economic development, the development of the construction sector is directly related to the external economic environment of the region [65]. As a result, the degree of regional economic prosperity has a significant impact on the adoption of green technologies in the construction sector. The level of the region's economic development is expressed by the gross regional product.

3.4.5. Scientific and Technological Innovation Level

A strong basis for growth and innovation is provided by the level of scientific and technological innovation, which is a reflection of the region's overall capacity for invention and creativity. It is the industry's main engine for innovation in green technologies. The level of regional innovation in science and technology is assessed by the turnover of the technology market.

3.4.6. Urbanization Level

Urbanization is a process of gathering talent and labor force capital against the backdrop of sustainable development. In addition, it creates a conducive atmosphere for innovation to promote the advancement of green industrial technologies while funding scientific and technological advancement. The level of urbanization is determined by dividing the population of cities by the overall population.

3.4.7. Foreign Direct Investment

Capital, cutting-edge machinery and technology, seasoned management expertise, and other factors will be brought to the region via foreign direct investment [43]. It is conducive to making up for industrial deficiencies through green technology innovation and creating a good environment. The real amount of foreign capital utilized in that year is used to calculate regional foreign direct investment.

3.4.8. Environmental Regulation

Environmental regulation belongs to social regulation, which is the power to restrict the behavior of economic subjects by means of tangible rules and regulations or intangible consciousness. It is mainly aimed at improving the environment by realizing the coordination between the rational utilization of resources and economic and social development [66]. The comprehensive intensity indicator of sulfur dioxide emission, smoke dust emission, and industrial wastewater emission of each prefecture-level city is used to measure environmental control while taking data accessibility into account. The specific calculation steps of environmental regulation are as follows:

(1) Calculate the relative intensity of environmental pollution emission:

$$ER_{n,it} = \frac{e_{n,it}}{z_{it}} / \sum_{i=1}^{16} \frac{e_{n,it}}{z_{it}} \tag{7}$$

where $ER_{n,it}$ (i = 1, 2, ..., 16) is the relative intensity of environmental pollution emissions, $e_{n,it}$ is the emission of the n-th pollutant in the i-th municipality, and z_{it} denotes the total industrial output value of the i-th municipality.

(2) Calculate the comprehensive index of environmental regulation:

$$ER_{it} = \frac{1}{3} \sum_{n=1}^{3} ER_{n,it} \tag{8}$$

The sum of the relative intensities of pollution emissions from sulfur dioxide, smoke, and dust emissions, as well as industrial wastewater emissions, is represented by $\sum_{n=1}^{3} ER_{n,it}$, while ER_{it} stands for the comprehensive index of environmental regulation.

(3) Calculate the intensity index of environmental regulation:

$$ERY_{it} = 1/ER_{it} \tag{9}$$

where ERY_{it} is the environmental regulation intensity index.

3.4.9. Education Level

A key assurance for the growth of green technology innovation is the degree of a high-quality labor force, which is directly impacted by education levels. The proportion of full-time teachers in primary, middle, and high schools in the resident population at the end of the school year is used to gauge educational levels.

3.4.10. Public Environmental Concern

Public environmental concern is a nonmandatory regulatory measure. When the public is aware of the negative impact of environmental pollution on public welfare, its opinion will be formed through relevant measures and means. In this way, there will be a certain pressure on the relevant construction industry to protect its own interests and demands [67] and urge the industry or government departments to carry out green environmental protection behaviors. The number of searches on Baidu's index websites for selected relevant keywords is a measure of the public's environmental concern. The search times of each prefecture-level city in the Chengdu–Chongqing urban agglomeration in each year can be obtained by entering keywords related to environmental issues such as "environmental pollution", "environmental governance", and "carbon emission". The value obtained by accumulating the number of keyword searches is used as a proxy for the public's concern about environmental issues in that city.

Table 3. Influencing factors and classification of green technology innovation efficiency in the construction industry.

Characterization Type	Influence Factor	Variable Description	Symbolic Representation
Resource endowment of construction industry	Industrial scale [4]	Total output value of construction industry above designated size/number of construction enterprises (%)	X1
	Technical equipment rate [68]	Technical equipment coefficient or technical equipment degree (%)	X2
	Industrial agglomeration degree [69]	Location entropy of construction industry	X3
Social economic factors	Economic development level [50]	Regional GDP (10^8 CNY)	X4
	Scientific and technological innovation level [70]	Technology market turnover (10^4 CNY)	X5
	Urbanization level [71]	Urban population/total population (%)	X6
	Foreign direct investment [43]	Actual amount of foreign capital used in the current year (10^4 USD)	X7
Environmental awareness factors	Environmental regulation [48]	Reciprocal of the comprehensive index of environmental pollution emissions	X8
	Education level [51]	The sum of full-time teachers in primary, middle, and high schools/the proportion of the last permanent resident population (%)	X9
	Public environmental concern [72]	The number of searches of "environmental pollution", "environmental governance", and "carbon emissions" on Baidu index website (times)	X10

3.5. Data Sources

This study used a research sample of urban statistics data from 2011 to 2019 and 16 cities in the Chengdu–Chongqing urban agglomeration as its research target. The data for the study were obtained from the China City Statistical Yearbook, the Sichuan Statistical Yearbook, the Chongqing Statistical Yearbook, the China Energy Statistical Yearbook, and the statistical yearbooks of other cities in the region as well as statistics from the Sichuan Provincial Department of Science and Technology. For individual missing datasets, linear interpolation was performed in this paper to ensure the integrity and validity of the data.

4. Results

4.1. Space–Time Characteristics of Green Technology Innovation Efficiency

The MATLAB software was employed to determine the green technology innovation efficiency of the construction industry in 16 cities within the study area from 2011 to 2019 using the efficiency evaluation index system developed above. The efficiency value reflects the level of innovation in green technology and differences in the construction industry in each city. The mean and ranking of green technology innovation efficiency in cities from 2011 to 2019 are shown in Table 4 below. By referring to the dividing standards of efficiency measurement in existing studies, the essential values for green technology innovation efficiency are 0.25, 0.5, and 0.75, which are separated into four tiers. Ultra-low efficiency is defined as being less than or equal to 0.25, low efficiency is defined as being between 0.25 and 0.5 (including 0.5), medium efficiency is defined as being between 0.5 and 0.75 (including 0.75), and high efficiency is defined as being greater than 0.75. The geographical and temporal evolution traits of green technology innovation efficiency in the construction industry in the Chengdu–Chongqing urban agglomeration were further examined based on the grade division of efficiency value.

Table 4. Green technology innovation efficiency of 16 cities in the Chengdu–Chongqing urban agglomeration.

City	2011		2019		2011–2019	
	Scores	Rank	Scores	Rank	Average Scores	Rank
Chengdu	0.590	5	1.013	2	0.807	2
Zigong	0.235	15	0.664	5	0.458	12
Luzhou	0.329	9	0.591	7	0.473	10
Deyang	0.320	10	0.418	13	0.470	11
Mianyang	0.295	13	0.477	9	0.452	13
Suining	1.029	2	0.618	6	0.732	4
Neijiang	0.225	16	0.459	11	0.413	14
Leshan	0.300	12	0.341	15	0.318	16
Nanchong	0.601	4	1.003	4	0.709	5
Meishan	1.035	1	0.415	14	0.501	9
Yibin	0.256	14	0.311	16	0.365	15
Guang'an	0.313	11	1.055	1	0.791	3
Dazhou	0.466	7	0.462	10	0.502	8
Ya'an	0.358	8	0.571	8	0.657	6
Ziyang	0.579	6	0.435	12	0.621	7
Chongqing	0.680	3	1.005	3	0.936	1

4.1.1. Time Series Evolution Characteristics of Green Technology Innovation Efficiency

As can be seen from Table 4, in all the prefecture-level cities of the Chengdu–Chongqing urban agglomeration, the level of green technology innovation in the construction industry shows an overall upward trend from 2011 to 2019, while there are clear disparities between cities. From the perspective of the change in efficiency level, Chengdu, Zigong, Luzhou, Neijiang, Nanchong, Guang'an, Ya'an, and Chongqing are eight cities that have improved in the green technology innovation efficiency level. Among them, Guang'an saw a continual two-level increase in the level of green technology innovation, from low efficiency to high efficiency. It is the city with the largest increase in green technology innovation efficiency among the Chengdu–Chongqing urban agglomeration, with an annual growth rate of 26.34%. Zigong's green technology innovation has improved from an ultra-low efficiency level to a medium efficiency level, and its annual growth rate is second only to Guang'an, reaching 20.32%. This might be the result of Zigong having a low value for green technology innovation efficiency in 2011 and a lot of potential for development. The efficiency of green technology innovation has significantly increased with the growth of the city's green ecological economy. The green technology innovation efficiency of Chengdu and Chongqing improved from the medium efficiency level to the high-efficiency level. Chengdu and Chongqing's green technology innovation efficiency was at the forefront of the Chengdu–Chongqing metropolitan agglomeration, but the annual growth rate was low, at 7.95% and 5.31%, respectively. Compared to the average level of the Chengdu–Chongqing economic circle, green technology innovation is higher in Chengdu and Chongqing, and the relative improvement space is limited. They work toward comprehensive and integrated development while also serving as the metropolitan agglomeration's core development cities in the Chengdu–Chongqing region. A range of factors ought to be taken into account when enhancing the efficiency of green technology innovation. Consequently, this improvement has been relatively flat.

Based on the overall average, the green technology innovation efficiency of the construction industry in the study area is primarily at the low and medium efficiency levels, and the efficiency value gap between cities is significant. Overall, there is still room for improvement. Chongqing has the highest efficiency of green technology innovation, with an average efficiency of 0.936. The average efficiency of green technology innovation in Chengdu ranks second only to Chongqing, and its efficiency value is 0.807. Chengdu and Chongqing are the two central cities in Southwest China, and they are the backbone of promoting green technology innovation and development. They have established the groundwork for green technology innovation in the two cities' construction industries by relying on their solid economic bases and technological advantages. The green technology innovation efficiency of Guang'an, Suining, and Nanchong ranked third, fourth, and fifth, and the average value of their overall green technology innovation efficiency from 2011 to 2019 was 0.791, 0.732, and 0.709, respectively. They are the main drivers behind the advancement of efficient green technology innovation in the research domain. However, the green technology innovation efficiency of Yibin and Leshan is relatively low, lower than 0.4, and the efficiency value is in the least developed state among the Chengdu–Chongqing urban agglomeration, with low efficiency and serious solidification.

4.1.2. Spatial Evolution Characteristics of Green Technology Innovation Efficiency

Due to the different conditions of social and economic development levels, geographical location, construction industry resource endowment, and other aspects, the characteristics of spatial heterogeneity of green technology innovation efficiency are easily brought about within the study area. ArcGIS 10.7 software was used to draw spatial distribution maps of the green technology innovation efficiency of the Chengdu–Chongqing urban agglomeration in 2011, 2014, 2017, and 2019, as shown in Figure 2 below.

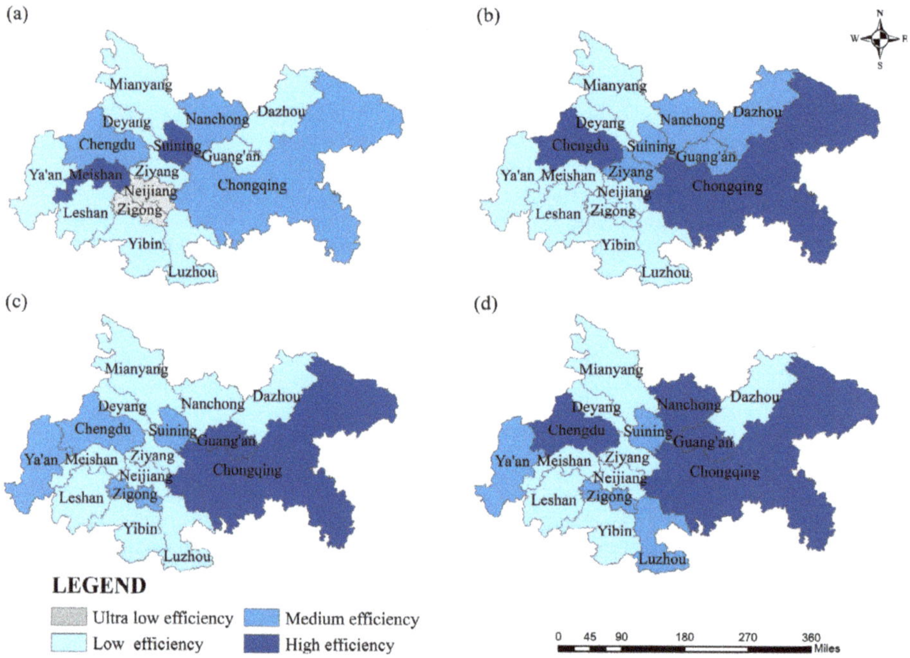

Figure 2. Spatial distribution pattern of green technology innovation efficiency in the Chengdu–Chongqing urban agglomeration. (**a**) 2011. (**b**) 2014. (**c**) 2017. (**d**) 2019.

On the whole, firstly, cities in the study area's prefecture level exhibit glaring discrepancies in the efficiency of green technology innovation. The number of areas with high and medium efficiency has improved, with the share increasing from 31.25% in 2011 to 50% in 2019. At the same time, places with ultra-low levels of efficiency just started to exist in 2011, and the number of areas with ultra-low efficiency levels and low efficiency levels showed a downward trend, accounting for 50% in 2019 from 68.75% in 2011. Secondly, the west and northeast are where green technology innovation in the building industry is quite efficient, whereas the southwest and northwest are where it is reasonably low. The spatial distribution pattern of Chengdu-Ya'an and Chongqing-Guang'an-Nanchong, which are the main axes of the high-efficiency level area, and Yibin-Leshan-Meishan and

Mianyang-Deyang, which are the two wings of the low-efficiency level area, are gradually formed. Finally, the regions with higher levels of efficiency primarily extended from the midwest and north central to the west and northeast, whereas the regions with lower levels of efficiency were concentrated in the southwest and northwest, demonstrating a tendency of narrowing regional reach.

This study created the spatial correlation strength matrix by applying the modified gravity model in order to further investigate and analyze the spatial distribution features of the green technology innovation efficiency of the construction sector in the Chengdu–Chongqing urban agglomeration. The geographical spillover network structure chart of green technology innovation efficiency in the construction sector was created using the ArcGIS software, and it efficiently and clearly highlighted the spillover characteristics of various spatial units. Five classes were used to categorize the spatial spillover intensity (V) of the green technology innovation efficiency in the construction industry: ultra-high ($V \geq 6$), high ($4.5 \leq V < 6$), medium ($3 \leq V < 4.5$), low ($1.5 \leq V < 3$), and ultra-low ($0 \leq V < 1.5$), and combined with the ArcGIS software for visual expression, as shown in Figure 3.

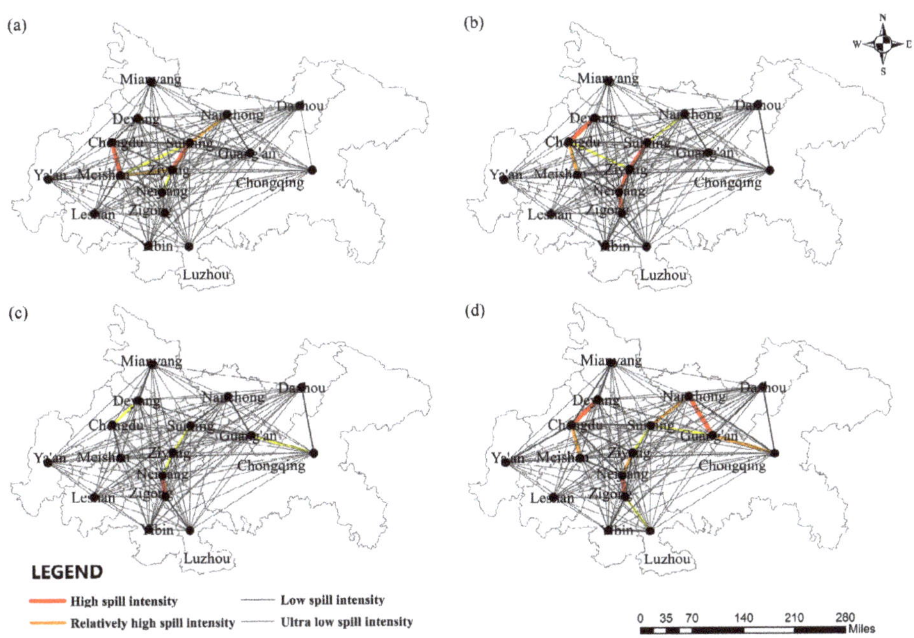

Figure 3. Spatial spillover network structure of green technology innovation efficiency in the Chengdu–Chongqing urban agglomeration. (**a**) 2011. (**b**) 2014. (**c**) 2017. (**d**) 2019.

The spatial network structure of green technology innovation efficiency exhibits the features of layer-level, accessibility, and multi-threaded spillover when viewed from the standpoint of the total spatial spillover network structure. It overcomes the conventional regional restrictions, broadens the radiation range, achieves the fusion of nearby and cross-regional radiation, enhances the spillover effect of green technology innovation on the construction sector of the Chengdu–Chongqing urban agglomeration, and narrows the regional divide even further. Chongqing in the east and Chengdu in the northwest form the key connections in the network spillover structure of green technology innovation efficiency in the construction sector. Among them, the network spillover structure of the main linkage objects in Chengdu is closely distributed. There is a significant amount of spatial spillover across cities, and there are more regional point pairs with high spatial correlation. Secondly, the spatial network with Chongqing as the main linkage object is relatively sparse. There are many regional point pairs with poor spatial correlation, and the spatial network structure made up of cities has a generally weak spillover link. The adoption of green technologies in Chongqing and the surrounding cities' construction sectors exhibits erratic growth, and the growth trend is typically slowing down, which has an unequal spatial spillover effect on a global scale.

From a local viewpoint, consider Chengdu and Chongqing the "leading goose" cities in the area, exploit the head goose effect to the fullest extent possible, and encourage the development of multi-wing radiation and diffusion to neighboring cities. Green technology innovation in the building sector gradually formed a space overflow area with high efficiency, including Chengdu-Meishan, Chengdu-Deyang, Luzhou-Zigong-Neijiang-Ziyang-Suining-Nanchong, and Chongqing-Guang'an-Nanchong, among others. These organizations make up the central network framework for the spatial spillover of green technology innovation efficiency in the Chengdu–Chongqing metropolitan agglomeration's construction sector. With the development of these groups, the structure is constantly strengthened, and the initial structure of the spatial overflow network is formed. The agglomeration effect of these groups will cause significant local and territorial differences in the efficiency of green technology innovation. In the short term, high-efficiency agglomeration has a "siphon effect" on surrounding cities, absorbing innovation resources from the surrounding areas, achieving rapid local development, and impeding technological innovation efficiency improvement in low-efficiency surrounding areas. However, high-efficiency agglomeration areas such as Chengdu-Meishan, Chongqing-Guang'an, and Nanchong show a strong spillover effect of space technology on surrounding cities in the long run. As the distance increases, there is a spillover strength limit in terms of external expansion, which mostly manifests as low-strength spatial links across regions.

4.2. Analysis on the Influencing Factors of Spatial and Temporal Differentiation of Green Technology Innovation Efficiency

This paper selects the factors that may affect the spatial differentiation of green technology innovation efficiency in the construction industry from three dimensions: resource endowment, social economy, and environmental awareness, and detects the dominant factors of spatial and temporal variation in green technology innovation efficiency in the construction industry and the interaction between related influencing factors by using geographic detectors.

4.2.1. Analysis of Influencing Factors

In this paper, we use the factor detector to obtain and rank the q-values for the three time points of 2011, 2015, and 2019. Every component chosen in the preceding research that had explanatory power passed the 1% significance level test. Table 5 displays the specific outcomes.

The impact of resource endowment, social and economic factors, and environmental awareness factors on the efficiency of green technology innovation in the construction industry can be seen as having substantial disparities from the perspective of factor classification. Overall, socioeconomic and environmental awareness factors have a greater effect on efficiency values than resource endowment in the construction industry, which has a relatively lesser effect on them. The q-values of all factors, taken in the context of the 10 driving factors, range from 0.14 to 0.84, demonstrating that there are clear distinctions between the effects of various driving factors on the effectiveness of green technology innovation in the construction sector. According to rankings and q-values for 2011, 2015, and 2019, environmental regulation, economic development level, public environmental concern, urbanization level, and foreign direct investment have a significant impact on the efficiency of green technology innovation in the construction industry. The total q-value is greater than 0.5, and these influencing factors are the main driving factors. Among the remaining influencing factors, the explanatory power of industry scale decreased slightly and then increased. The efficiency of green technology innovation moved up to the second spot in the construction sector in 2019, and its q-value was 0.75, indicating a persistent increasing trend. Therefore, the scale of the industry is chosen as a potential variable affecting the efficiency of green technology innovation in the construction sector. In addition, the influence of education level and technical equipment rate on green technology innovation efficiency in the construction industry is relatively weak.

Table 5. Factor detection results of space-time differentiation of green technology innovation efficiency in the Chengdu–Chongqing urban agglomeration.

Factor	2011		2015		2019	
	q	Rank	q	Rank	q	Rank
Industrial scale (X1)	0.39	6	0.14	10	0.75	2
Technical equipment rate (X2)	0.46	3	0.52	4	0.28	10
Industrial agglomeration degree (X3)	0.28	8	0.48	5	0.55	8
Economic development level (X4)	0.46	3	0.56	3	0.71	4
Scientific and technological innovation level (X5)	0.43	5	0.34	8	0.67	7
Urbanization level (X6)	0.5	2	0.32	9	0.74	3
Foreign direct investment (X7)	0.37	7	0.45	6	0.68	6
Environmental regulation (X8)	0.26	9	0.7	2	0.84	1
Education level (X9)	0.16	10	0.75	1	0.36	9
Public environmental concern (X10)	0.55	1	0.35	7	0.69	5

From the time dimension, the driving factors fluctuated to different degrees during 2011–2019. The explanatory power revealed a general rising trend that the influence of leading factors and potential variables on the spatial difference of green technology innovation efficiency in the construction industry gradually increased. Among them, there has been a noticeable increase in the degree of economic development, foreign direct investment, and environmental regulation. The industrial scale, urbanization level, and public environmental attention show V-shaped fluctuations, which first decline and then rise. The construction industry generally exhibits an increase in the explanatory power of green technology innovation efficiency. The education level and the rate of technical equipment both increased first and then decreased. The explanatory power of education level on the efficiency of green technology innovation in the construction industry is unstable and fluctuates greatly. However, in terms of explanatory power, the rate of technical equipment revealed a declining tendency.

4.2.2. Analysis on the Interaction of Spatial-Temporal Differentiation of Green Technological Innovation Efficiency

To further investigate the variation of explanatory power on green technology innovation efficiency in the construction industry when different driving factors interact, the dominant factors and potential factors after factor detection were selected to analyze their interaction mechanisms affecting the spatial divergence of green technology innovation efficiency in the construction industry. Due to the interaction between the two driving factors, it is not a simple linear addition [73]. Therefore, the q-value of the interaction influence of two drivers on green technology innovation in the construction industry is studied using the interaction detection of geographic detectors.

Table 6 displays the findings from the interactive detection investigation using the data from 2011 through 2019: There is a close relationship between the selected leading factors and the potential factors, and the q-value of each influencing factor shows different degrees of improvement after complex interaction. There are two modes of combination: double factor enhancement and nonlinear enhancement, and the interaction of all influencing variables has a greater impact on the geographic and temporal diversity of green technology innovation efficiency in the construction sector than any single factor. The interaction between industrial scale and urbanization level, foreign direct investment, and environmental regulation degree is explained by more than 60%, showing a stable growth trend over time. The level of economic development interacts with foreign direct investment, the degree of environmental regulation, and the public's environmental concern, respectively, and the explanatory force is above 50%. As for the interaction between foreign direct investment and public environmental concern, the explanation strength is more than 80%, and the impact degree on the spatial and temporal differentiation of green technology innovation efficiency in the construction industry shows a continuous increasing trend from 2011 to 2019. The interaction effect of industrial scale and public environmental concern with urbanization level and environmental regulation degree is obvious, and the explanatory power is relatively strong. The explanatory power of the interaction between environmental regulation and public environmental attention is significantly improved compared with the single factor, and its explanatory power is relatively stable and above 90%, which is larger than the pairwise interaction result between other types of factors. From 2011 to 2019, each influencing factor experienced a transition between double-factor enhancement and nonlinear enhancement. The last mode of action stabilized at twofold factor enhancement in order to strengthen the justification of interaction factors on the efficiency of green technology innovation in the construction sector.

Table 6. Interactive detection results of spatial differentiation of green technology innovation efficiency of construction industry.

Factor Interaction	2011		2015		2019	
	q	Type	q	Type	q	Type
X1 ∩ X4	0.802	DE	0.759	NE	0.781	DE
X1 ∩ X6	0.763	DE	0.774	NE	0.872	DE
X1 ∩ X7	0.611	DE	0.657	NE	0.939	DE
X1 ∩ X8	0.713	NE	0.888	NE	0.931	DE
X1 ∩ X10	0.778	DE	0.999	NE	0.873	DE
X4 ∩ X6	0.82	DE	0.698	DE	0.939	DE
X4 ∩ X7	0.507	DE	0.702	DE	0.786	DE
X4 ∩ X8	0.794	NE	0.947	DE	0.994	DE
X4 ∩ X10	0.574	DE	0.618	DE	0.901	DE
X6 ∩ X7	0.794	DE	0.787	NE	0.831	DE
X6 ∩ X8	0.781	NE	0.964	DE	0.888	DE
X6 ∩ X10	0.966	DE	0.473	DE	0.756	DE
X7 ∩ X8	0.807	NE	0.874	DE	0.992	DE
X7 ∩ X10	0.718	DE	0.782	DE	0.768	DE
X8 ∩ X10	0.944	NE	0.938	DE	0.916	DE

DE: Double enhancement; NE: nonlinear enhancement.

5. Discussion and Conclusions

5.1. Research Conclusions

The Chengdu–Chongqing urban agglomeration is a crucial growth pole for developing high-quality economic development in western China. This research assessed the green technology innovation efficiency of the construction sector in each city using data over the period of 16 cities in the Chengdu–Chongqing urban agglomeration from 2011 to 2019. Using a gravity model and a geographic detector, the geographical and temporal development characteristics of green technology innovation efficiency in the construction sector were explored. At the same time, the pertinent driving factors were identified, and the extent to which each driving element affects the efficiency of green technology innovation in the construction sector was explored for both single-factor and double-factor analyses. The ensuing conclusions were reached: (1) Within the Chengdu–Chongqing urban agglomeration, there are considerable regional variations in the efficiency of green technology innovation in the construction industry, and the overall trend is upward. (2) The research area exhibits spatially heterogeneous characteristics in terms of the efficiency of green technology innovation in the construction industry. Additionally, it demonstrates the tendency whereby the area with high efficiency levels gradually spreads to the surrounding areas with lower efficiency levels, and the area with low efficiency levels gradually decreases in scope. (3) The Chengdu–Chongqing urban agglomeration's geographical spillover impact is undoubtedly constrained by distance. Additionally, the western region's spatial spillover impact is superior to that of Chongqing's eastern region. The western portion of the Chengdu–Chongqing urban agglomeration has a better spatial spillover impact than the eastern portion, which is represented by Chongqing. Moreover, the spatial spillover effect is significantly limited by distance. (4) Environmental regulation, the level of economic development, public environmental concern, the level of urbanization, and foreign direct investment, as the dominant factors of green technology innovation efficiency in the construction industry, and the industry's scale as a potential factor, all have significant effects on the efficiency of green technology innovation in the construction industry. (5) In comparison to the single component, the interaction between the leading factor and the potential factor has a greater influence on the regional and temporal differentiation of green technology innovation efficiency in the construction sector.

5.2. Theoretical Contribution

This paper's theoretical contribution, as compared to previous studies, focuses primarily on three areas:

Firstly, prior to measuring the efficiency of green technology innovation in the research area's construction industry, the undesirable output is fully taken into account. It is discovered that the research area's overall innovation efficiency in green technology is notably different and exhibits an upward trend. This confirms the opinion of Qian et al. (2022) [74] that there is an imbalance in green technology innovation in inland areas and that there are obvious differences between regions. Additionally, it was discovered that places distant from the central cities were more likely to have severe solidification and ultra-low efficiency, which was in line with the findings of Xu et al. (2020) [75]. Based on these findings, this study investigates and analyzes the characteristics of the green technology innovation efficiency of the construction sector in the study area over time and space, as well as further examining the variations between cities and the degree of spatial connectivity.

Secondly, green technology innovation in the construction sector has had an optimistic spillover effect in the study area, gradually transferring from the high-efficiency-level to the neighboring low-efficiency-level areas, and the low-efficiency-level area's scope gradually exhibiting a trend of narrowing. The findings of Hu et al. (2022) [76], Wang et al. (2022) [77], and Zhao et al. (2021) [78] are in agreement with this finding. They believe that high-efficiency areas have radiation effects on low-efficiency areas and narrow the gap between cities. In order to intuitively reveal the spillover effect between different spatial units, this paper introduces the gravity model and utilizes the spatial spillover network structure diagram. As a result, the research findings on the efficiency of green technology innovation in the construction industry are further enhanced.

Thirdly, this research analyzes the factors that affect the efficiency of green technology innovation in the construction sector. The results are consistent with those of Zhao et al. (2022) [72], Li et al. (2022) [45], and Stucki et al. (2018) [79] and indicate that environmental regulation and economic development levels have a significant impact on green technology innovation efficiency in the sector. According to Porter's theory, environmental regulation, to some extent, has a favorable effect on the development of green technology [80]. High economic development locations typically have enough funding for green technology innovation activities, which can significantly encourage the improvement of green technology innovation efficiency. Existing studies consider the influencing factors to be thin and do not include multiple influencing factors in the same space for interaction impact analysis. In order to make up for these deficiencies, based on the characteristics of geographic detectors, factors of multicollinearity can be included in the same framework system for discussion. This paper expands the influencing factor system of green technology innovation efficiency in the construction industry and enriches the research findings by taking into account and examining the driving role of related influencing factors from the three aspects of the construction industry's resource endowment, social economy, and environmental awareness.

5.3. Management Inspiration

The Chengdu–Chongqing City cluster is situated at the intersection of the "Belt and Road" and the Yangtze River Economic Belt, which has considerable regional advantages and serves as an essential platform for the development of the western province. With the continuous promotion of the "double carbon" policy, the construction industry is in urgent need of green and low-carbon transformation. Therefore, the following suggestions are put forward:

Firstly, develop differentiated environmental regulation policies to enhance the institutional environment for the development of green technology innovation. The Chengdu–Chongqing region's construction industry's use of green technology innovation is best explained by environmental regulation, which has the strongest overall impact. Increase government involvement, bolster the administration's commitment to environmental protection, develop local conditions-specific environmental regulation laws, enhance the relevance and efficiency of environmental regulation, and facilitate the balanced development of green technology innovation efficiency.

Secondly, focus on bringing in top-notch foreign funding and promoting the advancement of green technology innovation in the construction industry. The spillover impact of technology, funding, resources, and knowledge delivered by foreign direct investment is fully utilized through the infusion of high-quality foreign investment by the government. This is a significant technique to increase the efficiency of green technology innovation in the research domain and is conducive to accelerating the transition of green technology innovation accomplishments in the construction sector.

Thirdly, encourage the public's excitement about environmental issues and fully utilize the public's oversight role. Develop policies to support and encourage public participation in environmental governance while continuously improving and standardizing the format of letters and media reports. This supports modernizing and scientifically validating environmental governance, ensures the timely and efficient implementation of public supervision and management, and is a crucial building block for attaining green, sustainable, and healthy development in the construction industry.

Finally, strengthen coordinated development among regions to narrow the imbalance. An efficient method of coordinating and promoting the growth of green technology innovation in the construction sector in each prefecture-level city in the Chengdu–Chongqing region is to improve the level of technical openness among cities. Give large cities such as Chengdu and Chongqing their due as "leading goose", radiate these cities' advantages in cutting-edge technology and resources to neighboring cities with low rates of green technology innovation, and encourage the integration and sustainable growth of the construction sector in this area.

5.4. Limitations and Deficiencies

In this study, the efficiency of green technology innovation in the construction sector is evaluated. The gravity model and geographic detector are used to investigate the characteristics of the spatial and temporal evolution of efficiency and its affecting elements. It expands and enriches the research theory of green technology innovation in the construction industry and helps to promote the green and low-carbon transformation of the construction industry. This study may have several shortcomings, which should be addressed and resolved in further studies. First of all, only the Chengdu–Chongqing urban agglomeration in China is used as the research region for this work, which focuses on the efficiency and impact variables of green technology innovation in the construction industry of 16 of those cities. However, there might be regional differences in the construction industry's use of green technology innovation in various metropolitan agglomerations. In the future, a comparison study of typical regions such as the Beijing-Tianjin-Hebei urban agglomeration and the Yangtze River Delta urban agglomeration will be necessary. Explore in further detail the regulations for green technology innovation in the construction industry in various urban agglomerations. Secondly, the research object is not sufficiently detailed. The construction industry of each city in the region is the research object of this paper, and the research scope is broad. It can be refined further in future research, and the city can be refined further for each construction enterprise or county for more in-depth research. Last but not least, this essay primarily focuses on the effects of resource abundance, social economics, and environmental consciousness in light of the influencing variables of green technology innovation and efficiency in the construction industry. It might also be impacted by factors such as the energy consumption structure, ancillary industries, and management levels, among others, which will require more investigation and in-depth debate in the future.

Author Contributions: Conceptualization, B.W.; methodology, H.C.; software, H.C.; formal analysis, Y.A.; investigation, F.L.; resources, B.W.; data curation, F.L. and B.W.; writing—original draft preparation, H.C.; writing—review and editing, F.L. and B.W.; supervision, Y.A. All authors have read and agreed to the published version of the manuscript.

Funding: This research was funded by National Social Science Foundation of China (22BJY142).

Institutional Review Board Statement: Not applicable.

Informed Consent Statement: Not applicable.

Data Availability Statement: Not applicable.

Conflicts of Interest: The authors declare no conflict of interest.

Abbreviations

SBM: slacks-based measure. SFA: stochastic frontier analysis. DEA: data envelopment analysis. BCC: Banker-Charnes-Cooper. CCR: Charnes-Cooper-Rhodes. OECD: Organization for Economic Co-operation and Development. DMU: decision-making units. R&D: research and development. GDP: gross domestic product.

References

1. Zhang, L.L. Environmental Regulation, Government R & D Funding and Green Technology Innovation. Master's Thesis, Anhui University of Finance and Economics, Bengbu, China, 2020.
2. Friedlingstein, P.; O'Sullivan, M.; Jones, M.W.; Andrew, R.M.; Hauck, J.; Olsen, A.; Peters, G.P.; Peters, W.; Pongratz, J.; Sitch, S.; et al. Global Carbon Budget 2020. *Earth Syst. Sci. Data* **2020**, *12*, 3269–3340. [CrossRef]
3. Du, K.; Li, P.Z.; Yan, Z.M. Do green technology innovations contribute to carbon dioxide emission reduction? Empirical evidence from patent data. *Technol. Forecast. Soc. Chang.* **2019**, *146*, 297–303. [CrossRef]
4. Hong, M.; Li, Z.; Drakeford, B. Do the Green Credit Guidelines Affect Corporate Green Technology Innovation? Empirical Research from China. *Int. J. Environ. Res. Public Health* **2021**, *18*, 1682. [CrossRef] [PubMed]
5. Sun, S.S. Research on the Spatial Network Structure Effect of the Construction Industry and Its Carbon Dioxide Emissions. Ph.D. Thesis, Xi'an University of Architecture and Technology, Xi'an, China, 2017.
6. Xiao, L.; Gong, Y.C. Spatial agglomeration analysis of gross output value of construction industry in East China. *Coop. Econ. Technol.* **2013**, *3*, 24–25. [CrossRef]
7. Suki, N.M.; Sharif, A.; Afshan, S.; Jermsittiparsert, K. The role of technology innovation and renewable energy in reducing environmental degradation in Malaysia: A step towards sustainable environment. *Renew. Energy* **2022**, *182*, 245–253. [CrossRef]
8. Lu, Y.Z. Green energy-saving buildings are extremely urgent in China. *China Real Estate* **2011**, *06*, 29–32.
9. Wurlod, J.D.; Noailly, J. The impact of green innovation on energy intensity: An empirical analysis for 14 industrial sectors in OECD countrie. *Energy Econ.* **2018**, *71*, 47–61. [CrossRef]
10. Sun, X.T.; Gao, J.H.; Fan, D. Analysis on Regional Differences and Efficiency Improvement of China's Green Technology Innovation. *Sci. Technol. Dev.* **2018**, *14*, 1092–1098.
11. Wei, W.D.; Cai, W.Q.; Guo, Y.; Bai, C.Q.; Yang, L.Z. Decoupling relationship between energy consumption and economic growth in China's provinces from the perspective of resource security. *Resour. Policy* **2020**, *68*, 101693. [CrossRef]
12. Feng, Y.; Chen, H.; Chen, Z.J.; Wang, Y.N.; Wei, W.D. Has environmental information disclosure eased the economic inhibition of air pollution? *J. Clean. Prod.* **2021**, *284*, 125412. [CrossRef]

13. Chen, H.; Guo, W.; Feng, X.; Wei, W.D.; Liu, H.B.; Feng, Y.; Gong, W.Y. The impact of low-carbon city pilot policy on the total factor productivity of listed enterprises in China. *Resour. Conserv. Recycl.* **2021**, *169*, 105457. [CrossRef]
14. Wei, W.D.; Hao, S.J.; Yao, M.T.; Chen, W.; Wang, S.S.; Wang, Z.Y.; Wang, Y.; Zhang, P.F. Unbalanced economic benefits and the electricity-related carbon emissions embodied in China's interprovincial trade. *J. Environ. Manag.* **2020**, *263*, 110390. [CrossRef] [PubMed]
15. Ren, K.; Kong, Y.S.; Imran, M.; Bangash, A.K. The Impact of the Voluntary Environmental Agreements on Green Technology Innovation: Evidence from the Prefectural-Level Data in China. *Front. Environ. Sci.* **2022**, *10*, 833724. [CrossRef]
16. Wang, D. Research on the Motivation and Economic Consequences of Green Technological Innovation. Ph.D. Thesis, Zhejiang Business University, Hangzhou, China, 2022.
17. Braun, E.; Wield, D. Regulation as a means for the social control of technology. *Technol. Anal. Strateg. Manag.* **1994**, *6*, 259–272. [CrossRef]
18. Sun, Y.; Anwar, A.; Razzaq, A.; Liang, X.; Siddique, M. Asymmetric role of renewable energy, green innovation, and globalization in deriving environmental sustainability: Evidence from top-10 polluted countries. *Renew. Energy* **2022**, *185*, 280–290. [CrossRef]
19. Tolliver, C.; Fujii, H.; Keeley, A.R.; Managi, S. Green Innovation and Finance in Asia. *Asian Econ. Policy Rev.* **2021**, *16*, 67–87. [CrossRef]
20. Du, K.; Li, J.L. Towards a green world: How do green technology innovations affect total-factor carbon productivity. *Energy Policy* **2019**, *131*, 240–250. [CrossRef]
21. Acemoglu, D.; Aghion, P.; Bursztyn, L.; Hemous, D. The Environment and Directed Technical Change. *Am. Econ. Rev.* **2012**, *102*, 131–166. [CrossRef]
22. Bai, C.Q.; Feng, C.; Yan, H.; Yi, X.; Chen, Z.J.; Wei, W.D. Will income inequality influence the abatement effect of renewable energy technological innovation on carbon dioxide emissions? *J. Environ. Manag.* **2020**, *264*, 110482. [CrossRef]
23. Yang, F.T. Institutional Research on Green Technological Innovation—Based on the Perspective of Ecological Civilization. Ph.D. Thesis, Party School of the CPC Central Committee, Beijing, China, 2014.
24. Peneder, M.; Arvanitis, S.; Rammer, C.; Stucki, T.; Wörter, M. Policy instruments and self-reported impacts of the adoption of energy saving technologies in the DACH region. *Empirica* **2022**, *49*, 369–404. [CrossRef]
25. Conceição, P.; Heitor, M.V.; Vieira, P.S. Are environmental concerns drivers of innovation? Interpreting Portuguese innovation data to foster environmental foresight. *Technol. Forecast. Soc. Chang.* **2004**, *73*, 266–276. [CrossRef]
26. VandeWerf, F. Management of acute myocardial infarction in patients presenting with ST-segment elevation. *Eur. Heart J.* **2003**, *24*, 28–66. [CrossRef] [PubMed]
27. Hellström, T. Dimensions of environmentally sustainable innovation: The structure of eco-innovation concepts. *Sustain. Dev.* **2007**, *15*, 148–159. [CrossRef]
28. Manral, L. An evolutionary theory of demand-side determinants of strategy dynamics. *Manag. Res. Rev.* **2018**, *41*, 314–344. [CrossRef]
29. Fernando, Y.; Chiappetta Jabbour, C.J.; Wah, W.X. Pursuing green growth in technology firms through the connections between environmental innovation and sustainable business performance: Does service capability matter? *Resour. Conserv. Recycl.* **2019**, *141*, 8–20. [CrossRef]
30. Zhang, M.L.; Liu, Y. Influence of digital finance and green technology innovation on China's carbon emission efficiency: Empirical analysis based on spatial metrology. *Sci. Total Environ.* **2022**, *838*, 156463. [CrossRef]
31. Wang, X.Y.; Wang, Q. Research on the impact of green finance on the upgrading of China's regional industrial structure from the perspective of sustainable development. *Resour. Policy* **2021**, *74*, 102436. [CrossRef]
32. Wang, M.Y.; Li, Y.M.; Li, J.Q.; Wang, Z.T. Green process innovation, green product innovation and its economic performance improvement paths: A survey and structural model. *J. Environ. Manag.* **2021**, *297*, 113282. [CrossRef]
33. Sun, L.Y.; Miao, C.L.; Yang, L. Ecological-economic efficiency evaluation of green technology innovation in strategic emerging industries based on entropy weighted TOPSIS method. *Ecol. Indic.* **2017**, *73*, 554–558. [CrossRef]
34. Liu, L.; Zhang, Z.S.; Wang, Z.; Xu, J.G. Health evaluation and key influencing factor analysis of green technological innovation system. *Environ. Sci. Pollut. Res.* **2022**, *29*, 77482–77501. [CrossRef]
35. Zhu, L.; Wang, Y.; Shang, P.P.; Qi, L.; Yang, G.C.; Wang, Y. Improvement path, the improvement potential and the dynamic evolution of regional energy efficiency in China: Based on an improved nonradial multidirectional efficiency analysis. *Energy Policy* **2019**, *133*, 110883. [CrossRef]
36. Lin, S.F.; Sun, J.; Marinova, D.; Zhao, D.T. Evaluation of the green technology innovation efficiency of China's manufacturing industries: DEA window analysis with ideal window width. *Technol. Anal. Strateg. Manag.* **2018**, *30*, 1166–1181. [CrossRef]
37. Zhang, L.Y.; Ma, X.; Ock, Y.S.; Qing, L.L. Research on Regional Differences and Influencing Factors of Chinese Industrial Green Technology Innovation Efficiency Based on Dagum Gini Coefficient Decomposition. *Land* **2022**, *11*, 122. [CrossRef]
38. Tone, K. A slacks-based measure of efficiency in data envelopment analysis. *Eur. J. Oper. Res.* **2001**, *130*, 498–509. [CrossRef]
39. Feng, Z.J. Research on Green Innovation Efficiency of Chinese Industrial Enterprises. *China Sci. Technol. Forum* **2013**, *2*, 82–88. [CrossRef]
40. Shen, T.T.; Li, D.J.; Jin, Y.Y.; Li, J. Impact of Environmental Regulation on Efficiency of Green Innovation in China. *Atmosphere* **2022**, *13*, 767. [CrossRef]

41. Li, J.; Du, Y.X. Spatial effect of environmental regulation on green innovation efficiency: Evidence from prefectural-level cities in China. *J. Clean. Prod.* **2021**, *286*, 125032. [CrossRef]
42. Miao, C.L.; Fang, D.B.; Sun, L.Y.; Luo, Q.L. Natural resources utilization efficiency under the influence of green technological innovation. *Resour. Conserv. Recycl.* **2017**, *126*, 153–161. [CrossRef]
43. Liu, C.Y.; Gao, X.Y.; Ma, W.L.; Chen, X.T. Research on regional differences and influencing factors of green technology innovation efficiency of China's high-tech industry. *J. Comput. Appl. Math.* **2020**, *369*, 112597. [CrossRef]
44. Behera, P.; Sethi, N. Nexus between environment regulation, FDI, and green technology innovation in OECD countries. *Environ. Sci. Pollut. Res.* **2022**, *29*, 52940–52953. [CrossRef]
45. Li, M.Y.; Gao, X. Implementation of enterprises' green technology innovation under market-based environmental regulation: An evolutionary game approach. *J. Environ. Manag.* **2022**, *308*, 114570. [CrossRef] [PubMed]
46. Nazli, T.; Janice, C.; Vedat, V. Strategic supply chain decisions under environmental regulations: When to invest in end-of-pipe and green technology. *Eur. J. Oper. Res.* **2020**, *283*, 601–613. [CrossRef]
47. Sung, B. Do government subsidies promote firm-level innovation? Evidence from the Korean renewable energy technology industry. *Energy Policy* **2019**, *132*, 1333–1344. [CrossRef]
48. Rubashkina, Y.; Galeotti, M.; Verdolini, E. Environmental regulation and competitiveness: Empirical evidence on the Porter Hypothesis from European manufacturing sectors. *Energy Policy* **2015**, *83*, 288–300. [CrossRef]
49. Wang, P.; Dong, C.; Chen, N.; Qi, M.; Yang, S.; Nnenna, A.B.; Li, W. Environmental Regulation, Government Subsidies, and Green Technology Innovation—A Provincial Panel Data Analysis from China. *Int. J. Environ. Res. Public Health* **2021**, *18*, 11991. [CrossRef]
50. Adewale Alola, A.; Ozturk, I.; Bekun, F.V. Is clean energy prosperity and technological innovation rapidly mitigating sustainable energy-development deficit in selected sub-Saharan Africa? A myth or reality. *Energy Policy* **2021**, *158*, 112520. [CrossRef]
51. Muhammad, A.N.; Toan, L.D.H.; Huong, T.X.T. Role of financial development, economic growth & foreign direct investment in driving climate change: A case of emerging ASEAN. *J. Environ. Manag.* **2019**, *242*, 131–141. [CrossRef]
52. Zhang, D.Y.; Vigne, S.A. How does innovation efficiency contribute to green productivity? A financial constraint perspective. *J. Clean. Prod.* **2021**, *280*, 124000. [CrossRef]
53. Lei, S.Y.; Pan, Z.L.; Niu, C.Q. *Analysis of the China-Eurasian Economic Union Trade Potential Based on Trade Gravity Model*; Springer International Publishing: Cham, Switzerland, 2022; pp. 288–297. [CrossRef]
54. Michele, F.; Francesco, M. Trade Costs and Economic Development. *Econ. Geogr.* **2012**, *88*, 137–163. [CrossRef]
55. Wu, G.C.; Li, J.; Chong, D.; Niu, X. Analysis on the Housing Price Relationship Network of Large and Medium-Sized Cities in China Based on Gravity Model. *Sustainability* **2021**, *13*, 4071. [CrossRef]
56. Lourenço, D.; Sá, C. Estimating the Effect of Distance on the Migration of Higher Education Candidates. *B. E. J. Econ. Anal. Policy* **2022**, *22*, 739–761. [CrossRef]
57. Wang, J.F.; Xu, C.D. Geographic detector: Principles and prospects. *J. Geogr.* **2017**, *72*, 116–134.
58. Xin, L.; Sun, H.; Wang, H.; Xiao, H.Y. Research on Spatial and Temporal Differentiation and Driving Forces of Green Economic Efficiency Based on Geographic Detector. *China's Popul. Resour. Environ.* **2020**, *30*, 128–138.
59. Zhang, J.X.; Ouyang, Y.; Ballesteros-Pérez, P.; Li, H.; Philbin, S.P.; Li, Z.L.; Skitmore, M. Understanding the impact of environmental regulations on green technology innovation efficiency in the construction industry. *Sustain. Cities Soc.* **2021**, *65*, 102647. [CrossRef]
60. Shan, Y.L.; Guan, D.B.; Liu, J.H.; Mi, Z.F.; Liu, Z.; Liu, J.R.; Schroeder, H.; Cai, B.F.; Chen, Y.; Shao, S.; et al. Methodology and applications of city level CO_2 emission accounts in China. *J. Clean. Prod.* **2017**, *161*, 1215–1225. [CrossRef]
61. Shen, T.; Chen, H.H.; Zhao, D.H.; Qiao, S. Examining the impact of environment regulatory and resource endowment on technology innovation efficiency: From the microdata of Chinese renewable energy enterprises. *Energy Rep.* **2022**, *8*, 3919–3929. [CrossRef]
62. Xie, C.; Wang, Q. Analysis on the spatio-temporal pattern evolution and influencing factors of China's new energy industry's technological innovation capability. *Geogr. Res.* **2022**, *41*, 130–148.
63. Ye, G.; Fu, Y.; Wang, Y.H.; Mou, P.; Tang, X.Y. Overview of Research on Total Factor Productivity Measurement in the Construction Industry. *Constr. Econ.* **2019**, *40*, 24–28. [CrossRef]
64. Peng, Y.F.; Fan, Y.Y.; Liang, Y. A Green Technological Innovation Efficiency Evaluation of Technology-Based SMEs Based on the Undesirable SBM and the Malmquist Index: A Case of Hebei Province in China. *Sustainability* **2021**, *13*, 11079. [CrossRef]
65. Deng, Y.L.; You, D.M.; Wang, J.J. Research on the nonlinear mechanism underlying the effect of tax competition on green technology innovation—An analysis based on the dynamic spatial Durbin model and the threshold panel model. *Resour. Policy* **2022**, *76*, 102545. [CrossRef]
66. Zhang, X.M.; Wu, N.; Wu, J.; Feng, Q.; Fu, Z.Q. Review on the connotation, characterization and application of environmental regulation. *J. Environ. Eng. Technol.* **2021**, *11*, 1250–1257.
67. Khosla, R.; Sagar, A.; Mathur, A. Deploying Low-carbon Technologies in Developing Countries: A view from India's buildings sector. *Environ. Policy Gov.* **2017**, *27*, 149–162. [CrossRef]
68. Hua, J.N.; Wang, Y. Analysis of green total factor productivity of China's construction industry based on panel data of 30 provinces. *Pract. Understand. Math.* **2020**, *50*, 297–305. (In Chinese)
69. Li, X.; Lai, X.D.; Zhang, F.C. Research on green innovation effect of industrial agglomeration from perspective of environmental regulation: Evidence in China. *J. Clean. Prod.* **2021**, *288*, 125583. [CrossRef]

70. Mormina, M. Science, Technology and Innovation as Social Goods for Development: Rethinking Research Capacity Building from Sen's Capabilities Approach. *Sci. Eng. Ethics* **2018**, *25*, 671–692. [CrossRef]
71. Dong, X.; Fu, W.; Yang, Y.; Liu, C.; Xue, G. Study on the Evaluation of Green Technology Innovation Efficiency and Its Influencing Factors in the Central Plains City Cluster of China. *Sustainability* **2022**, *14*, 11012. [CrossRef]
72. Hunt, K.; Gruszczynski, M. The influence of new and traditional media coverage on public attention to social movements: The case of the Dakota Access Pipeline protests. *Inf. Commun. Soc.* **2021**, *24*, 1024–1040. [CrossRef]
73. Han, J.J.; Wang, J.P.; Chen, L.; Xiang, J.Y.; Ling, Z.Y.; Li, Q.K.; Wang, E.L. Driving factors of desertification in Qaidam Basin, China: An 18-year analysis using the geographic detector model. *Ecol. Indic.* **2021**, *124*, 107404. [CrossRef]
74. Wang, Q.; Ren, S.M. Evaluation of green technology innovation efficiency in a regional context: A dynamic network slacks-based measuring approach. *Technol. Forecast. Soc. Chang.* **2022**, *182*, 121836. [CrossRef]
75. Xu, S.R.; Wu, T.; Zhang, Y. The spatial-temporal variation and convergence of green innovation efficiency in the Yangtze River Economic Belt in China. *Environ. Sci. Pollut. Res. Int.* **2020**, *27*, 26868–26881. [CrossRef]
76. Hu, B.; Yuan, K.; Niu, T.Y.; Zhang, L.; Guan, Y.Q. Study on the Spatial and Temporal Evolution Patterns of Green Innovation Efficiency and Driving Factors in Three Major Urban Agglomerations in China—Based on the Perspective of Economic Geography. *Sustainability* **2022**, *14*, 9239. [CrossRef]
77. Wang, X.; Wang, S.; Zhang, Y. The Impact of Environmental Regulation and Carbon Emissions on Green Technology Innovation from the Perspective of Spatial Interaction: Empirical Evidence from Urban Agglomeration in China. *Sustainability* **2022**, *14*, 5381. [CrossRef]
78. Zhao, N.; Liu, X.J.; Pan, C.F.; Wang, C.Y. The performance of green innovation: From an efficiency perspective. *Soc.-Econ. Plan. Sci.* **2021**, *78*, 101062. [CrossRef]
79. Stucki, T.; Woerter, M.; Arvanitis, S.; Peneder, M.; Rammer, C. How different policy instruments affect green product innovation: A differentiated perspective. *Energy Policy* **2018**, *114*, 245–261. [CrossRef]
80. Porter, M.E.; Linde, C. Toward a New Conception of the Environment-Competitiveness Relationship. *J. Econ. Perspect.* **1995**, *9*, 97–118. [CrossRef]

buildings MDPI

Review

Application of Artificial Neural Networks in Construction Management: A Scientometric Review

Hongyu Xu [1], Ruidong Chang [2], Min Pan [3], Huan Li [1], Shicheng Liu [1], Ronald J. Webber [4], Jian Zuo [2] and Na Dong [1,*]

1 College of Architecture and Environment, Sichuan University, Chengdu 610065, China; xuhongyu@stu.scu.edu.cn (H.X.); lihuan@stu.edu.cn (H.L.); liushicheng@stu.edu.cn (S.L.)
2 School of Architecture and Built Environment, The University of Adelaide, Adelaide 5005, Australia; ruidong.chang@adelaide.edu.au (R.C.); jian.zuo@adelaide.edu.au (J.Z.)
3 Sichuan Kaiyuan Engineering Project Management Consulting Co., Ltd., Chengdu 610041, China; sckyzx@sckyzx.com
4 Department of Mining-Built Environment, Central Queensland University, Rockhampton 4701, Australia; r.webber@cqu.edu.au
* Correspondence: dongna@scu.edu.cn; Tel.: +86-85407049

Abstract: As a powerful artificial intelligence tool, the Artificial Neural Network (ANN) has been increasingly applied in the field of construction management (CM) during the last few decades. However, few papers have attempted to draw up a systematic commentary to appraise the state-of-the-art research on ANNs in CM except the one published in 2000. In the present study, a scientometric analysis was conducted to comprehensively analyze 112 related articles retrieved from seven selected authoritative journals published between 2000 and 2020. The analysis identified co-authorship networks, collaboration networks of countries/regions, co-occurrence networks of keywords, and timeline visualization of keywords, together with the strongest citation burst, the active research authors, countries/regions, and main research interests, as well as their evolution trends and collaborative relationships in the past 20 years. This paper finds that there is still a lack of systematic research and sufficient attention to the application of ANNs in CM. Furthermore, ANN applications still face many challenges such as data collection, cleaning and storage, the collaboration of different stakeholders, researchers and countries/regions, as well as the systematic design for the needed platforms. The findings are valuable to both the researchers and industry practitioners who are committed to ANNs in CM.

Keywords: artificial neural network (ANN); construction management; scientometric analysis; future trends

Citation: Xu, H.; Chang, R.; Pan, M.; Li, H.; Liu, S.; Webber, R.J.; Zuo, J.; Dong, N. Application of Artificial Neural Networks in Construction Management: A Scientometric Review. *Buildings* **2022**, *12*, 952. https://doi.org/10.3390/buildings12070952

Academic Editors: Yongjian Ke, Jingxiao Zhang, Simon P. Philbin and David J. Edwards

Received: 14 April 2022
Accepted: 30 June 2022
Published: 4 July 2022

1. Introduction

The characteristics of high investment, long period and high uncertainty make construction management an indispensable part of the modern construction industry [1,2]. The urgent need of upgrading and transformation of the construction industry also drives the renewal of construction management concepts and methods [3–5]. In this data-intensive industry, data, which can significantly improve the performance of CM, is becoming the key resource [6,7]. Nevertheless, data application in CM has been considered relatively conservative [1]. It is difficult to analyze and process the large volume data of the construction industry with traditional technologies, so that a large amount of data is shelved and wasted [8]. The digitization report released by McKinsey indicated that the construction industry is one of the worst-performing digitally at the moment, which maybe is the main reason for decades of persistently low productivity in the construction industry [9,10]. Therefore, in the digitalization era, it is of great significance for the construction industry to use intelligent technology to process large volume data in CM and obtain knowledge hidden in the data for assist decision making [11]. Furthermore, one of the most promising technologies is the artificial neural network (ANN) [12].

ANN, as a mathematical model inspired by and imitating the biological brain, can be used to extract knowledge hidden in large historical data, and to process it productively [13]. As an importance branch of artificial intelligence, largely due to its good self-learning, self-organizing function and high-speed computing ability, ANN does not need to assume the relationship between variables and performs well in dealing with complex nonlinear problems [14,15]. Even with incomplete or previously unknown data acceptable results can be obtained [16]. Accordingly, ANN is extremely suitable for dealing with practical CM problems that are difficult to solve by mathematical methods and traditional modeling [13] and has been used in the Architecture, Engineering and Construction industry (AEC) since the early 1990s [17]. Previous studies have shown that ANN can play a major role in prediction, optimization, classification, and decision-making in CM practice [18–20]. It has successfully aided in solving specific problems throughout the project's life cycle from the planning stage to the operation and maintenance stage.

Because of the considerable use of ANN in CM [12], the literature related to ANN has proliferated and several available literature reviews on ANN in AEC have been put forward. For instance, Rajesh systematically reviewed the literature related to ANN in energy analysis [20], and Sony et al. [21] reviewed the related applications of convolution neural network (CNN) from the perspective of structural state assessment. However, these two literatures on ANN had specific perspectives, which only focused on the energy or the structural state assessment. Pan et al. [22] provided a comprehensive review on AI in construction engineering and management in which ANN is only mentioned to a limited extent in some paragraphs throughout the review, rather than in terms of a detailed vision. Adeli [23] reviewed the literature on the application of ANN in CM published from 1989 to 2000, but now this work is limited by timeliness. It is not difficult to conclude that the existing literature on ANNs in CM has rarely been comprehensively and systematically reviewed in the last 20 years. If the literature review of ANN in CM is not updated, this may lead to the following problems. Firstly, due to the lack of a comprehensive review, it is difficult for the beginner to learn about the authoritative authors, outlets, publications and the active countries/regions to serve as an example. Secondly, a comprehensive application profile is needed to show current research progress including what topics have been focused on and what progress has been made in the field of ANN in CM. Finally, without a summary of research evolution and current breakthroughs and limitations, researchers interested in this field will spend more time ascertaining current research status, future trends and possible research directions.

Therefore, it is necessary to make a comprehensive and systematic review of the application of ANN in CM. This paper intends to achieve the following objectives: (1) Identify main research authors, institutions and countries/regions that are active in the field of ANN in CM over the past 20 years and their cooperation relationships; (2) Present the main research interests on ANN in CM over the past two decades; (3) Uncover the relationships among different research interests and their evolution tendency; (4) Summarize benefits and challenges of ANN in CM and propose promising research directions.

2. Materials and Methods

Different methods are available for reviewing literatures [24] among which the scientometric analysis is good at visualizing significant structure and trends based on author, keyword, and reference in a large body of literature data [25]. The scientometric analysis can meet all of the research objectives mentioned above. Therefore, it is adopted in this paper. A three-stage process was carried out including: data collection, scientometric analysis, and discussion and conclusion. The outline of research methodology is shown as Figure 1.

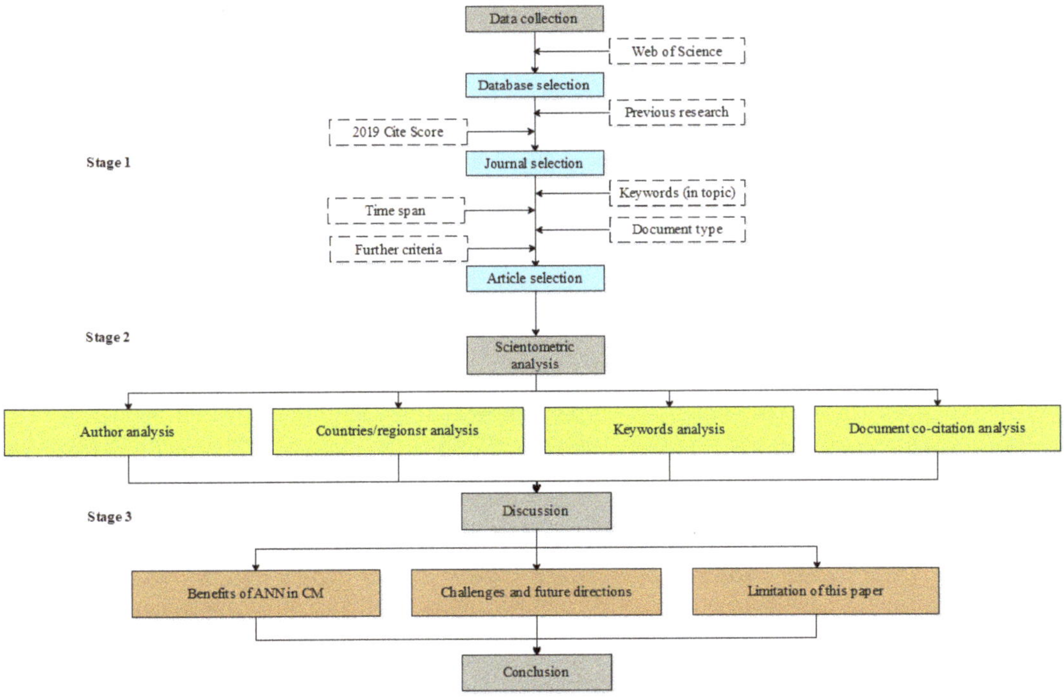

Figure 1. Outline of research methodology.

2.1. Data Collection

Due to the interdisciplinary nature of CM, a comprehensive academic database, Web of Science, known for its comprehensiveness, organizational structure, and scientific robustness, was chosen in this study [26]. It is a consensus in the industry that articles in high-ranking journals usually have high influence [25]. Considering this fact, the authors selected the journals that have an important impact and top quality in CM based on the 2019 Scopus journal metrics (CiteScore is not less than 2) and the ranking of CM journals [27]. Moreover, in view of timeliness, these journals must have published at least three papers related to ANN in CM between 2000 and 2020.

To summarize the research progress in the past 20 years, the retrieval time of the journal article is limited to 2000 and 2020. In order to ensure that all relevant papers are captured, different search keywords are applied. The following retrieval code was adopted, and the search was conducted using the 'topic' in the Web of Science.

(neural network) AND (construction management), (neural network) AND (engineering management)

Four hundred potential articles were initially identified and then filtered based on the criteria that ANN emerged as the main technology or played an important role rather than just a comparison. A two-stage selection strategy was adopted to meet the above criteria. Firstly, title, abstract, and keywords in each article are examined to exclude unrelated articles. Secondly, the entire paper content is analyzed in detail to ensure that all the selected articles are closely related to the research objectives. Finally, 112 papers were selected for later scientometric review; the different journals in which the selected papers were published can be seen in Table 1. These articles provide a representative sampling of existing studies on the ANN in CM and form the dataset utilized in the current research.

Table 1. Distribution of the selected papers among different journals.

No.	Journal	Cite Score 2019	Literature List	No. of Paper
1	Journal of Construction Engineering and Management (JCEM)	5.8	[15,28–64]	38
2	Automation in Construction (AC)	9.5	[65–98]	34
3	Journal of Civil Engineering and Management (JCiEM)	4.7	[99–111]	13
4	Engineering, Construction and Architectural Management (ECAM)	2.5	[112–119]	9
5	Journal of Management in Engineering (JME)	6.7	[120–127]	8
6	International Journal of Project Management (IJPM)	13.0	[128–133]	6
7	Journal of Computing in Civil Engineering (JCCE)	7.6	[134–137]	4
total				112

2.2. Introduction and Process of Scientometric Analysis

The earliest definition of scientometrics is "quantitative research on scientific development research" [138]. The purpose of scientometric analysis is to help literature reviews overcome subjective issues in content analysis [139]. A scientometric analysis consists of the text-mining and citation analysis which helps researchers find systematic literature-related findings, by finding literature information that may be ignored in manual review studies [140]. There are several available tools to realize the goal of scientometric analysis such as VOSviewer and Citespace. Citespace is an advantageous application for analyzing and visualizing networks, and is specialized in analyzing what the major research interests are and how they are evolved and linked [141]. VOSviewer offers the basic functionality required for producing, visualizing, and exploring bibliometric networks, and also has special text-mining features [142]. Each tool has its own strengths, and it is necessary to appropriately use different tools for different kinds of analysis [143]. The combination of these two tools is widely accepted for reviews, such as that of AI in the Architecture, Engineering and Construction industry [12] and computer vision applications for construction [144]. Therefore, VOSviewer and Citespace were selected for scientometric analysis in this paper. VOSviewer was used to implement keyword co-occurrence analysis and Citespace was used for other tasks.

Whether VOSviewer or Citespace, its main process can be summarized as follows: (1) importing literature data from WOS into software for visualization; (2) Figures presentation and optimization of different aspects according to software functions; (3) In-depth scientometric analysis of the adjusted figures. After the screening mentioned above, each piece of data is downloaded from WOS as a full-record text format. In total, 112 valid publications were extracted. Then, the output file format after renaming is 'download_*.txt' and is imported into VOSviewer and Citespace for format conversion. Visualizations are generated and can be converted to different views through different functions of the software. After adjusting different visualizations of input data to make them easier to read and analyze, the scientometric analysis begins, including the following four aspects considered in this paper. First, through co-author network analysis, core research groups and their cooperative relationships were identified. Second, with network analysis of participating countries/regions, the most influential countries/regions which are particularly active on ANN in CM and the collaborations among them were described. Third, taking note of the keywords, network analysis is conducted to discern the main research interests and the hot topics on ANN in CM. Fourth, with the timeline visualization and citation bursts, the keywords evolution network shows the trends and changes. Finally, network analysis of the co-citation references was carried out to mine the most representative literature in this filed.

3. Results of Scientometric Analysis

3.1. Author Analysis

Co-authorship network analysis of current research in ANN in CM can promote access to specialists and expand research productivity [25]. The core group of authors and their cooperative relationships in this field can be determined by analyzing the structural characteristics of the corresponding authors and their cooperation networks. In this paper, the co-authorship network was generated in Citespace and the collaboration map is presented in Figure 2. The size of the node indicates the number of the articles published by the author, and the connecting line indicates the collaboration relationship among them, and the color of the line indicates the authors in the same cluster. Publication dates from past to present are shown in a transition from cool to warm color. As can be seen from Figure 2, there are 253 nodes, 287 links, and the network density is 0.0101. The typical value for network density is between 0 and 1. Especially low network density, even close to 0, indicates that the authors in the network are not closely connected [145]. The most productive author on ANN in CM was *MINYUAN CHENG* with 8 articles, followed by *HSINGCHIH TSAI* (4), and *XUEFENG ZHAO* (4). It can be noted that many authors tend to collaborate with a relatively stable group of collaborators, so there are several major groups of authors. Among them, the cluster with *MINYUAN CHENG* is the core research team. They represent the important scholars in the application of ANN in CM and can offer highly individualized scientific research information to other researchers in this field.

Figure 2. Network of co-authorship in research on ANN in CM [15,28–137].

3.2. *Countries/Regions Analysis*

The leading countries/regions in research on ANN in CM can be identified through network analysis. The results are useful for interested scholars to help them identify leading countries with high potential for cooperation. At the same time, the results can also provide top management with macro data to facilitate policy decisions on industry digitization. This section presents the countries/regions contributing to the 112 research articles extracted for the study. Figure 3 shows the network of citing countries/regions, which contains 29 nodes and 33 links. The size of a node represents the total number of published articles in the 112 articles, and the thickness of the links indicates the levels of the cooperative relationships. As a result, PEOPLES R CHINA (23 articles), TAIWAN (21 articles) and USA (17 articles) top the list, demonstrating that the considerable number of related articles in these countries/regions have made significant contributions to research in this field. However, compared with other emerging technology such as AI and BIM, as a growing new technology, ANN in CM has not yet attracted the global attention it deserves. It is believed that in the future, more and more countries/regions will pay attention to and promote research in this field. The betweenness centrality is an important index in

Citespace. Freeman [146] noted that the betweenness centrality could be calculated by the ratio of the shortest path between two nodes to the sum of all such shortest paths. The greater the betweenness centrality, the higher its importance. From the perspective of centrality, Citespace identified a collaborative pattern, and the network reveals that countries such as PEOPLES R CHINA (centrality = 0.43), USA (centrality = 0.35) and TURKEY (centrality = 0.13) were the key infrastructure nodes in the network. Researchers from these countries collaborate more actively than others. The centrality for TURKEY is only 0.13 which indicates insufficient collaboration, while Taiwan, as the second prolific region, has a centrality of 0.00. In general, these results imply that strengthening academic exchanges and contacts to expand current research productivity may be a subject worth more attention.

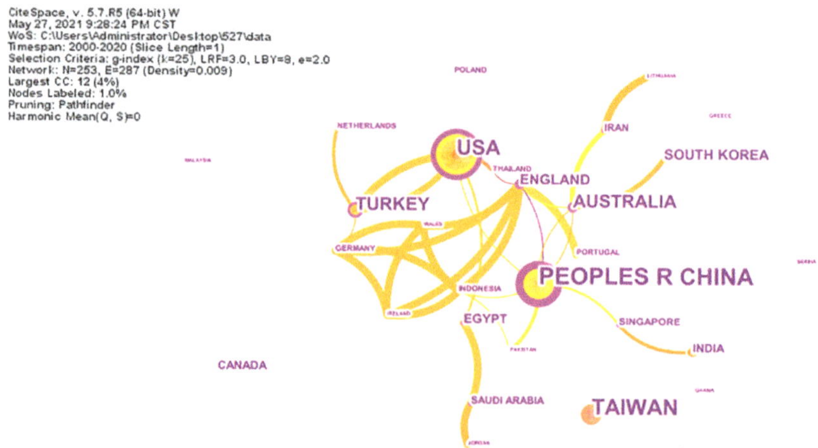

Figure 3. Collaboration network of countries/regions in the research on ANN in CM.

3.3. Keywords Analysis

3.3.1. Co-Occurrence Network of Keywords

Keywords are representative and concise descriptions of research paper content, and analyzing keywords provides an opportunity to identify major research interests in this field [147]. A network of keywords offers a good picture of a knowledge domain, which help to identify the interests over a specific timespan and provides an understanding of how they are connected and organized [138]. Identical terms (e.g., cost estimate, construction cost estimation and cost prediction; genetic algorithm and GA; regression and regression analysis) were merged (as cost estimate, genetic algorithm and regression analysis, respectively) and generic keywords related to research areas, etc. (e.g., management, model) were omitted during analysis because they do not reflect the current related research trend and have an impact on the clustering accuracy of analysis results [148].

To construct and map the knowledge domain on ANN in CM, keyword co-occurrence in the research area was obtained using VOSviewer. Main research interests on ANN in CM is shown as Figure 4. The network of co-occurring keywords has 122 nodes, 342 links, and a total link strength of 432. In this network, each node represents a different keyword, and the link between the two nodes is the co-relationship between the connected keywords, the node size is determined by the frequency of the words appeared in the 112 articles.

The frequency with which keywords are cited indicates the main research interests in the research field [149]. Table 2 summarizes the keyword occurrences and node strength of each. The links are the number of linkages between a given node and others, while the total link strength reflects the total strength linked to a specific item.

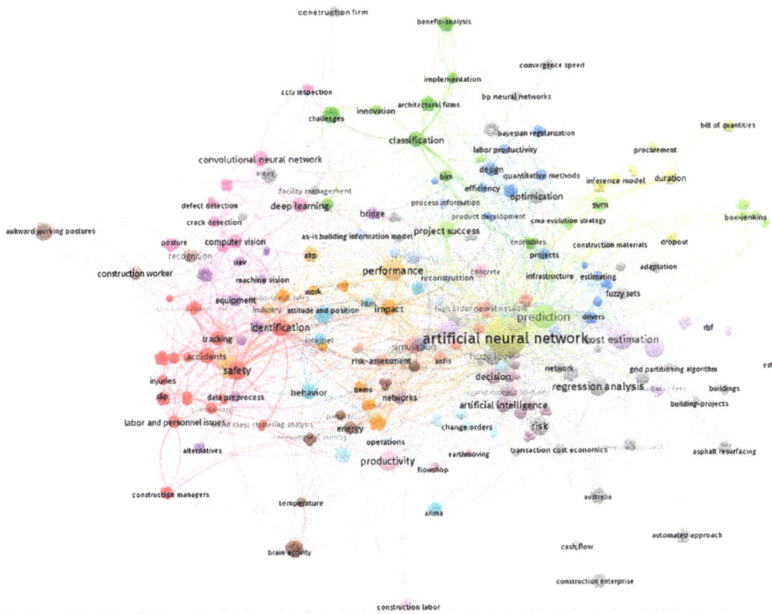

Figure 4. Main research interests on ANN in CM (co-occurrence network of keywords).

Table 2. Top keywords of existing research interests on ANN in CM.

Keyword	Occurrences	Links	Total Link Strength	Keyword	Occurrences	Links	Total Link Strength
artificial neural network	75	333	556	data mining	6	33	38
prediction	20	113	157	deep learning	6	59	66
cost estimation	16	79	127	impact	6	49	55
performance	16	102	138	svm	6	48	55
construction cost	14	70	101	computer vision	5	31	41
genetic algorithm	13	78	115	construction worker	5	50	58
regression analysis	13	73	106	cost and schedule	5	29	41
productivity	12	86	102	design	5	45	48
risk	11	77	99	duration	5	25	29
algorithm	10	80	96	machine learning	5	45	48
safety	10	72	106	networks	5	35	40
fuzzy logic	9	42	71	recognition	5	54	60
identification	9	63	88	tracking	5	53	61
optimization	8	57	71	bridge	4	17	19
project success	8	50	62	cluster-analysis	4	28	33
simulation	8	62	75	contractor	4	28	31
accidents	7	50	75	data	4	34	39
artificial intelligence	7	37	49	disputes	4	26	31
decision	7	54	70	energy	4	42	47
behavior	6	51	67	equipment	4	43	49
cbr	6	32	45	fuzzy sets	4	18	23
classification	6	42	45	labor and personnel issues	4	37	43
convolutional neural network	6	48	54	risk-assessment	4	34	38

From Table 2, it is revealed that ANN was the keyword with the highest frequency, and was used as the keyword in 75 of 112 articles, which further verifies the rationale of the literature selection. Besides ANN, prediction is the most frequent keyword and the total link strength is 113. Prediction is in the highest level of the keywords indicating the strong inter-relatedness between ANN and prediction. The analysis result that prediction has received considerable attention could be interpreted by the fact that as a powerful algorithm for AI, ANN is an effective tool for prediction [150]. ANN is typically applied in prediction models for knowledge discovery from a large quantity of information and documents which are generated in the process of construction management, and the result can provide reliable assistance for decision-making [117] and optimization [33]. In addition, ANN can also be used for recognition and classification, such as defect classification [85] and construction activity recognition [87]. Compared with prediction, research on recognition and classification attracts relatively insufficient attention and deserves further exploration in the future.

Except for the main functions of ANN in CM, the range of problems or tasks ANN has been applied to solve in CM is another important issue. It can be seen from Figure 4 and Table 2 that cost estimation, performance, productivity, risk, safety, project success and duration represent other important types of nodes in the network, which are key tasks of construction management [151]. The results indicate that ANN has gradually indeed become an effective tool for CM and is gradually replacing the traditional mainstream methods due to its advantages [79]. It is worth mentioning that behavior, construction worker, contractor, labor and personnel issues are an emerging type of research topic, and the importance of personnel management is further highlighted in project management [59]. There is, however, a conspicuous absence of interest in the topic of environment in the network, which needs further attention.

Finally, there are many keywords related to algorithms such as genetic algorithm, regression analysis, algorithms, fuzzy logic, data mining etc. It indicates that in order to better complete project management tasks, a variety of methods have become more widely applied together with ANN to improve the efficiency and precision of the model.

3.3.2. Timeline Visualization and Citation Bursts of Keywords

Cluster analysis is used to identify the semantic themes hidden in the textual data. Figure 5 shows a timeline visualization of cluster analysis of keywords which was created by Citespace 5.7.R1. There are three text mining algorithms that can be used to label clusters in CiteSpace. Log-Likelihood ratio (LLR) clustering technique was implemented in this study because of its good clustering results [80]. The network is divided into 9 major co-citation clusters (with cluster IDs #0, #1, etc.). CiteSpace automatically selects a label for each cluster based on titles, keywords, and abstracts of the articles in each cluster. Usually, the Modularity (Q value) and Mean Silhouette (MS) value are used to evaluate the clustering effect. Generally speaking, Q value is within the interval of [0, 1). $Q > 0.3$ means that the community structure is significantly divided. A cluster's silhouette value ranges from -1 to 1 and assesses the uncertainty involved in defining the cluster's nature. A value of 1 signifies that the cluster is perfectly isolated [12]. As shown in Figure 5, the network has high modularity ($Q = 0.581 > 0.3$), which shows that the network is divided into clusters with dense links amid nodes within clusters. The MS value is 0.363, indicating that the homogeneity of the clusters is not high. This MS value shows that although the studies in the network in each cluster may be consistent in exploring somewhat similar issues, they address different issues in fact [12].

The largest cluster (#0) has 25 members and is labeled as 'machine learning' by LLR, which includes deep learning, modeling, LSTM, machine learning, framework, hybrid intelligence, back-propagation neural network, prediction, etc. This result means that the enhancement and optimization of ANN is one of the most popular research interests in recent years and is in line with earlier observations; development of ANN-based deep learning (DL) model is a representative direction. DL affords a machine learning technique in which computers are taught to perform what comes naturally to humans by training [150]. Zhou, Xu, Ding, Wei and Zhou [98] combined a wavelet transform noise filter, CNN, and long short-term memory predictor to propose a DL method. The DL method proposed by Rafiei and Adeli [56], including unsupervised deep Boltzmann machine learning and BPNN. The actual data is used to verify the proposed DL algorithm, and the results show that the effectiveness and accuracy of a single ANN are improved.

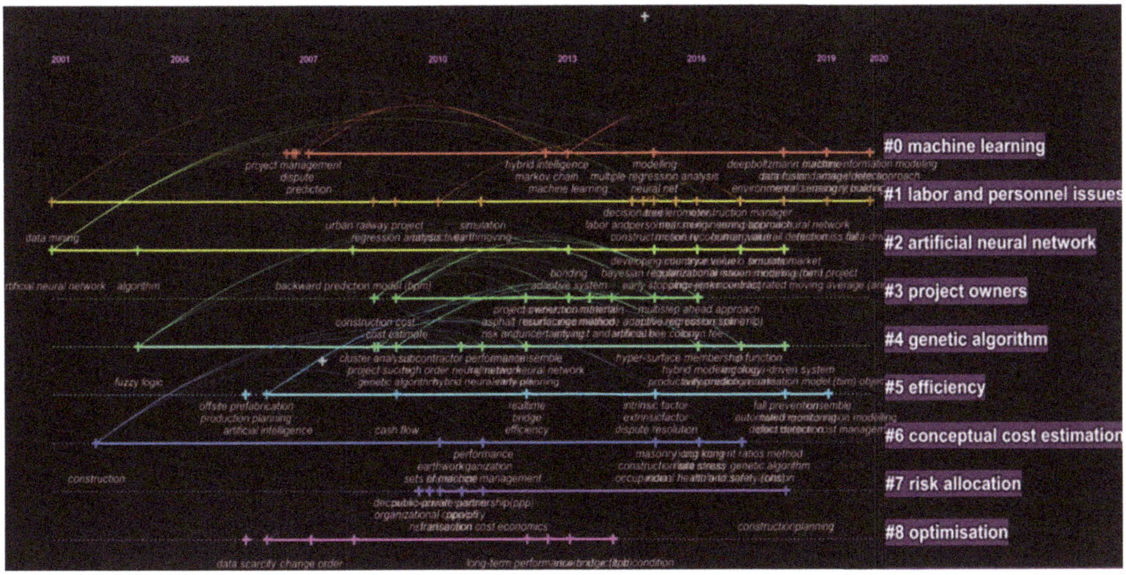

Figure 5. Timeline visualization of keywords (clustering structure).

The second largest cluster (#1), 'labor and personnel issues', mainly includes research on management, behavior, safety climate, worker, health, etc. The clustering results show that worker safety is another important topic. Especially since 2015, there has been a node explosion in Figure 5 (#1). This result shows that CM has paid great attention to safety and ANN has been widely used in the safety field in recent years. Emerging technologies such as laser scanning and smart sensors have made massive data acquisition a reality, which has provided tremendous support for this development [65]. Yi, Chan, Wang and Wang [94] proposed a system which could be automated by integrating smart sensor technology, location tracking technology and ANN to protect the wellbeing of those who have to work in hot and humid conditions.

In addition, cost estimation (#6) and risk allocation (#7), as the main tasks of traditional building management, were important topics before 2017, but the attention has been gradually weakened in recent years. Optimization (#8) is labeled as the smallest cluster indicating that the research on ANN applications in this area needs to be strengthened, as shown in Figure 5.

Citation bursts reflect the dynamics and evolution of the field by citing articles with a sharp increase in citations [152]. The higher the suddenness of a keyword, the more attention is paid to it in the time interval considered, and to some extent, it represents the research frontier and hotspot in the subject area [153]. Figure 6 shows the top 25 keywords with strongest citations bursts from 2000 to 2020. The light green line indicates the range of literature years reviewed, while the orange line indicates the duration of a citation burst event.

As shown in Figure 6, keywords with citation bursts can be divided into two categories including methods and issues. As to the methods appearing in the last 20 years, data mining first burst between 2001 and 2006, followed by fuzzy logic, case-based reasoning, genetic algorithm, and regression analysis. Meanwhile, hybrid neural network, machine learning and artificial intelligence began to appear in the last 10 years indicating that as an important representative of machine learning and artificial intelligence, ANN is attracting more and more attention. ML had the strongest of the citation bursts (3.77) and the bursts began from 2018 up until 2020 implying that ML based on ANN represent emerging themes in research on ANN in CM.

Figure 6. Top 25 keywords with the strongest citation bursts in the selected literature.

As to the issues, ANN has been applied to bridge management, project success, risk allocation, organizational capability, transaction and construction cost, bidding, dispute, safety, workers and productivity, etc., which have been hot research topics in the past 20 years. Before 2010, risk allocation had strong citation bursts due to the continuous development of the PPP model [134]. After 2010, Figure 6 shows that cost estimate (burst strength, 1.87), cost and schedule (1.86), had strong citation bursts in the literature. This implies that these were hot topics in the respective years.

3.4. *Document Co-Citation Analysis*

The references of frontier manuscripts can represent the knowledge base in a field [154]. Document co-citation analysis (DCA) studies a network of co-cited references. Thus, through analyzing the co-cited references, DCA objectively explored the underlying knowledge base of the ANN research in CM. CiteSpace was used to analyze the documents cited in 112 records. Figure 7 shows the detailed outcome of the document co-citation analysis, i.e., a co-citation network including 497 nodes and 1485 links. Each link represents the co-citation relationship between the two corresponding articles while the font size represents the co-citation frequency of these documents. The node documents were among cited documents and were not necessarily included in the 112 retrieved articles.

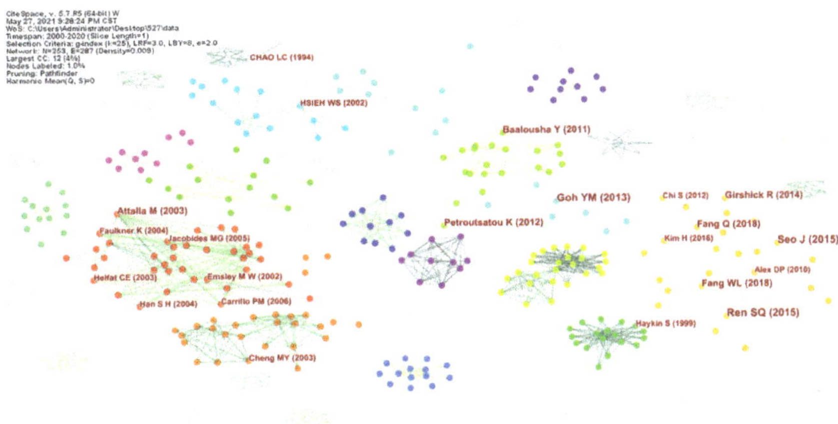

Figure 7. Document co-citation network of ANN in CM [15,28–137].

The top 10 cited documents are listed in Table 3. These articles were widely recognized by peers and had high value for research on ANN in CM. A systematic review of these 10 high-quality articles reveals the following findings: (1) Except for two studies on the improvement of CNN methods, the remaining research topics are divided equally into cost and safety for CM. (2) It can be seen from the publication dates of top cited documents that cost and safety have been the main research interests of the last two decades. Cost boomed in the last decade, but now the focus has turned to safety.

Four important cost documents cover almost all the research directions of ANN. First, the research on ANN: for example, articles compare traditional prediction methods (regression analysis, etc.) and ANN [155], develop different types of ANN algorithms (MLFN and GRNN) [55], other algorithms (FL) improve ANN [35], and establish a database for ANN [99]. Second, research on different topics: for example, articles include the cost for different types of construction projects, such as road tunnel construction cost [55], and the cost of reconstruction projects [155]. In addition, there is discussion of price, total cost, maintenance cost and other cost predictions from different perspectives.

For safety, the top cited literature shows ANN is mainly used for object safety detection and accident analysis. Specific techniques such as object detection, tracking and action recognition can be used to effectively identify unsafe acts and conditions. A large number of related researches in computer vision technology provide conditions for CNN to further realize accurate object detection. Fang Q developed a CNN model for automatic non-hardhat-use detection technology [156]. Fang WL proposed an improved and faster approach with CNN features to detect the presence of workers and equipment in real-time [157].

Table 3. Top 10 highly co-cited papers.

No.	Author	Article	Topic	Year	Total Citations	Source
1	Seo, et al. [158]	Computer vision techniques for construction safety and health monitoring	safety	2015	5	Advanced Engineering Informatics
2	Ren, et al. [159]	Faster R-CNN: towards real-time object detection with region proposal networks	CNN	2017	5	IEEE Transactions on Pattern Analysis and Machine Intelligence
3	Goh, et al. [160]	Neural network analysis of construction safety management systems: a case study in Singapore	safety	2013	5	Construction Management and Economics
4	Petroutsatou, Georgopoulos, Lambropoulos and Pantouvakis [55]	Early cost estimating of road tunnel construction using neural networks	cost	2012	4	Journal of Construction Engineering and Management
5	Girshick, et al. [161]	Rich feature hierarchies for accurate object detection and semantic segmentation	CNN	2014	4	Conference on Computer Vision and Pattern Recognition
6	Fang, Ding, Zhong, Love and Luo [157]	Automated detection of workers and heavy equipment on construction sites: A convolutional neural network approach	safety	2018	4	Advanced Engineering Informatics
7	Fang, Li, Luo, Ding, Luo, Rose and An [156]	Detecting non-hardhat-use by a deep learning method from far-field surveillance videos	safety	2018	4	Automation in Construction
8	Baalousha and Celik [99]	An integrated web-based data warehouse and artificial neural networks system for unit price analysis with inflation adjustment	cost	2011	4	Journal of Civil Engineering and Management
9	Attalla and Hegazy [155]	Predicting cost deviation in reconstruction projects: artificial neural networks versus regression	cost	2003	4	Journal of Construction Engineering and Management
10	Cheng and Ko [35]	Object-oriented evolutionary fuzzy neural inference system for construction management	cost	2003	3	Journal of Construction Engineering and Management

4. Discussion

4.1. Benefits of ANN in CM

Specific applications of ANN in dealing with CM problems were represented by the above literature analysis. The specific benefits of ANN in CM should be summarized so as to have a clear understanding of what value has been achieved.

Firstly, compared with traditional CM, data-driven CM based on ANN makes the construction and management activity more intelligent. As a powerful instrument to intelligently discover hidden knowledge from the mass of accumulated data in completed construction projects, ANN visualizes the tacit knowledge in the project experience and provides reliable advice for automated analysis and decision-making [162]. Table 2 proves that ANN has been successfully applied to the intelligent solution of problems such as prediction, success, cost, performance, productivity, risk and safety throughout the whole life cycle of construction projects without too much manual intervention. For instance, ANN has realized intelligent decision-making by quickly predicting the key indicators of project success in the project planning stage [75]. In the construction phase, current ANN models can provide real-time performance evaluation [49] and dynamic monitoring of on-site operations [15], so as to provide early warning. In order to ensure that the projects can be accomplished successfully, such an intelligent monitoring system is undoubtedly a great boon for on-site project managers. The intelligent maintenance system based on ANN has realized the automatic and remote assessment of structural condition [89]. In general, the intelligent function with ANN in CM is gradually replacing the traditional manual-led pattern, which tends to be time-consuming, subject to personal judgment and experience, and prone to error.

Secondly, improving the efficiency of CM is another prominent benefit of ANN being applied. Accurate estimation and reliable optimization are provided by the ANN model to make the project more effective and smoother. Experience has proved that the predictive performance of ANN is indeed better than traditional methods such as multiple regression analysis [120], and the recently optimized ANN model has made great progress in prediction performance [97]. Reliable results can avoid potential errors and reduce unnecessary waste, which is essential to the success of the construction project. The optimization effect is reflected in generating valuable suggestions for improvement in complex CM tasks under conflicting requirements and limitations, such as cash flow control [74], capital allocation plan [101], optimization control of schedule plan [69], time allocation of material processing [67], construction quality inspection [136], and adjustment of construction machinery posture and position [98]. Such findings can guide the optimization of the construction execution process, enabling timely adjustments at an early stage. Therefore, unnecessary steps, re-working, conflicts and potential delays can be effectively avoided and efficiency is greatly improved.

Thirdly, as shown in Table 3, ANN has great potential and value in reducing risks in the current complex and uncertain environments in CM. The ANN model can identify and evaluate risks in the new environment by capturing the interdependence between accidents and their causes in historical data, which effectively avoids the limitations of traditional risk analysis, such as the vagueness and subjectivity of expert experience. In the case of high uncertainty, ANN has been predominantly adopted for risk analysis in terms of finance [64], safety [77], contract [46], and quality [136]. Research on project dispute claim risk prediction, optimal risk allocation in PPP projects, risk analysis for BOT project contracts, early-warning for site work risk and bid selection has been carried out to improve the level of risk management. Therefore, ANN-based risk analysis can provide auxiliary and predictive insights on key issues, helping project managers to quickly determine the priority of possible risks and to determine positive actions, such as simplifying the work site operation, adjusting personnel arrangements, and ensuring that the project is carried out on time and on budget, rather than relying on risk mitigation measures.

4.2. Challenges and Future Directions for ANN in CM

In the future, CM will experience rapid digital transformation and ANN will also attract more and more attention, both in academic research or in practical applications. Despite the potential opportunities and benefits accruing from ANN in CM, there are still some challenging issues that deserve further study. Only by clarifying these challenges can effective countermeasures be proposed more targeted and effectively. The following four important challenges and directions to further tackle a diversity of existing issues in laborious, complex, or even dangerous tasks in CM are put forward.

4.2.1. More Collaboration Is Essential for Rapid Progress in ANN in CM

Lack of collaboration is a symptom of lower research productivity [25] and strong collaborative relationships should be fostered to make better research progress [163]. According to the co-authorship analysis (Figure 2) and the collaboration network of countries/regions (Figure 3), although more and more researches have been conducted on ANN in CM in the past 20 years, as a new technology in the digital transformation of the construction industry, collaboration is still in its infancy. Except for a few scholars such as Minyuan Cheng, Hsingchih Tsai, Xuefeng Zhao and Mingyuan Zhang shown in Figure 2 and countries/regions such as the People's Republic of China and USA, which are active in this field and have moderate cooperation, the research in this field needs extensive attention, promotion and cooperation from more scholars from all over the world and different domains such as mathematics and computing as well as CM. Cooperation among them should be strengthened to enable progress in this cross-disciplinary area.

4.2.2. The System Design and the Platform Establishing for ANN in CM Has Not Yet Begun

According to Table 2 and Figure 4, among the existing 112 articles, some articles focus on algorithm optimization which can be verified by the keywords including genetic algorithm, regression analysis, fuzzy logic, etc., some focus on function realization with keywords such as prediction, estimation, identification, decision, cluster analysis etc., while some focus on problem solving with keywords such as cost, performance, safety, productivity, risk etc. These findings indicate that the current research is mainly focused on specific applications, and that research on system design has not attracted enough attention. Systematic design and platform building are crucial to the spread and application of new technologies [37]. For example, as to BIM, there are many researches on framework and platforms [164], and for manufacturing industry, the system optimization based on AI has gradually become a prominent research trend [165]. However, systematic research and platform building research on ANN in CM have not been started yet. In the future, data preparation, model optimization and application, system design and platform setup deserve further discussion which is critical to digital transformation.

4.2.3. Research Focused on Different Stakeholders as Well as the Data Sharing among Them Is Still Missing

There are many participants in the construction projects, and each have different data types and application demands [126]. Therefore, the research on ANN in CM aimed at specific stakeholders such as the owners and the contractors as well as the consultants has not yet been addressed sufficiently in practical application [107]. Meanwhile, among the construction project's key five objectives, although cost, schedule, quality, environment and safety have attracted some specific research, research on the whole life cycle of the project is still insufficient [68]. For example, systematic research is needed on what data stakeholders will have access to, and what can they do with it separately targeting the government, the owner, the contractor as well as consultants. Furthermore, the data sharing among all the stakeholders should be discussed to maximize the value of the data and minimize the harm of information asymmetry [107]. Therefore, data collection, processing, storage and application targeting specific stakeholder are worth systematic research in the future.

4.2.4. Data Collection Is the Key of ANN in CM

Current research has focused on the development and application of ANN models but ignored the input data preparation. According to Bilal et al. [166], the whole data mining work usually takes about 80% of the time to prepare the data. There are many data sources for ANN application in CM, and Figure 8 shows six types of data sources and frequency distribution of all articles. The most two extensive data sources are historical project files and online databases, both of which are important in mining historical engineering data and have the disadvantages of one-time data collection. Only a few scholars realize the importance of establishing a database that can be dynamically updated in real time for a long time, such as Baalousha and Celik [99] who created a data warehouse so as to integrate all kinds of cost related information for cost estimation. Meanwhile, because of the complexity of CM, errors in data collection, processing and application are inevitable in most cases. High quality data is the prerequisite for the success of big data projects. Problems such as null value, misleading value, outlier value and non-standard value make the application of ANN extremely challenging. Although previous scholars did consider this problem and have completed data standardization [55], they did not conduct comprehensive integration and structured data quality analysis. Future research could focus on efficient and standardized data collection from massive historical engineering files through information technology, and strengthen the development and application of databases to realize sustainable data updating and data mining utilization.

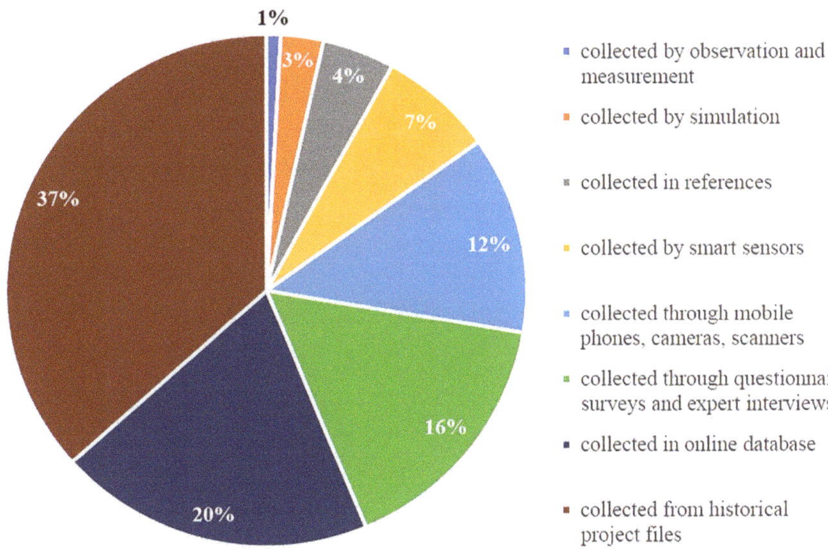

Figure 8. Methods of data collection of ANN application in CM.

4.3. Limitations of This Paper

Although some interesting findings have been made, there are still some shortcomings in this article that will be discussed in this section. Firstly, only a scientometric analysis is carried out in this paper. In order to get more detailed information on current research on ANN in CM, a content analysis can be carried out in the future. Furthermore, since the results and discussions are based on the findings of previous studies, the resulting theoretical framework should be validated and tested in future empirical studies. Moreover, although the comprehensiveness of the selected literature has been ensured to the extent possible, more searches of other database and additional keywords can be added in the future researches.

5. Conclusions

As an adaptive, model-free data mining method, ANN is one of the most promising data processing techniques in AEC, and has been increasingly used in CM. However, in the past 20 years, few papers have attempted to provide a comprehensive review of the existing literature on ANNs in CM. This study has analyzed a selection of 112 articles published in 7 high-quality journals between 2000 and 2020, and has conducted a comprehensive and structured review of ANN in CM. Through scientometric analysis, the review visualized the authors and countries/regions, main research interests and trends, providing a basis for further understanding of the application of ANN in CM. Challenges and future directions were put forward to provide references for future research. At present, there is still a lack of systematic research and sufficient attention to the application of ANN in CM. ANN applications still face many challenges such as data collection, cleaning and storage, the collaboration of different stakeholders, researchers and countries/regions, as well as the systematic design for the platform. More research is still needed in these fields so as to truly achieve intelligent CM based on ANN. The uniqueness of this paper is that it limited the research subject to research on ANN in CM rather than a broader field which is very important to clarify the current research status in the field of construction management. Despite all the contributions, this review has some limitations. Future directions should focus more on content analysis, example validation, and increased retrieval of databases and keywords.

Author Contributions: Methodology, H.X. and S.L.; software, H.X. and H.L.; validation, R.C. and M.P.; formal analysis, H.X.; writing—original draft preparation, H.X; writing—review and editing, R.C.; visualization, H.L. and S.L.; supervision, M.P. and R.J.W.; project administration, N.D. and J.Z. All authors have read and agreed to the published version of the manuscript.

Funding: This research received no external funding.

Institutional Review Board Statement: Not applicable.

Informed Consent Statement: Not applicable.

Data Availability Statement: Some or all data, models, or code that support the findings of this study are available from the corresponding author upon reasonable request.

Conflicts of Interest: The authors declare no conflict of interest.

References

1. Busta, H. KPMG Report: Construction Industry Slow to Adopt New Technology. 2016. Available online: https://www.constructiondive.com/news/kpmg-report-construction-industry-slow-to-adopt-new-technology/426268/ (accessed on 14 May 2022).
2. Yifan, T. Application of BIM Technology in Housing Construction Engineering Management. In Proceedings of the 2021 International Conference on Management Science and Software Engineering (ICMSSE), Chengdu, China, 9–11 July 2021.
3. Gao, Q.; Shi, R.; Wang, G. Construction of Intelligent Manufacturing Workshop Based on Lean Management. In Proceedings of the 9th International Conference on Digital Enterprise Technology (DET)—Intelligent Manufacturing in the Knowledge Economy Era, Nanjing, China, 29–31 March 2016; pp. 599–603.
4. Yang, X.; Zhao, Q. Influence of virtual reality and 3D printing on architectural innovation evaluation based on quality of experience evaluation using fuzzy logic. *J. Intell. Fuzzy Syst.* **2021**, *40*, 8501–8509.
5. Zhao, X. Exploration of Transformation and Upgrading Path of Construction Industry under the Background of Digital Information. In Proceedings of the International Conference on Construction and Real Estate Management (ICCREM), Beijing, China, 16–17 October 2021; pp. 501–507.
6. You, Z.; Wu, C. A framework for data-driven informatization of the construction company. *Adv. Eng. Inform.* **2019**, *39*, 269–277. [CrossRef]
7. Yu, T.; Liang, X.; Wang, Y. Factors Affecting the Utilization of Big Data in Construction Projects. *J. Constr. Eng. Manag.* **2020**, *146*, 04020032. [CrossRef]
8. Deng, H.; Hong, H.; Luo, D.H.; Deng, Y.C.; Su, C. Automatic Indoor Construction Process Monitoring for Tiles Based on BIM and Computer Vision. *J. Constr. Eng. Manag.* **2020**, *146*, 04019095. [CrossRef]
9. Manyika, S.L.J.; Bughin, J.; Woetzel, J.; Stamenov, K.; Dhingra, D. *Digital Globalization: The New Era of Global Flows*; DhingraMcKinsey & Company: New York, NY, USA, 2016.
10. Mostafa, K.; Hegazy, T. Review of image-based analysis and applications in construction. *Automat. Constr.* **2021**, *122*, 103516. [CrossRef]
11. Hang, Y.A.; Nan, Y.B.; Yi, P.C.; Yr, D. Data mining in the construction industry: Present status, opportunities, and future trends. *Autom. Constr.* **2020**, *119*, 103331.
12. Darko, A.; Chan, A.P.C.; Adabre, M.A.; Edwards, D.J.; Hosseini, M.R.; Ameyaw, E.E. Ameyaw. Artificial intelligence in the AEC industry: Scientometric analysis and visualization of research activities. *Autom. Constr.* **2020**, *112*, 103081. [CrossRef]
13. Lesniak, A.; Juszczyk, M. Prediction of site overhead costs with the use of artificial neural network based model. *Arch. Civ. Mech. Eng.* **2018**, *18*, 973–982. [CrossRef]
14. Kim, G.H.; Yoon, H.E.; An, S.H.; Cho, H.H.; Kang, K.I. Neural network model incorporating a genetic algorithm in estimating construction costs. *Build. Environ.* **2004**, *39*, 1333–1340. [CrossRef]

15. Patel, D.A.; Jha, K.N. Neural Network Model for the Prediction of Safe Work Behavior in Construction Projects. *J. Constr. Eng. Manag.* **2015**, *141*, 04014066. [CrossRef]
16. Wang, Y.-R.; Gibson, G.E., Jr. A study of preproject planning and project success using ANNs and regression models. *Autom. Constr.* **2010**, *19*, 341–346. [CrossRef]
17. Chao, L.C.; Skibniewski, M.J. Neural-Network Method of Estimating Construction Technology Acceptability. *J. Constr. Eng. Manag.-ASCE* **1995**, *121*, 130–142. [CrossRef]
18. Amiruddin, A.A.A.M.; Zabiri, H.; Taqvi, S.A.; Tufa, L.D. Neural network applications in fault diagnosis and detection: An overview of implementations in engineering-related systems. *Neural Comput. Appl.* **2020**, *32*, 495–518. [CrossRef]
19. Moayedi, H.; Mosallanezhad, M.; Rashid, A.S.A.; Jusoh, W.A.W.; Muazu, M.A. A systematic review and meta-analysis of artificial neural network application in geotechnical engineering: Theory and applications. *Neural Comput. Appl.* **2020**, *322*, 495–518. [CrossRef]
20. Kumar, R.; Aggarwal, R.K.; Sharma, J.D. Energy analysis of a building using artificial neural network: A review. *Energy Build.* **2013**, *65*, 352–358. [CrossRef]
21. Sony, S.; Dunphy, K.; Sadhu, A.; Capretz, M. A systematic review of convolutional neural network-based structural condition assessment techniques. *Eng. Struct.* **2021**, *226*, 111347. [CrossRef]
22. Pan, Y.; Zhang, L.M. Roles of artificial intelligence in construction engineering and management: A critical review and future trends. *Automat. Constr.* **2021**, *122*, 103517. [CrossRef]
23. Adeli, H. Neural networks in civil engineering: 1989–2000. *Comput-Aided Civ. Inf.* **2001**, *16*, 126–142. [CrossRef]
24. Hu, X.; Xia, B.; Skitmore, M.; Chen, Q. The application of case-based reasoning in construction management research: An overview. *Autom. Constr.* **2016**, *72*, 65–74. [CrossRef]
25. Hosseini, M.R.; Martek, I.; Zavadskas, E.K.; Aibinu, A.A.; Arashpour, M.; Chileshe, N. Critical evaluation of off-site construction research: A Scientometric analysis. *Autom. Constr.* **2018**, *87*, 235–247. [CrossRef]
26. Neto, D.d.C.e.S.; Cruz, C.O.; Rodrigues, F.; Silva, P. Bibliometric Analysis of PPP and PFI Literature: Overview of 25 Years of Research. *J. Constr. Eng. Manag.* **2016**, *142*, 10.
27. Siraj, N.B.; Fayek, A.R. Risk Identification and Common Risks in Construction: Literature Review and Content Analysis. *J. Constr. Eng. Manag.* **2019**, *145*, 9. [CrossRef]
28. Awad, A.; Fayek, A.R. Adaptive Learning of Contractor Default Prediction Model for Surety Bonding. *J. Constr. Eng. Manag.* **2013**, *139*, 694–704. [CrossRef]
29. Ayhan, B.U.; Tokdemir, O.B. Accident Analysis for Construction Safety Using Latent Class Clustering and Artificial Neural Networks. *J. Constr. Eng. Manag.* **2020**, *146*, 04019114. [CrossRef]
30. Bayram, S.; Al-Jibouri, S. Efficacy of Estimation Methods in Forecasting Building Projects' Costs. *J. Constr. Eng. Manag.* **2016**, *142*, 05016012. [CrossRef]
31. Cao, M.T.; Cheng, M.Y.; Wu, Y.W. Hybrid Computational Model for Forecasting Taiwan Construction Cost Index. *J. Constr. Eng. Manag.* **2015**, *141*, 04014089. [CrossRef]
32. Chao, L.C.; Chien, C.F. Estimating Project S-Curves Using Polynomial Function and Neural Networks. *J. Constr. Eng. Manag.* **2009**, *135*, 169–177. [CrossRef]
33. Chaovalitwongse, W.A.; Wang, W.; Williams, T.P.; Chaovalitwongse, P. Data Mining Framework to Optimize the Bid Selection Policy for Competitively Bid Highway Construction Projects. *J. Constr. Eng. Manag.* **2012**, *138*, 277–286. [CrossRef]
34. Cheng, M.Y.; Chang, Y.H.; Korir, D. Novel Approach to Estimating Schedule to Completion in Construction Projects Using Sequence and Nonsequence Learning. *J. Constr. Eng. Manag.* **2019**, *145*, 04019072. [CrossRef]
35. Cheng, M.Y.; Ko, C.H. Object-oriented evolutionary fuzzy neural inference system for construction management. *J. Constr. Eng. Manag.* **2003**, *129*, 461–469. [CrossRef]
36. Cirilovic, J.; Vajdic, N.; Mladenovic, G.; Queiroz, C. Developing Cost Estimation Models for Road Rehabilitation and Reconstruction: Case Study of Projects in Europe and Central Asia. *J. Constr. Eng. Manag.* **2014**, *140*, 04013065. [CrossRef]
37. Delgado, J.M.D.; Oyedele, L.; Bilal, M.; Ajayi, A.; Akanbi, L.; Akinade, O. Big Data Analytics System for Costing Power Transmission Projects. *J. Constr. Eng. Manag.* **2020**, *146*, 05019017. [CrossRef]
38. Dikmen, I.; Birgonul, M.T.; Kiziltas, S. Prediction of organizational effectiveness in construction companies. *J. Constr. Eng. Manag.* **2005**, *131*, 252–261. [CrossRef]
39. Dursun, O.; Stoy, C. Conceptual Estimation of Construction Costs Using the Multistep Ahead Approach. *J. Constr. Eng. Manag.* **2016**, *142*, 04016038. [CrossRef]
40. El-Gohary, K.M.; Aziz, R.F.; Abdel-Khalek, H.A. Engineering Approach Using ANN to Improve and Predict Construction Labor Productivity under Different Influences. *J. Constr. Eng. Manag.* **2017**, *143*, 04017045. [CrossRef]
41. Elazouni, A.M. Classifying construction contractors using unsupervised-learning neural networks. *J. Constr. Eng. Manag.* **2006**, *132*, 1242–1253. [CrossRef]
42. ElMousalami, H.H.; Elyamany, A.H.; Ibrahim, A.H. Predicting Conceptual Cost for Field Canal Improvement Projects. *J. Constr. Eng. Manag.* **2018**, *144*, 04018102. [CrossRef]
43. Goh, Y.M.; Sa'adon, N.F.B. Cognitive Factors Influencing Safety Behavior at Height: A Multimethod Exploratory Study. *J. Constr. Eng. Manag.* **2015**, *141*, 04015003. [CrossRef]

44. Heravi, G.; Eslamdoost, E. Applying Artificial Neural Networks for Measuring and Predicting Construction-Labor Productivity. *J. Constr. Eng. Manag.* **2015**, *141*, 04015032. [CrossRef]
45. Hu, Q.J.; Bai, Y.; He, L.P.; Cai, Q.J.; Tang, S.; Ma, G.L.; Tan, J.; Liang, B.W. Intelligent Framework for Worker-Machine Safety Assessment. *J. Constr. Eng. Manag.* **2020**, *146*, 04020045. [CrossRef]
46. Jin, X.H. Model for Efficient Risk Allocation in Privately Financed Public Infrastructure Projects Using Neuro-Fuzzy Techniques. *J. Constr. Eng. Manag.* **2011**, *137*, 1003–1014. [CrossRef]
47. Kale, S.; Karaman, E.A. Evaluating the Knowledge Management Practices of Construction Firms by Using Importance-Comparative Performance Analysis Maps. *J. Constr. Eng. Manag.* **2011**, *137*, 1142–1152. [CrossRef]
48. Kim, B.; Lee, H.; Park, H.; Kim, H. Framework for Estimating Greenhouse Gas Emissions Due to Asphalt Pavement Construction. *J. Constr. Eng. Manag.* **2012**, *138*, 1312–1321. [CrossRef]
49. Ko, C.H.; Cheng, M.Y. Dynamic prediction of project success using artificial intelligence. *J. Constr. Eng. Manag.* **2007**, *133*, 316–324. [CrossRef]
50. Lim, T.K.; Park, S.M.; Lee, H.C.; Lee, D.E. Artificial Neural Network-Based Slip-Trip Classifier Using Smart Sensor for Construction Workplace. *J. Constr. Eng. Manag.* **2016**, *142*, 04015065. [CrossRef]
51. Liu, M.; Ling, Y.Y. Modeling a contractor's markup estimation. *J. Constr. Eng. Manag.* **2005**, *131*, 391–399. [CrossRef]
52. Liu, Y.C.; Yeh, I.C. Building Valuation Model of Enterprise Values for Construction Enterprise with Quantile Neural Networks. *J. Constr. Eng. Manag.* **2016**, *142*, 04015075. [CrossRef]
53. Marzouk, M.; Amin, A. Predicting Construction Materials Prices Using Fuzzy Logic and Neural Networks. *J. Constr. Eng. Manag.* **2013**, *139*, 1190–1198. [CrossRef]
54. Pereira, E.; Ali, M.; Wu, L.; Abourizk, S. Distributed Simulation-Based Analytics Approach for Enhancing Safety Management Systems in Industrial Construction. *J. Constr. Eng. Manag.* **2020**, *146*, 04019091. [CrossRef]
55. Petroutsatou, K.; Georgopoulos, E.; Lambropoulos, S.; Pantouvakis, J.P. Early Cost Estimating of Road Tunnel Construction Using Neural Networks. *J. Constr. Eng. Manag.* **2012**, *138*, 679–687. [CrossRef]
56. Rafiei, M.H.; Adeli, H. Novel Machine-Learning Model for Estimating Construction Costs Considering Economic Variables and Indexes. *J. Constr. Eng. Manag.* **2018**, *144*, 04018106. [CrossRef]
57. Said, H.M.; Kandimalla, P. Performance Measurement of Building Sheet-Metal Ductwork Prefabrication under Batch Production Settings. *J. Constr. Eng. Manag.* **2018**, *144*, 04017107. [CrossRef]
58. Shiha, A.; Dorra, E.M.; Nassar, K. Neural Networks Model for Prediction of Construction Material Prices in Egypt Using Macroeconomic Indicators. *J. Constr. Eng. Manag.* **2020**, *146*, 04020010. [CrossRef]
59. Wang, D.; Arditi, D.; Damci, A. Construction Project Managers' Motivators and Human Values. *J. Constr. Eng. Manag.* **2017**, *143*, 04016115. [CrossRef]
60. Wilmot, C.G.; Mei, B. Neural network modeling of highway construction costs. *J. Constr. Eng. Manag.* **2005**, *131*, 765–771. [CrossRef]
61. Zhang, M.Y.; Cao, T.Z.; Zhao, X.F. Using Smartphones to Detect and Identify Construction Workers' Near-Miss Falls Based on ANN. *J. Constr. Eng. Manag.* **2019**, *145*, 04018120. [CrossRef]
62. Zhang, M.Y.; Cao, Z.Y.; Yang, Z.; Zhao, X.F. Utilizing Computer Vision and Fuzzy Inference to Evaluate Level of Collision Safety for Workers and Equipment in a Dynamic Environment. *J. Constr. Eng. Manag.* **2020**, *146*, 04020051. [CrossRef]
63. AbouRizk, S.; Knowles, P.; Hermann, U.R. Estimating labor production rates for industrial construction activities. *J. Constr. Eng. Manag.-ASCE* **2001**, *127*, 502–511. [CrossRef]
64. Lhee, S.C.; Issa, R.R.A.; Flood, I. Prediction of Financial Contingency for Asphalt Resurfacing Projects using Artificial Neural Networks. *J. Constr. Eng. Manag.-ASCE* **2012**, *138*, 22–30. [CrossRef]
65. Antwi-Afari, M.F.; Li, H.; Yu, Y.T.; Kong, L.L. Wearable insole pressure system for automated detection and classification of awkward working postures in construction workers. *Autom. Constr.* **2018**, *96*, 433–441. [CrossRef]
66. Bang, S.; Baek, F.; Park, S.; Kim, W.; Kim, H. Image augmentation to improve construction resource detection using generative adversarial networks, cut-and-paste, and image transformation techniques. *Autom. Constr.* **2020**, *115*, 103198. [CrossRef]
67. Benjaoran, V.; Dawood, N. Intelligence approach to production planning system for bespoke precast concrete products. *Autom. Constr.* **2006**, *15*, 737–745. [CrossRef]
68. Callow, D.; Lee, J.; Blumenstein, M.; Guan, H.; Loo, Y.C. Development of hybrid optimisation method for Artificial Intelligence based bridge deterioration model—Feasibility study. *Autom. Constr.* **2013**, *31*, 83–91. [CrossRef]
69. Chao, L.C.; Chien, C.F. A Model for Updating Project S-curve by Using Neural Networks and Matching Progress. *Autom. Constr.* **2010**, *19*, 84–91. [CrossRef]
70. Chen, J.H.; Hsu, S.C. Hybrid ANN-CBR model for disputed change orders in construction projects. *Autom. Constr.* **2007**, *17*, 56–64. [CrossRef]
71. Cheng, J.C.P.; Chen, W.W.; Chen, K.Y.; Wang, Q. Data-driven predictive maintenance planning framework for MEP components based on BIM and IoT using machine learning algorithms. *Autom. Constr.* **2020**, *112*, 103087. [CrossRef]
72. Cheng, J.C.P.; Wang, M.Z. Automated detection of sewer pipe defects in closed-circuit television images using deep learning techniques. *Autom. Constr.* **2018**, *95*, 155–171. [CrossRef]
73. Cheng, M.Y.; Tsai, H.C.; Hsieh, W.S. Web-based conceptual cost estimates for construction projects using Evolutionary Fuzzy Neural Inference Model. *Autom. Constr.* **2009**, *18*, 164–172. [CrossRef]

74. Cheng, M.Y.; Tsai, H.C.; Liu, C.L. Artificial intelligence approaches to achieve strategic control over project cash flows. *Autom. Constr.* **2009**, *18*, 386–393. [CrossRef]
75. Cheng, M.Y.; Tsai, H.C.; Sudjono, E. Evolutionary fuzzy hybrid neural network for dynamic project success assessment in construction industry. *Autom. Constr.* **2012**, *21*, 46–51. [CrossRef]
76. Chou, J.S.; Lin, C.W.; Pham, A.D.; Shao, J.Y. Optimized artificial intelligence models for predicting project award price. *Autom. Constr.* **2015**, *54*, 106–115. [CrossRef]
77. Fang, Q.; Li, H.; Luo, X.C.; Ding, L.Y.; Luo, H.B.; Li, C.Q. Computer vision aided inspection on falling prevention measures for steeplejacks in an aerial environment. *Autom. Constr.* **2018**, *93*, 148–164. [CrossRef]
78. Hola, B.; Schabowicz, K. Estimation of earthworks execution time cost by means of artificial neural networks. *Autom. Constr.* **2010**, *19*, 570–579. [CrossRef]
79. Hong, Y.; Hammad, A.W.A.; Akbarnezhad, A.; Arashpour, M. A neural network approach to predicting the net costs associated with BIM adoption. *Autom. Constr.* **2020**, *119*, 103306. [CrossRef]
80. Kolar, Z.; Chen, H.N.; Luo, X.W. Transfer learning and deep convolutional neural networks for safety guardrail detection in 2D images. *Autom. Constr.* **2018**, *89*, 58–70. [CrossRef]
81. Lee, J.; Sanmugarasa, K.; Blumenstein, M.; Loo, Y.C. Improving the reliability of a Bridge Management System (BMS) using an ANN-based Backward Prediction Model (BPM). *Autom. Constr.* **2008**, *17*, 758–772. [CrossRef]
82. Leu, S.S. Data mining for tunnel support stability: Neural network approach. *Autom. Constr.* **2001**, *10*, 429–441. [CrossRef]
83. Lu, Q.C.; Lee, S.; Chen, L. Image-driven fuzzy-based system to construct as-is IFC BIM objects. *Autom. Constr.* **2018**, *92*, 68–87. [CrossRef]
84. McGlinn, K.; Yuce, B.; Wicaksono, H.; Howell, S.; Rezgui, Y. Usability evaluation of a web-based tool for supporting holistic building energy management. *Autom. Constr.* **2017**, *84*, 154–165. [CrossRef]
85. Meijer, D.; Scholten, L.; Clemens, F.; Knobbe, A. A defect classification methodology for sewer image sets with convolutional neural networks. *Autom. Constr.* **2019**, *104*, 281–298. [CrossRef]
86. Mirahadi, F.; Zayed, T. Simulation-based construction productivity forecast using Neural-Network-Driven Fuzzy Reasoning. *Autom. Constr.* **2016**, *65*, 102–115. [CrossRef]
87. Slaton, T.; Hernandez, C.; Akhavian, R. Construction activity recognition with convolutional recurrent networks. *Autom. Constr.* **2020**, *113*, 103138. [CrossRef]
88. Sousa, V.; Matos, J.P.; Matias, N. Evaluation of artificial intelligence tool performance and uncertainty for predicting sewer structural condition. *Autom. Constr.* **2014**, *44*, 84–91. [CrossRef]
89. Turkan, Y.; Hong, J.; Laflamme, S.; Puri, N. Adaptive wavelet neural network for terrestrial laser scanner-based crack detection. *Autom. Constr.* **2018**, *94*, 191–202. [CrossRef]
90. Wang, C.H.; Kao, C.H.; Lee, W.H. A new interactive model for improving the learning performance of back propagation neural network. *Autom. Constr.* **2007**, *16*, 745–758. [CrossRef]
91. Wang, N.N.; Zhao, X.F.; Zhao, P.; Zhang, Y.; Zou, Z.; Ou, J.P. Automatic damage detection of historic masonry buildings based on mobile deep learning. *Autom. Constr.* **2019**, *103*, 53–66. [CrossRef]
92. Wang, W.; Chen, J.Y.; Hong, T.Z. Occupancy prediction through machine learning and data fusion of environmental sensing and Wi-Fi sensing in buildings. *Autom. Constr.* **2018**, *94*, 233–243. [CrossRef]
93. Wei, Y.J.; Akinci, B. A vision and learning-based indoor localization and semantic mapping framework for facility operations and management. *Autom. Constr.* **2019**, *107*, 102915. [CrossRef]
94. Yi, W.; Chan, A.P.C.; Wang, X.Y.; Wang, J. Development of an early-warning system for site work in hot and humid environments: A case study. *Autom. Constr.* **2016**, *62*, 101–113. [CrossRef]
95. Yu, W.D.; Lin, H.W. A VaFALCON neuro-fuzzy system for mining of incomplete construction databases. *Autom. Constr.* **2006**, *15*, 20–32. [CrossRef]
96. Yip, H.L.; Fan, H.Q.; Chiang, Y.H. Predicting the maintenance cost of construction equipment: Comparison between general regression neural network and Box-Jenkins time series models. *Autom. Constr.* **2014**, *38*, 30–38. [CrossRef]
97. Yu, W.D.; Liu, Y.C. Hybridization of CBR and numeric soft computing techniques for mining of scarce construction databases. *Autom. Constr.* **2006**, *15*, 33–46. [CrossRef]
98. Zhou, C.; Xu, H.C.; Ding, L.Y.; Wei, L.C.; Zhou, Y. Dynamic prediction for attitude and position in shield tunneling: A deep learning method. *Autom. Constr.* **2019**, *105*, 102840. [CrossRef]
99. Baalousha, Y.; Celik, T. Integrated Web-Based Data Warehouse and Artificial Neural Networks System for Unit Price Analysis with Inflation Adjustment. *J. Civ. Eng. Manag.* **2011**, *17*, 157–167. [CrossRef]
100. Bayram, S.; Ocal, M.E.; Laptali Oral, E.; Atis, C.D. Comparison Of Multi Layer Perceptron (MLP) and Radial Basis Function (RBF) FOR Construction Cost Estimation: The Case OF Turkey. *J. Civ. Eng. Manag.* **2016**, *22*, 480–490. [CrossRef]
101. Cheng, M.Y.; Su, C.W.; Tsai, M.H.; Lin, K.S. Data Preprocessing for Artificial Neural Network Applications in Prioritizing Railroad Projects—A Practical Experience in Taiwan. *J. Civ. Eng. Manag.* **2012**, *18*, 483–494. [CrossRef]
102. Chou, J.S.; Tsai, C.F.; Lu, Y.H. Project Dispute Prediction by Hybrid Machine Learning Techniques. *J. Civ. Eng. Manag.* **2013**, *19*, 505–517. [CrossRef]
103. Gerek, I.H.; Erdis, E.; Mistikoglu, G.; Usmen, M. Modelling masonry crew productivity using two artificial neural network techniques. *J. Civ. Eng. Manag.* **2015**, *21*, 231–238. [CrossRef]

104. Han, S.; Hong, T.; Kim, G.; Lee, S. Technical Comparisons of Simulation-Based Productivity Prediction Methodologies by Means of Estimation Tools Focusing on Conventional Earthmovings. *J. Civ. Eng. Manag.* **2011**, *17*, 265–277. [CrossRef]
105. Juan, Y.K.; Lai, W.Y.; Shih, S.G. Building Information Modeling Acceptance and Readiness Assessment in Taiwanese Architectural Firms. *J. Civ. Eng. Manag.* **2017**, *23*, 356–367. [CrossRef]
106. Juszczyk, M.; Zima, K.; Lelek, W. Forecasting of Sports Fields Construction Costs Aided by Ensembles of Neural Networks. *J. Civ. Eng. Manag.* **2019**, *25*, 715–729. [CrossRef]
107. Shahrara, N.; Celik, T.; Gandomi, A.H. Risk Analysis of Bot Contracts Using Soft Computing. *J. Civ. Eng. Manag.* **2017**, *23*, 232–240. [CrossRef]
108. Sonmez, R.; Ontepeli, B. Predesign Cost Estimation of Urban Railway Projects with Parametric Modeling. *J. Civ. Eng. Manag.* **2009**, *15*, 405–409. [CrossRef]
109. Wang, E.D.; Shen, Z.G. Lifecycle Energy Consumption Prediction of Residential Buildings by Incorporating Longitudinal Uncertainties. *J. Civ. Eng. Manag.* **2013**, *19*, S161–S171. [CrossRef]
110. Wang, W.C.; Bilozerov, T.; Dzeng, R.J.; Hsiao, F.Y.; Wang, K.C. Conceptual Cost Estimations Using Neuro-Fuzzy and Multi-Factor Evaluation Methods for Building Projects. *J. Civ. Eng. Manag.* **2017**, *23*, 1–14. [CrossRef]
111. Yousefi, V.; Yakhchali, S.H.; Khanzadi, M.; Mehrabanfar, E.; Saparauskas, J. Proposing a Neural Network Model to Predict Time and Cost Claims In Construction Projects. *J. Civ. Eng. Manag.* **2016**, *22*, 967–978. [CrossRef]
112. Hassim, S.; Muniandy, R.; Alias, A.H.; Abdullah, P. Construction tender price estimation standardization (TPES) in Malaysia: Modeling using fuzzy neural network. *Eng. Constr. Archit. Manag.* **2018**, *25*, 443–457. [CrossRef]
113. Bai, L.B.; Wang, Z.G.; Wang, H.L.; Huang, N.; Shi, H.J. Prediction of multiproject resource conflict risk via an artificial neural network. *Eng. Constr. Archit. Manag.* **2021**, *28*, 2857–2883. [CrossRef]
114. Meng, J.N.; Yan, J.H.; Xue, B.; Fu, J.; He, N. Reducing construction material cost by optimizing buy-in decision that accounts the flexibility of non-critical activities. *Eng. Constr. Archit. Manag.* **2018**, *25*, 1092–1108. [CrossRef]
115. Mensah, I.; Adjei-Kumi, T.; Nani, G. Duration determination for rural roads using the principal component analysis and artificial neural network. *Eng. Constr. Archit. Manag.* **2016**, *23*, 638–656. [CrossRef]
116. Nasirzadeh, F.; Kabir, H.M.D.; Akbari, M.; Khosravi, A.; Nahavandi, S.; Carmichael, D.G. ANN-based prediction intervals to forecast labour productivity. *Eng. Constr. Archit. Manag.* **2020**, *27*, 2335–2351. [CrossRef]
117. Utama, W.P.; Chan, A.P.C.; Zahoor, H.; Gao, R.; Jumas, D.Y. Making decision toward overseas construction projects An application based on adaptive neuro fuzzy system. *Eng. Constr. Archit. Manag.* **2019**, *26*, 285–302. [CrossRef]
118. Wu, H.Q.; Shen, G.; Lin, X.; Li, M.L.; Zhang, B.Y.; Li, C.Z.D. Screening patents of ICT in construction using deep learning and NLP techniques. *Eng. Constr. Archit. Manag.* **2020**, *27*, 1891–1912. [CrossRef]
119. Yuan, F.N.; Tang, M.H.; Hong, J.K. Efficiency estimation and reduction potential of the Chinese construction industry via SE-DEA and artificial neural network. *Eng. Constr. Archit. Manag.* **2020**, *27*, 1533–1552. [CrossRef]
120. Abdelaty, A.; Shrestha, K.J.; Jeong, H.D. Estimating Preconstruction Services for Bridge Design Projects. *J. Manag. Eng.* **2020**, *36*, 04020034. [CrossRef]
121. Bai, Y.; Huan, J.; Kim, S. Measuring Bridge Construction Efficiency Using the Wireless Real-Time Video Monitoring System. *J. Manag. Eng.* **2012**, *28*, 120–126. [CrossRef]
122. Cheung, S.O.; Tam, C.M.; Harris, F.C. Project dispute resolution satisfaction classification through neural network. *J. Manag. Eng.* **2000**, *16*, 70–79. [CrossRef]
123. Han, S.; Ko, Y.; Kim, J.; Hong, T. Housing Market Trend Forecasts through Statistical Comparisons based on Big Data Analytic Methods. *J. Manag. Eng.* **2018**, *34*, 04017054. [CrossRef]
124. Hyari, K.H.; Al-Daraiseh, A.; El-Mashaleh, M. Conceptual Cost Estimation Model for Engineering Services in Public Construction Projects. *J. Manag. Eng.* **2016**, *32*, 04015021. [CrossRef]
125. Jang, Y.J.; Jeong, I.; Cho, Y.K. Business Failure Prediction of Construction Contractors Using a LSTM RNN with Accounting, Construction Market, and Macroeconomic Variables. *J. Manag. Eng.* **2020**, *36*, 04019039. [CrossRef]
126. Lam, K.C.; Oshodi, O.S. Using Univariate Models for Construction Output Forecasting: Comparing Artificial Intelligence and Econometric Techniques. *J. Manag. Eng.* **2016**, *32*, 04016021. [CrossRef]
127. Patel, D.A.; Jha, K.N. Neural Network Approach for Safety Climate Prediction. *J. Manag. Eng.* **2015**, *31*, 05014027. [CrossRef]
128. Chaphalkar, N.B.; Iyer, K.C.; Patil, S.K. Prediction of outcome of construction dispute claims using multilayer perceptron neural network model. *Int. J. Proj. Manag.* **2015**, *33*, 1827–1835. [CrossRef]
129. Cheng, M.Y.; Tsai, H.C.; Sudjono, E. Evaluating subcontractor performance using evolutionary fuzzy hybrid neural network. *Int. J. Proj. Manag.* **2011**, *29*, 349–356. [CrossRef]
130. Costantino, F.; Di Gravio, G.; Nonino, F. Project selection in project portfolio management: An artificial neural network model based on critical success factors. *Int. J. Proj. Manag.* **2015**, *33*, 1744–1754. [CrossRef]
131. Jin, X.H.; Zhang, G.M. Modelling optimal risk allocation in PPP projects using artificial neural networks. *Int. J. Proj. Manag.* **2011**, *29*, 591–603. [CrossRef]
132. Pal, R.; Wang, P.; Liang, X.P. The critical factors in managing relationships in international engineering, procurement, and construction (IEPC) projects of Chinese organizations. *Int. J. Proj. Manag.* **2017**, *35*, 1225–1237. [CrossRef]
133. Wang, Y.R.; Yu, C.Y.; Chan, H.H. Predicting construction cost and schedule success using artificial neural networks ensemble and support vector machines classification models. *Int. J. Proj. Manag.* **2012**, *30*, 470–478. [CrossRef]

134. Jin, X.H. Neurofuzzy Decision Support System for Efficient Risk Allocation in Public-Private Partnership Infrastructure Projects. *J. Comput. Civ. Eng.* **2010**, *24*, 525–538. [CrossRef]
135. Tam, C.M.; Leung, A.W.T.; Liu, D.K. Nonlinear models for predicting hoisting times of tower cranes. *J. Comput. Civ. Eng.* **2002**, *16*, 76–81. [CrossRef]
136. Tinoco, J.; Correia, A.G.; Cortez, P.; Toll, D.G. Data-Driven Model for Stability Condition Prediction of Soil Embankments Based on Visual Data Features. *J. Comput. Civ. Eng.* **2018**, *32*, 04018027. [CrossRef]
137. Zhang, M.Y.; Zhu, M.; Zhao, X.F. Recognition of High-Risk Scenarios in Building Construction Based on Image Semantics. *J. Comput. Civ. Eng.* **2020**, *34*, 04020019. [CrossRef]
138. Zhong, B.; Wu, H.; Li, H.; Sepasgozar, S.M.E.; Luo, H.; He, L. A scientometric analysis and critical review of construction related ontology research. *Autom. Constr.* **2019**, *101*, 17–31.
139. Ke, Y.; Wang, S.; Chan, A.P.C.; Cheung, E. Research trend of public-private partnership in construction journals. *J. Constr. Eng. Manage.-ASCE* **2009**, *135*, 1076–1086. [CrossRef]
140. Su, H.; Lee, P. Mapping knowledge structure by keyword co-occurrence: A first look at journal papers in Technology Foresight. *Scientometrics* **2010**, *85*, 65–79. [CrossRef]
141. Liu, M.Q.; Le, Y.; Hu, Y.; Xia, B.; Skitmore, M.; Gao, X.Y. System Dynamics Modeling for Construction Management Research: Critical Review and Future Trends. *J. Civ. Eng. Manag.* **2019**, *25*, 730–741. [CrossRef]
142. Araujo, A.G.; Carneiro, A.M.P.; Palha, R.P. Sustainable construction management: A systematic review of the literature with meta-analysis. *J. Clean. Prod.* **2020**, *256*, 120350. [CrossRef]
143. Cobo, M.J.; Lopez-Herrera, A.G.; Herrera-Viedma, E.; Herrera, F. Science Mapping Software Tools: Review, Analysis, and Cooperative Study among Tools. *J. Am. Soc. Inf. Sci. Technol.* **2011**, *62*, 1382–1402. [CrossRef]
144. Martinez, P.; Al-Hussein, M.; Ahmad, R. A scientometric analysis and critical review of computer vision applications for construction. *Automat. Constr.* **2019**, *107*, 102947. [CrossRef]
145. Hu, W.; Li, C.-h.; Ye, C.; Wang, J.; Wei, W.-w.; Deng, Y. Research progress on ecological models in the field of water eutrophication: CiteSpace analysis based on data from the ISI web of science database. *Ecol. Model.* **2019**, *410*, 108779. [CrossRef]
146. Freeman, L.C. A Set of Measures of Centrality Based on Betweenness. *Sociometry* **1977**, *40*, 35–41. [CrossRef]
147. Shrivastava, R.; Mahajan, P. Artificial Intelligence Research in India: A Scientometric Analysis. *Sci. Technol. Libr.* **2016**, *35*, 136–151. [CrossRef]
148. He, Q.; Wang, G.; Luo, L.; Shi, Q.; Xie, J.; Meng, X. Mapping the managerial areas of Building Information Modeling (BIM) using scientometric analysis. *Int. J. Proj. Manag.* **2017**, *35*, 670–685. [CrossRef]
149. Jayantha, W.M.; Oladinrin, O.T. Bibliometric analysis of hedonic price model using CiteSpace. *Int J Hous Mark Anal.* **2019**, *13*, 357–371. [CrossRef]
150. Ben Chaabene, W.; Flah, M.; Nehdi, M.L. Machine learning prediction of mechanical properties of concrete: Critical review. *Constr. Build. Mater.* **2020**, *260*, 119889. [CrossRef]
151. Jin, R.Y.; Zuo, J.; Hong, J.K. Scientometric Review of Articles Published in ASCE's Journal of Construction Engineering and Management from 2000 to 2018. *J. Constr. Eng. M.* **2019**, *145*, 8. [CrossRef]
152. Ekanayake, E.M.A.C.; Shen, G.Q.; Kumaraswamy, M.M. Mapping the knowledge domains of value management: A bibliometric approach. *Eng. Constr. Archit. Manag.* **2019**, *26*, 499–514. [CrossRef]
153. Li, Q.; Long, R.; Chen, H.; Chen, F.; Wang, J. Visualized analysis of global green buildings: Development, barriers and future directions. *J. Clean. Prod.* **2020**, *245*, 118775. [CrossRef]
154. Chen, C. CiteSpace II: Detecting and visualizing emerging trends and transient patterns in scientific literature. *J. Assoc. Inf. Sci. Technol.* **2006**, *57*, 359–377. [CrossRef]
155. Attalla, M.; Hegazy, T. Predicting cost deviation in reconstruction projects: Artificial neural network versus regression. *J. Constr. Eng. Manag.-ASCE* **2003**, *129*, 405–411. [CrossRef]
156. Fang, Q.; Li, H.; Luo, X.; Ding, L.; Luo, H.; Rose, T.M.; An, W. Detecting non-hardhat-use by a deep learning method from far-field surveillance videos. *Autom. Constr.* **2018**, *85*, 1–9. [CrossRef]
157. Fang, W.; Ding, L.; Zhong, B.; Love, P.E.D.; Luo, H. Automated detection of workers and heavy equipment on construction sites: A convolutional neural network approach. *Adv. Eng. Inform.* **2018**, *37*, 139–149. [CrossRef]
158. Seo, J.; Han, S.; Lee, S.; Kim, H. Computer vision techniques for construction safety and health monitoring. *Adv. Eng. Inform.* **2015**, *29*, 239–251. [CrossRef]
159. Ren, S.; He, K.; Girshick, R.; Sun, J. Faster R-CNN: Towards Real-Time Object Detection with Region Proposal Networks. *IEEE Trans. Pattern Anal. Mach. Intell.* **2017**, *39*, 1137–1149. [CrossRef] [PubMed]
160. Goh, Y.M.; Chua, D. Neural network analysis of construction safety management systems: A case study in Singapore. *Constr. Manag. Econ.* **2013**, *31*, 460–470. [CrossRef]
161. Girshick, R.; Donahue, J.; Darrell, T.; Malik, J. Rich feature hierarchies for accurate object detection and semantic segmentation. In Proceedings of the 27th IEEE Conference on Computer Vision and Pattern Recognition (CVPR), Columbus, OH, USA, 23–28 June 2014; pp. 580–587.
162. Roman, N.D.; Bre, F.; Fachinotti, V.D.; Lamberts, R. Application and characterization of metamodels based on artificial neural networks for building performance simulation: A systematic review. *Energy Build.* **2020**, *217*, 109972. [CrossRef]

163. Hosseini, M.R.; Maghrebi, M.; Akbarnezhad, A.; Martek, I.; Arashpour, M. Analysis of Citation Networks in Building Information Modeling Research. *J. Constr. Eng. Manag.* **2018**, *144*, 04018064. [CrossRef]
164. Wen, Q.J.; Ren, Z.J.; Hui, L.; Wu, J.F. The progress and trend of BIM research: A bibliometrics-based visualization analysis—ScienceDirect. *Autom. Constr.* **2021**, *124*, 103558. [CrossRef]
165. Nti, I.K.; Adekoya, A.F.; Weyori, B.A.; Nyarko-Boateng, O. Applications of artificial intelligence in engineering and manufacturing: A systematic review. *J. Intell. Manuf.* **2021**, *33*, 1581–1601. [CrossRef]
166. Bilal, M.; Oyedele, L.O.; Qadir, J.; Munir, K.; Ajayi, S.O.; Akinade, O.O.; Owolabi, H.A.; Alaka, H.A.; Pasha, M. Big Data in the construction industry: A review of present status, opportunities, and future trends. *Adv. Eng. Inform.* **2016**, *30*, 500–521. [CrossRef]

buildings

MDPI

Article

Risk Propagation Model and Simulation of an Assembled Building Supply Chain Network

Yingchen Wang [1], Ran Sun [1], Liyuan Ren [1,*], Xiaoxiao Geng [2], Xiangmei Wang [1] and Ling Lv [1]

1 School of Management Engineering and Business, Hebei University of Engineering, Handan 056038, China
2 School of Architecture and Art, Hebei University of Engineering, Handan 056038, China
* Correspondence: renliyuan@hebeu.edu.cn

Abstract: In recent years, the prefabricated building supply chain has received strong support from the government and has developed rapidly, but there are various risks in the operation process. In this paper, on the basis of considering asymptomatic infections and relapse, this paper establishes a risk transmission model that considers a recurrent Susceptible–Exposed–Asymptomatic–Infectious–Recovered (abbr. SEAIR) model, systematically analyses the risks in the supply chain, and calculates the risk balance point to conclude that the risks can exist in the supply chain for a long time. By drawing a causal circuit diagram, the relationship between the influencing factors in the process of risk transmission is found, establishing a stock flow map to explore the law of risk propagation. The simulation results using Vensim PLE software show that the five influencing factors of infection rate, transmission rate, government financial support, government policy supervision, and immunity loss ratio have an important impact on the number of risk-unknown enterprises, risk-latent enterprises, risk transmission enterprises, and infection rehabilitation enterprises in risk transmission, and relevant countermeasures to deal with risk transmission in the supply chain are proposed. Theoretically, this paper broadens the ideas for improving infectious disease models. From the management point of view, it reveals how the prefabricated building supply chain enables enterprises to improve their ability to deal with risks through the risk propagation model, providing reference and helping to manage the risks faced by the prefabricated building supply chain.

Keywords: supply chain; risk transmission; recurrent; SEAIR; emulation

Citation: Wang, Y.; Sun, R.; Ren, L.; Geng, X.; Wang, X.; Lv, L. Risk Propagation Model and Simulation of an Assembled Building Supply Chain Network. *Buildings* **2023**, *13*, 981. https://doi.org/10.3390/buildings13040981

Academic Editor: Tarek Zayed

Received: 10 March 2023
Revised: 26 March 2023
Accepted: 3 April 2023
Published: 7 April 2023

1. Introduction

In recent years, the supply chain of prefabricated buildings has been widely considered by society in shortening the construction period, saving labour, and improving the quality of construction enterprises. At the same time, states have successively issued relevant policies to encourage and support the development of the supply chain of prefabricated buildings, which has brought prefabricated construction industrialization to a new level [1]. According to the statistics of the prefabricated building industry, the market size of China's prefabricated buildings in 2011 was about CNY 4.3 billion, reached CNY 262.3 billion in 2017, and soared to CNY 1.02 trillion by 2022, with a compound growth rate of more than 100%, and the national prefabricated buildings accounted for more than 22%. Supply chain management is chosen as a management mode for prefabricated buildings to enhance their strength, integrating the advantages of various related enterprises, connecting all the enterprise nodes in the supply chain, and forming a chain network with the owner's demand as the guide and the general contractor as the core. Through the commercial interaction among the enterprises of each node, information flow, resource flow, cash flow, and logistics run through the design and construction units, material purchasing units, logistics and transportation units, assembly and construction units, and structural design units, thus promoting the functional operation of the overall construction chain [2]. However, with the continuous improvements of the technology of prefabricated building supply chains, the number of node enterprises is increasing, and the types of nodes are increasing from the initial single type to the present full coverage. The prefabricated links involved are more complicated, and the suppliers are more diverse, which leads to increasing risks in the supply chain. External environmental factors, such as natural environmental changes, policies and regulations, and internal factors, such as the planning stage, purchasing stage, manufacturing stage, transportation stage, and assembly stage, all affect the change in risks [3]. While the risks in the supply chain increase, the stability of the supply chain worsens, and the changes in these risks affect the normal construction and transportation of the node enterprises in the supply chain [4]. On the other hand, after being affected by the risks, the node enterprises will further spread the risks through the propagation path of the supply chain, which will have an impact on the normal operation of the whole supply chain. Therefore, the risk problem in the supply chain of prefabricated buildings has become an urgent problem in supply chain management.

Risk research on supply chains was first influenced by enterprise risk, and the connection between enterprises increased, so risk arose. A supply chain is a network composed of many enterprises to meet the needs of customers, and the risks among enterprises are bound to affect the risks of the whole supply chain. However, the prefabricated building supply chain has many participants, complicated strategic relationships, and opaque information among enterprises, so it is difficult for the traditional supply chain to meet the development of the prefabricated building supply chain. Therefore, research on the risk of prefabricated building supply chains has gradually attracted people's attention, and many scholars have explored and studied the risks of prefabricated building supply chains from different perspectives. Considering the risks in all stages of the whole life cycle of prefabricated building projects, in Zhang and Qiao [5] an index system was constructed, a risk evaluation model is formed, and relevant suggestions are put forward for the risk management of the prefabricated building supply chain. Chan et al. used a literature review, structured discussion, and multi-attribute group decision models to explore the benefits and challenges of modular integrated architecture [6]. Wang et al.'s factor analysis was used to identify the key points of risk control in the supply chain of prefabricated buildings, and the relationship and weight of risk factors in each stage are quantitatively analysed. It is concluded that the risk mainly exists in the manufacturing stage, and the risk in the delivery stage is the lowest [7]. Zhang refers to the information dissemination model, identifies the relationship among 19 risk dissemination factors, and introduces blockchain technology to construct the information flow model of the assembled supply chain [8]; Gu identified and analysed the risk factors for the supply chain based on the Supply Chain Operations Reference model (abbr. SCOR) theory, reduced the dimension of risk factors by principal component analysis, and simulated and analysed it by system dynamics [9]; Zhang evaluated the risk of a prefabricated concrete supply chain by using a cloud model under an engineering procurement construction (abbr. EPC) general contracting project [10]; An et al. established a Structural Equation Model (abbr. SEM) by multivariate data analysis and obtained the order of influencing factors for the research objective of supply chain integration [11]; Wang et al. introduced the idea of interface management, analysed the risk mechanism of supply chain integration interface, and used a Combination Ordered Weighted Averaging (abbr. C-OWA) operator weighting and grey clustering evaluation method to build a risk evaluation model [12]; Al-Hussein M. H. Al-Aidrous analysed the relationship between the influencing factors of ground-floor housing in prefabricated buildings, and proposed relevant measures to promote the development of prefabricated buildings in Malaysia [13]; Kristopher Orlowski analysed the manufacturing principles of special weatherproof seals for prefabricated buildings and evaluated their influencing factors [14]; Luo et al. analysed the supply chain risk network of stakeholders by using social network, to explore the supply chain network risks of prefabricated houses in Hong Kong and help employees to deal with these risks more effectively [15]; Ibrahim Yahaya Wuni et al. identified the risk factors of modular integrated buildings by using the fuzzy comprehensive evaluation method [16]; Syed Saad examined the key factors for success in Malaysia's construction industry from five areas: stakeholder understanding, resource availability, process management, issues and perceptions, and future needs [17]. Most of these studies adopted quantitative research methods and explored the risk generation mechanism in different ways, defined the risk propagation angle, and constructed a risk propagation model.

The infectious disease model is a typical transmission dynamics model in mathematical engineering, and the research on the infectious disease dynamics model has great practical significance, mainly including the following three aspects: establishing a mathematical model to simulate the process of infectious disease transmission, analysing the spread trend of the infectious disease model, and studying targeted prevention and control strategies. Typical infectious disease models include the Susceptible–Infected (abbr. SI) model, the Susceptible–Infected–Susceptible (abbr. SIS) model, the Susceptible–Infected–Recovered (abbr. SIR) model [18], and the Susceptible–Exposed–Infected–Recovered (abbr. SEIR) model [19]. Since then, many scholars have improved on the basis of the above classical models based on different considerations: Yiping Tan built a stochastic Susceptible–Infected–Susceptible (abbr. SIS) dynamics model from the perspective of media coverage and used dynamics for research [20]; Ayse Peker Dobie studied a Susceptible–Infectious–Susceptible (SIS) model with virus mutation in a variable population size, calculating equilibrium points [21]; Wonhyung Choi studied a spatial Susceptible—nfected–Susceptible epidemic model with a free boundary [22]; Jianhua Chen considered that the possibility of infected people becoming removed after treatment is very small, and improved the Susceptible–Infected–Recovered (abbr. SIR) model to analyse the emergency supply chain transmission mechanism [23]; Di Liang studied supply chain risk propagation based on the Susceptible–Infected–Recovered (abbr. SIR) epidemic model [24]. Based on the Susceptible–Exposed–Infected–Recovered (abbr. SEIR) model, Guanhua Ni proposed a Susceptible–Contacts–Exposed–Infected–Recovered (abbr. SCEIR) model that incorporates close contacts (C) and self-protectors (P) into a Parameters Sensitivity Analysis of COVID-19 [25]; Isa Abdullahi Baba studied the transmission dynamics of the disease by incorporating the saturated incidence rate into the model, and the Caputo sense was constructed for studying the risk balance point [26]; Mauro Aliano proposed a time delay differential system describing risk diffusion among companies inside an economic sector by means of a Susceptible–Infected–Recovered (abbr. SIR) dynamics [27]; based on the Bo Li epidemiological model, the stability of discrete time and local bifurcation were considered, and the dynamic behaviour of the infectious disease model was analysed [28]. Tchavdar T. Marinov proposed an adaptive susceptible infection-removal (abbr. A-SIR) epidemic model with time-dependent transmission and clearance rates and applied it to address COVID-19 in Latin America [29].

At present, research on prefabricated building supply chains mainly focuses on risk evaluation and risk generation mechanisms. Most of these studies regard risk factors as independent individuals and identify and evaluate related risk factors by using factor analysis and grey cluster evaluation methods, but there are few studies on the spread of risk factors in the supply chain. At the same time, the wide spread of risks in the supply chain can easily break the safe operation of the supply chain, making it unable to achieve the expected goal of supply chain management, resulting in a decrease in supply chain efficiency and an increase in the cost of each subject in the supply chain. Therefore, it is crucial to study risk transmission from the perspective of the prefabricated building supply chain. Due to the similarities between virus transmission and risk transmission in terms of spread object, spread process, and spread environment, scholars have applied basic virus transmission models such as SI, SIS, SIR, SEIR, and so on to the field of risk transmission. However, the traditional virus model fails to take into account the symptoms of enterprises in the supply chain after being eroded by risk, and asymptomatic infected enterprises have no obvious characteristics after risk erosion and are easy to ignore, thereby underestimating the harm brought by risk and making it difficult to comprehensively solve the corresponding problems caused by risk transmission. At the same time, the performance of diseases caused by virus transmission after enterprises in the supply chain are eroded by risk is similar, and some disease problems cannot be completely solved at one time and recur after treatment; enterprises in the supply chain may not be able to completely solve the risk problem once after being eroded by risk, and may reappear after a period of time, forming a secondary diffusion. Focusing on the problem of risk propagation in the prefabricated building supply chain, this paper uses the Vensim PLE software of system dynamics to identify the causes and consequences of the prefabricated building supply chain risk, determine the key risk factors, innovatively introduce the SEAIR model in the complex network infectious disease transmission dynamics model to establish the risk transmission model, and study the risk transmission process. Table 1 shows the relationship between the existing literature on risk in the prefabricated building supply chain and the virus model improvement perspective and this study.

Table 1. Shows a summary of the existing research on risk in the prefabricated building supply chain and virus model improvement perspectives.

Au (Year)	Research Perspective	Research Method/Used Models	Whether to Consider Asymptomatic Infection	Whether Recurrent Is Considered
Zhang and Qiao [5]	Risk assessment of prefabricated building supply chain	The whole life cycle of a construction project	NO	NO
Chan Tsz Wai [6]	Benefits and critique of modular integrated architecture	Multi-attribute group decision making	NO	NO
Wang et al. [7]	The relationship between risk factors in the prefabricated building supply chain	Factor analysis	NO	NO
Zhang [8]	The relationship between risk transmission factors in the prefabricated building supply chain	Block chain technology	NO	NO
Gu [9]	Supply chain risk factor identification and analysis	SCOR, Principal component analysis	NO	NO
Zhang [10]	Risk assessment of prefabricated concrete supply chain	Cloud model	NO	NO
An et al. [11]	Rank supply chain influencers	SEM	NO	NO
Wang et al. [12]	Risk assessment of prefabricated building supply chain	C-OWA operator weighting and grey clustering evaluation methods	NO	NO
Al-Hussein M. H. Al-Aidrous [13]	The relationship between the influencing factors of ground-floor housing in prefabricated buildings	Statistical Package of Social Science (SPSS)	NO	NO
Kristopher Orlowski [14]	Evaluation of influencing factors of special weatherproof seals for prefabricated buildings	Neo-Hookean Model	NO	NO
Luo et al. [15]	Explore the supply chain cyber risks of prefabricated housing projects	Social network analysis	NO	NO
Ibrahim Yahaya Wuni [16]	Identify risk factors for modular integrated construction [16]	Fuzzy comprehensive evaluation	NO	NO
Syed Saad [17]	Analysis of the key factors required for success in the Malaysian construction industry	Data analysis methods	NO	NO
Yiping Tan [20]	Risk propagation	A stochastic SIS dynamics model	NO	NO
Ayse Peker Dobie [21]	Risk propagation, risk balancing points	SIS model with virus mutation in a variable population size	NO	NO

Table 1. *Cont.*

Au (Year)	Research Perspective	Research Method/Used Models	Whether to Consider Asymptomatic Infection	Whether Recurrent Is Considered
Wonhyung Choi [22]	Risk propagation	A spatial SIS epidemic model with a free boundary	NO	NO
Jianhua Chen [23]	Emergency supply chain transmission mechanism	Improve SIR	NO	NO
Di Liang [24]	Supply chain risk propagation	SIR	NO	NO
Guanhua Ni [25]	Parameters Sensitivity Analysis of COVID-19	Propose a SCEIR model	NO	NO
Isa Abdullahi Baba [26]	Risk balancing point	The transmission dynamics of the disease are studied	NO	NO
Mauro Aliano [27]	Dynamic risk propagation model	Time delay is considered on the basis of the SIR model	NO	NO
Bo Li [28]	Analyse the dynamic behaviour of infectious disease models	A discrete-time SIR epidemic model	NO	NO
Tchavdar T. Marinov [29]	Infectious disease models applied to COVID-19 in Latin America.	Presents an Adaptive Susceptible–Infected–Removed (A-SIR) epidemic model with time-dependent transmission and removal rates	NO	NO
This paper	Risk propagation in the supply chain of prefabricated buildings	Consider the recurrent SEAIR model	Yes	Yes

The contributions of this article are as follows:

1. Focusing on the problem of supply chain risk transmission, this paper establishes a supply chain network model and innovatively introduces the recurrent SEAIR model when considering asymptomatic infections and relapses, adding two situations: asymptomatic characteristics and symptomatic characteristics.
2. Considering both of these scenarios, the asymptomatic infection enterprise and the symptomatic infection enterprise are jointly defined as risk transmission enterprises based on the actual supply chain, and are regarded as one of the main bodies of the risk transmission model.
3. The SEAIR supply chain risk propagation model considering recurrence is established, and the computer simulation method based on system dynamics is combined with the propagation dynamics idea to simulate and analyse the risk propagation process, seek risk mitigation strategies, and provide decision support for better reducing supply chain risks.

The remainder of this paper is organized as follows: Section 2 identifies risk factors associated with the prefabricated building supply chain. Section 3 establishes a SEAIR model that considers relapse. In Section 4, based on the recurrent SEAIR model, a supply chain risk propagation model was established and simulated for research. Finally, in Section 5, the relevant conclusions provide some directions for further research.

2. Identification and Dissemination of Prefabricated Building Supply Chain Risk

The assembled building supply chain is guided by the owner's demand, and the general contractor is responsible for the coordination and scheduling of each node enterprise, such as the designer, the manufacturer, the transporter, and the assembly formula, to meet the needs of the owner, the general contractor, and the government. Different stakeholders are faced with different risks in the five working stages of design, procurement, manufacturing, transportation, and assembly. At the same time, the risks are mutually restricted, and the problems of one party will drive the whole supply chain to produce a linkage effect.

2.1. Assembled Supply Chain Risk Identification

The prefabricated building supply chain is guided by the needs of the owner, and the general contractor is responsible for the coordination and scheduling of various node enterprises, such as the designer, manufacturer, transportation, and assembly party, to meet the needs of the owner, the general contractor, and the government. Different stakeholders face different risks in the five stages of design, procurement, manufacturing, transportation, and assembly, and at the same time, each risk is mutually restricted, and the problem of one problem will drive the entire supply chain to have a linkage effect. As the first step in the construction activities of various stakeholders, the formulation of the plan directly affects the work arrangement of the subsequent procurement, manufacturing, transportation and assembly stages; the raw material standards procured by the owner and general contractor also affect the component manufacturing and later assembly work; the manufacturer's control of the delivery time and quality of prefabricated components will affect the transportation arrangement and assembly quality; as transportation is an indispensable part of construction activities, its safety will also affect the progress of engineering assembly. Based on this, the relevant literature at home and abroad and prefabricated building accidents are analysed and summarized, and the risk factors of the five stages of design, procurement, manufacturing, transportation, and assembly and the perspective of the external environment are preliminarily formed, and then experts in the field of prefabricated building are invited to summarize the relevant risks based on the characteristics of the prefabricated building supply chain [30–34]:

1. From the perspective of the external environment, the risk factors are changes in government policies, sudden changes in the natural environment, and the impact of the economic environment.
2. From the perspective of planning, the planning objectives are inaccurate; the difference between the change plan and the plan information and the actual situation is too large.
3. From the purchasing point of view, the cost of purchased materials is too high, the quality of purchased materials is not up to standard, and the purchasing time is delayed.
4. From the manufacturing point of view, the process of manufacturing products does not meet the specification requirements, the delivery time of manufactured products is delayed, and the capacity of manufacturing products is not up to standard.
5. From the perspective of transportation, materials are damaged during transportation, transportation time is delayed, transportation schemes are not perfect, and the transportation company is reliable.
6. From the perspective of assembly, the risk factors are changes in the assembly period, safety accidents at assembly sites, a low technical operation level of assembly personnel, and unqualified quality of assembly parts.

2.2. Risk Transmission Mechanism of the Prefabricated Building Supply Chain

At the planning level, the accuracy of the target of the plan affects the time of purchasing materials and the transportation time of products, and the standards for manufacturing will also be limited. At the purchasing level, the quality grade of purchased materials limits the assembly results and the quality of manufactured products. At the manufacturing level, the time of manufacturing products and the delivery time limit the transportation of products, and the qualified rate of manufactured products has an impact on assembly. At the transportation level, unreasonable transportation planning leads to changes to the supply chain structure and delays the progress of the project. At the level of assembly, the quality of assembly parts is unqualified, which cannot meet the assembly standards and delays the construction period; at the same time, it may lead to accidents in production safety. At the external risk level, sudden weather changes delay the progress of supply chain projects, and some policies issued by the government restrict parts of the products, which will have an impact on the subsequent manufacturing of products and assembly materials. The risk transmission mechanism of the prefabricated building supply chain is shown in Figure 1.

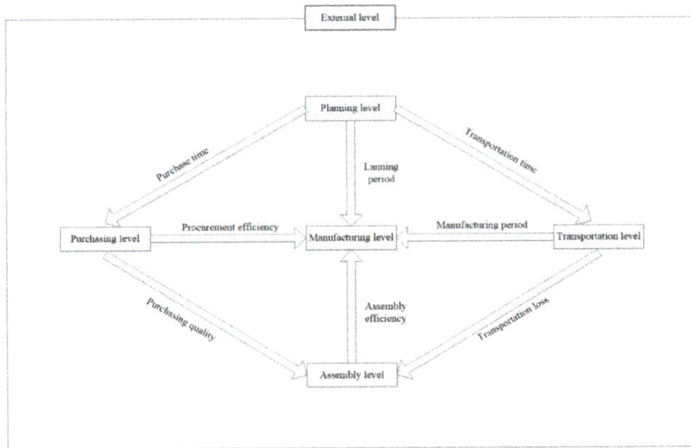

Figure 1. Risk transmission mechanism of a prefabricated building supply chain.

3. Establishment of the Model

3.1. Establishment of Recurrent SEAIR Models

The recurrent SEAIR model divides the population into five categories: S—susceptible, E—latent, A—asymptomatic, I—symptomatic, and R—recovering. In the recurrent SEAIR model, the population always keeps a constant Q, and at time T, the number of susceptible people is S(t), the number of latent people is E(t), the number of asymptomatic people is A(t), the number of symptomatic people is I(t), and the number of recovered people is R(t). A class of SEAIR models with asymptomatic infection and secondary recurrence was established.

The law of infection is as follows: (1) After contacting the infected person, the susceptible person will be infected with a certain probability, from a healthy state to a latent state and then to an infected state. (2) The state of infected people includes two types: one is symptomatic infection and the other is asymptomatic infection. (3) After these two infected people recover their health, they will develop their own immunity. However, over time, it is possible that this immunity will disappear and the person will again be vulnerable to infection. The mechanism of infection of the recurrent SEAIR risk transmission model is shown in Figure 2.

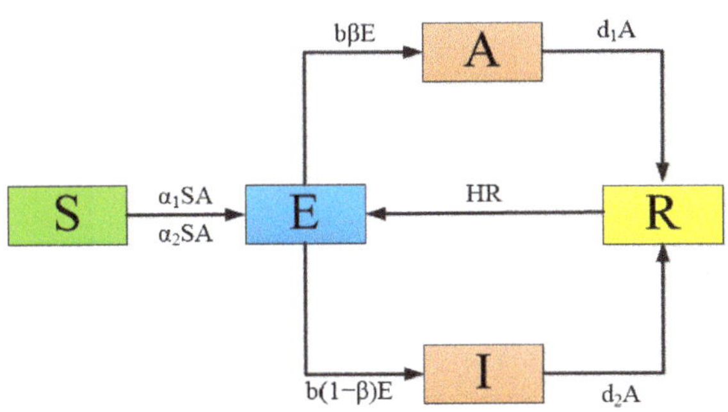

Figure 2. Recurrent SEAIR model.

In the recurrent SEAIR model, it is assumed that (1) the natural mortality is μ, regardless of factors such as birth population and floating population; (2) the infection rate of symptomatic infected persons is α_1 and that of asymptomatic infected persons is α_2; (3) the probability of transmission from latent to infected persons is b; (4) the mortality rate of asymptomatic infected persons is c_1 and that of symptomatic infected persons is c_2;

and (5) the recovery rate of asymptomatic infected persons is d_1 and that of symptomatic infected persons is d_2; (6) the probability of latent patients entering the asymptomatic period is β, and the probability of latent patients entering the symptomatic period is $1-\beta$. Those who recover will recover to latent patients with the probability of β; (7) H represents the recurrence rate of the disease.

The infectious disease model is as follows:

$$\begin{cases} \frac{dS(t)}{dt} = Q - \alpha_1 S(t)A(t) - \alpha_2 S(t)I(t) - \mu S(t) \\ \frac{dE(t)}{dt} = \alpha_1 S(t)A(t) + \alpha_2 S(t)I(t) - \mu E(t) - b\beta E(t) - b(1-\beta)E(t) + HR(t) \\ \frac{dA(t)}{dt} = b\beta E(t) - (c_1+\mu)A(t) - d_1 A(t) \\ \frac{dI(t)}{dt} = b(1-\beta)E(t) - (c_2+\mu)I(t) - d_2 I(t) \\ \frac{dR(t)}{dt} = d_1 A(t) + d_2 I(t) - \mu R(t) - HR(t) \end{cases} \quad (1)$$

3.2. Analysis of the Applicability of the Network Model and Virus Model of a Prefabricated Building Supply Chain

1. The objects of diffusion are similar: In the process of risk propagation in the supply chain of prefabricated buildings, the main enterprises in the supply chain suffer from risk erosion, and each enterprise is an independent whole with its own ability to resist risks. Enterprises with a strong ability to resist risks are ultimately less affected by risks; in contrast, enterprises with a weak ability to resist risks are greatly affected by risks. For virus transmission, the infected individual is the object of transmission, and the transmission effect is affected by individual immunity [23].
2. The process of diffusion is similar: In the supply chain risk of prefabricated buildings, the spread of risk is disorderly. When the manufacturer receives the erosion of risk, the affected risk will spread in the supply chain network, and the upstream raw material suppliers, downstream transporters, and packaging formulas will be affected. In the supply chain, one node enterprise affected by the risk will spread to another node enterprise in contact with it, and the spread results will worsen. In the process of virus transmission, the virus that comes into contact with the infected virus is the first to be affected, and then it spreads in all directions in a radial way. Without being stopped, the impact is even more serious [35].
3. The diffusion environment is similar: The risk of the prefabricated building supply chain is diffused through the business dealings between node enterprises. When the quality of the products produced by the manufacturer does not meet the requirements of the owner, the downstream transportation faces the risk of delay in transportation time, which leads to the failure to complete the equipment installation on schedule, resulting in compensation for a contract breach by the general contractor. The main body of virus transmission is the cells in the body of the virus, and the virus carried in the cells is spread to other places through the flow of blood [36].

3.3. The Existence of a Risk Balance Point

In the infectious disease model, there are two modes: disease-free equilibrium and endemic equilibrium. The disease-free equilibrium is the extinction of the disease, and the endemic equilibrium means that the disease will develop in the population for a long time, forming a normalization [37]. Because this paper studies and analyses risk transmission based on the infectious disease model, it redefines the disease-free equilibrium point as the risk extinction point and the endemic equilibrium point as the long-term risk existence point. The risks in the supply chain can only be controlled but do not completely disappear. Therefore, this paper only explores the spread of the endemic equilibrium point, that is, the long-term existence point of risks, defined according to the supply chain:

In Model (1), there is always a disease-free equilibrium $P_0\left(\frac{\mu}{Z},0,0,0,0\right)$. According to the method of the regeneration matrix in the literature [38], the basic regeneration number is obtained:

$$R_0 = \frac{\alpha_2 Qb\beta}{\mu M_1 M_2} + \frac{\alpha_1 Zb(1-\beta)}{\mu M_0 M_2} + \frac{Hb\beta M_2 d_1 + Hb(1-\beta)M_1 d_2}{M_0 M_1 M_2 (H+\beta)}$$

among $M_0 = b + \mu, M_1 = \mu + d_1 + c_1, M_2 = \mu + d_2 + c_2$. There are the following conclusions about the existence of the equilibrium point of the model:

Theorem 1. *When $R_0 > 1$, Model (1) has a unique endemic equilibrium $P^*(S^*, E^*, A^*, I^*, R^*)$*

It is proven that if the right side of Model (1) is equal to 0 and an endemic equilibrium P^* exists, the algebraic equations are satisfied:

$$\begin{cases} Q - \alpha_1 S^* A^* - \alpha_2 S^* I^* - \mu S^* = 0 \\ \alpha_1 S^* A^* + \alpha_2 S^* I^* - \mu E^* - b\beta E^* - b(1-\beta)E^* + HR^* = 0 \\ b\beta E^* - (c_1+\mu)A^* - d_1 A^* = 0 \\ b(1-\beta)E^* - (c_2+\mu)I^* - d_2 I^* = 0 \\ d_1 A^* + d_2 I^* - \mu R^* - HR^* = 0 \end{cases} \tag{2}$$

The following can be obtained from the third and fourth formulas of the equations:

$$b\beta E^* = (c_1+\mu)A^* + d_1 A^* = M_1 A^*$$
$$b(1-\beta)E^* = (c_2+\mu)I^* + d_2 I^* = M_2 A^*$$

Therefore,

$$A^* = \frac{b\beta}{M_1}E^*, I^* = \frac{b(1-\beta)}{M_2}E^*. \tag{3}$$

Substituting (3) into (2) yields

$$\frac{d_1 b\beta}{M_1}E^* + \frac{d_2 b(1-\beta)}{M_2}E^* = (H+\mu)R^*,$$

Therefore,

$$R^* = \frac{1}{H+\mu}\left(\frac{d_1 b\beta}{M_1} + \frac{d_2 b(1-\beta)}{M_2}\right). \tag{4}$$

Combining Equation (4) and bringing the second equation in equation group (2) into the first equation, we obtain $Q - M_0 E^* + \frac{H}{H+\mu}\left(\frac{d_1 b\beta}{M_1} + \frac{d_2 b(1-\beta)}{M_2}\right)E^* = \mu S^*$

So

$$S^* = \frac{Q}{\mu} - \frac{(H+\mu)M_0 M_1 M_2 - d_1 b\beta H M_2 - d_2 b(1-\beta)HM_1}{\mu(H+\mu)M_1 M_2}E^* \tag{5}$$

Substituting (3), (4), and (5) into the second equation set of equation set (2) gives

$$\alpha_1 S^* A^* + \partial_2 S^* I^* - \mu E^* - b\beta E^* - b(1-\beta)E^* + HR^* = 0$$

$$\left(\frac{\alpha_1 b(1-\beta)}{M_2} + \frac{\alpha_2 b\beta}{M_1}\right)E^* \times \left\{\frac{Q}{\mu} - \frac{(H+\mu)M_0 M_1 M_2 - d_1 b\beta H M_2 - d_2 b(1-\beta)HM_1}{\mu(H+\mu)M_1 M_2}E^*\right\}$$
$$-M_0 E^* + \frac{H}{H+\mu}\left(\frac{d_1 b\beta}{M_1} + \frac{d_2 b(1-\beta)}{M_2}\right)E^* = 0.$$

Therefore,

$$E^* = \frac{\mu M_0 M_1^2 M_2^2 (H+\mu)}{\{(H+\mu)M_0 M_1 M_2 - [Hb_1\beta M_2 d_1 + Hb_2(1-\beta)M_2 d_2]\}[\alpha_1 b_1 M_1(1-\beta) + \alpha_2 \mathrm{b}_2 M_2 \beta]}(R_0 - 1)$$

Therefore, this paper needs to prove
$\{(H+\mu)M_0 M_1 M_2 - [Hb_1\beta M_2 d_1 + Hb_2(1-\beta)M_2 d_2]\} > 0$ Just do it.

The specific steps are as follows:

$$\begin{aligned}
&\{(H+\mu)M_0M_1M_2-[Hb_1\beta M_2d_1+Hb_2(1-\beta)M_2d_2]\}\\
&> HbM_1M_2-[Hb_1\beta M_2d_1+Hb_2(1-\beta)M_2d_2]\}\\
&= Hb[(\beta M_1M_2-\beta d_1M_1M_2)+(1-\beta)M_1M_2-(1-\beta)d_2M_1]\\
&> 0
\end{aligned}$$

It can be proven that there is an endemic equilibrium point, so this situation will form a long-term risk in enterprises, that is, the risk that a supply chain will exist for a long time.

4. Based on the Establishment and Simulation Analysis of Supply Chain Risk Propagation Model Considering the Recurrent SEAIR Model

4.1. Identification of the Causes and Results of Supply Chain Risks of Prefabricated Buildings

From the perspective of system theory, the prefabricated building supply chain is a complex and huge system, and this paper analyses the risk factors affecting the prefabricated building supply chain from two aspects: the external system and the internal system. The stakeholders involved in the prefabricated construction supply chain include planners, purchasers, fabricators, transporters, and assemblers. This paper classifies these five stakeholders as internal systems that affect the risk propagation of the prefabricated building supply chain, and the external environment as external systems. System dynamics are used to analyse the influencing factors of external and internal systems in combination with Vensim PLE software, and identify the key factors affecting risk, that is, the main causes of risk.

4.1.1. Causality Diagram Analysis

The causal circuit diagram includes six systems, namely, external environment stage system, planning stage subsystem, procurement stage subsystem, manufacturing stage subsystem, transportation stage subsystem, and assembly stage subsystem; these six systems affect each other, interconnect, and ultimately affect the spread of risk in the entire prefabricated building supply chain. A causal analysis of the supply chain risk of prefabricated buildings is shown in Figure 3.

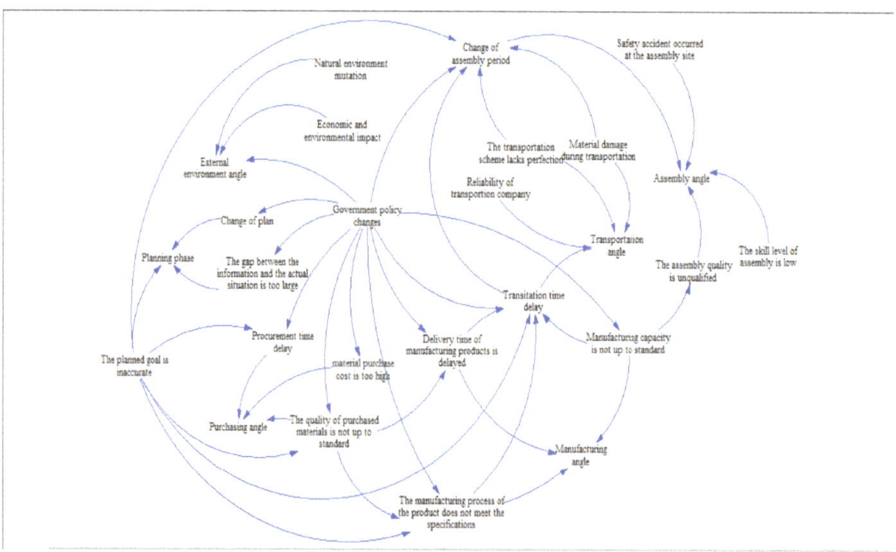

Figure 3. Causal analysis of the supply chain risk of prefabricated buildings.

4.1.2. Result Tree Analysis

Set each risk factor as the current state and use the analysis tool uses tree to analyse the results brought by each risk factor. It is concluded that when the government policy changes, it has the widest impact on other stages. The result trees of other factors are not introduced here. The result tree of government policy changes is shown in Figure 4.

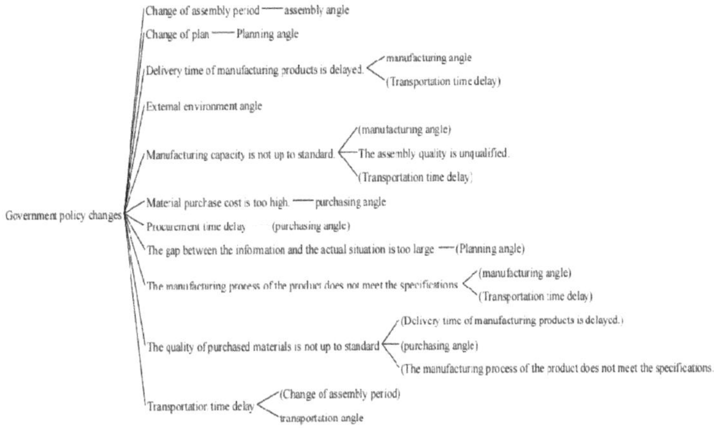

Figure 4. Result tree of government policy changes.

4.2. Based on the Establishment of the Supply Chain Risk Propagation Model Considering the Recurrent SEAIR Model

Through the above analysis, this paper concludes that changes in government policies have the greatest impact on the other stages of the supply chain of prefabricated buildings. Therefore, the government is regarded as the main body of the supply chain of prefabricated buildings, and it is introduced into the SEAIR model to explore the impact of relevant government policies on the supply chain of prefabricated buildings.

4.2.1. Definition of the System Boundary

In the management and operation of the prefabricated building supply chain, one or more enterprises will be involved, which will influence each other. One of them will be affected by risks, and the whole supply chain may be affected. Under the supply chain of prefabricated buildings, the risk spread can be divided into unknown risk enterprises (susceptible to infection), latent risk enterprises (latent persons), risk transmission enterprises (symptomatic and asymptomatic infection), and infection rehabilitation enterprises (recovered persons). When the node enterprises in the assembled supply chain are disturbed by risks, they change from unknown risk enterprises to latent risk enterprises. Under the influence of a certain transfer rate, latent enterprises turn into risk transmission enterprises, and risk transmission enterprises turn into infection rehabilitation enterprises under the influence of the government. When the node enterprises become infection rehabilitation enterprises, they may lose their immunity to risks and then continue to become unknown risk enterprises. It is found that the system in this process includes six subjects: enterprises with unknown risks, enterprises with latent risks, enterprises with risk transmission, enterprises with infection rehabilitation, enterprises without risk immunity, and the government.

4.2.2. Causality Analysis of Risk Communication

After determining the boundary of system research, through the analysis of the relationship between the internal factors of the boundary, the system causality diagram can be obtained. The risk communication system is no longer a single linear relationship but an intricate nonlinear system, which is a dynamic system under the action of many factors. Through the analysis of the risk propagation mechanism and the risk system of the prefabricated building supply chain, the system causal loop diagram is constructed based on the SEAIR model considering recurrence, and the feedback analysis theory of system dynamics is applied to describe the system structure of the risk propagation of the prefabricated building supply chain by the feedback loop, as shown in Figure 5:

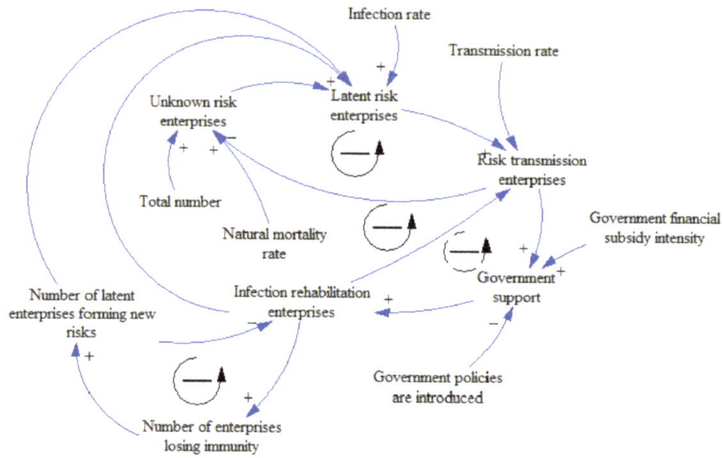

Figure 5. Causal circuit diagram of risk propagation.

The main circuits included in the figure are explained as follows:

1. Positive feedback loop R1: enterprises with latent risks → enterprises with risk transmission → government support → enterprises with infection and rehabilitation → enterprises with lost immunity → enterprises with new latent risks →enterprises with latent risks. The increase in the number of enterprises with latent risks leads to an increase in the number of risk transmission enterprises. When the number of communication enterprises is large, the government is urged to take measures to reduce the number of unstable enterprises affected by risks in the industry so that the number of infection and rehabilitation enterprises will increase, and the number of enterprises losing immunity in the later period will also rise, becoming new enterprises with latent risks.
2. Negative feedback loop B1: latent risk enterprise → risk-spreading enterprise → infection rehabilitation enterprise → latent risk enterprise.
3. Negative feedback loop B2: enterprises with unknown risks → enterprises with latent risks → enterprises with risk transmission → government support → enterprises with infection rehabilitation → enterprises with unknown risks. An increase in the number of enterprises with latent risks thus increases the number of enterprises with risk transmission. The increase in the number of communication enterprises is bound to attract the attention of the government, and the enterprises with risk communication should be guided and planned. Under the government's measures, the number of infection and rehabilitation enterprises in the industry has increased. When the number of enterprises in the industry is fixed, the number of enterprises with latent risks eventually decreases.
4. Negative feedback loop B3: risk communication enterprises → government support → infection rehabilitation enterprises → risk communication enterprises.
   With an increase in the number of risk communication enterprises, the government has strengthened risk control, and some risk communication enterprises will become risk rehabilitation enterprises, which will increase the number of risk rehabilitation enterprises and reduce the number of risk communication enterprises to a certain extent.
5. Negative feedback loop B4: infection rehabilitation enterprises → number of enterprises losing immunity → number of enterprises with latent new risks → infection

rehabilitation enterprises. With an increase in the number of infection rehabilitation enterprises, to some extent, the number of enterprises that have lost their immunity increases; thus, the number of enterprises with latent new risks will increase, and the number of infection rehabilitation enterprises will be reduced to a certain degree.

4.2.3. Flow Chart Analysis of Risk Communication

According to the causality diagram, a complete system flow diagram is constructed by using the system dynamics Vensim software, as shown in Figure 6:

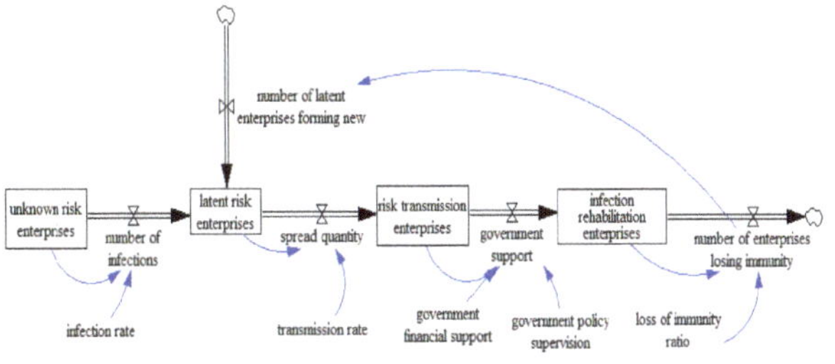

Figure 6. Flow chart of risk propagation.

As shown in Figure 6, the model contains four horizontal variables: enterprises with unknown risks, enterprises with latent risks, enterprises with risk transmission, and enterprises with infection rehabilitation. There are five rate variables: the number of unknown enterprises forming new risks, the number of infections, the number of transmissions, government support, and the number of enterprises losing immunity. There are five constants: infection rate, transmission rate, government financial support, government policy supervision, and the proportion of lost immunity. Through consultation with actual construction personnel and researchers in the field of prefabricated buildings, combined with the research literature in related fields [24,39], each constant is assigned to determine the parameter expression, as shown in Table 2:

Table 2. Variable expression and constant assignment in the table flow graph.

Serial Number	Expression Assignment
1	Unknown enterprise at risk = INTEG (30 − Number of infections, 30)
2	Latent risk enterprises = INTEG (Number of infections-Spread quantity + number of latent enterprises forming new, 0)
3	Risk transmission enterprises = INTEG (spread quantity − government support, 0)
4	Infection rehabilitation enterprises = INTEG (Government support − number of enterprises losing immunity, 0)
5	Number of latent enterprises forming new = number of enterprises losing immunity.
6	Number of infections = unknown risk enterprises × infection rate
7	Spread quantity = latent risk enterprises × Transmission rate
8	Government support = risk transmission enterprises × (Government financial support + government policy supervision)
9	Number of enterprises losing immunity = Loss of immunity ratio × infection rehabilitation enterprises
10	Infection rate = 0.10
11	Transmission rate = 0.15
12	Government financial support = 0.25
13	Government policy supervision = 0.25
14	Loss of immunity ratio = 0.20

4.3. Analysis of Model Simulation Results

As seen from Figure 5, the four main bodies involved in risk communication are risk-unknown enterprises, risk-latent enterprises, risk communication enterprises, and infection rehabilitation enterprises, and the number of the two main bodies of risk communication enterprises and infection rehabilitation enterprises directly reflects the intensity of risk communication. Among them, enterprises with unknown risks are affected by the number of infections, and the number of infections is determined by the infection rate. Latent enterprises are affected by the number of infections, the number of transmissions, and the number of latent enterprises forming new risks, and the number of transmissions is determined by the transmission rate. The number of risk communication enterprises is affected by the amount of communication and government support, and government support is determined by financial support and policy supervision. The number of rehabilitation enterprises is affected by the number of enterprises that have lost their immunity and government support, while the number of enterprises that have lost their immunity is determined by the proportion of those that have lost their immunity. Therefore, in the process of risk spread, the infection rate, spread rate, government financial support, government policy supervision, and the value of the proportion of lost immunity play an important role. Through the analysis of the above factors, combined with the actual participants in the supply chain, the change in the number of risk-spreading enterprises in the component manufacturers and the change in the number of risk rehabilitation enterprises in the transporters are taken as examples to study. Vensim software was used to simulate this, and the influence of the infection rate, transmission rate, government financial support, government policy supervision, and the change in the proportion of losing immunity on risk transmission is explored.

4.3.1. Impact of Change in Infection Rate on Risk Transmission

Set the initial infection rate at 10%, increase the infection rate by 5% and increase by 10%, and observe the change in the number of companies affected by risks. Figure 7 shows that, under the premise of infection rate as an influencing factor, the number of enterprises in the process of risk transmission shows significant changes.

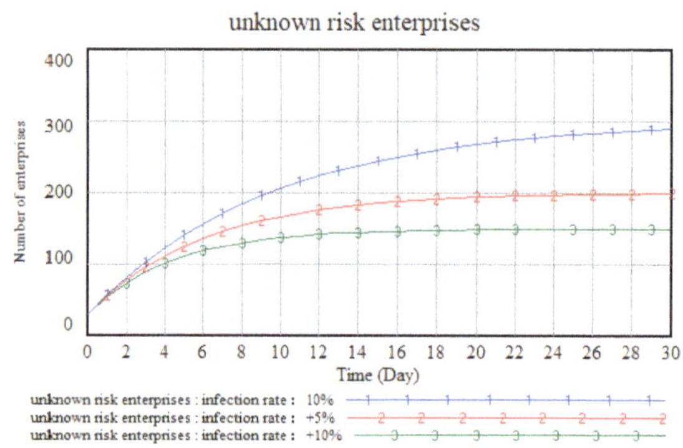

(**a**)

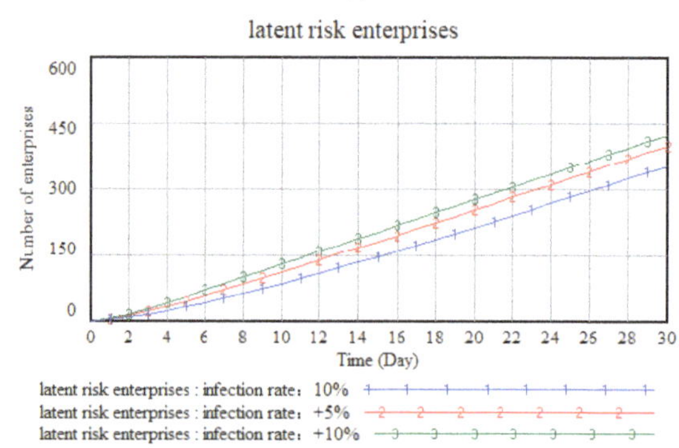

(**b**)

Figure 7. *Cont.*

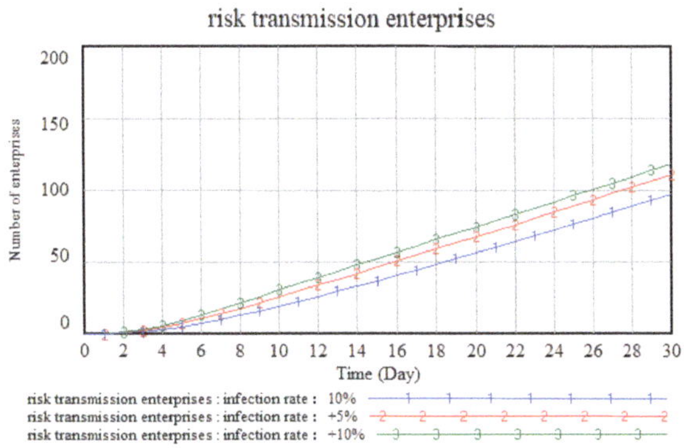

(c)

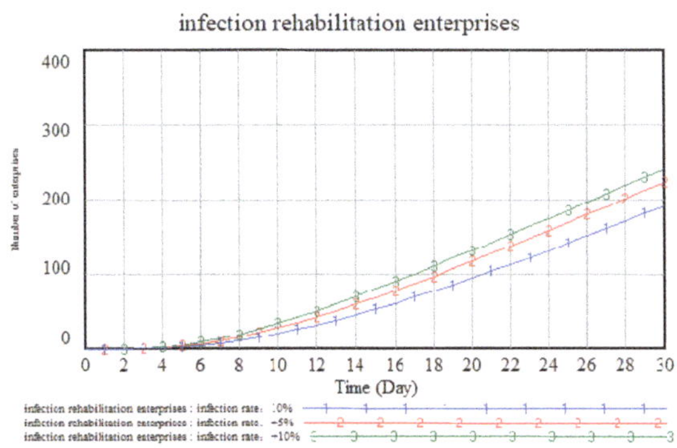

(d)

Figure 7. Impact of changes in infection rates on the number of enterprises in the process of risk transmission: (**a**) the impact of infection rates on businesses with unknown risks; (**b**) the impact of infection rates on businesses with latent risks; (**c**) the impact of infection rates on businesses where risk spreads; (**d**) impact of infection rate on infection recovery businesses.

With the increase in infection rate, the number of enterprises with unknown risks gradually increased, showing an overall increasing trend. The greater the initial infection rate, the smaller the number of businesses with unknown risk and the greater the number of businesses affected by the risk.

With the increase in infection rate, the number of risk-latent enterprises gradually increases, the infection rate is greater, and the number of risk-latent enterprises is greater; the greater the initial infection rate is, the greater the number of businesses with latent risk. On the thirtieth day, the number of risk-latent companies was as high as about 400.

The number of companies that spread the risk gradually increases with the increase in the infection rate, and the greater the infection rate, the greater the number of enterprises affected by the risk and the faster the growth rate. The number of enterprises showed a steady growth trend overall.

With the increase in infection rate, under the influence of risks, the number of infection rehabilitation enterprises gradually showed a slow growth trend from the fifth day, and the growth trend accelerated after that; in general, the greater the infection rate, the greater the infection, and the greater the number of rehabilitation enterprises. The study by Chen, J. et al. found a similar problem [23].

4.3.2. Impact of Change in Transmission Rate on Risk Transmission

The initial transmission rate is set to 15%, and the change in the number of companies in the process of risk transmission is observed under the premise that the transmission rate is observed as an influencing factor by 5% in Figure 8.

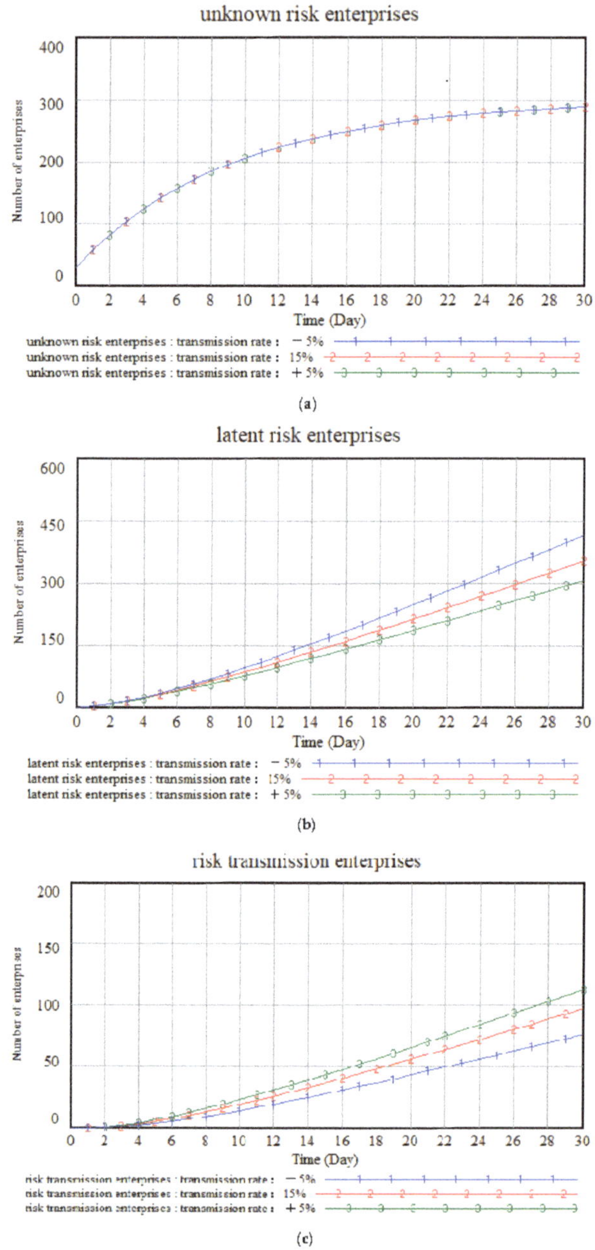

Figure 8. *Cont.*

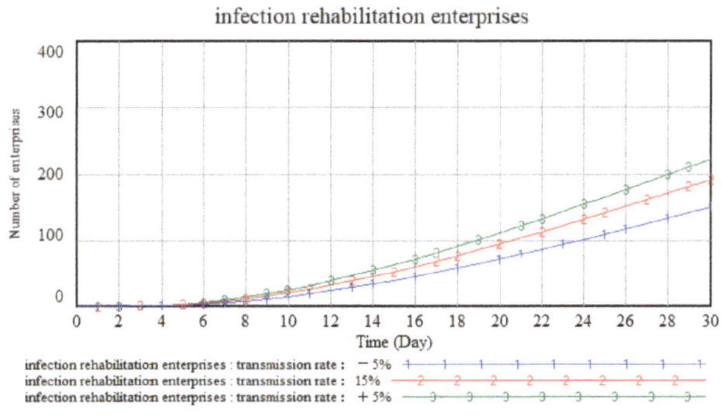

(d)

Figure 8. The impact of changes in transmission rate on the number of enterprises in the process of risk transmission: (**a**) the impact of transmission rates on businesses with unknown risks; (**b**) the impact of transmission rates on businesses with latent risks; (**c**) the impact of transmission rates on businesses where risk spreads; (**d**) impact of transmission rate on infection recovery businesses.

Under the influence of the transmission rate, the number of enterprises with unknown risk gradually increases, but a certain increase or decrease in the transmission rate of 15% will not affect the image transformation of enterprises with unknown risk. This is because in the context of considering relapse, infection rehabilitation enterprises become risk-latent enterprises after losing immunity to risks, and then continue to become risk transmission enterprises and infection rehabilitation enterprises with the spread of risks, and will not become risk-unknown enterprises. In this context of risk transmission, changes in transmission rate will have an impact on risk-latent enterprises, risk transmission enterprises, and infection recovery enterprises, and will not affect the change in the number of enterprises with unknown risks.

With the increase in transmission rate, the number of risk-latent enterprises has shown an increasing trend. Under different transmission rates, the number of latent enterprises increased in the first two days in line with the level of growth; the greater the transmission rate, the smaller the number of companies with latent risks; the smaller the transmission rate, the greater the number of companies with latent risks. Research by Subrata Paul et al. found a similar problem [40].

With the increase in the transmission rate, the number of risk communication companies showed an increasing trend from the fourth day. The greater the transmission rate, the greater the number of businesses affected by the risk; under the influence of different transmission rates, the initial difference in the number of risk transmission enterprises is small, and the number gap gradually increases over time.

With the increase in the transmission rate, the number of recovered enterprises showed an increasing trend from the sixth day. The transmission rate is proportional to the number of infected recovery enterprises, and the larger the transmission rate, the higher the number of infection rehabilitation enterprises; under the influence of different transmission rates, the initial difference in the number of risk transmission enterprises is small, and the quantitative gap gradually increases with the change in time.

4.3.3. Impact of Changes in Government Financial Support on Risk Communication

The initial government financial support is set at 25%, and the probability of increasing or decreasing by 10% is based on this basis, and the number of companies in the process of risk transmission is changed by the government support obtained in Figure 9.

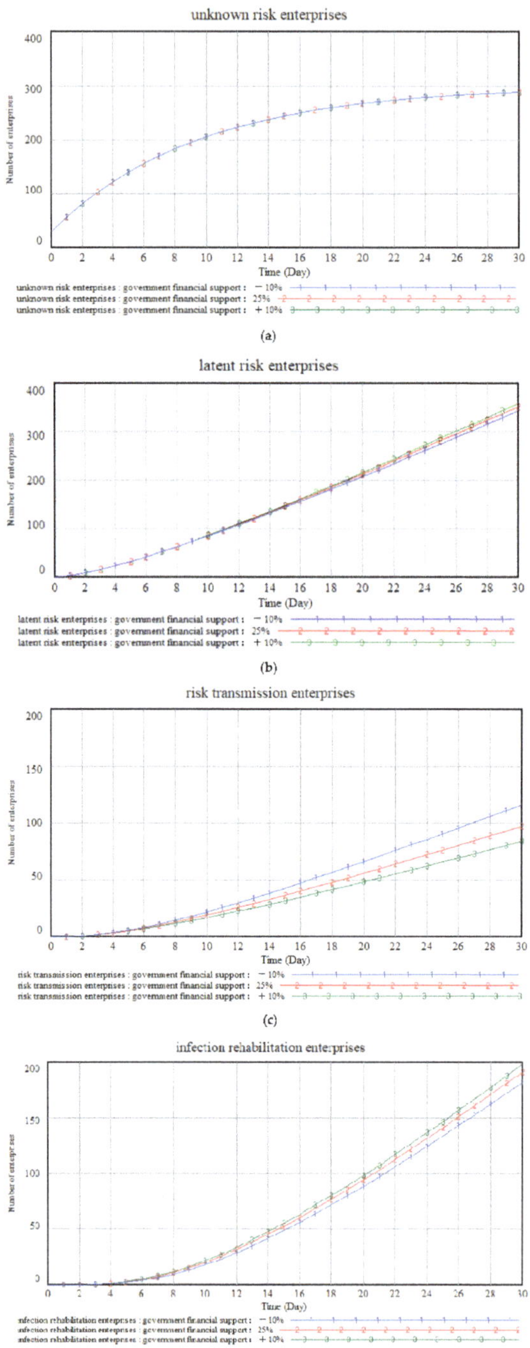

Figure 9. The impact of changes in government financial support on the number of enterprises in the process of risk propagation: (**a**) the impact of government financial support on enterprises with unknown risks; (**b**) the impact of government financial support on enterprises with latent risks; (**c**) the impact of government financial support on risk communication enterprises; (**d**) the impact of government financial support on enterprises recovering from infection.

Under the influence of government financial support, the number of enterprises with unknown risks gradually increases, but a certain increase or decrease in the ratio of 20% will not affect the image transformation of enterprises with unknown risks. This is because in the context of considering relapse, infection rehabilitation enterprises become risk-latent enterprises after losing immunity to risks, and then continue to become risk transmission enterprises and infection rehabilitation enterprises with the spread of risks and will not become risk-unknown enterprises. In the context of this risk transmission, changes in government financial support will have an impact on risk-latent enterprises, risk transmission enterprises, and infection recovery enterprises, and will not affect the change in the number of enterprises with unknown risks.

Under the change of government support, the number of enterprises with latent risks has shown an increasing trend. With different government support, the number of latent enterprises increased in the first 14 days in line with the level of growth; the greater the government support, the greater the number of companies with latent risks; the smaller the government support, the smaller the number of companies with latent risks.

Under the change in government support, the number of risk communication enterprises has shown an increasing trend. The financial support of the government is inversely proportional to the number of risk communication enterprises, and the greater the government support, the smaller the number of risk communication enterprises under the state of being affected by risk; with the passage of time, the initial difference in the number of risk-propagating enterprises under different government support is small, and the number gap gradually increases over time.

The degree of government support is directly proportional to the number of enterprises recovering from infection, and the number of rehabilitation enterprises has increased. With the increase in government financial support, the number of enterprises recovering from infection first showed a slow growth, and then increased the range of change.

4.3.4. Impact of Changes in Government Policy Supervision on Risk Communication

The initial value of government policy supervision is set to 25%, and on this basis, a change of 15% is made, and Figure 10 shows the change in the number of enterprises in the process of risk transmission under government supervision.

Under the influence of government policies and supervision, the number of enterprises with unknown risks gradually increases, but a certain increase or decrease in the ratio of 15% will not affect the image transformation of enterprises with unknown risks. This is because in the context of considering relapse, infection rehabilitation enterprises become risk-latent enterprises after losing immunity to risks, and then continue to become risk transmission enterprises and infection rehabilitation enterprises with the spread of risks, and will not become risk-unknown enterprises. Under this background of risk transmission, changes in government policies and supervision will have an impact on risk-latent enterprises, risk transmission enterprises, and infection recovery enterprises, and will not affect the change in the number of enterprises with unknown risks.

The impact of government policies and supervision on the number of enterprises with latent risks is to increase slowly at first, and then increase the growth range. However, on the whole, the impact of changes in government supervision on the number of enterprises with latent risks has not changed much. Under different policies and supervision, the number of enterprises affected by risk all increased from the first 0, and then with the increase in supervision, the number of enterprises with latent risks gradually increased.

The stronger the government supervision, the greater the number of risk transmission enterprises and infection rehabilitation enterprises, but in general, the impact of government supervision on the number of risk transmission enterprises and rehabilitation enterprises gradually increases over time. Under the influence of government policies and supervision, the number of risk transmission enterprises and infection recovery enterprises gradually showed an upward trend since the previous days, and the number growth changed little, and then the quantitative gap gradually increased.

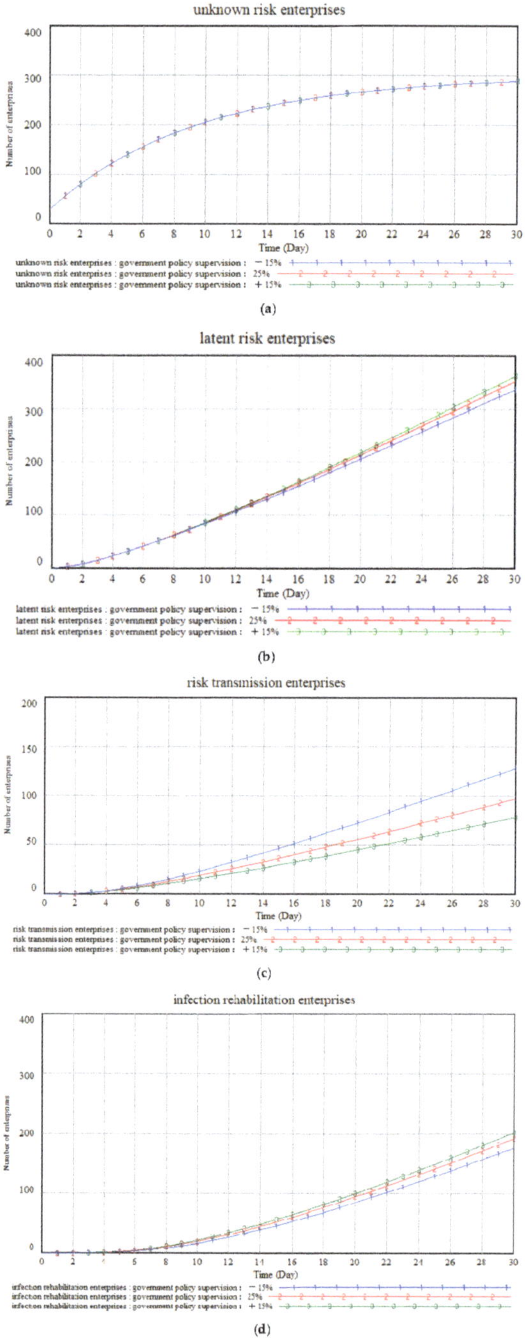

Figure 10. The impact of changes in government policy supervision on the number of enterprises in the process of risk propagation: (**a**) the impact of government policy support on enterprises with unknown risks; (**b**) the impact of government policy support on enterprises with latent risks; (**c**) the impact of government policy support on risk communication enterprises; (**d**) the impact of government policy support on enterprises recovering from infection.

4.3.5. Impact of Change in the Proportion of Lost Immunity on Risk Transmission

The initial immunity loss ratio is set to 20%, the increase is set to 25%, and the decrease is set to 15%, and the change in the number of companies in the process of risk transmission is obtained from Figure 11.

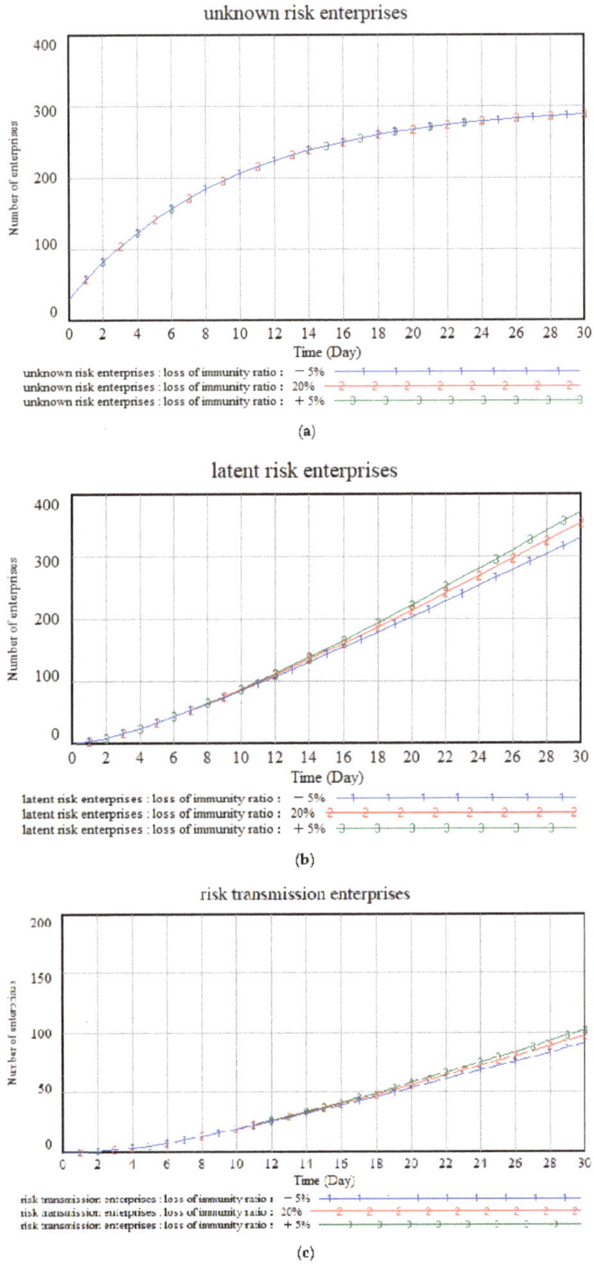

Figure 11. *Cont.*

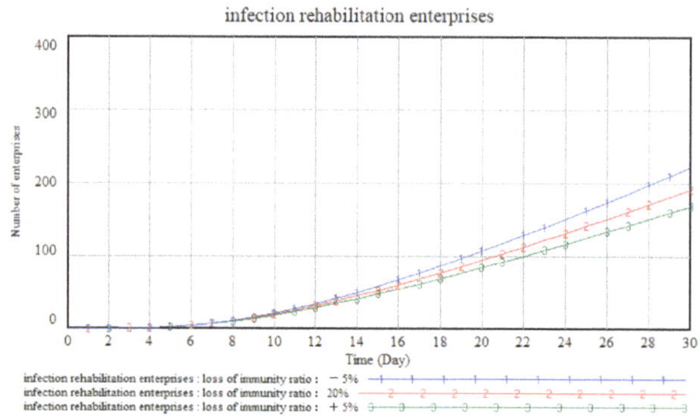

(**d**)

Figure 11. The impact of changes in the proportion of lost immunity on the number of enterprises in the process of risk transmission: (**a**) the impact of the proportion of immunity loss on enterprises with unknown risk; (**b**) the impact of the proportion of loss of immunity on risk-latent enterprises; (**c**) the impact of the proportion of immunity loss on enterprises that spread risk; (**d**) impact of the proportion of lost immunity on enterprises recovering from infection.

Under the influence of government policies and supervision, the number of enterprises with unknown risks gradually increases, but a certain increase or decrease in the ratio of 20% will not affect the image transformation of enterprises with unknown risks. This is because in the context of considering relapse, infection rehabilitation enterprises become risk-latent enterprises after losing immunity to risks, and then continue to become risk transmission enterprises and infection rehabilitation enterprises with the spread of risks, and will not become risk-unknown enterprises. In the context of this risk transmission, the change in the proportion of loss of immunity will have an impact on risk-latent enterprises, risk transmission enterprises, and infection recovery enterprises, and will not affect the change in the number of risk-unknown enterprises.

Enterprises with lurking risks are generally showing a growth trend. The change in the proportion of loss of immunity has a direct impact on the number of risk-latent enterprises, and the larger the proportion of loss of immunity, the greater the number of risk-latent enterprises.

With the increase in the proportion of loss of immunity, the number of risk transmission enterprises gradually increases, which is directly proportional to the two; a lower number of recovered enterprises is inversely proportional to the two.

5. Conclusions

Based on the consideration of the SEAIR model, this paper established the risk propagation model of the prefabricated building supply chain and studied the impact of each influencing factor on the supply chain risk propagation by simulating each enterprise, increase and decrease in the virus transmission model, and used Vensim simulation software to set the initial value of related factors. Based on the above results, relevant suggestions are made:

1. With the increase in infection rate, the number of enterprises in the supply chain that are not affected by risk will be greatly reduced, and the number of risk transmission enterprises and the number of risk-latent enterprises will increase significantly. With the increase in transmission rate, the change trend of risk transmission enterprises, risk-latent enterprises, and infection recovery enterprises in the supply chain is roughly the same as the change in the number of enterprises brought about by the increase or decrease in infection rate, but the change in transmission rate will not affect the change in the number of enterprises with unknown risk. Through the above simulation, it can be seen that this is because, while considering the recurrence, some of the risk rehabilitation enterprises may continue to form new risk-latent enterprises,

and with the spread of risks, new risk-latent enterprises become risk transmission enterprises and infection rehabilitation enterprises, which will not affect the number of risk-unknown enterprises. In order to effectively control the number of enterprises affected by risks, it is necessary to take certain measures to reduce the infection rate, which requires enterprises to actively seek the cooperation of other manufacturers when producing products, find enterprises eroded by risks, avoid risks in time, and reduce infection from the root. The enterprise itself actively takes anti-risk measures, understands the enterprises it contacts to prevent itself from being infected, understands the potential risk factors of the department of the enterprise affected by the risk in the production process of the product, assesses its risk level, prepares emergency plans in advance, reduces the business dealings between the department affected by the risk and other departments that are not affected, and avoids the expansion of the risk infection rate.

2. With the strengthening of government financial support and government policy supervision, the number of risk-spreading enterprises in some manufacturers affected by risks in the supply chain began to decline, and the ability of enterprises to eliminate risks was enhanced. Therefore, to reduce the number of risk-spreading enterprises in manufacturers affected by risks, the government can appropriately introduce relevant policies, such as giving appropriate financial subsidies to help the trapped manufacturing enterprises get rid of the risks as soon as possible and resume production. A series of targeted measures can be taken to reduce the tax ratio and reduce the burden on enterprises.
3. With the increase in the proportion of enterprises losing immunity in the supply chain, the ability of enterprises to mitigate risks decreases, and they are easily eroded by risks, resulting in an increase in the number of risk-spreading enterprises affected by risks. Therefore, to reduce the spread of risks in the supply chain, enterprises should gradually improve their ability to cope with risks, avoid the impact of risks, strengthen contact with other enterprises in the supply chain, and enhance their ability to cope with risks. As the information flow, capital flow, logistics, and other aspects among manufacturing enterprises in the supply chain are interconnected and closely related, one party will be implicated in many parties when it receives the risk impact, and the enterprises affected by the risk should actively inform the relevant business enterprises of their own information affected by the risk to provide a reference for the risk prevention of other enterprises. In the process of dealing with risks, each node enterprise should help each other and avoid fighting alone. Enterprises that have not been affected by risks should actively face risks while assisting those affected by risks to reduce the impact of risk erosion on the supply chain. This paper studies the mechanism of supply chain risk transmission from a macro perspective. In this study, enterprises without obvious symptoms are easily ignored after risk erosion, so asymptomatic infected enterprises are increased, a virus transmission model is established, and the impact of risk transmission in the supply chain on enterprises is analysed. Taking the risk impact of unknown risk enterprises, risk-latent enterprises, risk transmission enterprises, and infection recovery enterprises as examples, suggestions for dealing with supply chain risks are put forward from three perspectives: reducing the infection rate and transmission rate, strengthening government support, and coping with risk erosion. At the same time, because the SEAIR model is a typical model for studying virus transmission, with the development of society, human beings have more diverse ways to deal with virus transmission, and the model for studying virus transmission also needs to be improved. The research process of this paper does not consider the time lag between the measures taken by enterprises and their effects after suffering from risk erosion, that is, the transmission process of enterprises affected by risks and solving risk problems has a time transmission process. The length of the time transmission process has a certain impact on the number of risk transmission

enterprises and infection recovery enterprises. Therefore, future studies will improve on the basis of the SEAIR model and increase the time delay for analysis.

Author Contributions: Conceptualization, X.G. and R.S.; methodology, Y.W.; software, Y.W.; validation L.R. and Y.W.; formal analysis, L.L. and X.W.; investigation. L.R.; resources. Y.W.; data curation, L.R.; writing—original draft preparation, R.S.; writing—review and editing, X.G. and Y.W. visualization, L.R.; supervision L.L. and X.W. All authors have read and agreed to the published version of the manuscript.

Funding: This research was funded by the State Key Program of National Social Science Foundation of China (Grant no. 19AGL030) and Research Subject of Social Science Development of Hebei (Grant no. 20210201118).

Data Availability Statement: We declare that all data, models, and code generated or used during the study appear in the submitted article.

Acknowledgments: In this study, I would like to thank Wang Yingchen for his careful guidance and strong support and the students on the team for listening and exchanging ideas so that the research paper could be successfully published.

Conflicts of Interest: The authors declare no conflict of interest.

References

1. Masood, R.; Lim, J.B.P.; González, V.A.; Roy, K.; Khan, K.I.A. A Systematic Review on Supply Chain Management in Prefabricated House-Building Research. *Buildings* **2022**, *12*, 40. [CrossRef]
2. Zhu, T.; Liu, G. A Novel Hybrid Methodology to Study the Risk Management of Prefabricated Building Supply Chains: An Outlook for Sustainability. *Sustainability* **2023**, *15*, 361. [CrossRef]
3. Zhang, K.; Tsai, J.-S. Identification of Critical Factors Influencing Prefabricated Construction Quality and Their Mutual Relationship. *Sustainability* **2021**, *13*, 11081. [CrossRef]
4. Liu, K.; Su, Y.; Zhang, S. Evaluating Supplier Management Maturity in Prefabricated Construction Project-Survey Analysis in China. *Sustainability* **2018**, *10*, 3046. [CrossRef]
5. Zhang, C.; Qiao, M. Risk assessment of prefabricated building supply chain based on BP neural network. *Proj. Manag. Technol.* **2022**, *20*, 28–33. (In Chinese)
6. Wai, C.T.; Yi, P.W.; Olanrewaju, O.I.; Abdelmageed, S.; Hussein, M.; Tariq, S.; Zayed, T. A critical analysis of benefits and challenges of implementing modular integrated construction. *Int. J. Constr. Manag.* **2021**, *23*, 656–668. [CrossRef]
7. Wang, X.; Zhou, S.; Guo, Y.; Duan, Y. Research on supply chain risk of prefabricated buildings based on EFA and CFA. *Pract. Underst. Math.* **2021**, *51*, 286–299. (In Chinese)
8. Zhang, M. Research on the Risk of Prefabricated Building Supply Chain from the Perspective of Information Dissemination. Master's Thesis, Tianjin University of Technology, Tianjin, China, 2021. (In Chinese).
9. Gu, L. Research on the Supply Chain Risk Management of Prefabricated Concrete Buildings Based on System Dynamics. Master's Thesis, Yangzhou University, Yangzhou, China, 2020. (In Chinese).
10. Zhang, F. Study on the Risk Assessment of EPC Prefabricated Concrete Building Supply Chain Based on Cloud Model. Master's Thesis, Jiangxi University of Science and Technology, Ganzhou, China, 2020. (In Chinese).
11. An, H.; Kuang, Y.; Yang, W.; Song, L. Analysis on the motivation of supply chain integration of prefabricated buildings based on SEM. *J. Civ. Eng. Manag.* **2020**, *37*, 50–56. (In Chinese)
12. Wang, J.; Zhu, X.; Song, Z. Risk identification and evaluation of supply chain integration of prefabricated buildings based on interface management. *J. Qingdao Univ. Technol.* **2020**, *41*, 41–49.
13. Al-Aidrous, A.-H.M.H.; Shafiq, N.; Al-Ashmori, Y.Y.; Al-Mekhlafi, A.-B.A.; Baarimah, A.O. Essential Factors Enhancing Industrialized Building Implementation in Malaysian Residential Projects. *Sustainability* **2022**, *14*, 11711. [CrossRef]
14. Orlowski, K.; Shanaka, K.; Mendis, P. Manufacturing, Modeling, Implementation and Evaluation of a Weatherproof Seal for Prefabricated Construction. *Buildings* **2018**, *8*, 120. [CrossRef]
15. Luo, L.; Shen, G.Q.; Xu, G.; Liu, Y.; Wang, Y. Stakeholder-Associated Supply Chain Risks and Their Interactions in a Prefabricated Building Project in Hong Kong. *J. Manag. Eng.* **2019**, *35*, 05018015. [CrossRef]
16. Wuni, I.Y.; Shen, G.Q.; Osei-Kyei, R.; Agyeman-Yeboah, S. Modelling the critical risk factors for modular integrated construction projects. *Int. J. Constr. Manag.* **2022**, *22*, 2013–2026. [CrossRef]
17. Saad, S.; Alaloul, W.S.; Ammad, S.; Altaf, M.; Qureshi, A.H. Identification of critical success factors for the adoption of Industrialized Building System (IBS) in Malaysian construction industry. *Ain Shams Eng. J.* **2021**, *13*, 101547. [CrossRef]
18. Daley, D.J.; Kendall, D.G. Epidemics andrumours. *Nature* **1964**, *204*, 11–18. [CrossRef] [PubMed]
19. Anderson, R.; May, R. *Infectious Diseases of Humans: Dynamics and Control*; Oxford University Press: Oxford, UK, 1991.

20. Tan, Y.; Cai, Y.; Wang, X.; Peng, Z.; Wang, K.; Yao, R.; Wang, W. Stochastic dynamics of an SIS epidemiological model with media coverage. *Math. Comput. Simul.* **2023**, *204*, 1–27. [CrossRef]
21. Dobie, A.P. Susceptible-infectious-susceptible (SIS) model with virus mutation in a variable population size. *Ecol. Complex.* **2022**, *50*, 101004. [CrossRef]
22. Choi, W.; Lin, Z.; Ahn, I. SIS reaction–diffusion model with risk-induced dispersal under free boundary. *Nonlinear Anal. Real World Appl.* **2022**, *67*, 103605. [CrossRef]
23. Chen, J.; Yin, T. Transmission Mechanism of Post-COVID-19 Emergency Supply Chain Based on Complex Network: An Improved SIR Model. *Sustainability* **2023**, *15*, 3059. [CrossRef]
24. Liang, D.; Bhamra, R.; Liu, Z.; Pan, Y. Risk Propagation and Supply Chain Health Control Based on the SIR Epidemic Model. *Mathematics* **2022**, *10*, 3008. [CrossRef]
25. Ni, G.; Wang, Y.; Gong, L.; Ban, J.; Li, Z. Parameters Sensitivity Analysis of COVID-19 Based on the SCEIR Prediction Model. *COVID* **2022**, *2*, 1787–1805. [CrossRef]
26. Baba, I.A.; Humphries, U.W.; Rihan, F.A. A Well-Posed Fractional Order Cholera Model with Saturated Incidence Rate. *Entropy* **2023**, *25*, 360. [CrossRef] [PubMed]
27. Aliano, M.; Cananà, L.; Cestari, G.; Ragni, S. A Dynamical Model with Time Delay for Risk Contagion. *Mathematics* **2023**, *11*, 425. [CrossRef]
28. Li, B.; Eskandari, Z.; Avazzadeh, Z. Dynamical Behaviors of an *SIR* Epidemic Model with Discrete Time. *Fractal Fract.* **2022**, *6*, 659. [CrossRef]
29. Marinovg, T.T.; Marinovag, R.S. Inverse problem for adaptive SIR model: Application to COVID-19 in Latin America. *Infect. Dis. Model.* **2022**, *7*, 134–148. [CrossRef]
30. Karmaker, C.L.; Al Aziz, R.; Palit, T.; Bari, A.M. Analyzing supply chain risk factors in the small and medium enterprises under fuzzy environment: Implications towards sustainability for emerging economies. *Sustain. Technol. Entrep.* **2023**, *2*, 100032. [CrossRef]
31. Kumar, S.; Barua, M.K. Modeling and investigating the interaction among risk factors of the sustainable petroleum supply chain. *Resour. Policy* **2022**, *79*, 102922. [CrossRef]
32. Arshad, H.; Zayed, T. Critical influencing factors of supply chain management for modular integrated construction. *Autom. Constr.* **2022**, *144*, 104612. [CrossRef]
33. Sunmola, F.; Burgess, P.; Tan, A.; Chanchaichujit, J.; Balasubramania, S.; Mahmud, M. Prioritising Visibility Influencing Factors in Supply Chains for Resilience. *Procedia Comput. Sci.* **2023**, *217*, 1589–1598. [CrossRef]
34. Xu, M.; Cui, Y.; Hu, M.; Xu, X.; Zhang, Z.; Liang, S.; Qu, S. Supply chain sustainability risk and assessment. *J. Clean. Product.* **2019**, *225*, 857–867. [CrossRef]
35. Yi, C.-Q.; Meng, S.-D.; Zhang, D.-M. Studies on the Supply Chain Risk Management Using Complex Network. In *Liss 2012*; Springer: Berlin/Heidelberg, Germany, 2013; pp. 181–187.
36. Sun, Q.; Yang, G.; Zhou, A. An Entropy-Based Self-Adaptive Node Importance Evaluation Method for Complex Networks. *Complexity* **2020**, *2020*, 4529429. [CrossRef]
37. Batistela, C.M.; Ramos, M.M.; Cabrera, M.A.; Dieguez, G.M.; Piqueira, J.R. Vaccination and social distance to prevent COVID-19. *IFAC-PapersOnLine* **2021**, *54*, 151–156. [CrossRef]
38. Liu, P.; Tan, X. Dynamics Analysis of a Class of Stochastic SEIR Models with Saturation Incidence Rate. *Symmetry* **2022**, *14*, 2414. [CrossRef]
39. Wu, Z.; Wang, Y.; Gao, J.; Song, J.; Zhang, Y. A Multistage Time-Delay Control Model for COVID-19 Transmission. *Sustainability* **2022**, *14*, 14657. [CrossRef]
40. Paul, S.; Mahata, A.; Ghosh, U.; Roy, B. Study of SEIR epidemic model and scenario analysis of COVID-19 pandemic. *Ecol. Genet. Genom.* **2021**, *19*, 100087. [CrossRef] [PubMed]

buildings

MDPI

Article

The Current Risk Management Practices and Knowledge in the Construction Industry

Rami A. Bahamid [1], Shu Ing Doh [1], Muhamad Azry Khoiry [2,*], Mukhtar A. Kassem [3,*] and Mohammed A. Al-Sharafi [4]

1 Faculty of Civil Engineering Technology, Universiti Malaysia Pahang, Gambang, Kuantan 26300, Pahang, Malaysia; ra2007mi2007@gmail.com (R.A.B.); dohsi@ump.edu.my (S.I.D.)
2 Department of Civil Engineering, Faculty of Engineering and Built Environment, Universiti Kebangsaan Malaysia, Bangi 43600, Selangor, Malaysia
3 Department of Quantity Surveying, Faculty of Built Environment & Surveying, Universiti Teknologi Malasia, Skudai 81310, Malaysia
4 Department of Information Systems, Azman Hashim International Business School, Universiti Teknologi Malaysia, Skudai 81310, Malaysia; alsharafi@ieee.org
* Correspondence: azrykhoiry@ukm.edu.my (M.A.K.); mukhtarkas@gmail.com (M.A.K.)

Abstract: Construction is a critical sector of any economy in terms of value production, labor, and contributing to the gross national product. Managing risk is a relatively young area in Yemen's construction sector, but it is gaining traction as building activity and competition rise. Construction firms mitigate risk by using a variety of risk management methods. Therefore, there is a need to assess these procedures in order to detect shortcomings. This research aims to establish the existing risk management strategies used in Yemeni building projects. Survey questionnaires were used to collect data. Respondents were drawn from Yemeni construction businesses. Risk management is not executed systematically, intentionally, or continuously, and most firms' risk management procedures are reactive, semipermanent, informal, and unstructured, with no or few dedicated resources to address risks. This strategy is inconsistent with generally accepted risk management principles. Nonetheless, the findings suggest a general understanding of risk management and a willingness to learn from previous errors. The study of the findings suggests that risk identification approaches such as judgment and historical data are employed for risk analysis, and that the industry typically attempts to avoid or transfer risks in Yemeni building projects. The results shed light on the shortcomings of Yemen's project management practices. To guarantee that construction projects obtain maximum value for money, project managers of big construction businesses in Yemen need a strong understanding of and training in globally accepted systematic risk management procedures. Finally, this study can help future stakeholders determine how to work together to manage risk.

Keywords: risk management; risk management techniques; construction projects; systematic risk management; Yemen

Citation: Bahamid, R.A.; Doh, S.I.; Khoiry, M.A.; Kassem, M.A.; Al-Sharafi, M.A. The Current Risk Management Practices and Knowledge in the Construction Industry. *Buildings* **2022**, *12*, 1016. https://doi.org/10.3390/buildings12071016

Academic Editors: Yongjian Ke, Jingxiao Zhang and Simon P. Philbin

Received: 9 June 2022
Accepted: 7 July 2022
Published: 14 July 2022

1. Introduction

The construction sector contributes significantly to the growth of any nation's economy by creating value, labor, and contributing to the gross national product (GDP) [1–5]. Risk management is critical to attaining project goals owing to the complexity of construction projects, which include a plethora of hazards affecting a variety of stakeholders [6]. Since risk management is a new concept in Yemen's construction sector, it has received little investigation. The aims of project risk management are to exploit the possibility and effect of good occurrences and to reduce the probability and influence of negative events [7]. Thus, project risk management would enhance project performance by ensuring that project objectives are met and by seeking chances to maximize positive effects on goals. Risk is a random incidence or condition that, if it occurs, has an influence on one or more project goals either positively or negatively [8]. Risk management is the procedure of evaluating and managing these risk occurrences to reduce or eliminate the incidence and effect of negative impacts and promote positive benefits. As project risk grows, management and control become more challenging [9]. Numerous failures occur over the course of the project as a result of the risk and unpredictability inherent in the environment and in the project structure. However, because the risk is hard to measure, a full and complete definition has yet to be developed [10].

Owoo et al. [2] defined construction as the process of constructing, repairing, maintaining, altering, and deconstructing structures, highways, streets, bridges, roads, sewers, trains, and communication networks. The industry's significance stems from its close ties to other sectors of the economy since it generates considerably more significant effects through strong linkages with other industries [2,9,11]. However, no construction activity is risk-free due to the industry's demanding, dynamic, and complicated character. According to research, both practitioners and scholars in the sector believe that construction projects face higher risks than other industries due to their complexity [12–15].

Crispim et al. [16] said that an incapability to manage risk is the primary reason for project budgets, timelines, and other project goals being exceeded. Numerous scholars have examined underperformance in construction projects, including cost overruns, schedule delays, and lengthy and expensive conflicts that result in unnecessary claims [17–19]. Construction industries in developing countries handle risk management in construction by relying on a collection of often inadequate techniques resulting in substandard outcomes and restricting project performance [13,20]. Crispim et al. [16] carried out research to ascertain in what way project managers handle risk in a business; 6.8% of the respondents queried in the research worked in civil construction, while the remainder worked in information technology (IT), finance, telecommunications, and manufacturing. As a result, their conclusions were not customized for the construction sector in particular. Serpell et al. [13] also examined how to assess the ability of construction companies to manage risk effectively and how to assist them in improving. Their research did not examine whether the methodical processes advocated by effective risk management methods were being followed but rather examined organizational maturity. As a result, this article tries to examine existing risk management procedures in the construction sector via sequential steps, thus systematizing the process.

Yemen and other developing nations must develop and build further development projects to boost economic growth and create more sophistication and employment possibilities since these projects need significant investment. Therefore, in this scenario, any expenditures (delays, assets, and budget) will result in a significant financial loss, despite the fact that the numerous risk factors connected with these projects will have a considerable influence on the project's completion time, without outweighing the expenses [21,22]. Yemen's construction projects are the fourth most significant source of employment, employing roughly 9–10% of the workforce and growing at an annual average pace of approximately 5.4 percent, effectively contributing to Yemen's economic development. Yemen's construction industry continues to be a critical component of the country's development process and an income source for all organizations and individuals, including the construction sector, which has a significant influence on domestic earnings and average employee wages [21,23]. Yemen, similar to other developing nations, faces grave hazards in the building business [5,18,24–27].

Additionally, based on [23], statistics indicate that 47% of finished projects in Yemen were late, while 40% of overall projects ran over budget. Construction risk has emerged as a significant problem for investors, requiring careful analysis and thorough study to manage. To identify gaps in the adoption of a formal risk management process, the primary goal of this study is to assess the existing risk management techniques and their implementation in the Yemeni construction sector. The objective is to assist stakeholders in assessing their current and prospective projects, emphasizing risk management strategies. This research investigated these topics and adds to the field of risk management by outlining essential risk management strategies.

2. Literature Review

In the context of construction management, risk management refers to a detailed and systematic method for determining, assessing, and mitigating risks to accomplish project objectives [28]. El-Karim et al. [29] described risk management as the systematic method of recognizing, assessing, and reacting to project risk, which involves increasing and decreasing the likelihood and effects of positive and negative qualities, respectively. Serpell et al. [13] also described risk management as a planned and proactive approach aimed at minimizing the chance for unsatisfactory outcomes throughout the many phases of a construction project, including design, construction, and operation. According to Taofeed et al. [15], Risk management is a process that entails identifying hazards, assessing them qualitatively and quantitatively, and reacting with an effective strategy for control and treatment. Banaitiene and Banaitis [28] discovered that to ensure effective and efficient risk management, it is essential to have a sound and systematic methodology in place and breadth of knowledge and expertise. Nevertheless, how risks are addressed on the majority of building projects in Yemen, which reflect the environment of impoverished nations, is unknown. Risk management throughout the project delivery value chain has the potential to help the industry. This may have a cascading impact on a construction project's performance, productivity, budget, and quality [30]. Enshassi and Mosa [31] also said that in order to accomplish a company's goals and expectations, it is important to build a system of time management, cost, safety, and quality to manage a project well.

In order to best position it to satisfy customers' expectations, the growth of the construction industry requires an understanding of risk management procedures [32]. The industry has a bad reputation for risk management since it is burdened with several uncertainties as a result of expert assessment's subjective opinion, resulting in projects missing their cost and schedule objectives [27,33,34]. To ensure that both project customers and contractors receive value for money, project managers must use sound risk management procedures [13]. Identifying the risk is the first and most important and dangerous step in risk management processes, as indicated by many researchers [28,35,36]; nevertheless, according to [8], the entire risk management process must be planned, and thus the risk management plan must begin before identifying and classifying risks and their sources. Risk identification attempts to classify responsibilities [28]; identify potential risks for any project, including sources and potential consequences [37,38]; document characteristics that help in analyzing those risks; formulate appropriate responses to them [8]; and thus build an understanding of the circumstances and events that lead to risks [39].

Risk analysis is a process that occurs between risk identification and risk management [40]. This approach assigns a value to recognized sources of risk and uncertainty about the project's objectives and is often accompanied by an evaluation of the likelihood of occurrence and severity of risk consequences [38,41,42]. A qualitative and/or quantitative analysis of the identified hazards is performed. Qualitative risk analysis quantifies the likelihood of hazards occurring and the severity of their associated repercussions in a project [37]. Banaitiene and Banaitis [28] noted that qualitative risk analysis enables the identification of critical risk elements. It is not necessary to quantify all the dangers that have been identified qualitatively. As Project Management Institute PMI [8] explains, it is not even essential for all projects. The body of knowledge is broadly in agreement over what defines quantitative risk analysis. According to Banaitiene and Banaitis [28], quantitative risk management entails the use of increasingly advanced tools and procedures for investigating and analyzing risk, including predicting the likelihood of occurrence of identified risk factors and their possible effects on projects.

Risk response is the process through which plans are developed to address hazards that have been recognized and evaluated as they happen [37]. According to PMI [8], these expressed approaches are chosen among produced alternatives and actions addressing the total project risk exposure and the treatment of agreed-upon project hazards. This agreement is founded on PMI's belief that the plan must be agreed upon by all members of the project team and important stakeholders. This is the stage at which a purchasing strategy for reacting to recognized and evaluated hazards when they arise is executed. According to Hansen-Addy and Fekpe [37], this is the ultimate procedure in systematic risk management. They continue to assess and react to present and new threats throughout this period. Residual risks, that is, risks that persist after risk measures are implemented, are monitored, as are secondary risks, which occur directly due to risk remedies being implemented [8]. Rezakhain [35] also mentioned that this stage is responsible for implementing the risk mitigation plan, which PIM [8] classified as risk response, as well as tracking identified risks, monitoring residual risks, identifying new risks, and evaluating the overall risk management process effectiveness. This notion is also strongly supported by PIM [8]. Prudent risk management does not mean avoiding it, but rather recognizing it and making decisions concerning all the opportunities and risks that come with it [38,42].

Risks in the construction industry have been identified by numerous studies in different countries [9,43–52]. While there are several books on risk management, there is scant data available on its use in practice [53]. Additionally, risk management is significantly impacted by the distinctiveness of a country's building sector [46]. The implementation of project risk management varies significantly by industrial area [54,55]. Moreover, [8,9,49,50,56–58] have examined a number of risk management techniques. Nevertheless, most of these techniques may not be appropriate in local environments. Zou et al. [59] identified significant risks in construction projects in Australia and measured their possibility of occurrence. In addition, various risks that cause delays in construction projects have been identified in Malaysia, Libya, and Ghana [18,60,61]. Perera et al. [62] implemented the Delphi method in Sri Lanka to classify serious risks on a lifecycle basis and clarify how such risks were shared and controlled by the parties involved in the projects. Pawar and Pagey [63] identified risks, examined severity, and surveyed risk management actions in India, while El-Karim et al. [29] measured the consequence of factors affecting time and cost contingencies. These earlier studies only considered risk identification, apportionment, and effects, but not how contractors handle risk.

It is important to note that many types of research, including those reported in [32,64–68], have examined risk management in construction project management. However, the majority of these studies place a premium on risk management from the standpoint of the project client, with slight consideration given to the viewpoint of contractors. Expanding risk management by including other viewpoints from contractors is required to provide a fair picture of the construction risk literature. Furthermore, because construction projects are country- and sector-specific [69], and risk management practices differ significantly across sectors and nations [70], attempting to apply tactics from one economic system to another would be futile.

Within organizations, the project management technique now in use does not effectively support the growing need for risk management [71]. As a consequence, many projects lack risk management procedures [7,72,73]. A risk management system that is effective must be more dynamic than the risk itself. Otherwise, it is likely that it will struggle to integrate with the corporate culture and other business procedures. Regardless of the effort and care with which a system is prepared and deployed, it may fail to fulfill its aim upon the first installation and will need continual calibration. This demands management's leadership, patience, direction, time, and resources.

3. Materials and Methods

The first stage was to identify the critical activities involved in the construction risk management process. This was accomplished mostly through a survey of the literature. A questionnaire was developed based on the literature study, and a pilot study was performed to determine the questionnaire's applicability in the Yemeni context. The study questionnaire was developed specifically for the Yemeni construction sector and was adapted from those used previously [9,49,50,74]. Two parts comprised the questionnaire. The first portion was designed to elicit information about the respondents' demographic characteristics. Section 2 was designed to elicit impressions of existing risk management procedures and expertise in Yemeni building projects. A five-point Likert-type scale was integrated into the questionnaire, allowing for statistical analysis of the data. To elicit replies, questionnaire survey forms were delivered to construction experts affiliated with Yemen's building sector. The profile of the responders is shown in Table 1.

Table 1. Respondents profile.

Item	Frequency	Percent
Profession of Respondents		
Engineer	80	46.00
Quantity surveyor	25	14.40
Architect	49	28.20
Others	20	11.50
Total	174	100
Role of Respondents		
Chairman/director	7	4.00
Project manager	39	22.40
Project consultant	27	15.50
Site engineer/Architects	65	37.40
Site supervisor	17	9.80
Others	19	10.90
Total	174	100
Years of experience		
1–5 years	36	20.70
6–10 years	52	29.90
11–15 years	49	28.20
16–20 years	17	9.80
Above 20 years	20	11.50
Total	174	100

The study sampled reputable companies registered with Yemen's Ministry of Public Works and Highways. Although statistics were obtained from just four major cities in Yemen (Sana'a, Aden, Taiz, and Hadramout) due to the current situation and war in Yemen, these areas account for a significant portion of the country's construction sector owing to the current crisis and conflict in Yemen. A total of 400 questionnaires were distributed, and 174 (43.5 percent) of the responses were analyzed. The respondents were distributed geographically as follows: 27 in Sana'a, 47 in Aden, 39 in Taiz, and 61 in Hadramout. Because of the respondents' geographical dispersion, their size variances, their contribution to industrial growth, and their broad expertise in a range of building projects, the data gathered were comprehensive and assumed to be representative of the construction industry. The majority of the respondents are engineers (46 percent) who are there for guiding projects in the structural and stability aspects of the construction venture; architects (28.20 percent), who are focused more on the aesthetics and visual aspects of a project; quantity surveyors (14.40 percent), who are there from the start of the project and using their expertise to evaluate the value and space of a project; and other respondents (11.50 present). Most of the respondents are site engineers/architects and project managers (50 percent). The gathered data were analyzed using the Statistical Package for Social Science (SPSS) software, produced by IBM in USA.

4. Results and Discussions

This section is divided into subheadings. It provides a concise and precise description of the experimental results, their interpretation, and the experimental conclusions that can be drawn.

4.1. Current State of Risk Management Practices

To ascertain respondents' knowledge and practice of risk management in construction projects in Yemen, participants were asked to react to questions using a Likert scale ranging from 1 to 5, with 1 indicating strong disagreement and 5 indicating strong agreement [9,66,75,76]. These questions were added to ascertain the current state of an organization's risk management system by eliciting broad views regarding the formality of their risk management processes (see Table 2). Risk management is not often used in Yemeni building projects. The Yemeni construction professionals agree that they encounter risks in their projects (mean of 4.01), but when they were asked to evaluate their knowledge and the implementation of risk management, most of them do not implement risk management in their projects—the mean score was 2.34, although almost all participants recognize the necessity of its implement (mean 3.99). Assigning reasons for these results have been answered in the following question, which over half of the respondents do not have knowledge of risk management (mean of 2.70). Moreover, the findings indicate that many respondents believe that risks in Yemen's construction projects are not allocated to the most appropriate party and other many allocations, which implies that risk needs to be allocated to the most appropriate party, either the owner or the contractors (mean 2.68). The majority of assessed firms' risk management systems and practices are responsive, semipermanent, unplanned, and unorganized, with few to none committing assets to control the risks. Nonetheless, there is an awareness of potential dangers and a willingness to learn from previous errors. These outcomes contradict those of [37], who found that construction experts consider risk management earnestly and are proactive instead of reactive in their approach. The study's results indicate that the scenario is much different in Yemen's building sector.

Table 2. Current practices and knowledge of risk management in Yemen.

Questions	Overall	
Risk management is generally implemented in your projects	2.34	Disagree
It is necessary to implement risk management in construction projects	3.99	Agree
Risks in Yemen construction are allocated to the most appropriate party	2.68	Disagree
You have knowledge of risk management in construction projects	2.70	Disagree
You encounter risks in your projects (e.g., war and political instability, etc.)	4.01	Agree
Risk Identification is implemented in a systematic manner throughout the project	2.58	Disagree
Risk Analysis is implemented in a systematic manner throughout the project	2.48	Disagree
Risk Response is implemented in a systematic manner throughout the project	2.72	Disagree

To assess current risk management methods in Yemen's construction sector, respondents were asked to identify whether risk identification, risk analysis, and risk response are carried out in a systematic way throughout the project (Table 2). From the results, respondents claimed that the risk management process does not apply systematically in their projects, with a mean of 2.58 for risk identification, 2.48 for risk assessment, and 2.72 for risk response. The findings show that most construction projects in Yemen do not apply systematic risk management identification, assessment, or response. As a result of this, it is reasonable to infer that risk management was not conducted in a continuous, deliberate, and systematic way. Risk management practices were not given sufficient attention, time, or resources, and their execution might best be described as chaotic and ad hoc. The aspect of risk management does not end at the identification phase—it continues through the subsequent steps of analysis and reaction. These findings contradict those of [77], who concluded that medium- to large-sized businesses in the United States of America primarily practice all the risk management process components in their projects. This is substantiated by [64], who argue that formal risk management is practiced in Malaysia by firms with strong reputations, unwavering financial standing, and involvement in large construction projects.

Notably, this research examined individuals in the Yemeni construction business as well as seasoned specialists at the top echelon. If they show that systematic risk management is not practiced, it is reasonable to assume that the construction sector in Yemen has missed the critical nature of systematic risk management and so pushed it to the sidelines. As a result, it is not surprising that building projects in underdeveloped nations, such as Yemen, face delays, cost overruns, substandard quality, and, in some cases, full abandonment. Consequently, project goals are not met [78,79].

4.2. Current State of Risk Management Practices

The preceding section discussed construction businesses' risk management techniques. This section discusses the tools and procedures that are currently being used in building projects. These methods and procedures are classified into three categories: risk identification, risk analysis, and risk response. On a scale of 1–5, respondents were asked to rate their frequency of use of risk management approaches, with 1 being never used and 5 representing always used.

4.2.1. Risk Identification Techniques

Risk identification is a critical step in the risk management process for construction projects. Unidentified hazards cannot be effectively controlled. Identification of risks is highly dependent on the expertise and insight of key project workers. Risk identification is a demanding undertaking, made more difficult by the increasing complexity of projects and the rapid rate of technological progress [54]. Table 3 shows many ways and strategies for detecting risks, along with how often they are used. According to the results in Table 2, intuition/judgment (mean = 3.30) is the frequently used method for identifying risk in Yemeni construction projects, followed by historical data (mean = 3.14), consulting experts (mean = 2.79), checklists (mean = 2.46), brainstorming (mean = 2.17), and interviewing (mean = 2.10).

Table 3. Ranking of risk identification techniques.

Risk Identification Techniques	Mean	Ranking
Historical Data	3.14	2
Checklists	2.46	4
Brainstorming	2.17	5
Intuition/judgment	3.30	1
Interviewing	2.10	6
Consulting expert	2.79	3

The findings indicated that the various techniques are used improperly. The majority of respondents rely on personal experience, and they were unfamiliar with reactive and proactive risk identification methods and their use in proper risk management process or system, despite it being the most significant (Mills, 2001). Moreover, the use of these tools are contingent upon the project's nature, the policy of the organization, the project management approach, the availability of resources, and the risk tolerance of project team members. This may reflect the fact that there is no one-size-fits-all strategy to risk identification methods. The findings indicate that the different strategies are used in an ad hoc manner. The majority of parties depend on personal experience and knowledge. Although interviewing is a primary strategy for identifying risks [80], it is placed sixth (mean 2.10). These findings reveal that the primary approaches to identifying risks in Yemen are based on prior experiences, personal knowledge, and information rather than extensive project analysis. This indicates that risk identification does not take place in accordance with the results of [37,53,81], which recognized consulting experts and brainstorming as effective techniques for risk identification. This study's conclusion elucidates the main reasons for building project failure. It is worth noting that building projects continue to fall short of their cost, schedule, and quality expectations [79].

4.2.2. Risk Analysis Techniques

Qualitative and quantitative risk assessments are both used in risk analysis. Risk analysis is possibly the most challenging aspect of project risk management [82]. Numerous frequently used qualitative and quantitative risk assessment methodologies and approaches have been found via a survey of relevant literature [7,80,83–85]. The participants were asked to rate the frequency with which they used three risk analysis methods. Table 4 summarizes the total rank of risk analysis methods using means: qualitative (mean = 2.20), quantitative (mean = 1.11), and semiquantitative (mean = 1.23). The low mean values show that analysis is rarely used to identify previously identified risks. The findings indicated that there are few documents of risks analyzed by any method, which is best viewed as a casual and trivial effort. Regardless of the fact that some project managers identify their risk analysis software, computers and utility are rarely used in combination with project management software. Advanced quantitative risk analysis techniques are rarely used. An additional matter is the lack of dependable data for quantitative risk analysis as the majority of organizations lack the necessary capacity, expertise, and system to track completed and ongoing construction projects.

Table 4. Risk analysis techniques ranking.

Risk Analysis Techniques	Mean	Ranking
Qualitative analysis	2.72	1
Quantitative analysis	1.10	3
Semiquantitative analysis	1.56	2

The low mean values indicate that analysis is rarely used to identify previously known dangers and that these groups are unfamiliar with their usefulness. Advanced quantitative risk analysis approaches are seldom used. Another further difficulty is the lack of trustworthy data for quantitative risk assessments since the majority of businesses lack the necessary infrastructure, skills, and capability to track current and finished construction projects. Additionally, the results indicated that risk analysis was not conducted in a systematic way throughout the project. However, it was discovered that the whole risk analysis procedure is not methodical. This additionally establishes the notion that risk management in construction is an ad hoc endeavor [86]. While [37,87,88] identified collaborative risk analysis among major project stakeholders as the most frequently employed risk analysis practice, this study's results imply otherwise. This indicates that, although this risk management approach was carried out in certain circumstances, it was not carried out consistently [89]. This makes sense since if the risk is not detected in a systematic manner, it cannot be assessed in a systematic manner. This is addressed by the absence of instruments for qualitative and/or quantitative risk analysis. As a result, risks are not evaluated for their likelihood and effect, priorities are not identified, and the whole influence of projects is not determined [8,28].

4.2.3. Risk Response Techniques

The overall mean rankings of response methods follow: avoid the risk (mean = 4.47), reduce the likelihood of occurrence (mean = 3.76), minimize the consequences (mean = 3.64), transfer the risk (mean = 3.46), risk-sharing (mean = 3.17), and retain the risk (mean = 2.80), as shown in Table 5. Avoiding risks, at the top of the ranking, implies missing out on significant business opportunities as a result of an overall cautious attitude. By taking risks, organizations earn money and increase their value. It is advisable to make enlightened decisions and to grip the advantage of opportunities that can be controlled efficiently while avoiding risks that are beyond the capabilities of the organization. Making enlightened decisions requires not only experience and sound judgment, but also familiarity with the risk management processes. Nonetheless, risk management is viewed as a function of the decision's quality. The decision maker's ability to make a bad or good decision is mostly determined by the quality of information he or she obtains. Information is the primary source of information during the risk identification and analysis steps. The ranking of risk response methods indicates that the construction industry has progressed far beyond risk-sharing and relies heavily on risk transfer. The findings indicate that no principles are followed when risk is transferred to a business partner; additionally, they lack the appropriate attitude toward risk, which results in conflicts and is frequently detrimental to the project objectives. The preceding findings demonstrate that there is no one-size-fits-all method to risk management and that a combination of techniques would be appropriate.

Table 5. Ranking of risk response techniques.

Risk Response Techniques	Mean	Ranking
Reduce the possibility of occurrence (probability)	3.76	2
Reduce the consequences (impact)	3.64	3
Avoid the risk	4.47	1
Transfer risk	3.46	4
Interviewing	2.10	6
Consulting expert	2.79	3

5. Conclusions and Recommendations

In the majority of developing nations, risk management research has concentrated on the identification of risk variables, allocation of risk, whether or not project teams possess the competence to appropriately evaluate risk and employ risk management assessment techniques. However, this article is intended to analyze the present risk management procedures in Yemen's construction sector. The techniques of project risk management in Yemen's construction sector were discussed. Because the majority of businesses do not conduct risk management, there is a need to enhance specific procedures and raise their frequency of usage.

The construction sector often depends on intuition and judgment to identify risks (mean = 3.30) and seldom does quantitative risk analysis (mean = 1.10). Risk is often avoided in the construction business (mean = 4.47). The majority of firms' risk management systems and practices are reactive, semipermanent, informal, and unstructured, with little or no committed resources to address risks. There is little recording of the risk management process, which is viewed as an informal and insignificant effort by all stakeholders. The research discovered that risk management is not conducted in a systematic manner due to a lack of risk identification, analysis, reaction, and monitoring tools. Generally, risk management procedures focus only on discovering, assessing, and reacting to hazards as they occur.

In the construction sector, a systematic risk management method must be used to limit the unfavorable effects of these risks, both collectively and individually, on the building project. The application and implementation of risk management approaches must be tailored to the local situation. Systematic development in the risk maturity level of local firms is required, particularly as risk management knowledge increases. Risk management rules for the country's industry must be created. International donor organizations interested in funding building projects across the country will indeed be educated on the industry's risk management techniques. Similarly, when foreign construction corporations decide to enter into a joint venture agreement with local contractors, they will be considerably assisted in their appraisal and risk assessment. Additionally, the results produce academic and professional dialogue that might be used to develop strategies for encouraging members to practice risk management in a systematic manner. Additionally, the government would be pushed to provide legislative support for risk management as a qualifying criterion for building project tendering. The findings indicate that throughout the nation, project management practitioners have put systematic risk management approaches in the background. The lack of adequate risk management methods results in substandard performance across several areas of the Yemeni construction industry. To address the low-performance phenomenon, the country's regular and scheduled periodic training of construction professionals has to be improved and made more frequent. This will improve and raise the bar for project managers' systematic risk management methods. The outcomes of this study paint a dismal picture for the business, implying the need for more research on how to pique practitioners' and stakeholders' interest in risk management implementation for the advantage of the sector. Recently, advocacy has shifted toward enterprise-level risk management. However, little information is available on enterprise risk management in Yemen's construction business.

The research findings enable designers, construction managers, and supervisors, together with other important project participants, to evaluate their current and prospective projects in light of the risk management approaches discussed in the study. The results contribute to a greater understanding of risk management in the Yemeni construction sector, hence adding to hitherto untapped information. Risk management assessment using stepwise techniques is new in Yemen's building business. Owing to the war and current situation in Yemen, the data were collected only from four cities in Yemen; for future research, a larger sample size should be obtained. Additional studies should concentrate on identifying the impediments to enterprise risk management adoption in poor countries such as Yemen. Further study will need to include the creation of a decision support system to facilitate enterprise risk management adoption. Finally, a quantitative study is proposed to identify the financial and other losses sustained by the construction sector as a consequence of contractors' poor risk management.

Author Contributions: Conceptualization, R.A.B., S.I.D., M.A.K. (Muhamad Azry Khoiry), M.A.K. (Mukhtar A. Kassem) and M.A.A.-S.; methodology, R.A.B., S.I.D. and M.A.K. (Mukhtar A. Kassem); validation, R.A.B., S.I.D., M.A.K. (Muhamad Azry Khoiry) and M.A.A.-S.; formal analysis, R.A.B., S.I.D., M.A.K. (Muhamad Azry Khoiry), M.A.K. (Mukhtar A. Kassem) and M.A.A.-S.; writing—original draft preparation, R.A.B., S.I.D. and M.A.A.-S.; writing—review and editing M.A.K. (Muhamad Azry Khoiry), M.A.K. (Mukhtar A. Kassem) and M.A.A.-S.; visualization, R.A.B., S.I.D., M.A.K. (Muhamad Azry Khoiry), M.A.K. (Mukhtar A. Kassem) and M.A.A.-S.; supervision, S.I.D., M.A.K. (Muhamad Azry Khoiry) and M.A.K. (Mukhtar A. Kassem); project administration, S.I.D., M.A.K. (Muhamad Azry Khoiry) and M.A.A.-S. All authors have read and agreed to the published version of the manuscript.

Funding: The research was supported by Universiti Teknologi Malaysia (UTM) Research Grant Vot No. 05E79.

Institutional Review Board Statement: Not applicable.

Informed Consent Statement: Not applicable.

Data Availability Statement: The data sets during and/or analyzed during the current study are available from the corresponding author upon reasonable request.

Acknowledgments: Authors are grateful to the University of Technology Malaysia (UTM) for supporting this research and providing research facilities.

Conflicts of Interest: The authors declare no conflict of interest.

References

1. Schoonwinkel, S.; Fourie, C.J.; Conradie, P.D.F. A risk and cost management analysis for changes during the construction phase of a project. *J. South Afr. Inst. Civ. Eng.* **2016**, *58*, 21–28. [CrossRef]
2. Owoo, N.S.; Lambon-Quayefio, M.P. The Construction Sector in Ghana. 2018. Available online: https://www.econstor.eu/handle/10419/190166 (accessed on 25 May 2022).
3. Urbański, M.; Haque, A.U.; Oino, I. The moderating role of risk management in project planning and project success: Evidence from construction businesses of Pakistan and the UK. *Eng. Manag. Prod. Serv.* **2019**, *11*, 23–35. [CrossRef]
4. Quigley, Q.J.; Dickson, A.; Kirytopoulos, K. Project Complexity and Risk Management (ProCRiM): Towards modelling project complexity driven risk paths in construction projects. *Int. J. Proj. Manag.* **2016**, *34*, 1183–1198. [CrossRef]

5. Bahamid, R.A.; Doh, S.I.; Al-Sharafi, M.A.; Rahimi, A.R. Risk Factors Influencing the Construction Projects in Yemen from Expert's Perspective. *IOP Conf. Ser. Mater. Sci. Eng.* **2020**, *712*, 12007. [CrossRef]
6. Wu, J.; Zhou, Y. Risk assessment of urban rooftop distributed PV in energy performance contracting (EPC) projects: An extended HFLTS-DEMATEL fuzzy synthetic evaluation. *Sustain. Cities Soc.* **2019**, *47*, 101524. [CrossRef]
7. Choudhry, R.M.; Iqbal, K. Identification of risk management system in construction industry in Pakistan. *J. Manag. Eng.* **2013**, *29*, 42–49. [CrossRef]
8. PMI. *A Guide to the Project Management Body of Knowledge (PMBOK Guide)*, 5th ed.; Project Management Institute, Inc.: Newton Square, PA, USA, 2017.
9. Boateng, S.; Ameyaw, A.; Mensah, C. Assessment of systematic risk management practices on building construction projects in Ghana. *Int. J. Constr. Manag.* **2020**, 1–10. [CrossRef]
10. Amani, N.; Safarzadeh, K. Project risk management in Iranian small construction firms. *J. Eng. Appl. Sci.* **2022**, *69*, 1–14. [CrossRef]
11. Elghdban, M.G.; Azmy, N.B.; Zulkiple, A.B.; Al-Sharafi, M.A. *A Systematic Review of the Technological Factors Affecting the Adoption of Advanced IT with Specific Emphasis on Building Information Modeling*; Springer: Berlin/Heidelberg, Germany, 2021; Volume 295. [CrossRef]
12. Mills, A. A systematic approach to risk management for construction. *Struct. Surv.* **2001**, *19*, 245–252. [CrossRef]
13. Serpell, A.; Ferrada, X.; Rubio, L.; Arauzo, S. *Evaluating Risk Management Practices in Construction Organizations*; Elsevier: Amsterdam, The Netherlands, 2015.
14. Shojaei, S.A.S.; Haeri, P. Development of supply chain risk management approaches for construction projects: A grounded theory approach. *Comput. Ind. Eng.* **2019**, *128*, 837–850. [CrossRef]
15. Taofeeq, K.; Adeleke, D.M.; Hassan, A.Q. Factors Affecting Contractors risk attitude from Malaysia construction industry perspective. *Soc. Sci. Humanit. J.* **2019**, *3*, 1281–1298.
16. Crispim, J.; Silva, L.H.; Rego, N. Project risk management practices: The organizational maturity influence. *Int. J. Manag. Proj. Bus.* **2019**, *12*, 187–210. [CrossRef]
17. Odeh, H.T.; Battaineh, A.M. Causes of construction delay: Traditional contracts. *Int. J. Proj. Manag.* **2002**, *20*, 67–73. [CrossRef]
18. Sambasivan, Y.W.; Soon, M. Causes and effects of delays in Malaysian construction industry. *Int. J. Proj. Manag.* **2007**, *25*, 517–526. [CrossRef]
19. Wang, C.; Tang, Y.; Kassem, M.A.; Ong, H.Y.; Yap, J.B.H.; Ali, K.N. Novel Quality-Embedded Earned Value Performance Analysis Tool for Sustainable Project Portfolio Production. *Sustainability* **2022**, *14*, 8174. [CrossRef]
20. Adeleke, M.W.; Bahaudin, A.Q.; Kamaruddeen, A.Y.; Bamgbade, A.M.; Ali, J.A. An empirical analysis of organizational external factors on construction risk management. *Int. J. Suppl. Chain Manag.* **2019**, *8*, 932.
21. Kassem, M.A.; Khoiry, M.A.; Hamzah, N. Using Relative Importance Index Method for Developing Risk Map in Oil and Gas Construction Projects. *J. Kejuruter.* **2020**, *32*, 85–97. [CrossRef]
22. Renuka, S.M.; Umarani, C.; Kamal, S. A Review on Critical Risk Factors in the Life Cycle of Construction Projects. *J. Civ. Eng. Res.* **2014**, *4*, 31–36. [CrossRef]
23. Ahmad, S.A.; Issa, U.H.; Farag, M.A.; Abdelhafez, L.M. Evaluation of Risk Factors Affecting Time and Cost of Construction Projects in Yemen. *Int. J. Manag. IJM* **2013**, *4*, 11.
24. El-Sayegh, R.; Manjikian, S.M.; Ibrahim, S.; Abouelyousr, A.; Jabbour, A. Risk identification and assessment in sustainable construction projects in the UAE. *Int. J. Constr. Manag.* **2021**, *21*, 327–336. [CrossRef]
25. Issa, U.H.; Farag, M.A.; Abdelhafez, L.M.; Ahmed, A.S. A risk allocation model for construction projects in Yemen. *Civ. Environ. Res.* **2015**, *7*, 78–89.
26. Mukhtar, N.H.; Kassem, A.; Khoiry, M.A. Theoretical review on critical risk factors in oil and gas construction projects in Yemen. *Eng. Constr. Archit. Manag.* 2020; *in press*. [CrossRef]
27. Bahamid, R.A.; Doh, S.I.; Al-Sharaf, M.A. Risk factors affecting the construction projects in the developing countries. *IOP Conf. Ser. Earth Environ. Sci.* **2019**, *244*, 012040. [CrossRef]
28. Banaitiene, N.; Banaitis, A. Risk Management in Construction Projects. In *Risk Management—Current Issues and Challenges*; InTech Open: London, UK, 2012. [CrossRef]
29. El-Karim, M.S.B.A.A.; El Nawawy, O.A.M.; Abdel-Alim, A.M. Identification and assessment of risk factors affecting construction projects. *HBRC J.* **2017**, *13*, 202–216. [CrossRef]
30. Kassem, M.A. Risk Management Assessment in Oil and Gas Construction Projects Using Structural Equation Modeling (PLS-SEM). *Gases* **2022**, *2*, 33–60. [CrossRef]
31. Enshassi, A.; Mosa, J.A. Risk Management in Building Projects: Owners ' Perspective. *IUG J. Nat. Eng. Stud.* **2008**, *16*, 95–123.
32. Augustine, E.; Ajayi, J.R.; Ade, B.A.; Edwin, A.A. Assessment of risk management practices in Nigerian construction industry: Toward establishing risk management index. *Int. J. Pure Appl. Sci. Technol.* **2013**, *16*, 20–31.
33. Bahamid, R.A.; Doh, S.I. A review of risk management process in construction projects of developing countries. *IOP Conf. Ser. Mater. Sci. Eng.* **2017**, *271*, 012042. [CrossRef]
34. Chatterjee, K.; Zavadskas, E.K.; Tamošaitien, J.; Adhikary, K.; Kar, S. A hybrid MCDM technique for risk management in construction projects. *Symmetry* **2018**, *10*, 46. [CrossRef]
35. Rezakhani, P. Classifying Key Risk Factors in Construction Projects. *Bull. Polytech. Inst. Jassy Constr. Archit. Sect.* **2012**, *62*, 27–38.
36. Dutta, E. Study of risk in construction contracts. *Int. J. Res. Eng. Technol.* **2014**, *2*, 21–26.

37. Hansen-Addy, E.; Fekpe, A. Risk Management Knowledge and Practices in the Ghanaian Construction Industry. 2015. Available online: http://www.westeastinstitute.com/wp-content/uploads/2015/07/Andrew-Hansen-Addy.pdf (accessed on 26 May 2022).
38. Szymański, P. Risk management in construction projects. *Procedia Eng.* **2017**, *208*, 174–182. [CrossRef]
39. Fan, Y.; Stevenson, M. Reading on and between the lines: Risk identification in collaborative and adversarial buyer–supplier relationships. *Supply Chain. Manag.* **2018**, *23*, 351–376. [CrossRef]
40. Wang, S.Q.; Dulaimi, M.F.; Aguria, M.Y. Risk management framework for construction projects in developing countries. *Constr. Manag. Econ.* **2004**, *22*, 237–252. [CrossRef]
41. Zavadskas, E.K.; Turskis, Z.; Tamošaitiene, J. Risk assessment of construction projects. *J. Civ. Eng. Manag.* **2010**, *16*, 33–46. [CrossRef]
42. Wang, C.; Tang, Y.; Kassem, M.A.; Li, H.; Hua, B. Application of VR technology in civil engineering education. *Comput. Appl. Eng. Educ.* **2022**, *30*, 335–348. [CrossRef]
43. Ismael, D.; Shealy, T. Sustainable construction risk perceptions in the Kuwaiti construction industry. *Sustainability* **2018**, *10*, 1854. [CrossRef]
44. Kassem, M.A.; Khoiry, A.; Hamzah, N. Evaluation of Risk Factors Affecting on Oil and Gas Construction Projects in Yemen. *Int. J. Eng. Technol.* **2019**, *8*, 6–14. [CrossRef]
45. Siang, L.C.; Ali, A.S. Implementation of Risk Management in the Malaysian Construction Industry. *J. Surv. Constr. Prop.* **2012**, *3*, 1–15. [CrossRef]
46. Andi. The importance and allocation of risks in Indonesian construction projects. *Constr. Manag. Econ.* **2006**, *24*, 69–80. [CrossRef]
47. Iqbal, S.; Choudhry, R.M.; Holschemacher, K.; Ali, A.; Tamošaitienė, J. Risk management in construction projects. *Technol. Econ. Dev. Econ.* **2015**, *21*, 65–78. [CrossRef]
48. Kartam, N.A.; Kartam, S.A. Risk and its management in the Kuwaiti construction industry: A contractors' perspective. *Int. J. Proj. Manag.* **2001**, *19*, 325–335. [CrossRef]
49. Li, X.; Wang, C.; Kassem, M.A.; Alhajlah, H.H.; Bimenyimana, S. Evaluation Method for Quality Risks of Safety in Prefabricated Building Construction Using SEM–SDM Approach. *Int. J. Environ. Res. Public Health* **2022**, *19*, 5180. [CrossRef] [PubMed]
50. El-Sayegh, S.; El-Sayegh, S.M. Project risk management practices in the UAE construction industry. *Int. J. Proj. Organ. Manag.* **2014**, *6*, 121–137. [CrossRef]
51. Hiyassat, M.A.; Alkasagi, F.; El-Mashaleh, M.; Sweis, G.J. Risk allocation in public construction projects: The case of Jordan. *Int. J. Constr. Manag.* **2022**, *22*, 1478–1488. [CrossRef]
52. Alaghbari, W.; Salim, A.; Dola, K.; Ali, A.A.A. Identification of significant factors influencing housing cost in Yemen. *Int. J. Hous. Mark. Anal.* **2012**, *5*, 41–52. [CrossRef]
53. Lyons, M.; Skitmore, T. Project risk management in the Queensland engineering construction industry: A survey. *Int. J. Proj. Manag.* **2004**, *22*, 51–61. [CrossRef]
54. El-Sayegh, S.M.; Mansour, M.H. Risk assessment and allocation in highway construction projects in the UAE. *J. Manag. Eng.* **2015**, *31*, 04015004. [CrossRef]
55. Hatem, Z.M.; Kassem, M.A.; Ali, K.N.; Khoiry, M.A. A New Perspective on the Relationship Between the Construction Industry Performance and The Economy Outcome-A Literature Review. 2022. Available online: https://www.researchgate.net/profile/Zaid-Hatem/publication/359402109_A_New_Perspective_on_the_Relationship_Between_the_Construction_Industry_Performance_and_The_Economy_Outcome-A_Literature_Review/links/623a2e3a2708166c05437280/A-New-Perspective-on-the-Relationship-Between-the-Construction-Industry-Performance-and-The-Economy-Outcome-A-Literature-Review.pdf (accessed on 28 May 2022).
56. Loosemore, M.; Raftery, J.; Reilly, C. *Risk Management in Projects*; Routledge: London, UK, 2006.
57. Forbes, M.; Smith, D.; Horner, S. Tools for selecting appropriate risk management techniques in the built environment. *Constr. Manag. Econ.* **2008**, *26*, 1241–1250. [CrossRef]
58. Jaafari, A. Management of risks, uncertainties and opportunities on projects: Time for a fundamental shift. *Int. J. Proj. Manag.* **2001**, *19*, 89–101. [CrossRef]
59. Zou, P.X.W.; Zhang, G.; Wang, J. Understanding the key risks in construction projects in China. *Int. J. Proj. Manag.* **2007**, *25*, 601–614. [CrossRef]
60. Tumi, H.K.; Omran, S.A.H.; Pakir, A. Causes of delay in construction industry in Libya. In Proceedings of the International Conference on Economics and Administration, Bucharest, Romania, 14–15 November 2009; pp. 265–272.
61. Fugar, B.; Agyakwah-Baah, F.D. Delays in building construction projects in Ghana. *Australas. J. Constr. Econ. Build.* **2010**, *10*, 128–141. [CrossRef]
62. Perera, A.K.S.; Rameezdeen, R.; Chileshe, N.; Hosseini, M.R. Enhancing the effectiveness of risk management practices in Sri Lankan road construction projects: A Delphi approach. *Int. J. Constr. Manag.* **2014**, *14*, 1–14. [CrossRef]
63. Pawar, S.; Pagey, A. Survey and analysis of risk management in building construction work. *Int. J. Res. Eng. Technol.* **2017**, *4*, 2297–2299.
64. Mohamed, O.; Abd-Karim, B.; Roslan, H.; Suhaimi, M.; Danuri, M.; Zakaria, N. Risk management: Looming the modus operandi among construction contractors in Malaysia. *Int. J. Constr. Manag.* **2015**, *15*, 82–93. [CrossRef]

65. Perez, M.; Gray, D.; Skitmore, J. Perceptions of risk allocation methods and equitable risk distribution: A study of medium to large Southeast Queensland commercial construction projects. *Int. J. Constr. Manag.* **2017**, *17*, 132–141. [CrossRef]
66. Renault, N.; Agumba, B.; Ansary, J. An exploratory factor analysis of risk management practices: A study among small and medium contractors in Gauteng. *Acta Structilia* **2018**, *25*, 1–39. [CrossRef]
67. Al Mhdawi, M.K.; Motawa, I.; Rasheed, H.A. Assessment of Risk Management Practices in Construction Industry. In *Lecture Notes in Mechanical Engineering*; Springer: Berlin/Heidelberg, Germany, 2020; pp. 421–433. [CrossRef]
68. Jepson, J.; Kirytopoulos, K.; Chileshe, N. Isomorphism within risk-management practices of the Australian construction industry. *Int. J. Constr. Manag.* **2020**, *22*, 1508–1524. [CrossRef]
69. Alfen, H.; Kalidindi, S.; Ogunlana, S.; Wang, S. *Public-Private Partnership in Infrastructure Development: Case Studies from Asia and Europe*; Verlag der Bauhaus-Universität: Weimar, Germany, 2009.
70. Khodeir, L.M.; Mohamed, A.H.M. Identifying the latest risk probabilities affecting construction projects in Egypt according to political and economic variables. From January 2011 to January 2013. *HBRC J.* **2015**, *11*, 129–135. [CrossRef]
71. Baarimah, A.O.; Alaloul, W.S.; Liew, M.S.; Kartika, W.; Al-Sharafi, M.A.; Musarat, M.A.; Alawag, A.M.; Qureshi, A.H. A Bibliometric Analysis and Review of Building Information Modelling for Post-Disaster Reconstruction. *Sustainability* **2022**, *14*, 393. [CrossRef]
72. Afila, D.; Smith, N.J. Risk management and value management in project appraisal. *Proc. Inst. Civ. Eng. Manag. Procure. Law* **2007**, *160*, 63–67. [CrossRef]
73. Ismail, Y.A.B.; Khoiry, M.A. A Conceptual Model of Delay Factors Affecting Road Construction Projects in Libya. 2017. Available online: http://jestec.taylors.edu.my/Vol%2012%20issue%2012%20December%202017/12_12_12.pdf (accessed on 1 June 2022).
74. Tang, W.; Qiang, M.; Duffield, C.F.; Young, D.M.; Lu, Y. Risk Management in the Chinese Construction Industry. *J. Constr. Eng. Manag.* **2007**, *133*, 944–956. [CrossRef]
75. Amoah, P.; Ahadzie, D.K.; Dansoh, A. The factors affecting construction performance in Ghana: The perspective of small-scale building contractors. *Ghana Surv.* **2011**, *4*, 41–48.
76. Zafar, T.; Wuni, I.; Shen, I.Y.; Ahmed, G.Q.; Yousaf, S. A fuzzy synthetic evaluation analysis of time overrun risk factors in highway projects of terrorism-affected countries: The case of Pakistan. *Int. J. Constr. Manag.* **2019**, *22*, 732–750. [CrossRef]
77. Azhar, R.U.; Ginder, S.; Farooqui, W.C. An Assessment of Risk Management Practices in the Alabama Building Construction Industry. 2008. Available online: https://www.academia.edu/5097554/An_Assessment_of_Risk_Management_Practices_in_the_Alabama_Building_Construction_Industry (accessed on 20 May 2022).
78. Marzouk, T.I.; El-Rasas, M.M. Analyzing delay causes in Egyptian construction projects. *J. Adv. Res.* **2014**, *5*, 49–55. [CrossRef]
79. Asiedu, H.W.; Alfen, R.O. Understanding the underlying reasons behind time overruns of government building projects in Ghana. *KSCE J. Civ. Eng.* **2016**, *20*, 2103–2111. [CrossRef]
80. PMBOK. *A Guide to the Project Management Body of Knowledge*, 3rd ed.; American National Standard: Newtown Square, PA, USA, 2004.
81. Kassem, M.A.; Khoiry, M.A.; Hamzah, N. Using probability impact matrix (PIM) in analyzing risk factors affecting the success of oil and gas construction projects in Yemen. *Int. J. Energy Sect. Manag.* **2020**, *14*, 527–546. [CrossRef]
82. Thomas, R. A General Inductive Approach for Analyzing Qualitative Evaluation Data. *Am. J. Eval.* **2006**, *27*, 237–246. [CrossRef]
83. Del Caño, A.; De La Cruz, M.P. Integrated Methodology for Project Risk Management. *J. Constr. Eng. Manag.* **2002**, *128*, 473–485. [CrossRef]
84. Goh, C.S.; Abdul-Rahman, H. The identification and management of major risks in the Malaysian construction industry. *J. Constr. Dev. Ctries.* **2013**, *18*, 19.
85. Kassem, M.A.; Khoiry, M.A.; Hamzah, N. Assessment of the effect of external risk factors on the success of an oil and gas construction project. *Eng. Constr. Archit. Manag.* **2020**, *27*, 2767–2793. [CrossRef]
86. Mhetre, K.; Konnur, B.A.; Landage, A.B. Risk Management in Construction Industry Article in. *Int. J. Eng. Res.* **2016**, *1*, 153–155. [CrossRef]
87. Alfadil, M.O.; Kassem, M.A.; Ali, K.N.; Alaghbari, W. Construction Industry from Perspective of Force Majeure and Environmental Risk Compared to the COVID-19 Outbreak: A Systematic Literature Review. *Sustainability* **2022**, *14*, 1135. [CrossRef]
88. Ghaleb, H.; Alhajlah, H.H.; Abdullah, A.A.b.; Kassem, M.A.; Al-Sharafi, M.A. A Scientometric Analysis and Systematic Literature Review for Construction Project Complexity. *Buildings* **2022**, *12*, 482. [CrossRef]
89. Ali, K.N.; Alhajlah, H.H.; Kassem, M.A. Collaboration and Risk in Building Information Modelling (BIM): A Systematic Literature Review. *Buildings* **2022**, *12*, 571. [CrossRef]

buildings

MDPI

Article

Safety Risk Assessment of Prefabricated Buildings Hoisting Construction: Based on IHFACS-ISAM-BN

Junwu Wang [1,2,†], Feng Guo [1,2,†], Yinghui Song [1], Yipeng Liu [1,2], Xuan Hu [1,2] and Chunbao Yuan [3,*]

1 School of Civil Engineering and Architecture, Wuhan University of Technology, Wuhan 430070, China; 267544@whut.edu.cn (J.W.); guofeng777@whut.edu.cn (F.G.); songyinghui@whut.edu.cn (Y.S.); liuyipeng@whut.edu.cn (Y.L.); xuanhu17@whut.edu.cn (X.H.)
2 Hainan Institute, Wuhan University of Technology, Sanya 572025, China
3 China Construction Seventh Engineering Division Corp. Ltd., Shenzhen 518000, China
* Correspondence: 262895@whut.edu.cn
† These authors contributed equally to this work.

Abstract: Prefabricated buildings that are more environmentally friendly have been vigorously promoted by the Chinese government because of the reduced waste and carbon emissions during the construction process. Most of the construction processes of prefabricated buildings are completed in the prefabricated component factory, but the safety risks during the hoisting process cannot be ignored. In this paper, the initial framework of a Bayesian Network (BN) is obtained from the combination of the improved Human Factors Analysis and Classification System Model (HFACS) and BN. The improved similarity aggregation method (SAM) is used to calculate the prior probability of BN, which can better summarize and deal with the fuzzy judgment of experts on risk accidents. The improved SAM can consider both the weight of experts and the relative consistency of their opinions, which is of great significance for improving the reliability of BN inputted data. This paper uses the construction project in Sanya, Hainan Province, to verify the validity of the model. The results show that the calculation results of the model are basically consistent with the actual situation. The safety risk of this project is relatively low, and the premise of unsafe behaviors and unsafe supervision are the key risk factors of the project. In addition to maintaining good construction conditions and workers' healthy states, it is also necessary to carefully check the performance of tower cranes and equipment such as spreaders. During the operation process of the tower crane, workers should avoid walking or staying within the hoisting range.

Keywords: prefabricated buildings; improved human factor analysis and classification system; improved similarity aggregation method; Bayesian Network

Citation: Wang, J.; Guo, F.; Song, Y.; Liu, Y.; Hu, X.; Yuan, C. Safety Risk Assessment of Prefabricated Buildings Hoisting Construction: Based on IHFACS-ISAM-BN. *Buildings* **2022**, *12*, 811. https://doi.org/10.3390/buildings12060811

Academic Editors: Yongjian Ke, Jingxiao Zhang and Simon P. Philbin

Received: 6 May 2022
Accepted: 9 June 2022
Published: 12 June 2022

1. Introduction

Prefabricated buildings have been vigorously promoted by the Chinese government because of the advantages of producing less construction waste [1] and carbon emissions [2], and faster construction [3]. In the past few years, the areas of newly built prefabricated buildings have been increasing [4], and the proportion of newly built prefabricated buildings in 2021 has reached 20.5% of the new buildings in China [5]. The "Opinions on Promoting the Green Development of Urban and Rural Construction" issued by the General Office of the Central Committee of the Communist Party of China and the General Office of the State Council pointed out [6] that it is necessary to vigorously develop prefabricated buildings and focus on promoting the construction of steel-structure prefabricated buildings, so as to continuously improve the standardization level of components and promote the formation of a complete industrial chain, thus increasing the coordinated development of intelligent construction and building industrialization. Therefore, local governments have actively responded to the call of the central government through the issues of various planning policies for the development of prefabricated buildings [7,8].

For example, Beijing clearly required steel structures to be used in new public buildings [9]. By 2022, the area of prefabricated buildings will account for more than 35% of that of the new constructions; Guangdong Province clearly pointed out that by 2025, the proportion of urban prefabricated buildings in the Pearl River Delta will account for more than 35% of the new construction area, and more than 30% of the prefecture level downtown areas in eastern and northwestern Guangdong with a permanent resident population of more than 3 million, and more than 20% in other areas. Hainan Province has higher requirements: by the end of 2025, prefabricated buildings should account for more than 80% of new constructions, and two national-level prefabricated building demonstration cities should be built. It is also vital to balance the supply and demand of the annual production capacity of prefabricated components. The introduction of these policies demonstrates that prefabricated buildings will be an important and even main construction method in China [10].

The construction process of prefabricated buildings can be roughly divided into five stages: component production [11], transportation [12], on-site storage, hoisting, splicing, and installation [13,14]. The component production is mainly carried out in the prefabricated component factory, in which the corresponding prefabricated components are produced according to the design instructions and the production standards [15]. Then, the prefabricated components will be transported to the construction site for storage. When the construction starts, the construction unit will transport the prefabricated components to the corresponding position through the tower crane. During the hoisting process, the tower crane driver and the ground workers (signalmen and riggers) must cooperate with each other to better promote construction safety because of their different sights [16]. After the prefabricated components are hoisted to the corresponding position, the tower crane driver and the installers also need to cooperate with each other. The installers set up temporary supports to completely fix the prefabricated components, then the connection between the components and the tower crane can be removed [17].

By repeating the above process, a prefabricated building can be built [18]. It is obvious to find that the main process of the prefabricated building at the construction site is to hoist the prefabricated components. A more specific description of the hoisting process of the prefabricated components is shown in Figure 1 (referring to the actual prefabricated construction project, the solid line frame is the basic process of hoisting the prefabricated components; the dotted line frame is the preparation and perfection).

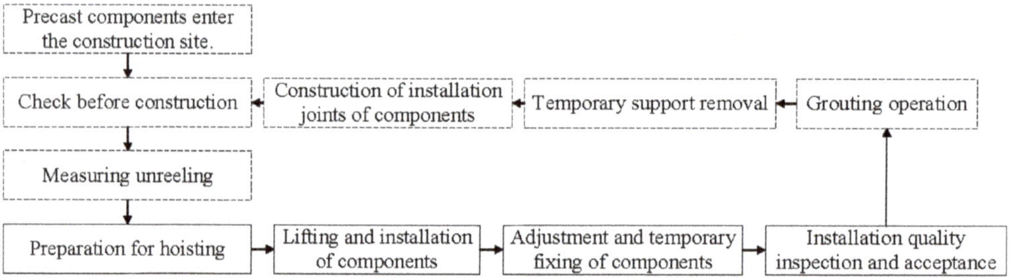

Figure 1. Specific construction procedures for hoisting of prefabricated components.

In Figure 1, "component lifting and installation" and "component adjustment and temporary fixation" are the two links with the longest construction time and are the most likely to lead to safety accidents [19]. Before hoisting the prefabricated components, the positioning traction ropes should be fastened onto the components to ensure safety and firmness; the special spreader shall be installed and hung on the hook of the tower crane and connected with the hanging point on the component. The workers should check whether it is firm. After the prefabricated components are hoisted and before installation, the sling should be kept balanced, and the components should be lowered when it is safe. When lowering the prefabricated components, the installers should use the traction rope to control the position and direction to make the whole process smooth and slow. After the installation is completed, the position without force or unbalanced force shall be adjusted in time [20]. In summary, the hoisting construction of prefabricated buildings is very complex. It is necessary to analyze the construction risks in hoisting construction to avoid safety accidents.

The rest of this paper is structured as follows: Section 2 is the relevant literature. Section 3 introduces the methodology and establishes a model for the problem of this paper. Section 4 validates the model with real cases and conducts sensitivity analysis. Section 5 discusses the model and puts forward relevant management suggestions. Section 6 summarizes the full text and gives an outlook for future research.

2. Literature Review

2.1. Safety Risk Analysis of Construction Project

Construction safety risk analysis has always been the focus of academic research. The analysis can be divided into two aspects: the overall construction risks analysis and the risk analysis in the specific construction process.

From the overall construction risks, many scholars are keen to study the construction risks of subway projects. Zhou [21] established an intelligent model based on random forest for the risk prediction of subway construction. The prediction model can be used as the basis for the implementation of subway foundation pit safety risk prediction, helping to implement emergency measures in advance. Based on the BP neural network, Li [22] has carried out intelligent identification of the safety risks of subway construction from the aspects of human factors and managements, which can ensure that the construction unit finds the risks and takes measures in time. In order to determine the success factors of construction safety management, Liu [23] established an interpretative structural model (ISM) through a literature review and questionnaire to find out the relationship between different factors, which is helpful to improve the safety performance in the process of subway construction and reduce the safety accidents. Many scholars have also conducted research on prefabricated buildings. Through the analytic hierarchy process (AHP) and entropy weight method, Liu [24] proposed an evaluation method of prefabricated buildings' construction safety based on a cloud model, which provides a new perspective to objectively evaluate the safety of prefabricated projects. Based on an ISM and the analytical network process (ANP), Xu [25] evaluated the safety factors of prefabricated building construction, which is of great significance to reduce the safety risks in prefabricated building construction.

From the perspective of specific construction risks, many scholars believe that the hoisting is the most risky. Liu [26] thought that some of the existing pieces of research do not consider the interaction of risk factors in the hoisting stage, so he proposed a security risk analysis method that integrates the Internet of Things, a building information model, the Apriori algorithm, and a complex network, in order to achieve effective security management and decision-making. Lu [27] established a comprehensive prefabricated construction site layout model, which integrated the hoisting efficiency, construction risks, and transportation costs of the prefabricated components, and obtained the Pareto optimal solution by a genetic algorithm. This model helps to solve the site layout problem of prefabricated building construction.

In summary, the hoisting construction is the key point in the safety management of prefabricated buildings. This paper will further analyze the relationship between the safety risks of prefabricated building hoisting construction and propose new management suggestions.

2.2. HFACS Model

The traditional theory of accident causes is mostly analyzed separately from the four aspects of human, object, management, and environment, without considering the internal relationship between the factors [28]. The HFACS frame takes into account the transitive impact of organizational factors on unsafe behaviors at the individual level, which is more comprehensive and scientific than the independent analysis. The HFACS model was first proposed by scholars such as SHAPPELL in 2000 [29], and it is still used by researchers in various academic fields. The specific analysis of the model includes four levels: unsafe behavior, the premise of unsafe behavior, unsafe supervision, and organizational impact. As shown in Figure 2.

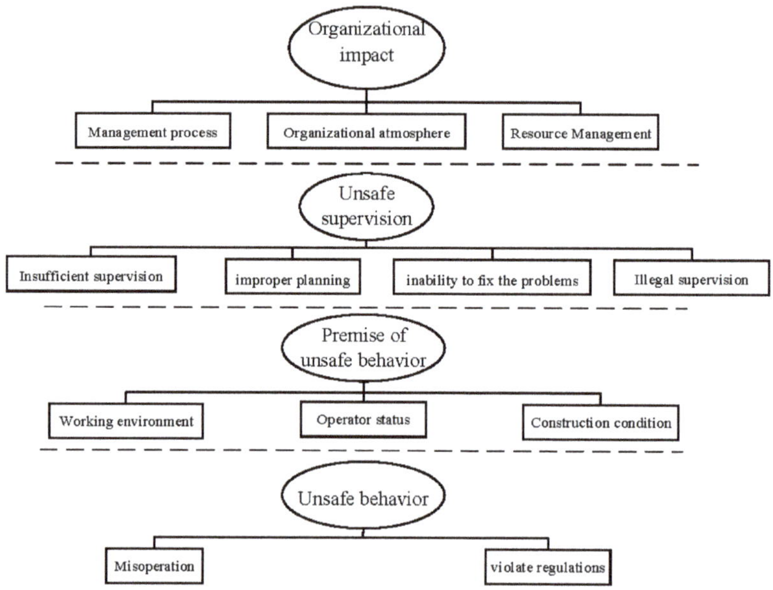

Figure 2. HFACS frame.

Since 2000, many scholars have promoted and applied the HFACS model. For example, Reinach and Viale [30] firstly modified the HFACS frame by adding external factors and established the HFACS-RR frame suitable for railway accidents. Chauvin [31] constructed a HFACS-Cloo frame for marine collision accidents and proved its rationality by analyzing 27 typical marine collision accidents. Spiess [32] applied the HFACS frame to the analysis of medical malpractice by adding health education and proved that it can improve the health condition of patients. Patterson and Shappell [33] introduced the HFACS frame into the safety analysis of the coal mining industry and verified the applicability and rationality of the frame with 508 typical accidents in the coal mining industry. In conclusion, the HFACS frame has strong extensibility and practicability. The extension and application of HFACS to the safety risk analysis of prefabricated building hoisting construction will contribute to different conclusions from the previous research. The specific extended application in this paper is detailed in Chapter 3 Methodology and Chapter 4 Case Analysis.

2.3. Fuzzy Bayesian Network

The HFACS frame can determine the specific risk factors of safety accidents and their interrelationships, but it cannot confirm the weight and control focus of each risk factor in detail [34]. Therefore, in order to further sort out and determine the impact of risk factors, the author mapped the HFACS frame to a BN. Liu [35] constructed the HFACS-CM frame of coal mine accidents and analyzed it combined with structural equation model (SEM), thus obtaining the main risk factors that will lead to safety accidents for miners. Xia [36] constructed the HFACS-BN model to actively predict the safety performance of construction projects and provide some suggestions for the safety risk management. Rostamabadi [37] proposed an accident analysis model that combines BN and the fuzzy best worst method (fuzzy BWM) into the HFACS frame. This method can effectively analyze and predict the safety risks in accidents. Based on the literature analysis and the characteristics of high-altitude crashes during construction, Luo [38] established the HFACS-BN model and put forward management opinions on high-altitude crash events. When constructing a BN and analyzing the risk probability, the prior probability of the accident must be determined by integrating the opinions of experts. There are many methods of integrating experts' opinions, such as the arithmetic mean method of reserve calculation [39], the Delphi method [40], the similarity aggregation method (SAM) [41], and the fuzzy analytic hierarchy process (FAHP) [42]. Among them, the arithmetic mean method of reserve calculation is just a simple arithmetic average of experts' opinions; the Delphi method considers the maximum uncertainty of experts; and FAHP is an extension of traditional AHP, which uses fuzzy language to deal with the experience and knowledge of each expert, so as to obtain the objective weight; while SAM can comprehensively consider the weight of each expert and the consistency between different experts.

The above-mentioned experts' opinions and methods have their own characteristics and scopes of application, and most of them pay attention to the evaluation value of high-weight experts, while ignoring the opinions of low-weight experts. If the opinions of most low-weight experts are similar, the results will be biased, because authoritative experts may also make inaccurate judgments. Among the above methods, only SAM considers both the expert weight and consistency. However, the traditional SAM method ignores the influence of expert weight on consistency; therefore, this paper improves the traditional SAM by integrating the influence of expert weight on consistency. The improved SAM method can make the aggregation results more scientific. Using this method to calculate the prior probability of a BN can reduce the uncertainty and identify the probabilities of key accidents more reliably.

3. Methodology

3.1. HFACS Frame for Hoisting Construction of Prefabricated Building Components

The traditional HFACS frame is mostly used in the aviation industry [43]. Compared with the hoisting construction of prefabricated components studied in this paper, the working conditions, workers, management, and many other factors are different, which brings different transmission process of risks. Therefore, it is not suitable to directly apply the HFACS frame for the aviation industry to the hoisting construction of prefabricated components. This paper extracts the process and causes of the accidents from the investigation report and cases of safety accidents in hoisting construction of prefabricated components in recent years [44] and modifies the original HFACS frame to adapt to the environmental characteristics of prefabricated component hoisting construction accidents. The revised HFACS frame is shown in Figure 3. With reference to literature [44,45] and specific hoisting construction accidents of prefabricated buildings, and combined with the construction characteristics of prefabricated buildings, the specific causative factors are obtained and shown in Table 1.

Considering the current situation of China's construction industry and the characteristics of prefabricated building construction, the following improvements are made on the basis of the original HFACS frame [46,47]:

(1) External environment factor that includes policy factors and industry management is added. Policy factors include two aspects: imperfect technical safety standards for hoisting constructions and imperfect management methods of special operation workers. If the government cannot issue perfect management methods, it is not conducive to standardizing the behaviors of the construction unit and construction safety.

(2) In terms of unsafe supervision, improper planning, and inability to fix the problems are combined into insufficient supervision. Improper planning corresponds to the behavior before supervision, and unable to fix the problems corresponds to the behaviors after the accidents, both of which can be summarized in insufficient supervision. Insufficient dynamic supervision is added. Tower crane construction

needs to consider the cooperation between workers in construction space and on the ground. The tower crane hook visualization system and safety monitoring system for collision avoidance of tower crane can effectively reduce the probability of mishook and collision accidents.

(3) In terms of the premise of unsafe behavior, the operators' states and personnel factors are merged into the states of workers. As mentioned above, tower crane construction workers mainly include tower crane drivers and ground workers whose situation should be taken into account. The factor of construction conditions is added. The natural environment is uncontrollable, and the construction conditions are artificially determined, including unreasonable stacking of prefabricated components on the ground and the height of tower crane.

(4) Unsafe behaviors are divided into two major aspects: professional skill errors such as invalid communication between tower crane drivers and ground workers, and errors in normal operations such as multitasking.

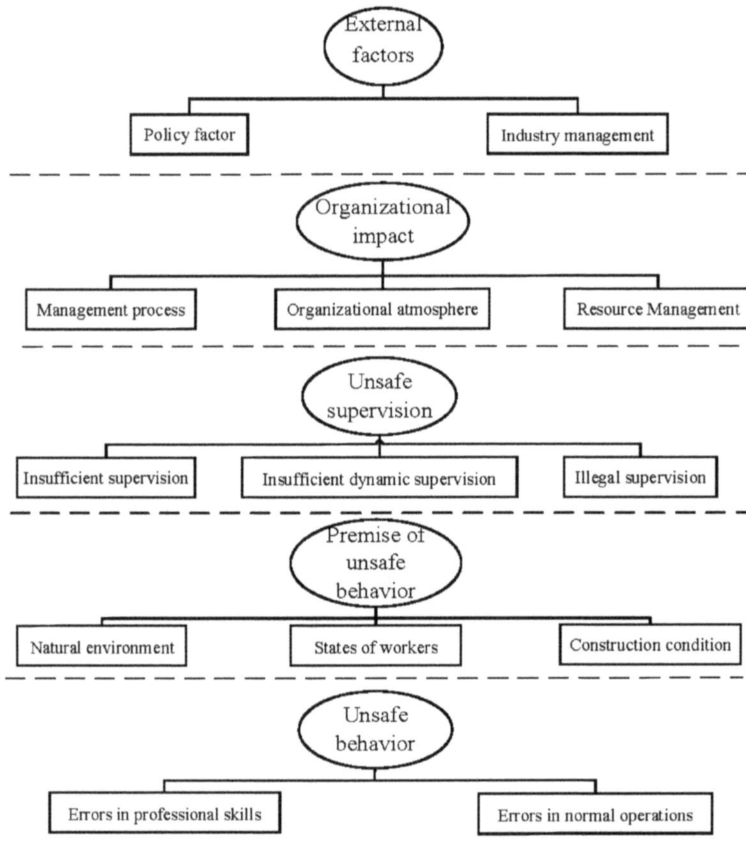

Figure 3. Prefabricated building hoisting construction HFACS frame.

Table 1. Causation factors of prefabricated building hoisting construction accidents based on improved HFACS frame.

Level	Category	Causation Factor
External factors A	Policy factor A_1	A_{11} Imperfect technical safety standards for hoisting constructions. A_{12} Imperfect management methods of special operation workers.
	Industry management A_2	A_{21} Lack of registration management and inspection of tower cranes. A_{22} No qualification of workers involved in special type of work.
Organizational impact B	Management process B_1	B_{11} Incomplete safety plans for tower crane assembly, operation, and separation. B_{12} Unreasonable construction schedule.
	Organizational atmosphere B_2	B_{21} Insufficient safety awareness of tower crane installation and demolition workers. B_{22} Insufficient safety awareness of tower crane drivers and ground workers.
	Resource Management B_3	B_{31} Safety defects of imported the tower cranes, spreaders, slings, hoisting baskets, and claps. B_{32} The lack of safety education and training. B_{33} The lack of warning signs in the isolation area for tower crane installation, demolition, and hoisting. B_{34} No safety protection measures such as scaffolds, skids, and locks.
Unsafe supervision C	Insufficient supervision C_1.	C_{11} Insufficient safety supervision of tower crane operation by contractors. C_{12} Incomplete safety inspection of tower cranes by maintainers and drivers.
	Insufficient dynamic supervision C_2.	C_{21} Tower crane hook visualization system is unused. C_{22} Safe load indicator and graphic display of crane operation are unused. C_{23} Safety monitoring system for collision avoidance of tower crane is unused.
	Illegal supervision C_3.	C_{31} Authorization of extra-served tower cranes. C_{32} Workers such as tower crane drivers and signal workers are employed without certification. C_{33} Although the position of the tower crane does not match the working scope, the construction is still carried out.
Premise of unsafe behavior D	Natural environment D_1.	D_{11} Poor ground conditions at the construction site. D_{12} Low visibility at the construction site. D_{13} Quality changes of prefabricated components caused by ground water on the construction site.
	States of workers D_2.	D_{21} Physical condition of tower crane drivers and ground workers. D_{22} Tower crane drivers and ground workers' psychological pressure on construction progress. D_{23} Lack of practical skills of tower crane drivers and ground workers.
	Construction condition D_3.	D_{31} Cross work of multiple tower cranes. D_{32} Unreasonable stacking of prefabricated components on the ground. D_{33} Height of tower cranes.
Unsafe behavior E	Errors in professional skills E_1.	E_{11} Ineffective communication between tower crane drivers and ground workers. E_{12} Hoisting of prefabricated components does not use special spreaders and slings. E_{13} No temporary support is set up, which violates the requirements of the technical solution.
	Errors in normal operations E_2.	E_{21} In-site workers move or stay within the hoisting range. E_{22} Multitasking of tower crane drivers and ground workers.

3.2. Bayesian Networks (BN)

According to the above-established HFACS frame for hoisting construction of prefabricated buildings, the identified factors are converted into nodes in the BN, and the HFACS frame is mapped to the BN structure, as shown in Figure 4. H is the leaf node of BN, that is, the node where the accident happens.

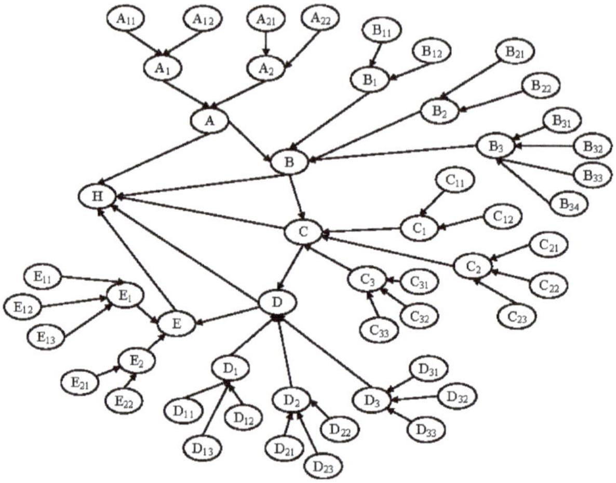

Figure 4. BN structure of prefabricated building hoisting construction.

Considering that the occurrence of variables requires certain conditions, the joint probability distribution $P(X)$ [48] of the variable $X = \{X_1, \cdots X_n\}$ in the BN can be expressed as:

$$P(X) = \prod_{i=1}^{n} P(X_i|Pa(X_i)) \tag{1}$$

In the above formula, $Pa(X_i)$ is the superset of X_i. When $\forall i \epsilon\ [1, n]$, the probability of X_i will be defined as:

$$P(X_i) = \sum_{X_i, j \neq i} P(X) \tag{2}$$

BN uses the observation result (defined as E) before the update of Bayesian theorem, that is, the prior probability of variables to produce a posterior probability [49]. As shown in formula (3):

$$P(X|E) = \frac{P(X,E)}{P(E)} = \frac{P(X,E)}{\sum_X P(X,E)} \tag{3}$$

The above prior probabilities are often obtained through expert interviews or questionnaires [42]. As described in the literature review, most scholars have adopted various methods to deal with experts' estimates, and different methods have different advantages and scope of application. Considering that experts will give different, or even opposite results, the reliability of the research may be greatly reduced. In order to consider the weight importance of experts and the relative consistency of the opinions, this paper intends to consider the use of an improved SAM method to aggregate experts' judgments, thus getting more reliable results.

3.3. Improved SAM

The method of aggregating experts' opinions adopted by previous scholars is to make the weight of experts the only indicator to show the reliability of the estimated values [50], that is, the opinions of experts with high weight tend to be more influential than those of low-weight experts. SAM can not only consider the relative importance of experts, but also the relative consistency of their opinions. However, the original SAM integrates these two factors only through simple linear addition. Therefore, the main goal of this chapter is to improve the original SAM and take the weight of experts and the consistency of their opinions into consideration (see details in formula (5)). In addition, it is unreasonable to judge experts' weights only by the educational background or professional title. This paper comprehensively considers the experts' professional title, work experience, educational background, and age [41,51]. The specific parameters and scores are shown in Table 2.

Table 2. Experts' weight and corresponding scores.

Indicator	Category	Score
Professional title	Senior engineer	10
	Professor	8
	Associate professor/Intermediate engineer	6
	Technician	4
	Worker	2
Work experience (years)	≥ 30	10
	20–29	8
	10–19	6
	6–9	4
	≤ 5	2
Educational background	Doctor	10
	Master	8
	Bachelor	6
	Junior college	4
	Middle school	2
Age	≥ 60	10
	50–59	8
	40–49	6
	30–39	4
	≤ 30	2

If there is a 45-year-old expert with the title of associate professor with a Ph.D. and 20 years of work experience, his score is 30 $(6 + 8 + 10 + 6)$. After synthesizing all the scores, his expert weight is the result of dividing his weight score by the scores of all experts. The specific steps of obtaining the prior probability by the improved SAM method are as follows:

Firstly, calculate the similarity $S(E_a, E_b)$ of opinions between each pair of experts. E_a and E_b represent the judgments of expert a and b on fuzzy events. The specific weight values of expert a and b are, respectively, defined as $E_a = (a_1, a_2, a_3, a_4)$, $E_b = (b_1, b_2, b_3, b_4)$. The calculation formula of $S(E_a, E_b)$ [49] is:

$$S(E_a, E_b) = 1 - 1/4 \sum_{i=1}^{4} |a_i - b_i| \tag{4}$$

The similarity of the two experts' opinions can be judged by calculating the differences between the professional titles, work experience, educational background, and age.

Secondly, calculate the weighted agreement degree $WA(E_a)$ of expert a. The weights of expert a and b are defined as $W(E_a)$ and $W(E_b)$. The calculation formula to define the weighted agreement $WA(E_a)$ of expert a is:

$$WA(E_a) = \frac{\sum_{b=1}^{N} W(E_b)\cdot S\left(\widetilde{R_a}, \widetilde{R_b}\right)}{\sum_{b=1}^{N} W(E_b)}, a \neq b \tag{5}$$

Then, calculate the degree of relative consistency (RA) of the experts [52], defined as:

$$RA(E_a) = \frac{WA(E_a)}{\sum_{a=1}^{N} WA(E_a)} \tag{6}$$

Then, calculate the Consensus Coefficient (CC) of each expert [52], defined as:

$$CC(E_a) = \beta \times W(E_a) + (1-\beta) \times RA(E_a) \tag{7}$$

$\beta(0 \leq \beta \leq 1)$ in the above formula is the relaxation coefficient, which is the key factor to balance the importance of $W(E_a)$ and $RA(E_a)$, so this needs to be decided by the decision makers.

Finally, the opinions of experts can be aggregated, and the final fuzzy number E can be obtained, which is defined as:

$$E = CC(E_1) \times E_1 + CC(E_2) \times E_2 + \cdots \tag{8}$$

In order to turn the fuzzy number E into a fuzzy possibility score (FPS), this paper adopts the CoA fuzzification technology. $E_a = (a_1, a_2, a_3, a_4)$ is a standard trapezoidal number, and its member function is defined as:

$$u(x) = \begin{cases} 0 & x < a_1 \\ \frac{x-a_1}{a_2-a_1} & a_1 \leq x < a_2 \\ 1 & a_2 \leq x < a_3 \\ \frac{x-a_4}{a_4-a_3} & a_3 \leq x < a_4 \\ 0 & x \geq a_4 \end{cases} \tag{9}$$

Defuzzification of trapezoidal fuzzy numbers is as follows:

$$\begin{aligned} FPS &= \frac{\int_{a_1}^{a_2} \frac{x-a_1}{a_2-a_1} x dx + \int_{a_2}^{a_3} x dx + \int_{a_3}^{a_4} \frac{a_4-x}{a_4-a_3} dx}{\int_{a_1}^{a_2} \frac{x-a_1}{a_2-a_1} dx + \int_{a_2}^{a_3} dx + \int_{a_1}^{a_2} \frac{a_4-x}{a_4-a_3} dx} \\ &= \frac{1}{3} \frac{(a_4+a_3)^2 - a_3a_4 - (a_1+a_2)^2 + a_1a_2}{(a_4+a_3-a_1-a_2)} \end{aligned} \tag{10}$$

In order to convert FPS into the corresponding fuzzy failure probability (FFP), this paper adopts the commonly used Onisawa function [53]. The conversion of fuzzy FPS into FFP is as follows:

$$FFP = \begin{cases} \frac{1}{10^K}\ if\ FPS \neq 0 \\ 0\ if\ FPS = 0 \end{cases} \quad K = \left[\left(\frac{1-FPS}{FPS}\right)\right]^{\frac{1}{3}} \times 2.301 \tag{11}$$

This paper defuzzifies the obtained fuzzy possibility to obtain FPS, and then converts it into FFP, so that a quantified probability value, that is, a prior probability value, can be obtained. Through the calculation of the formulas (4)–(11), the prior probability of the BN can be obtained, and by inputting the prior probability into the BN, the posterior probability, that is, the possibility of an accident, can be obtained. Validation studies with real cases will be analyzed in the next chapter.

4. Case Analysis

4.1. Project Overview and Data Sources

The Hongye Haitang Residential Community Project is located in the east of Sanya City, Hainan Province, China, with a total construction area of 36,911.4 m^2 and a prefabricated construction area of 21,438.92 m^2, including 11 six-story residential buildings and one commercial supporting building. The prefabricated components are prefabricated stairs, prefabricated laminated floor slabs, and prefabricated lightweight interior partition walls. The building height is 19.6m. The BN structure in this paper is shown in Figure 4. In order to figure out the prior probability of each accident during the construction, it is necessary to investigate with experts to determine the probability of the accidents. This paper selects an expert engaged in the construction of prefabricated buildings, an expert engaged in the research of tower crane construction in universities, and one safety manager on the construction site to collect their evaluation indicators of the project by means of telephone interview and questionnaire. The data are collected in the form of fuzzy numbers [51]. The specific fuzzy language terms are shown in Table 3. Taking "Very low" as an example, the fuzzy number is (0, 0, 0.1, 0.2), which corresponds to the judgment $E_a = (a_1, a_2, a_3, a_4)$ of expert a in formula (4). If expert a believes that the probability of the accident leading to the final result is very low, and then $E_a = (a_1, a_2, a_3, a_4) = (0, 0, 0.1, 0.2)$.

Table 3. Fuzzy number set.

Fuzzy Language	Fuzzy Number
Very low	(0, 0, 0.1, 0.2)
Low	(0.1, 0.2, 0.2, 0.3)
Lower	(0.2, 0.3, 0.4, 0.5)
Moderate	(0.4, 0.5, 0.5, 0.6)
Higher	(0.5, 0.6, 0.7, 0.8)
High	(0.7, 0.8, 0.8, 0.9)
Very high	(0.8, 0.9, 1, 1)

Attach: The larger the value in Table 3, the greater the security risk of the node.

The detailed data of the three experts involved in this study are shown in Table 4. Taking Expert One as an example, referring to the standard in Table 2, its weight score calculation formula is $36(\ 8 + 10 + 10 + 8)$. After the scores of all experts are obtained, the weight ratio of experts can be calculated by dividing the scores of individual experts by the sum of the total scores of experts.

Table 4. Experts' information and weight.

Expert	Professional Title	Work Experience	Educational Background	Age	Score	Weight Ratio
Expert1	Professor	30	Doctor	56	$8 + 10 + 10 + 8 = 36$	$36/84 = 0.43$
Expert2	Intermediate engineer	25	Master	47	$6 + 8 + 8 + 6 = 28$	$28/84 = 0.33$
Expert3	Technician	12	Bachelor	34	$4 + 6 + 6 + 4 = 20$	$20/84 = 0.24$
					84	1

After calculating the weight ratio of each expert and collecting the judgment of each expert on the node accident, the prior probability of each node accident can be calculated through the above formulas (4)–(11). Showing the process of calculating all 34 nodes in this study will lead to a cumbersome paper, so the researcher selected node C_{21}"without using the tower crane hook visualization system" for the example calculation. The judgments of the three experts on node C_{21} are (high, low, and lower). The detailed calculation process is shown in Table 5.

Table 5. The detailed calculation process of the prior probability of node C_{21}.

Indicator	Value	Calculation Process
$S(E_1, E_2)$	0.4	$S(E_a, E_b) = 1 - 1/4\sum_{i=1}^{4}\lvert a_i - b_i\rvert, S(E_1, E_2)$
$S(E_1, E_3)$	0.55	$= 1 - 1/4(\lvert 0.1 - 0.7\rvert + \lvert 0.2 - 0.8\rvert + \lvert 0.2 - 0.8\rvert + \lvert 0.3 - 0.9\rvert) = 0.4$
$S(E_2, E_3)$	0.85	$S(E_1, E_3) = 1 - 1/4(\lvert 0.7 - 0.2\rvert + \lvert 0.8 - 0.3\rvert + \lvert 0.8 - 0.4\rvert + \lvert 0.9 - 0.5\rvert) = 0.55$
$WA(E_1)$	0.46	$S(E_2, E_3) = 1 - 1/4(\lvert 0.2 - 0.1\rvert + \lvert 0.3 - 0.2\rvert + \lvert 0.4 - 0.2\rvert + \lvert 0.5 - 0.3\rvert) = 0.85$
$WA(E_2)$	0.56	$WA(E_a) = \frac{\sum_{b=1}^{N} W(E_b)\cdot S(\widetilde{R_a}, \widetilde{R_b})}{\sum_{b=1}^{N} W(E_b)}, a \neq b,$
$WA(E_3)$	0.68	
$RA(E_1)$	0.27	$WA(E_1) = \frac{W(E_2)\cdot S(E_1,E_2)+W(E_3)\cdot S(E_1,E_3)}{W(E_2)+W(E_3)} = \frac{0.33\cdot 0.4+0.24\cdot 0.55}{0.33+0.24} = 0.46$
$RA(E_2)$	0.33	$WA(E_2) = \frac{W(E_1)\cdot S(E_1,E_2)+W(E_3)\cdot S(E_2,E_3)}{W(E_1)+W(E_3)} = \frac{0.43\cdot 0.4+0.24\cdot 0.85}{0.43+0.24} = 0.56$
$RA(E_3)$	0.4	$WA(E_3) = \frac{W(E_1)\cdot S(E_1,E_3)+W(E_2)\cdot S(E_2,E_3)}{W(E_1)+W(E_2)} = \frac{0.43\cdot 0.55+0.33\cdot 0.85}{0.43+0.33} = 0.68$
$CC(E_1)$	0.35	
$CC(E_2)$	0.33	$RA(E_a) = \frac{WA(E_a)}{\sum_{a=1}^{N} WA(E_a)}, RA(E_1) = \frac{0.46}{0.46+0.56+0.68} = 0.27,$
$CC(E_3)$	0.32	$RA(E_2) = \frac{0.56}{0.46+0.56+0.68} = 0.33, RA(E_3) = \frac{0.68}{0.46+0.56+0.68} = 0.4$
E	(0.342, 0.442, 0.474, 0.574)	$CC(E_a) = \beta \times W(E_a) + (1-\beta) \times RA(E_a),$
FPS	0.458	$CC(E_1) = \beta \times W(E_1) + (1-\beta) \times RA(E_1) = 0.5 \times 0.43 + 0.5 \times 0.27 = 0.35$
FFP	0.003	$CC(E_2) = \beta \times W(E_2) + (1-\beta) \times RA(E_2) = 0.5 \times 0.33 + 0.5 \times 0.33 = 0.33$ $CC(E_3) = \beta \times W(E_3) + (1-\beta) \times RA(E_3) = 0.5 \times 0.24 + 0.5 \times 0.4 = 0.32$ $E = CC(E_1) \times E_1 \cdots,$ $E = 0.35 \times (0.7, 0.8, 0.8, 0.9) + 0.33 \times (0.1, 0.2, 0.2, 0.3) + 0.32 \times (0.2, 0.3, 0.4, 0.5) = (0.342, 0.442, 0.474, 0.574)$ $FPS = \frac{1}{3}\frac{(a_4+a_3)^2 - a_3a_4 - (a_1+a_2)^2 + a_1a_2}{(a_4+a_3-a_1-a_2)} = 0.458$ $FFP = \begin{cases} \frac{1}{10^K}\, if\ FPS \neq 0 \\ 0\, if\ FPS = 0 \end{cases} K = \left[\left(\frac{1-FPS}{FPS}\right)\right]^{\frac{1}{3}} \times 2.301,$ $K = \left[\left(\frac{1-0.458}{0.458}\right)\right]^{\frac{1}{3}} \times 2.301 = 2.43, FFP = 0.0037$

The calculation of other nodes is similar to the calculation process in Table 5, so it is not repeated. Through the above process, the prior probability of other nodes can be calculated. The prior probabilities of all nodes are shown in Table 6.

Summarize the opinions of experts and calculate the prior probability of each node. The posterior probability can be obtained by further analysis by establishing a hoisting construction safety evaluation model.

Table 6. Prior probability and ranking of all nodes.

Indicator	Prior Probability	Ranking	Indicator	Prior Probability	Ranking
A_{11}	0.004539	28	C_{31}	0.036522	5
A_{12}	0.001588	34	C_{32}	0.049201	3
A_{21}	0.002695	31	C_{33}	0.026871	9
A_{22}	0.007936	24	D_{11}	0.002445	32
B_{11}	0.010329	20	D_{12}	0.014106	30
B_{12}	0.005551	27	D_{13}	0.019146	12
B_{21}	0.032023	7	D_{21}	0.034773	6
B_{22}	0.023769	22	D_{22}	0.014044	16
B_{31}	0.097932	1	D_{23}	0.012839	18
B_{32}	0.014755	14	D_{31}	0.014044	17
B_{33}	0.009872	11	D_{32}	0.001707	33
B_{34}	0.007936	25	D_{33}	0.032023	8
C_{11}	0.011716	19	E_{11}	0.026871	10
C_{12}	0.056632	2	E_{12}	0.007936	26
C_{21}	0.003695	15	E_{13}	0.044988	4
C_{22}	0.009889	21	E_{21}	0.009872	23
C_{23}	0.004424	29	E_{22}	0.018226	13

4.2. Construction of Safety Evaluation Model for Hoisting Construction

According to the BN constructed in Section 3.2, the BN structure of the safety risk of the prefabricated building hoisting construction was established in the graphical view interface of GeNIe2.3. The forward and reverse reasoning are performed on the BN. Through forward reasoning, the prior probability of intermediate nodes and accident nodes can be obtained, that is, the probability of accidents during the hoisting process of prefabricated components. Through reverse reasoning, the posterior probability of the initial node can be obtained, and the key risk factors can be identified. Input the prior probability of each node obtained in Section 4.1 and calculate the data by GeNIe2.3. The result is shown in Figure 5.

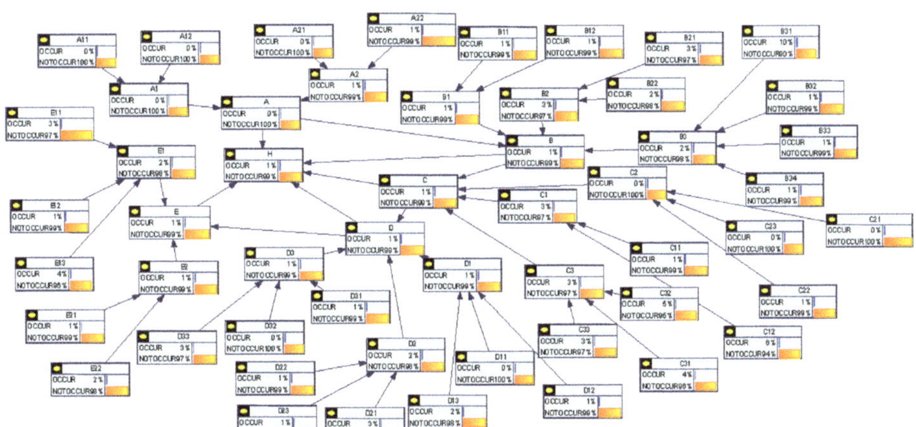

Figure 5. BN forward reasoning of safety accidents in hoisting construction of prefabricated components.

The results show that the probability of safety accidents in the hoisting construction of prefabricated components of the project is 1%, which is consistent with the operation of the project. No safety accidents occurred during the hoisting of the prefabricated components of the project, mainly because the buildings are not very high, and the construction is not difficult. In order to analyze the key points of control in the hoisting construction of prefabricated components, the author assumed that a safety accident has occurred (i.e., the probability of the occurrence of accident node H is 100%) and obtained the posterior probability of each risk factor through GeNIe2.3. The results are shown in Figure 6.

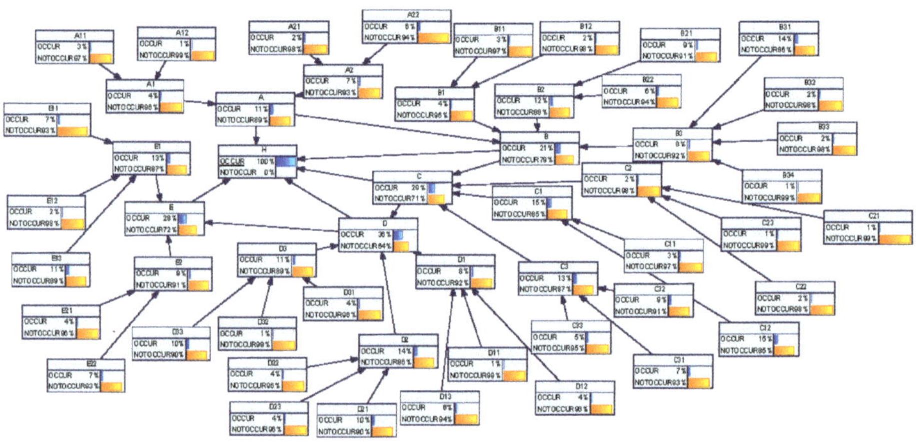

Figure 6. BN reverse reasoning of safety accidents in hoisting construction of prefabricated components.

It can be seen from the above that the posterior probabilities of A, B, C, D, and E are 0.11, 0.21, 0.29, 0.36, and 0.28, respectively, indicating that if a safety accident occurs in the hoisting construction of prefabricated components, the unsafe supervision, the premise of unsafe behaviors, and unsafe behaviors are most likely to have something wrong. This may because that the operation of the tower crane driver and the ground workers and their cooperation as well as the supervision are the necessary prerequisites for the safe construction of the tower crane. Further diagnosis and reasoning are carried out on the three factors to obtain the possible factors that need to be checked after the factors with a high posterior probability have problems. Let the occurrence probability of node C unsafe supervision, node D the premise of unsafe behavior, and node E unsafe behavior be 100%, respectively, and input them into the model. The risk probability of the factors affecting the root nodes of nodes C, D, and E is obtained as shown in Tables 7–9.

Table 7. Posterior probability of reasoning on node C.

Risk Factor	Posterior Probability	Ranking
Organizational impact *B*	0.113	3
Insufficient supervision C_1	0.487	1
Insufficient dynamic supervision C_2	0.07	4
Illegal supervision C_3	0.421	2

Table 8. Posterior probability of reasoning on node D.

Risk Factor	Posterior Probability	Ranking
Unsafe supervision C	0.227	3
Natural environment D_1	0.208	4
State of workers D_2	0.356	1
Construction conditions D_3	0.276	2

Table 9. Posterior probability of reasoning on node E.

Risk Factor	Posterior Probability	Ranking
Premise of unsafe behavior D	0.157	3
Errors in professional skills E_1	0.518	1
Errors in normal operations E_2	0.361	2

4.3. Sensitivity Analysis

In order to measure the importance of the root nodes in the BN structure, this paper calculates the sensitivity of the root node to the leaf nodes through the Ratio of Variation (RoV) of failure probability. The detailed calculation formula is as follows:

$$RoV(x_r) = \frac{P_L(x_r) - P_F(x_r)}{P_F(x_r)} \tag{12}$$

Among them, $P_L(x_r)$ and $P_F(x_r)$ are the posterior probability and prior probability of the root nodes. The larger the value of the RoV, the stronger the probability dependence of the leaf node on the root node. The author sets node H as the target node and calculates the sensitivity of each safety risk factor by formula (12). The results are shown in Tables 10–13.

Table 10. Risk factor sensitivity and ranking of each node (third-level indicators).

Indicator	Prior Probability	Posterior Probability	Sensitivity	Ranking
A_{11}	0.004539	0.033846	6.456709	9
A_{12}	0.001588	0.011872	−0.878777	24
A_{21}	0.002695	0.020128	0.364145	32
A_{22}	0.007936	0.059001	6.434639	10
B_{11}	0.010329	0.028590	1.896040	20
B_{12}	0.005551	0.015412	−0.686754	28
B_{21}	0.032023	0.085425	1.338991	22
B_{22}	0.023769	0.063744	5.445973	12
B_{31}	0.097932	0.139242	30.474179	3
B_{32}	0.014755	0.022596	5.115380	13
B_{33}	0.009872	0.015182	−0.435011	30
B_{34}	0.007936	0.012225	−0.728267	27

Table 10. *Cont.*

Indicator	Prior Probability	Posterior Probability	Sensitivity	Ranking
C_{11}	0.011716	0.032159	0.196787	34
C_{12}	0.056632	0.150956	18.021711	4
C_{21}	0.003695	0.007165	−0.873479	25
C_{22}	0.009889	0.019063	−0.404722	31
C_{23}	0.004424	0.008573	−0.639328	29
C_{31}	0.036522	0.070172	4.989439	14
C_{32}	0.049201	0.093437	8.046061	8
C_{33}	0.026871	0.052080	8.382063	7
D_{11}	0.002445	0.007520	−0.765171	26
D_{12}	0.014106	0.042728	0.228781	33
D_{13}	0.019146	0.057607	2.008810	19
D_{21}	0.034773	0.103335	6.325584	11
D_{22}	0.014044	0.042886	2.053663	18
D_{23}	0.012839	0.039266	1.795961	21
D_{31}	0.014044	0.042718	2.327203	16
D_{32}	0.001707	0.005275	2.090364	17
D_{33}	0.032023	0.095100	37.895755	2
E_{11}	0.026871	0.065590	2.598690	15
E_{12}	0.007936	0.019784	1.004101	23
E_{13}	0.044988	0.107527	12.549258	6
E_{21}	0.009872	0.037556	12.935310	5
E_{22}	0.018226	0.068905	42.391196	1

Table 11. Risk factor sensitivity and ranking of each node (second-level indicators).

Indicator	Prior Probability	Posterior Probability	Sensitivity	Ranking
Policy factor A_1	0.003064	0.042682	12.932486	1
Industry management A_2	0.005316	0.073903	12.903272	2
Management process B_1	0.007940	0.036082	3.544348	9
Organizational atmosphere B_2	0.027896	0.121332	3.349424	13
Resource Management B_3	0.017186	0.076546	3.454042	11
Insufficient supervision C_1	0.034174	0.150722	3.410432	12
Insufficient dynamic supervision C_2	0.004526	0.021276	3.700792	8
Illegal supervision C_3	0.029149	0.129996	3.459647	10
Natural environment D_1	0.009012	0.080985	7.986626	3
State of workers D_2	0.015690	0.138616	7.834901	5
Construction conditions D_3	0.012075	0.107675	7.917023	4
Errors in professional skills E_1	0.020389	0.132815	5.514145	7
Errors in normal operations E_2	0.014049	0.092511	5.584861	6

Table 12. Risk factor sensitivity and ranking of each node (first-level indicators).

Indicator	Prior Probability	Posterior Probability	Sensitivity	Ranking
External factor A	0.004190	0.112428	25.835758	5
Organizational impact B	0.007414	0.212063	27.601368	3
Unsafe supervision C	0.009851	0.289889	28.427518	2
Premise of unsafe behavior D	0.006027	0.359413	58.629575	1
Unsafe behavior E	0.010239	0.279783	26.325399	4

Table 13. Ranking of the sensitivity of each indicator at the three levels.

First Level Indicator	Ranking
Premise of unsafe behavior D	1
Unsafe supervision C	2
Organizational impact B	3
Unsafe behavior E	4
External factors A	5
Second level indicator	Ranking
Policy factor A_1	1
Industry management A_2	2
Natural environment D_1	3
Construction condition D_3	4
State of workers D_2	5
Third level indicator	Ranking
E_{22} Multitasking of tower crane drivers and ground workers	1
D_{33} Height of tower crane	2
B_{31} Safety defects of introduced tower cranes, spreaders, slings, hoisting baskets, and claps	3
C_{12} Incomplete safety inspection of tower cranes by maintainers and drivers	4
E_{21} In-site workers move or stay within the hoisting range	5

It can be seen from Table 10 that the key safety risk factors of prefabricated building hoisting construction mainly include: E_{22} Multitasking of tower crane drivers and ground workers, D_{33} Height of tower cranes, B_{31} Safety defects of imported the tower cranes, spreaders, slings, hoisting baskets, and claps, C_{12} Incomplete safety inspection of tower cranes by maintainers and drivers, E_{21} In-site workers move or stay within the hoisting range, and E_{13} No temporary supports.

This shows that the hoisting construction of prefabricated buildings should focus on the management of the tower crane drivers and ground workers on the construction site and improve their safety awareness. The imported tower cranes and the tools should be carefully checked, and perfect equipment management measures should be established. The training and assessment of construction workers are also of great importance, especially the training about technologies related to temporary support and the installation of prefabricated components. It can be seen from Table 12 that from a macro perspective, the risk factors affecting the hoisting and construction of prefabricated buildings are the unsafe supervision, organizational influence, unsafe behavior, external factors, and premise of unsafe behavior, which are consistent with the prior probability calculation. The same results indicate that the prefabricated building hoisting construction should pay attention to the premise of unsafe behavior at the macro level and improve the safety supervision system.

5. Discussion

5.1. Model Analysis

The prior probabilities of BN in this paper are calculated by the improved SAM. There is a relaxation coefficient β in the improved SAM, which is a key factor to balance the importance of $W(E_a)$ and $RA(E_a)$. In this paper, but the actual value is not necessarily 0.5. Take node C_{21} "without using the tower crane hook visualization system" an example calculation. Make β be 0.2, 0.5, and 0.8, and calculate the prior probability of node C_{21} again.

(1) Make $\beta = 0.2$, where the calculation of $S(E_a, E_b), WA(E_a),\ RA(E_a)$ are not affected.

$$\begin{gathered} CC(E_a) = \beta \times W(E_a) + (1-\beta) \times RA(E_a), CC(E_1) = \beta \times W(E_1) + (1-\beta) \times RA(E_1) \\ = 0.2 \times 0.43 + 0.8 \times 0.27 = 0.302 \\ CC(E_2) = \beta \times W(E_2) + (1-\beta) \times RA(E_2) = 0.2 \times 0.33 + 0.8 \times 0.33 = 0.33 \\ CC(E_3) = \beta \times W(E_3) + (1-\beta) \times RA(E_3) = 0.2 \times 0.24 + 0.8 \times 0.4 = 0.368 \\ E = CC(E_1) \times E_1 \cdots,\ E = 0.302 \times (0.7, 0.8, 0.8, 0.9) + 0.33 \times (0.1, 0.2, 0.2, 0.3) + 0.368 \times (0.2, 0.3, 0.4, 0.5) \\ = (0.319, 0.419, 0.455, 0.555) \end{gathered}$$

$$FPS = \frac{1}{3}\frac{(a_4+a_3)^2 - a_3a_4 - (a_1+a_2)^2 + a_1a_2}{(a_4+a_3-a_1-a_2)} = 0.437$$

$$FFP = \begin{cases} \frac{1}{10^K}\ if\ FPS \neq 0 \\ 0\ if\ FPS = 0 \end{cases} K = \left[\left(\frac{1-FPS}{FPS}\right)\right]^{\frac{1}{3}} \times 2.301,\ K = \left[\left(\frac{1-0.458}{0.458}\right)\right]^{\frac{1}{3}} \times 2.301 = 2.43, FFP = 0.0031$$

(2) Make $\beta = 0.5$

$$\begin{gathered} CC(E_a) = \beta \times W(E_a) + (1-\beta) \times RA(E_a), CC(E_1) = \beta \times W(E_1) + (1-\beta) \times RA(E_1) \\ = 0.5 \times 0.43 + 0.5 \times 0.27 = 0.35 \\ CC(E_2) = \beta \times W(E_2) + (1-\beta) \times RA(E_2) = 0.5 \times 0.33 + 0.5 \times 0.33 = 0.33 \\ CC(E_3) = \beta \times W(E_3) + (1-\beta) \times RA(E_3) = 0.5 \times 0.24 + 0.5 \times 0.4 = 0.32 \\ E = CC(E_1) \times E_1 \cdots,\ E = 0.35 \times (0.7, 0.8, 0.8, 0.9) + 0.33 \times (0.1, 0.2, 0.2, 0.3) + 0.32 \times (0.2, 0.3, 0.4, 0.5) \\ = (0.342, 0.442, 0.474, 0.574) \end{gathered}$$

$$FPS = \frac{1}{3}\frac{(a_4+a_3)^2 - a_3a_4 - (a_1+a_2)^2 + a_1a_2}{(a_4+a_3-a_1-a_2)} = 0.458$$

$$FFP = \begin{cases} \frac{1}{10^K}\ if\ FPS \neq 0 \\ 0\ if\ FPS = 0 \end{cases} K = \left[\left(\frac{1-FPS}{FPS}\right)\right]^{\frac{1}{3}} \times 2.301,\ K = \left[\left(\frac{1-0.458}{0.458}\right)\right]^{\frac{1}{3}} \times 2.301 = 2.43, FFP = 0.0037$$

(3) Make $\beta = 0.8$

$$\begin{gathered} CC(E_a) = \beta \times W(E_a) + (1-\beta) \times RA(E_a), CC(E_1) = \beta \times W(E_1) + (1-\beta) \times RA(E_1) \\ = 0.8 \times 0.43 + 0.2 \times 0.27 = 0.398 \\ CC(E_2) = \beta \times W(E_2) + (1-\beta) \times RA(E_2) = 0.8 \times 0.33 + 0.2 \times 0.33 = 0.330 \\ CC(E_3) = \beta \times W(E_3) + (1-\beta) \times RA(E_3) = 0.8 \times 0.24 + 0.2 \times 0.4 = 0.272 \\ E = CC(E_1) \times E_1 \cdots,\ E = 0.398 \times (0.7, 0.8, 0.8, 0.9) + 0.33 \times (0.1, 0.2, 0.2, 0.3) + 0.272 \times (0.2, 0.3, 0.4, 0.5) \\ = (0.366, 0.466, 0.493, 0.593) \end{gathered}$$

$$FPS = \frac{1}{3}\frac{(a_4+a_3)^2 - a_3a_4 - (a_1+a_2)^2 + a_1a_2}{(a_4+a_3-a_1-a_2)} = 0.480$$

$$FFP = \begin{cases} \frac{1}{10^K}\ if\ FPS \neq 0 \\ 0\ if\ FPS = 0 \end{cases} K = \left[\left(\frac{1-FPS}{FPS}\right)\right]^{\frac{1}{3}} \times 2.301,\ K = \left[\left(\frac{1-0.458}{0.458}\right)\right]^{\frac{1}{3}} \times 2.301 = 2.43, FFP = 0.0043$$

It can be seen from the above that $FFP|_{\beta=0.8} > FFP|_{\beta=0.5} > FFP|_{\beta=0.2}$, that is, if the value of β is increased, the value of FFP will be more inclined to the judgment of high-weight experts. While reducing the value of β and increasing $(1-\beta)$, the value of FFP will be more inclined to the result chosen by the majority of experts. Therefore, when decision makers use the above formulas to calculate the prior probabilities of BN, they should first clarify their risk preferences, whether they are willing to trust experts with a larger weight proportion and are more reliable, or to trust the choices of the majority.

5.2. Key Findings and Management Suggestions

In Section 4.3, the author conducts a sensitivity analysis on the different indicators of the three levels. The top five most sensitive indicators at each level are summarized in Table 13.

Based on the IHFACS frame, this paper establishes a safety risk evaluation indicator system for the hoisting construction of prefabricated buildings. The system includes 5 first-level indicators such as the premise of unsafe behaviors, 13 second-level indicators such as policy factors, industry management, and the natural environment, and 34 third-level indicators such as the multitasking of tower crane drivers and ground workers and tower cranes' height. By analyzing the sensitivity results of various security risk factors shown in Table 13, the author can obtain the following key findings and put forward some management suggestions:

From a macro perspective, the premise of unsafe behavior and unsafe supervision are the most critical factors, which are largely determined by the contractor. According to the IHFACS frame, the premise of unsafe behaviors will be affected by unsafe supervision, therefore the contractor should strengthen daily supervision and management to avoid illegal supervision and introduce dynamic supervision methods for hoisting construction. In addition, The Contractor shall fully consider the natural environment of the construction site to avoid the impact on the construction process, provide good construction conditions for the workers, pay attention to the health of the workers, and establish a safety technical support system. In addition, strengthen the management of tower cranes, drivers, and ground workers, avoid the situation of relevant personnel working without certificates, and prevent the tower crane from continuing to work in the presence of potential safety hazards.

At the medium level, in addition to paying attention to the natural environment, maintaining good construction conditions, and the state of workers, policy factors and industry management factors should also be considered. However, policy factors and industry managements are not determined by contractors. Incomplete policies and regulations and unreasonable industry management often make contractors put their own interests first without considering the potential safety risks of construction. The contractors often take actions that are not conducive to construction safety in order to catch up with the construction period and seek more benefits. Therefore, it is necessary for the government to strengthen the management of the industry and formulate relevant policies to ensure construction safety.

At the micro level, the key safety risk factors are the Multitasking of tower crane drivers and ground workers, the Height of tower crane, the Safety defects of introduced tower cranes, spreaders, slings, hoisting baskets, and claps, and so on. Contractors should attach importance to the management and safety education of the tower crane drivers and ground workers, and implement the safety responsibility distribution system step by step. A good safety awareness of the tower crane drivers and ground workers should be cultivated so that they can concentrate on the construction. During the contractors' daily supervision, the performance of the tower cranes' spreaders, slings, hanging baskets, and hooks should be carefully checked. The contractor should establish a daily maintenance and inspection system for tower cranes, which not only urges maintainers and tower crane drivers to conduct daily inspections and maintenance of tower cranes, but also needs to check maintenance records to avoid staffs' perfunctory effort. During the construction of the tower crane, workers should avoid walking or staying within the hoisting range.

6. Conclusions

Compared with the previous research on construction risk analysis, this paper adopts an improved HFACS model and considers the influence of external factors to integrate the factors of the original HFACS model, which makes the model more suitable for the research in this paper. In this paper, an improved SAM is proposed to calculate the prior probability in BN, and the improved SAM and BN are combined to evaluate the construction risks of prefabricated building hoisting. The improved SAM can better summarize the fuzzy opinions and truly reflect the judgments of the experts. Through the relaxation coefficient β, it can effectively balance the relationship between the experts' weight and consistency. The larger β is, the more it favors the opinions of high-weight experts, and the smaller β is, the more it favors the opinions chosen by the majority of experts. Compared with the previous SAM, the improved one reduces the weight of high-weight experts' opinions and increases the weight of the opinions selected by more experts, which can effectively avoid the judgment errors of high-weight experts.

For the cases selected in this paper, the overall safety risk probability level of the project is obtained through the forward reasoning of the BN. Through reverse reasoning, the key risk factors of the project were identified. The evaluation results of the Hongye Haitang Residential Community Project are basically consistent with the actual situation on the construction site, which proves the validity of the model. Since the buildings of the project selected in this paper are not very high and the construction is not difficult, it may be considered to use a project with more construction difficulty for case analysis in the future.

Through the analysis of this paper, the most sensitive factors are determined from the macro, medium, and micro levels. Macro level: D Premise of unsafe behavior, C Unsafe supervision, B Organizational impact, E Unsafe behavior, and A External factor; Medium level: A_1 Policy factors, A_2 Industry management, D_1 Natural environment, D_3 Construction conditions, and D_2 State of workers; Micro level: E_{22} Multitasking of tower crane drivers and ground workers, D_{33} Height of tower crane, B_{31} Safety defects of the tower cranes, spreaders, slings, hoisting baskets, and claps, C_{12} Incomplete safety inspection of tower cranes by maintainers and drivers, and E_{21} In-site workers move or stay within the hoisting range. By strengthening the management of the above-mentioned factors, it is possible to effectively avoid safety accidents during the hoisting construction of prefabricated buildings, which is beneficial for contractors to achieve optimal resources allocation with limited resources and carry out risk managements.

This study also has some limitations. Although the model constructed in this paper can summarize and deal with the fuzzy judgments of experts on risk accidents well, the subjectivity of experts' judgments still has an impact on the results. Subjective judgments combined with objective construction data can obtain more accurate results. In addition, for the purpose of acquiring more complete conclusion, it is necessary to consider more factors that affect prefabricated building hoisting construction accidents.

Author Contributions: Conceptualization, J.W.; Data curation, X.H. and C.Y.; Formal analysis, Y.S.; Funding acquisition, J.W. and C.Y.; Investigation, Y.S. and C.Y.; Methodology, Y.S. and F.G.; Software, F.G.; Supervision, J.W.; Validation, X.H. and F.G.; Visualization, Y.S.; Writing—original draft, F.G.; Writing— review and editing, Y.S. and Y.L. All authors have read and agreed to the published version of the manuscript.

Funding: This study was supported by the National Key R&D projects (grant number 2018YFC0704301); Science and Technology Project of Wuhan Urban and Rural Construction Bureau, China (201943); Research on theory and application of prefabricated building construction management (20201h0439); Wuhan Mo Dou Construction Consulting Co., Ltd. (20201h0414); and Preliminary Study on the Preparation of the 14th Five-Year Plan for Housing and Urban–Rural Development in Hubei Province (20202s0002).

Institutional Review Board Statement: Not applicable.

Informed Consent Statement: Informed consent was obtained from all subjects involved in the study.

Data Availability Statement: The case analysis data used to support the findings of this study are available from the corresponding author upon request.

Conflicts of Interest: The authors declare no conflict of interest. The funders had no role in the design of the study; in the collection, analyses, or interpretation of data; in the writing of the manuscript, or in the decision to publish the result.

References

1. Minunno, R.; O'Grady, T.; Morrison, G.M.; Gruner, R.L.; Colling, M. Strategies for Applying the Circular Economy to Prefabricated Buildings. *Buildings* **2018**, *8*, 125. [CrossRef]
2. Li, X.J.; Wang, C.; Alashwal, A.; Bora, S. Game analysis on prefabricated building evolution based on dynamic revenue risks in China. *J. Clean. Prod.* **2020**, *267*, 121730. [CrossRef]
3. Tavares, V.; Soares, N.; Raposo, N.; Marques, P.; Freire, F. Prefabricated versus conventional construction: Comparing life-cycle impacts of alternative structural materials. *J. Build Eng.* **2021**, *41*, 102705. [CrossRef]
4. Liu, S.; Li, Z.F.; Teng, Y.; Dai, L.R. A dynamic simulation study on the sustainability of prefabricated buildings. *Sustain. Cities Soc.* **2022**, *77*, 103551. [CrossRef]
5. Zhao, W.S.; Chen, Y.T. Study on Large-Scale Promotion of Prefabricated Buildings in Anhui Province Based on SEM and IoT. *Sci. Program.-Neth.* **2022**, *2022*, 6947365. [CrossRef]
6. General Office of the Central Committee of the Communist Party of China, General Office of the State Council. Opinions on Promoting the Green Development of Urban and Rural Construction. 2021. Available online: https://www.gov.cn/ (accessed on 21 October 2021).
7. Wang, Y.N.; Xue, X.L.; Yu, T.; Wang, Y.W. Mapping the dynamics of China's prefabricated building policies from 1956 to 2019: A bibliometric analysis. *Build. Res. Inf.* **2021**, *49*, 216–233. [CrossRef]
8. Du, H.; Han, Q.; Sun, J.; Wang, C.C. Adoptions of prefabrication in residential sector in China: Agent-based policy option exploration. *Eng. Constr. Archit. Manag.* **2022**. *ahead-of-print*. [CrossRef]
9. Beijing Municipal People's Government. Further Development of the Implementation of Assembled Buildings. 2022. Available online: https://www.beijing.gov.cn/ (accessed on 25 April 2022).
10. Du, Q.; Hao, T.T.; Huang, Y.D.; Yan, Y.Q. Prefabrication decisions of the construction supply chain under government subsidies. *Environ. Sci. Pollut. Res.* **2022**. *ahead-of-print*. [CrossRef]
11. Yazdani, M.; Kabirifar, K.; Fathollahi-Fard, A.M.; Mojtahedi, M. Production scheduling of off-site prefabricated construction components considering sequence dependent due dates. *Environ. Sci. Pollut. Res.* **2021**, 1–17. [CrossRef]
12. Zhang, H.; Yu, L. Dynamic transportation planning for prefabricated component supply chain. *Eng. Constr. Archit. Manag.* **2020**, *27*, 2553–2576. [CrossRef]
13. Han, Q.Y.; Chang, J.J.; Liu, G.W.; Zhang, H. The Carbon Emission Assessment of a Building with Different Prefabrication Rates in the Construction Stage. *Int. J. Environ. Res. Public Health* **2022**, *19*, 2366. [CrossRef] [PubMed]
14. Xiao, M.F.; Zhou, X.H.; Pan, X.X.; Wang, Y.A.; Wang, J.; Li, X.L.; Sun, Y.; Wang, Y. Simulation of emergency evacuation from construction site of prefabricated buildings. *Sci. Rep.* **2022**, *12*, 2732. [CrossRef]
15. Zhang, J.S.; Xiang, P.C.; Zhong, J.; Zhang, J.; Wu, Z.Z.; Antwi-Afari, M.F. Exploring Key Factors for Contractors in Opening Prefabrication Factories: A Chinese Case Study. *Front. Public Health* **2022**, *10*, 837350. [CrossRef] [PubMed]
16. Liu, Z.S.; Li, A.X.; Sun, Z.; Shi, G.L.; Meng, X.T. Digital Twin-Based Risk Control during Prefabricated Building Hoisting Operations. *Sensors* **2022**, *22*, 2522. [CrossRef] [PubMed]
17. Zhao, Y.H.; Cao, C.F.; Liu, Z.S. A Framework for Prefabricated Component Hoisting Management Systems Based on Digital Twin Technology. *Buildings* **2022**, *12*, 276. [CrossRef]
18. Zhao, W.S.; Zhang, B.B.; Yang, Y. Empirical study of comprehensive benefits for prefabricated buildings: A case study of Hefei city. *Int. J. Electr. Eng. Educ.* **2020**, 0020720920928465.
19. Zhang, Y.; Yi, X.; Li, S.; Qiu, H. Evolutionary game of government safety supervision for prefabricated building construction using system dynamics. *Eng. Constr. Archit. Manag.* **2022**. *ahead-of-print*. [CrossRef]
20. Fard, M.M.; Terouhid, S.A.; Kibert, C.J.; Hakim, H. Safety concerns related to modular/prefabricated building construction. *Int. J. Inj. Control. Saf. Promot.* **2017**, *24*, 10–23. [CrossRef]
21. Zhou, Y.; Li, S.Q.; Zhou, C.; Luo, H.B. Intelligent Approach Based on Random Forest for Safety Risk Prediction of Deep Foundation Pit in Subway Stations. *J. Comput. Civ. Eng.* **2019**, *33*, 05018004. [CrossRef]
22. Li, M.C.; Wang, J.C. Intelligent Recognition of Safety Risk in Metro Engineering Construction Based on BP Neural Network. *Math. Probl. Eng.* **2021**, *2021*, 1–10. [CrossRef]
23. Liu, P.; Li, Q.M.; Bian, J.; Song, L.L.; Xiahou, X.E. Using Interpretative Structural Modeling to Identify Critical Success Factors for Safety Management in Subway Construction: A China Study. *Int. J. Environ. Res. Public Health* **2018**, *15*, 1359. [CrossRef] [PubMed]
24. Liu, J.K.; Gong, E.Q.; Wang, D.; Teng, Y. Cloud Model-Based Safety Performance Evaluation of Prefabricated Building Project in China. *Wirel. Pers. Commun.* **2018**, *102*, 3021–3039. [CrossRef]
25. Xu, G.M. The Construction Site Management of Concrete Prefabricated Buildings by ISM-ANP Network Structure Model and BIM under Big Data Text Mining. *Int. J. Interact. Multimed. Artif. Intell.* **2020**, *6*, 138–145. [CrossRef]
26. Liu, Z.S.; Meng, X.T.; Xing, Z.Z.; Jiang, A.T. Digital Twin-Based Safety Risk Coupling of Prefabricated Building Hoisting. *Sensors* **2021**, *21*, 3583. [CrossRef]
27. Lu, Y.; Zhu, Y.Q. Integrating Hoisting Efficiency into Construction Site Layout Plan Model for Prefabricated Construction. *J. Constr. Eng. Manag.* **2021**, *147*, 04021130. [CrossRef]
28. Cheng, L.H.; Jiang, B.L.; Guo, H.M. Modeling the causes of accidental gas explosions from the perspective of safety information loss. *Process Saf. Prog.* **2022**. *ahead-of-print*. [CrossRef]

29. Diller, T.; Helmrich, G.; Dunning, S.; Cox, S.; Buchanan, A.; Shappell, S. The Human Factors Analysis Classification System (HFACS) Applied to Health Care. *Am. J. Med. Qual.* **2014**, *29*, 181–190. [CrossRef]
30. Reinach, S.; Viale, A. Application of a human error framework to conduct train accident/incident investigations. *Accid. Anal. Prev.* **2006**, *38*, 396–406. [CrossRef]
31. Chauvin, C.; Lardjane, S.; Morel, G.; Clostermann, J.P.; Langard, B. Human and organizational factors in maritime accidents: Analysis of collisions at sea using the HFACS. *Accid. Anal. Prev.* **2013**, *59*, 26–37. [CrossRef]
32. Spiess, B.D.; Rotruck, J.; McCarthy, H.; Suarez-Wincosci, O.; Kasirajan, V.; Wahr, J.; Shappell, S. Human factors analysis of a near-miss event:oxygen supply failure during cardiopulmonary bypass. *J. Cardiothorac. Vasc. Anesth.* **2015**, *29*, 204–209. [CrossRef]
33. Patterson, J.M.; Shappell, S.A. Operator error and system deficiencies: Analysis of 508 mining incidents and accidents from Queensland, Australia using HFACS. *Accid. Anal. Prev.* **2010**, *42*, 1379–1385. [CrossRef] [PubMed]
34. Yang, J.; Kwon, Y. Human factor analysis and classification system for the oil, gas, and process industry. *Process Saf. Prog.* **2022**. *ahead-of-print*. [CrossRef]
35. Liu, R.L.; Cheng, W.M.; Yu, Y.B.; Xu, Q.F.; Jiang, A.W.; Lv, T. An impacting factors analysis of miners' unsafe acts based on HFACS-CM and SEM. *Process Saf. Environ.* **2019**, *122*, 221–231. [CrossRef]
36. Xia, N.N.; Zou, P.X.W.; Liu, X.; Wang, X.Q.; Zhu, R.H. A hybrid BN-HFACS model for predicting safety performance in construction projects. *Saf. Sci.* **2018**, *101*, 332–343. [CrossRef]
37. Rostamabadi, A.; Jahangiri, M.; Zarei, E.; Kamalinia, M.; Banaee, S.; Samaei, M.R. A Novel Fuzzy Bayesian Network-HFACS (FBN-HFACS) model for analyzing Human and Organization Factors (HOFs) in process accidents. *Process Saf Environ.* **2019**, *132*, 59–72. [CrossRef]
38. Luo, X.X.; Liu, Q.L.; Qiu, Z.X. The Influence of Human-Organizational Factors on Falling Accidents From Historical Text Data. *Front. Public Health* **2022**, *9*, 783537. [CrossRef]
39. Detyniecki, M. Mathematical Aggregation Operators and Their Application to Video Querying. Ph.D. Thesis, Université Pierre et Marie Curie, Paris, France, 2000.
40. Ishikawa, A.; Amagasa, M.; Shiga, T.; Tomizawa, G.; Tatsuta, R.; Mieno, H. The max-min Delphi method and fuzzy Delphi method via fuzzy integration. *Fuzzy Sets Syst.* **1993**, *55*, 241–253. [CrossRef]
41. Hsu, H.-M.; Chen, C.-T. Aggregation of fuzzy opinions under group decision making. *Fuzzy Sets Syst.* **1996**, *79*, 279–285. [CrossRef]
42. Yazdi, M.; Kabir, S. A fuzzy Bayesian network approach for risk analysis in process industries. *Process Saf. Environ. Prot.* **2017**, *111*, 507–519. [CrossRef]
43. Wiegmann, D.A.; Shappell, S.A. Human error analysis of commercial aviation accidents: Application of the human factors analysis and classification system (HFACS). *Aviat. Space Environ. Med.* **2001**, *72*, 1006–1016.
44. Song, Y.H.; Wang, J.W.; Liu, D.H.; Guo, F. Study of Occupational Safety Risks in Prefabricated Building Hoisting Construction Based on HFACS-PH and SEM. *Int. J. Environ. Res. Public Health* **2022**, *19*, 1550. [CrossRef] [PubMed]
45. Sadeghi, S.; Soltanmohammadlou, N.; Rahnamayiezekavat, P. A systematic review of scholarly works addressing crane safety requirements. *Saf. Sci.* **2021**, *133*, 105002. [CrossRef]
46. Wu, J.S.; Zhang, L.L.; Bai, Y.P.; Reniers, G. A safety investment optimization model for power grid enterprises based on System Dynamics and Bayesian network theory. *Reliab. Eng. Syst. Saf.* **2022**, *221*, 108331. [CrossRef]
47. Vithanage, S.C.; Sing, M.C.P.; Davis, P.; Newaz, M.T. Assessing the Off-Site Manufacturing Workers' Influence on Safety Performance: A Bayesian Network Approach. *J. Constr. Eng. Manag.* **2022**, *148*, 04021185. [CrossRef]
48. Ali, M.S.; Kitali, A.E.; Kodi, J.; Alluri, P.; Sando, T. Quantifying the Safety Benefits of Transit Signal Priority Using Full Bayes Before-After Study. *J. Transp. Eng. Part A: Syst.* **2022**, *148*, 04021102. [CrossRef]
49. Sun, M.; Zhou, R.G.; Jiao, C.W.; Sun, X.D. Severity Analysis of Hazardous Material Road Transportation Crashes with a Bayesian Network Using Highway Safety Information System Data. *Int. J. Environ. Res. Public Health* **2022**, *19*, 4002. [CrossRef]
50. Ramzali, N.; Lavasani, M.R.M.; Ghodousi, J. Safety barriers analysis of offshore drilling system by employing Fuzzy Event Tree Analysis. *Saf. Sci.* **2015**, *78*, 49–59. [CrossRef]
51. Onisawa, T. An approach to human reliability in man-machine systems usingerror possibility. *Fuzzy Sets Syst.* **1988**, *27*, 87–103. [CrossRef]
52. Wang, Y.Z.; Yin, H.J.; Guo, X.W.; Zhang, W.G.; Li, Q.K. Distributed ANN-bi level two-stage stochastic fuzzy possibilistic programming with Bayesian model for irrigation scheduling management. *J. Hydrol.* **2022**, *606*, 127435. [CrossRef]
53. Otsubo, Y.; Otani, N.; Chikasue, M.; Nishino, M.; Sugiyama, M. Root cause estimation of faults in production processes: A novel approach inspired by approximate Bayesian computation. *Int. J. Prod. Res.* **2022**, *2022*, 1–19. [CrossRef]

buildings

MDPI

Article

Safety Risk Assessment Using a BP Neural Network of High Cutting Slope Construction in High-Speed Railway

Jianling Huang [1], Xiaoye Zeng [1], Jing Fu [2], Yang Han [1] and Huihua Chen [1,*]

[1] Department of Engineering Management, School of Civil Engineering, Central South University, Changsha 410083, China; hjl1201@csu.edu.cn (J.H.); xy.zeng@csu.edu.cn (X.Z.); hy1994csu@csu.edu.cn (Y.H.)
[2] Organization Department of the CPC Loudi Municipal, Loudi 417000, China; fujing6694@outlook.com
* Correspondence: chh@csu.edu.cn

Abstract: High-speed railway construction is extending to mountainous areas, and the harsh environment and complex climate pose various risks to the slope construction. This seriously threatens human lives and causes huge economic losses. The existing research results on the construction safety risks of high cutting slope construction in HSRs are limited, and a complete set of safety risk assessment processes and methods has not yet been formed. Therefore, in this study, we aimed to develop a safety risk assessment model, including factor identification and classification and assessment data processing, to help project managers evaluate safety risks in high cutting slope construction. In this study, comprehensive identification of high cutting slope construction safety risks was carried out from three dimensions, risk technical specification, literature analysis, and case statistical analysis, and a list of risk-influencing factors was formed. Based on the historical data, a high side slope risk evaluation model was established using a BP neural network algorithm. The model was applied to the risk evaluation of HF high cutting slopes. The results show that the risk evaluation level is II; the main risks are earthwork excavation method, scaffolding equipment, slope height, slope rate, groundwater, personnel safety awareness, and construction safety risk management system. Finally, a case study was used to verify the proposed model, and control measures for safety risks were proposed. Our findings will help conduct effective safety management, add to the knowledge of construction safety risk management in terms of implementation, and offer lessons and references for future construction safety management of HSR.

Keywords: high cutting slope; risk assessment; BP neural network

Citation: Huang, J.; Zeng, X.; Fu, J.; Han, Y.; Chen, H. Safety Risk Assessment Using a BP Neural Network of High Cutting Slope Construction in High-Speed Railway. *Buildings* **2022**, *12*, 598. https://doi.org/10.3390/buildings12050598

Academic Editors: Yongjian Ke, Jingxiao Zhang and Simon P. Philbin

Received: 18 March 2022
Accepted: 1 May 2022
Published: 5 May 2022

1. Introduction

The high-speed railway (HSR) has become a common solution to relieve the pressure on transportation systems worldwide, especially in China. HSR provides a fast and robust travel option that enhances the movement of people as a critical national infrastructure system. In China, the construction of HSR is extending to the mountainous areas; the poor construction environment and the complex and diverse climate make the construction personnel face various risks, some of which even threaten the safety of life [1]. When the HSR line passes through adverse geological sections, manual excavation and reinforcement are required [2]. Accidents such as slope instability, slump, gushing water, mechanical injury, electric shock, and falls are prone to occur during slope construction [3]. Especially in recent years, more and more slope disasters have caused extensive losses worldwide of human life and property [4]. Therefore, it is vital to develop a safety risk assessment system to avoid or mitigate those slope disasters.

The high cutting slope refers to a soil slope with a height greater than 20 m and less than 100 m or a rocky slope with a height greater than 30 m and less than 100 m [5]. The study of high cutting slope construction originated from the study of slope stability, which is important for the safety of slope construction. Slope stability is generally influenced by soil, hydrology, vegetation, earthquakes, climate, geological conditions, groundwater, slope height, slope rate, and other factors [6]. The study of risk in slope engineering started in the 1980s, and the initial stage of the research was mainly to clarify the existence of inherent uncertainty in geotechnical engineering and to try to characterize the uncertainty of geotechnical engineering by using probabilistic methods [7].

Current studies generally use reliability analysis, the limit equilibrium method, and the strength reduction method for slope stability. Reliability analysis was applied to the study of slope sexual stability using soil parameter data extracted from field and laboratory data, and the relative contribution of the uncertainty of different parameters to slope reliability varies [8]. The impact of uncertainty on the reliability and performance evaluation of slope design is often significant. Traditional safety factor-based slope practices can not explicitly address uncertainty and thus affect the adequacy of predictions [9]. For simple homogeneous soil slopes, the calculation results of the limit equilibrium method and strength reduction method are essentially the same [10].

For slope construction safety, the stability of slopes has also been analyzed using kinematic laws and digital elevation models. Ref. [11] used kinematic laws and digital elevation models for the study area to develop probabilistic risk maps for planar, tipping, and wedge damage. By comparing the actual fault distribution in the area with the probabilistic risk map prepared for the study area, it was found that the identified faults were located in the higher risk areas on the probabilistic risk map. Ref. [12] analyzed the fundamental changes of a particle subjected to flow dynamics, deposition, and erosion processes at high slope angles. Ref. [13] developed a nonlinear mathematical model for the degradation of sensitive clay soils after peak undrained shear strength based on experimental results.

In summary, the existing research mainly focuses on the stability analysis of slopes and the analysis of landslide hazard risk during the operation period of HSR, while the research on the safety risks occurring during the construction period of HSR is relatively rare. In actual engineering practice, the existing analysis of sliding stability of high cutting slopes mostly adopts the deterministic analysis method, which determines whether the damage will occur by calculating the safety coefficient of anti-sliding stability. The deterministic analysis method is more frequently used because the calculation is relatively simple and the result is more intuitive. However, there are a lot of uncertain parameters in the actual construction process of the high cutting slope of HSR, and the deterministic analysis method can not consider the influence of parameter uncertainty. Therefore, consideration of random factors and their construction dynamics in the analysis of the high cutting slope of HSR deserves further discussion and has important research value.

In 1943, the first neuronal M-P model was proposed by McCulloch-Pitts, and the research on neural networks began [14,15]. Since then, more derivative models of neural networks have emerged. After Rosenblatt's first perceptron model in 1957, many influential models have been proposed [16,17]. Neural networks rapidly developed and were used within different fields [18]. Neural networks have various functions, such as learning, training, simulation, storage, and error removal, which allow them to develop rapidly in many fields and achieve great success in signal processing, pattern recognition, etc. In recent years, more and more scholars have applied them in the research of artificial intelligence [19–25]. Scholars have applied BP neural networks to construction risk assessment and achieved a large number of results, which fully demonstrate the feasibility of BP neural networks for construction risk assessment [2,26–28]. A BP neural network is a nonlinear dynamical system that constructs a model to realize nonlinear analysis by learning and understanding historical data. Compared with the conventional linear analysis methods, a BP neural network has the following advantages [29,30]: (i) it can process data with ambiguous feature performance and logical relationships; (ii) it can process nonlinear characteristic random noisy data; and (iii) it does not require an in-depth understanding of the simulation process. In the case of having more uncertainty, the neural network model can fully demonstrate its superiority in processing data. Based on the above advantages, a BP neural network was selected as the model for risk evaluation in this study.

The paper is organized as follows. Section 2 introduces the data and methods, which includes factor identification and classification (Sections 2.1–2.3), construction safety risk assessment index system (Section 2.4), risk classification criteria (Section 2.5), and assessment model (Section 2.6). Section 3 presents a case study to verify the practicality of the proposed model and discusses the results, then, major conclusions and implications are drawn (Section 4).

2. Data and Methods

Many factors that are not interconnected are stimulated by a certain condition to produce a chain reaction leading to the existence of risk. Thus, the existence of accidents is not caused by a single factor. The construction safety risk of the high cutting slope of HSR has the characteristics of suddenness, damage, complexity, objectivity, and development. This also determines the unpredictability and diversity of its risk-influencing factors.

This study started with the technical specification of railway roadbed risk management to grasp the process of high cutting slope engineering of HSR and the common accidents and problems in the process of slope construction. Our goal was to make it easier to understand construction safety risks. Through literature review and practical research on slope stability, the main influencing factors of instability were studied. The occurrence of accidents is often related to people, materials, the environment, construction management, and other factors. Therefore, through the collection of relevant information and cases to supplement the construction safety risk factors of HSR, the personnel factors and construction management factors were fully considered to form a comprehensive list of construction safety risks.

2.1. Construction Safety Risk Identification Based on Technical Specifications

Technical code for risk management of railway subgrade engineering [31] gives the influencing factors of the construction safety risks of the high cutting slope, which play a reference role in the selection of risk factors in this study. Construction safety risk factors identified for the high cutting slope in HSR are shown in Table 1.

Table 1. Construction safety risk factors based on technical specifications.

Classification	Risk Factors
Natural factor	Topography, surface water, groundwater, scenic nature reserve, existing buildings (structures) and pipelines, rainstorms, floods, avalanches, thunder and lightning, etc.
Geological factor	Degree of rock weathering, landslides, cave strata such as karst and mined-out regions, regional subsidence, swelling rock (soil), permafrost, soft soil, collapsible loess, liquid formation, etc.
Technical factor	Improper classification and protection schemes of soil and rock, improper excavation methods, inadequate foundation treatment, staffing, mechanical equipment, material factors, disposal sites, and protection.
Social factor	Land acquisition and demolition, external influence, humanistic environment.

2.2. Construction Safety Risk Identification Based on Documentary Data

Through combing and screening statistics of related literature, the factors affecting the construction safety of the high cutting slope in HSR were derived, as shown in Table 2.

Table 2. Construction safety risk factors based on literature.

Literature Resources	Risk Factors
Seibold & Hinz [32]	Water movement, geological factors, slope construction process.
Li et al. [33]	Engineering geological conditions, construction methods, landslide treatment plans, slope construction remediation plans.
Cunsen et al. [34]	Topography, geomorphology, stratigraphic lithology, geological structure, hydrogeological conditions, slope design scheme, slope construction technology.
Jiang et al. [35]	Slope excavation methods, anchor cable arrangement, anchor section grouting, climatic factors, slope drainage facility.
Wei et al. [36]	Slope scouring, climatic factors, geological conditions, groundwater.
Park et al. [37]	Discontinuity of structural surface, lithological weathering and erosion, climatic conditions, groundwater.
Zhou et al. [38]	Implementation strength of construction organization design, construction wiring survey, earthwork excavation, surface drainage, anchoring engineering construction.
Wyllie & Mah [39]	Design of slope rate, reinforcement engineering, protection engineering, construction technology, construction wiring survey, geological conditions, groundwater displacement detection.
Scarpelli et al. [40]	Soil support structure, preliminary geological exploration, excavating sequence, terrain monitoring.
Fell & Hartford [41]	Geological conditions, climate, pre-reinforcement construction technology.
Tiwari & Upadhyaya [42]	Rainfall, groundwater, slope rate, adjacent buildings construction.
Zhang et al. [43]	Excavation height, slope rate, differences between stratigraphic and topographic
He et al. [44]	Slope height, slope shape, slope ratio, geological conditions, design plan of blasting, blasting environment, safety supervision, management factors.
Dahal et al. [45]	Excavation, surface water drainage system, anchor construction, planting bag construction, slope skeleton protection.
Abdulwahid & Pradhan [46]	Construction scale, geological conditions, construction environment, data integrity.

2.3. Construction Safety Risk Identification Based on Statistical Analysis of Cases

The construction safety risk identification was carried out through the high cutting slopes related to HK and HSH, two high-speed railways in China, and the risk identification table for the statistical analysis of the cases was formed as shown in Table 3.

Table 3. Construction safety risk identification based on statistical analysis of cases.

Cases	Risk Factors
HK DK438 + 355~ + 510	Soft rock high cutting slope. The main risk factors are slope excavation, geological environment, rainfall, personnel safety awareness, prestressing tensioning equipment, monitoring program reasonableness, mechanical excavation, and slope repair equipment.
HK DK830 + 955~DK831 + 020	Hard rock high cutting slope, smooth blasting. The main risk factors are slope excavation, rock blasting, blasting material, rockfall, falling objects, and the perfection of emergency rescue measures.
HSH DK35 + 309.83~ + 392.00	Soft rock high cutting slope, the groundwater is developed. The main risk factors are drainage tunnel technology and groundwater.
HSH DK97 + 006.89~DK99 + 227.52	Expansive soil high cutting slope. The main risk factors are slope excavation, rainfall, surface drainage system, and data integrity of monitoring program.
HSH DK123 + 612.32~DK124 + 312.50	Bedding high cutting slope. The main risk factors are slope structure, formation lithology, rainfall, slope excavation, and drilling rig equipment.
HSH DK269 + 378.81~ + 703.75	Expansive soil high cutting slope with buildings around. The main risk factors are anti-slip pile technology, rainfall, groundwater, geological condition, and surrounding buildings.
HSH DK270 + 262.11~ + 411.00	Bedding high cutting slope. The main risk factors are slope excavation, prestressing anchoring engineering technology, anchor cable material, geological condition, and scaffolding equipment.
HSH DK294 + 533.51~295 + 069.22	Soft rock high cutting slope. The main risk factors are slope excavation, slope rate, slope skeleton protection, mechanical excavation and slope repair equipment, implementation of safety management measures, and concrete material.
HSH DK318 + 237.71~ + 558.50	Bedding high cutting slope. The main risk factors are prestressed anchorage engineering technology, slope protection, personnel safety awareness, prestressed anchor anti-slip pile technology, and rainfall.

2.4. Construction Safety Risk Assessment Index System for High Cut Slope of HSR

Through SPSS software for analysis, combined with engineering practice, the construction safety risk factors were classified and summarized according to the principle of systemic feasibility. The secondary risk indicators include 5 items such as personnel risk, and the tertiary risk indicators include 39 items such as earth excavation. The risk assessment index system is formed in Table 4.

Table 4. Safety risk assessment index system for high cutting slope construction of HSR.

Goal Layer	Criterion Layer	Indicator Layer
Safety risk assessment indicator system for high cutting slope construction of HSR	Risk factors of construction technology (CT)	Earthwork excavation method CT_1 Rock blasting CT_2 Surface drainage system technology CT_3 Anti-slip retaining wall technology CT_4 Anti-slip pile technology CT_5 Prestressed anchorage technology CT_6 Prestressed anchor cable anti-slide pile technology CT_7 Grouting micro-pile technology CT_8 Tunnel drainage technology CT_9 Slope skeleton protection CT_{10} Slope surface protection CT_{11}
	Risk factors of material and equipment (ME)	Blasting materials ME_1 Anchor cable materials ME_2 Concrete materials ME_3 Mechanical excavation and repairment equipment ME_4 Drilling rig equipment ME_5 Scaffolding equipment ME_6 Prestressed tensioning equipment ME_7
	Risk factors of personnel (P)	Basic quality level of personnel P_1 Personnel working level P_2 Personnel operation error P_3 Personnel safety awareness P_4 Staffing level P_5
	Risk factors of the environment (E)	Slope height E_1 Slope rate E_2 Rainfall E_3 Surrounding buildings E_4 Falling objects E_5 Groundwater E_6 Geological conditions E_7
	Risk factors of construction management risk (CM)	Quality assurance measures CM_1 Quality testing standards CM_2 Perfection of emergency rescue measures CM_3 Degree of implementation of safety management fee CM_4 Degree of implementation of construction organization design CM_5 Reasonableness of monitoring program CM_6 Completeness of monitoring data CM_7 Communication and coordination among all parties involved in the project CM_8 Construction safety risk management system CM_9

2.5. Risk Classification Criteria of Safety in Construction

With a large number of risks and a variety of risk control measures in the construction of high cutting slopes in HSR., it is necessary to evaluate risks reasonably, effectively, and objectively, select appropriate risk control measures, and carry out scientific engineering decisions and risk evaluation. A key part of the evaluation is the need to study the corresponding risk acceptance criteria and control countermeasures. According to the Technical Code for Risk Management of Railway Construction Engineering [47], the risk acceptance criteria for high cutting slopes in HSR are shown in Table 5.

Table 5. Risk acceptance criteria.

Risk Level	Acceptance Level
Level I (lower risk)	Negligible
Level II (medium risk)	Acceptable
Level III (high risk)	Unexpected
Level IV (very high risk)	Unacceptable

2.6. *BP Neural Network Model*

2.6.1. Design of Network Topology

According to the risk assessment index system established above, combined with the BP network theory, three levels of indicators are used as the input layer, and risk levels are the output layer. Among them, the input layer has 39 indicators, and the output layer is the project risk level. The trial-and-error method was adopted to determine the number of hidden nodes corresponding to the fastest convergence and smallest error of the network. The operating mechanism diagram is shown in Figure 1.

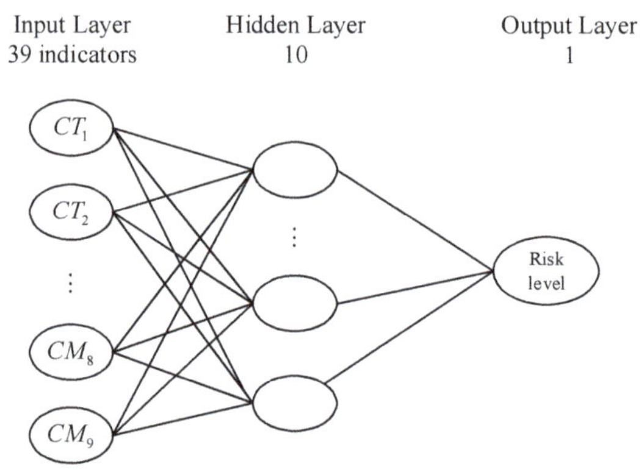

Figure 1. BP neural network operation structure.

2.6.2. Select the Activation Function of the Feedforward Neural Network

In this model, the maximum number of trainings is set to 20,000, after which the training is terminated. The error accuracy is set as 1×10^{-6}, and the operation is terminated when the accuracy is less than this error number. The model of this training is nonlinear, and the data need to be normalized to a certain extent before being brought in; therefore, the S-type function was chosen for this model.

2.6.3. Selection of Samples

The processing of samples directly affects the generalization mapping ability of BP networks [48]. When selecting the samples, we need to consider the engineering requirements and characteristics [49,50]. In this study, the sample parameters were obtained via a questionnaire survey; a total of 216 high cutting slopes in HSR of HK, HF, YG, and HSH were selected for the collection. The data were quantified by combining Table 7 and the method of expert interrogation. If the collected samples are evenly distributed, it is more conducive to the training of neural networks and can make a better prediction of the risk level of the cases.

2.6.4. Pretreatment of Input and Output Data

The collected data were divided into input data and output data, of which 80% were used as training samples and 20% as testing samples. The processed data of input nodes were regarded as the input layer, which was denoted by x. The processed data of output nodes were regarded as the output layer, which was denoted by y. In this study, the number of samples obtained was 216, denoted by m. The network inputs and outputs are shown in Table 6.

Table 6. Inputs and outputs data of network.

Input Data	Output Data
$x_{11}, x_{12}, \cdots x_{1n}$	y_1
$x_{21}, x_{22}, \cdots x_{2n}$	y_2
$\vdots$	$\vdots$
$x_{n1}, x_{n2}, \cdots x_{nn}$	y_n

Each indicator has a different dimension; if the initial data are used directly, it will be difficult to ensure that indicators are in the same dimension, which will make the BP network converge slowly. If the indicators have the same dimension, the S-shaped function in the BP network can be fully utilized and the saturation area of the function can be avoided, thus enhancing the sensitivity of the BP neural network to the indicators. Therefore, it is important to reduce the magnitude of the variation of the sample values and to lock the interval of the definition domain of the samples within a certain range. This ensures that the derivative values of the input function are within the appropriate interval and have an important role in the training of the neural network and the prediction of the samples. Therefore, in this study, the collected sample data were normalized to reduce the magnitude variation of the predicted values. Firstly, each risk indicator was evaluated by the table of construction safety risk assessment of high cutting slopes in HSR (Table 7) and the method of expert interrogation. The statistical data were then normalized to obtain the risk value of relevant indicators. For example, the transformation of construction technology risk (CT) is shown as Formulas (1)–(3). The expression of the input value of the whole indicators system is shown as Formula (4).

$$CT_i = \sum_{j=1}^{11} CT_{ij}/11 \; i = (1,2,\ldots,11) \tag{1}$$

$$CT_i = \frac{ct_i - X(\min)}{X(\max) - X(\min)} \; (i = 1,2,3,\cdots,11) \tag{2}$$

$$CT = (CT_1, CT_2, \ldots, CT_{11}) \tag{3}$$

$$All = (CT, ME, P, E, CM)^T \tag{4}$$

Table 7. Construction safety risk evaluation levels.

Comprehensive Risk Evaluation Value	[0~0.25]	(0.25~0.5]	(0.5~0.75]	(0.75~1]
Risk Level	Low risk	Medium risk	High risk	Extremely high risk

According to Table 5, combined with engineering practice, the construction safety risk evaluation levels can be divided into four levels, as shown in Table 7.

2.6.5. Selection of Initial Parameters

In this study, the BP neural network toolbox in MATLAB was used for simulation training, and the appropriate model parameters played a decisive role in the training quality and accuracy. The network creation function was debugged in the toolbox operation interface of the BP neural network, which is the first step to building the network object. Then, the transfer function was debugged. The transfer function represents the input and output objects with a correlation function using the logsig function. The learning function is to adjust the local error value size, and the training function is to adjust the global error value size; learngdm was chosen as the learning function, traingdx as the training function, and the network error performance function was the default MSE function of the toolbox. The network model parameters are set in Table 8.

Table 8. BP network model parameters.

Network Creation Function	Transfer Function	Learning Function	Training Function	Network Error Performance Function	Error Precision
newff	logsig	learngdm	traingdx	MSE	0.0003

2.6.6. Objective Function of Network Training

The initial expected predicted value of the objective function is $X_k(k = 1, 2, \cdots, N)$, the output value of the final function is $\overset{\wedge}{X_k} = (k = 1, 2, \cdots, N)$, and the prediction error e is $e = (e_1, e_2, \cdots, e_N) = \left(x_1 - \overset{\wedge}{x_1}, x_2 - x_2, \cdots, \overset{\wedge}{x_N} - \overset{\wedge}{x_N}\right)$. In the study, SSE was used to represent the evaluation value of neural network training maturity, which is shown in Formula (5).

$$SSE = \frac{1}{2}\sum\left(x_k - \overset{\wedge}{x_k}\right)^2 \tag{5}$$

where, x_k is the expected output and $\overset{\wedge}{x_k}$ is the actual output.

Simulation training was performed according to the neural network toolbox to achieve the minimum error criterion. If the trained neural network does not meet the error criterion, the BP neural network model parameters need to be readjusted to achieve a suitable neural network structure.

2.6.7. Selection of Output Node

The output node refers to the final information output, which is also the global control of the neural network training. The output nodes should be selected with reasonable and high credibility as often as possible to facilitate the collection of data and to normalize the sample data to obtain the output data that are beneficial to the neural network training. The sample data of the output nodes in this study were based on the risk levels corresponding to Table 7.

3. Results

3.1. Case Background

Hefei-Fuzhou HSR is an important part of Beijing-Fuzhou HRR; DK592 + 532.00~ + 562.00 is a section of the high cutting slope (Figure 2). The area is low and hilly with gentle terrain and developed vegetation. The groundwater is bedrock fracture water, which is more developed. The groundwater is not chemically erosive (judged by the chloride ion content, without chloride salt erosive).

Figure 2. The high cutting slope of section DK592 + 532.00~ + 562.00.

3.2. Establish BP Neural Network Model

The samples of this case came from a total of 216 high cutting slopes of the HK, HF, YG, and HSH high-speed railway. By collecting the geological situation, special construction plan, and minutes of each high cutting slope and considering the relevant indicator rating methods in risk guidance combined with engineering practices and expert interviews, the indicators in each sample were assigned values according to Table 7, and all data were normalized. The details of the data processing are shown in the Appendix A. The Ordered Weighted Averaging is used to determine the weights of indicators at all levels, and the weight determination method is shown in Formula (6).

$$w = \overline{w}_i \bigg/ \sum_{i=1}^{j} \overline{w}_i, i = 1, 2, \cdots, n\ j = i = 1, 2, \cdots, m \tag{6}$$

The weight values of CT, ME, P, E, and CM were 0.328, 0.104, 0.189, 0.258, and 0.121 respectively, as shown in Figure 3. From Figure 3, it can be seen that the risks of construction technology and environment have a relatively strong influence on the construction safety of high cutting slopes in HSR, and the risk of material and equipment, personnel, and construction management have a relatively weak influence on it.

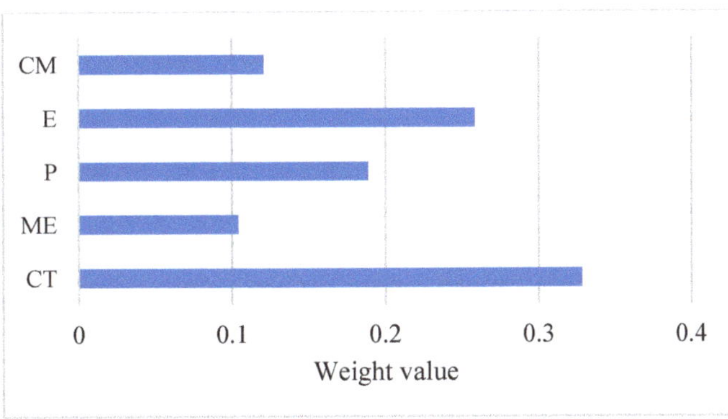

Figure 3. The weight values of the secondary risk indicators.

3.3. Training Simulation

According to the BP neural network algorithm, Matlab was used to create the neural network and run the results. After 11 iterations, the training MSE value of the simulation training was 0.000176, which is less than the target value of 0.0003. This met the predetermined accuracy requirements, and the BP neural network model achieved convergence. The simulation results are shown in Figure 4; it can be seen that the R-values of the training set and test set are 0.947 and 0.808, which indicate that the model fits the observed values well.

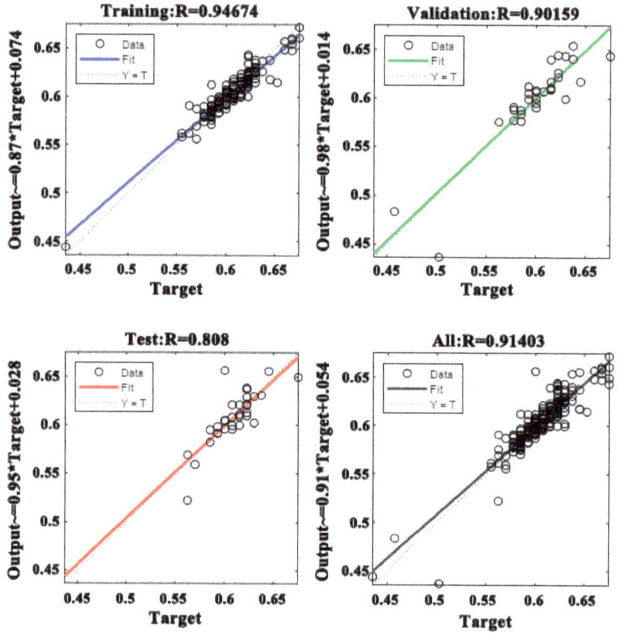

Figure 4. Schematic diagram of BP Neural network training fit degree.

The results of the error analysis for the test set data are shown in Figure 5. It can be seen that the maximum absolute value of the error is 8.22%; therefore, the training effect of the model is satisfactory.

3.4. Discussion

The trained BP artificial neural network was used to predict the risk of an HF grade eight high cutting slope construction project. After normalizing the data for each risk indicator (0.788, 0.394, 0.303, 0.303, 0.273, 0.455, 0.303, 0.303, 0.485, 0.273, 0.485, 0.273, 0.364, 0.485, 0.424, 0.212, 0.576, 0.278, 0.364, 0.389, 0.212, 0.697, 0.303, 0.818, 0.576, 0.273, 0.485, 0.424, 0.636, 0.394, 0.455, 0.364, 0.364, 0.394, 0.212, 0.394, 0.333, 0.424, 0.758), they were entered into the network to obtain the predicted value of 0.479. According to Table 7, the construction safety risk of the high cutting slope is predicted to be a medium risk, which is consistent with the risk level of the project; thus, the model fit is excellent.

According to the input value of each risk factor, the risk index greater than 0.5 is the main risk. It is known that the main risks of this high cutting slope construction are earthwork excavation method, scaffolding equipment, slope height, slope rate, groundwater, personnel safety awareness, and construction safety risk management system, and these influencing factors have a greater impact on the construction safety management of the project. Therefore, the safety control of this high cutting slope construction focuses on slope excavation risk control, slope reinforcement, waterproof measures, and construction safety management measures. The following control measures are proposed for the construction safety risks of this high cutting slope project: (1) choosing a reasonable earth excavation method; (2) setting up reinforcement protection measures; (3) strengthening of waterproof design; (4) conducting pre-reinforcement treatment; (5) monitoring stability; (6) strengthening construction management measures.

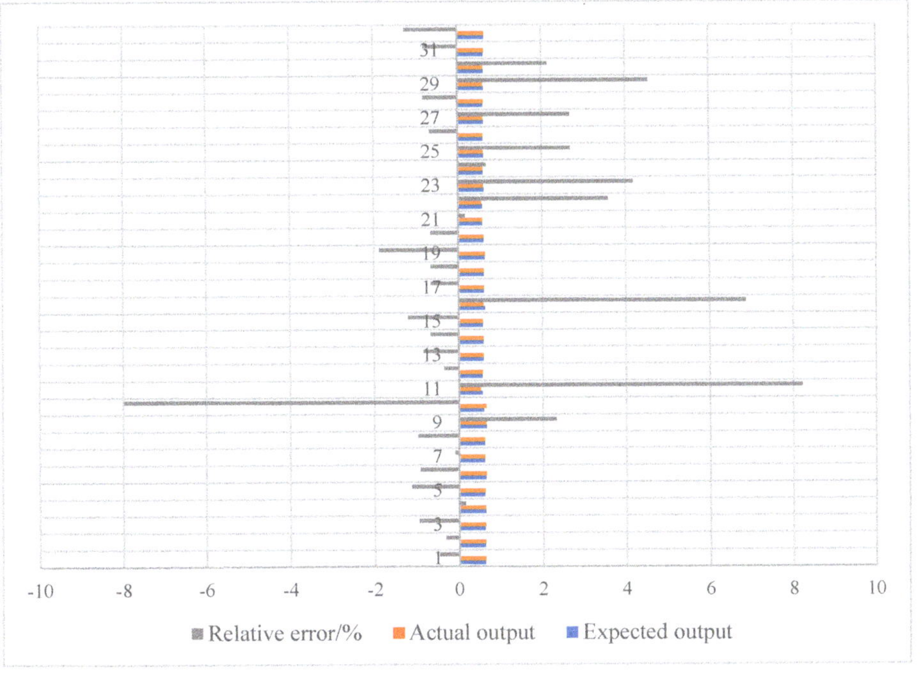

Figure 5. Error analysis for the test set data.

4. Conclusions

In this study, the safety risks of high cutting slope construction in HSR were identified in all aspects from three dimensions, risk technical specification, literature analysis, and case statistical analysis, and a list of risk influencing factors was formed. The evaluation indicator system was constructed by designing questionnaires and analyzing them with SPSS data statistical software. The assessment model was established by a BP neural network, and the pre-control measures were proposed for the risk factors. The construction safety risks of a high cutting slope of HF high-speed railway was analyzed and evaluated. The main findings of this study are (1) a list of construction safety risks of high cutting slopes in HSR was formed; (2) a risk assessment indicator system of high cutting slopes in HSR was constructed; (3) a construction safety risk assessment model based on a BP neural network was established; and (4) the feasibility of the assessment model was verified.

The limitation of this study is the identification and analysis of construction risk factors with a certain one-sidedness and subjectivity. Combined with the dynamic and difficult quantitative nature of construction risks, it needs to be further combined with engineering practice to refine and improve the construction safety impact factors. In addition, the number and authenticity of the learning samples directly ensure the feasibility of the trained neural network, and more samples need to be collected to improve the sample credibility. Finally, the BP neural network training process is related to set parameters, which will be combined with more intelligent algorithms for improvement to improve the accuracy of the training results in the future.

Author Contributions: Conceptualization, J.H. and H.C.; methodology, X.Z. and J.F.; software, X.Z.; validation, X.Z. and J.F.; formal analysis, X.Z.; investigation, J.H., X.Z., J.F., Y.H. and H.C.; resources, J.H. and H.C.; data curation, Y.H.; writing—original draft preparation, X.Z. and J.F.; writing—review and editing, X.Z.; visualization, X.Z.; supervision, J.H. and H.C.; project administration, J.H. and H.C.; funding acquisition, J.H. and H.C. All authors have read and agreed to the published version of the manuscript.

Funding: This study was funded by the National Key R&D Program of China (Grant Number 2021YFB2301802).

Institutional Review Board Statement: Not applicable.

Informed Consent Statement: Not applicable.

Data Availability Statement: All datasets generated for this study have been included in the article.

Conflicts of Interest: The authors declare no conflict of interest.

Appendix A

For the construction technology risk (CT), the collected assessment data are shown in Table A1.

Table A1. Raw data of construction technical risk (CT).

Indicator \ Expert	1	2	3	4	5	6	7	8	9	10	11
CT_1	2	2	3	3	2	3	2	3	3	2	2
CT_2	3	2	2	2	3	3	2	2	2	3	2
CT_3	2	2	1	1	2	2	2	2	3	2	1
CT_4	4	3	4	4	4	4	3	3	2	3	3
CT_5	2	3	3	3	2	3	2	2	3	3	3
CT_6	3	2	3	2	3	3	2	3	2	2	3
CT_7	3	3	2	3	2	2	3	3	4	3	4
CT_8	2	2	3	2	2	2	2	2	3	2	2
CT_9	2	2	1	1	2	2	1	2	2	2	2
CT_{10}	3	3	2	3	3	3	3	3	2	3	3
CT_{11}	2	2	1	2	2	2	2	2	1	2	2

By using Formula (1), data processing is shown in Table A2.

Table A2. Construction technical risk (CT) data processing.

Indicator	CT_1	CT_2	CT_3	CT_4	CT_5	CT_6
Data processing	2.455	2.364	1.818	3.364	2.636	2.545
Indicator	CT_7	CT_8	CT_9	CT_{10}	CT_{11}	
Data processing	2.909	2.182	1.727	2.818	1.818	

The output values were normalized according to Formula (2), and the results are shown in Table A3.

Table A3. Normalized data of construction technical risk (CT).

Indicator	CT_1	CT_2	CT_3	CT_4	CT_5	CT_6
Normalized data	0.485	0.455	0.273	0.788	0.545	0.515
Indicator	CT_7	CT_8	CT_9	CT_{10}	CT_{11}	
Normalized data	0.636	0.394	0.242	0.606	0.273	

For material and equipment risk (ME), the assessment data collected are shown in Table A4.

Table A4. Raw data of material and equipment risk (ME).

Expert / Indicator	1	2	3	4	5	6	7	8	9	10	11
ME_1	3	3	2	3	2	3	2	3	3	2	1
ME_2	2	1	1	2	1	1	1	1	1	1	2
ME_3	2	2	1	2	1	2	2	2	1	1	1
ME_4	2	1	1	1	1	1	1	1	2	2	2
ME_5	3	2	2	2	2	3	2	3	2	2	2
ME_6	4	3	3	3	3	3	3	3	3	3	3
ME_7	1	2	2	2	2	2	2	2	2	2	2

By using Formula (1), data processing is shown in Table A5.

Table A5. Data processing of material and equipment risk (ME).

Indicator	ME_1	ME_2	ME_3	ME_4	ME_5	ME_6	ME_7
Data processing	2.455	1.273	1.545	1.364	2.273	3.091	1.909

The output values are normalized according to Formula (2), and the results are shown in Table A6.

Table A6. Normalized data of material and equipment risk (ME).

Indicator	ME_1	ME_2	ME_3	ME_4	ME_5	ME_6	ME_7
Normalized data	0.485	0.091	0.182	0.121	0.424	0.697	0.303

For environmental risk (E), the assessment data collected are shown in Table A7.

Table A7. Raw data of environmental risk (E).

Expert / Indicator	1	2	3	4	5	6	7	8	9	10	11
E_1	2	2	1	1	2	1	2	2	3	2	2
E_2	2	1	1	1	1	2	2	2	2	2	2
E_3	3	2	3	3	2	3	3	2	2	3	2
E_4	2	3	3	3	3	2	3	2	3	3	2
E_5	1	2	1	1	1	1	2	1	2	2	2
E_6	2	1	2	1	2	2	2	2	2	2	1
E_7	3	3	3	2	3	2	2	3	2	2	2

By using the Formula (1), data processing is shown in Table A8.

Table A8. Data processing of environmental risk (E).

Indicator	E_1	E_2	E_3	E_4	E_5	E_6	E_7
Data processing	1.818	1.636	2.545	2.636	1.455	1.727	2.455

The output values were normalized according to Formula (2), and the results are shown in Table A9.

Table A9. Normalized data of environmental risk (E).

Indicator	E_1	E_2	E_3	E_4	E_5	E_6	E_7
Normalized data	0.273	0.212	0.515	0.545	0.152	0.242	0.485

For personnel risk (P), the assessment data collected are shown in Table A10.

Table A10. Raw data of personnel risk (P).

Indicator \ Expert	1	2	3	4	5	6	7	8	9	10	11
P_1	3	4	3	4	3	2	2	4	2	3	2
P_2	1	2	2	2	1	2	1	2	2	2	1
P_3	2	2	3	3	2	3	3	2	3	2	2
P_4	2	1	2	2	3	1	2	2	2	1	2
P_5	2	3	2	2	2	1	2	2	3	2	3

By using Formula (1), data processing is shown in Table A11.

Table A11. Data processing of Personnel risk (P).

Indicator	P_1	P_2	P_3	P_4	P_5
Data processing	2.909	1.636	2.455	1.818	2.182

The output values were normalized according to formula (2), and the results are shown in Table A12.

Table A12. Normalized data of personnel risk (P).

Indicator	P_1	P_2	P_3	P_4	P_5
Normalized data	0.636	0.212	0.485	0.273	0.394

For construction management risk (CM), the assessment data collected are shown in Table A13.

Table A13. Raw data of construction management risk (CM).

Indicator \ Expert	1	2	3	4	5	6	7	8	9	10	11
CM_1	3	2	3	2	2	2	3	2	2	2	3
CM_2	3	3	2	2	4	3	3	2	2	3	3
CM_3	4	3	3	3	2	3	3	2	3	3	3
CM_4	3	3	4	4	4	3	3	4	3	4	2
CM_5	4	3	2	2	2	2	3	3	2	3	3
CM_6	4	3	2	3	3	3	3	2	2	2	4
CM_7	2	3	3	2	3	3	3	2	3	2	2
CM_8	2	3	2	3	3	2	4	3	4	3	3
CM_9	3	3	3	3	3	3	2	3	3	2	2

By using Formula (1), data processing is shown in Table A14.

Table A14. Data processing of construction management risk (CM).

Indicator	CM_1	CM_2	CM_3	CM_4	CM_5	CM_6	CM_7	CM_8	CM_9
Data processing	2.364	2.727	2.909	3.364	2.636	2.818	2.545	2.909	2.909

The output values are normalized according to Formula (2), and the results are shown in Table A15.

Table A15. Normalized data of construction management risk (CM).

Indicator	CM_1	CM_2	CM_3	CM_4	CM_5	CM_6	CM_7	CM_8	CM_9
Normalized data	0.455	0.576	0.636	0.788	0.545	0.606	0.515	0.636	0.576

Due to space issues, only the first ten groups of case data processing results are listed in this study, as shown in Table A16.

Table A16. Input data for the first ten samples.

Indicator \ Sample	1	2	3	4	5	6	7	8	9	10
1	0.485	0.727	0.697	0.727	0.636	0.545	0.697	0.545	0.545	0.667
2	0.455	0.424	0.455	0.424	0.303	0.152	0.212	0.273	0.212	0.515
3	0.273	0.606	0.424	0.606	0.273	0.333	0.333	0.424	0.394	0.545
4	0.788	0.485	0.758	0.424	0.182	0.515	0.364	0.606	0.606	0.667
5	0.545	0.515	0.576	0.636	0.515	0.485	0.455	0.545	0.455	0.576
6	0.515	0.333	0.667	0.545	0.576	0.364	0.606	0.455	0.364	0.545
7	0.636	0.758	0.697	0.788	0.576	0.424	0.576	0.333	0.424	0.758
8	0.394	0.424	0.636	0.455	0.121	0.152	0.182	0.303	0.303	0.636
9	0.242	0.455	0.636	0.515	0.303	0.424	0.394	0.576	0.485	0.636
10	0.606	0.485	0.667	0.364	0.091	0.394	0.242	0.515	0.394	0.667
11	0.273	0.606	0.545	0.636	0.485	0.576	0.758	0.545	0.606	0.606
12	0.485	0.576	0.576	0.576	0.606	0.606	0.333	0.333	0.212	0.606
13	0.091	0.333	0.727	0.636	0.545	0.303	0.364	0.485	0.515	0.515
14	0.182	0.455	0.364	0.455	0.152	0.515	0.515	0.515	0.636	0.424
15	0.121	0.485	0.636	0.545	0.303	0.485	0.515	0.606	0.697	0.667
16	0.424	0.515	0.485	0.485	0.152	0.242	0.455	0.515	0.455	0.545
17	0.697	0.303	0.424	0.485	0.455	0.303	0.636	0.333	0.273	0.636
18	0.303	0.636	0.788	0.606	0.667	0.242	0.303	0.455	0.424	0.758
19	0.273	0.515	0.576	0.545	0.485	0.333	0.394	0.485	0.455	0.545
20	0.212	0.455	0.576	0.697	0.182	0.515	0.273	0.485	0.545	0.576
21	0.515	0.485	0.636	0.515	0.303	0.424	0.485	0.455	0.364	0.455
22	0.545	0.455	0.485	0.485	0.121	0.455	0.636	0.545	0.394	0.636
23	0.152	0.485	0.394	0.455	0.364	0.182	0.697	0.545	0.545	0.545
24	0.242	0.455	0.727	0.727	0.697	0.333	0.485	0.333	0.424	0.545
25	0.485	0.364	0.455	0.364	0.455	0.121	0.515	0.333	0.242	0.576
26	0.636	0.485	0.636	0.515	0.152	0.152	0.303	0.576	0.515	0.515
27	0.212	0.424	0.636	0.545	0.273	0.394	0.333	0.758	0.758	0.818
28	0.485	0.667	0.697	0.788	0.242	0.576	0.515	0.273	0.455	0.667
29	0.273	0.424	0.667	0.515	0.455	0.485	0.515	0.303	0.364	0.606
30	0.394	0.364	0.636	0.515	0.636	0.394	0.394	0.545	0.545	0.515
31	0.455	0.515	0.515	0.576	0.697	0.515	0.424	0.364	0.455	0.727
32	0.576	0.545	0.545	0.485	0.455	0.364	0.091	0.455	0.455	0.545
33	0.636	0.515	0.636	0.515	0.152	0.364	0.515	0.545	0.455	0.455
34	0.788	0.394	0.515	0.788	0.303	0.212	0.455	0.576	0.576	0.636
35	0.545	0.545	0.576	0.545	0.212	0.273	0.485	0.333	0.364	0.576
36	0.606	0.455	0.364	0.455	0.364	0.182	0.455	0.303	0.364	0.606
37	0.515	0.485	0.758	0.576	0.667	0.485	0.636	0.333	0.515	0.667
38	0.636	0.333	0.515	0.606	0.333	0.485	0.333	0.667	0.667	0.515
39	0.576	0.576	0.636	0.576	0.606	0.533	0.485	0.394	0.545	0.576

References

1. Lin, D.; Chen, P.; Ma, J.; Zhao, Y.; Xie, T.; Yuan, R.; Li, L. Assessment of slope construction risk uncertainty based on index importance ranking. *Bull. Eng. Geol. Environ.* **2019**, *78*, 4217–4228. [CrossRef]
2. Deng, X.; Xu, T.; Wang, R. Risk evaluation model of highway tunnel portal construction based on BP fuzzy neural network. *Comput. Intell. Neurosci.* **2018**, *2018*, 8547313. [CrossRef] [PubMed]
3. Alonso, E.E. Risk analysis of slopes and its application to slopes in Canadian sensitive clays. *Geotechnique* **1976**, *26*, 453–472. [CrossRef]
4. Yilmaz, I. Comparison of landslide susceptibility mapping methodologies for Koyulhisar, Turkey: Conditional probability, logistic regression, artificial neural networks, and support vector machine. *Environ. Earth Sci.* **2010**, *61*, 821–836. [CrossRef]
5. Fu, J. *Research on Construction Safety Risk Management of High Cut Slope of High Speed Railway Based on BP Neural Network*; Central South University: Changsha, China, 2019.
6. Sidle, R.C.; Pearce, A.J.; O'Loughlin, C.L. *Hillslope Stability and Land Use*; American Geophysical Union: Washington, DC, USA, 1985.
7. Brown, E.T. Risk assessment and management in underground rock engineering—An overview. *J. Rock Mech. Geotech. Eng.* **2012**, *4*, 193–204. [CrossRef]
8. Christian, J.T.; Ladd, C.C.; Baecher, G.B. Reliability applied to slope stability analysis. *J. Geotech. Eng.* **1994**, *120*, 2180–2207. [CrossRef]
9. El-Ramly, H.; Morgenstern, N.R.; Cruden, D.M. Cruden, Probabilistic slope stability analysis for practice. *Can. Geotech. J.* **2002**, *39*, 665–683. [CrossRef]
10. Cheng, Y.M.; Lansivaara, T.; Wei, W.B. Two-dimensional slope stability analysis by limit equilibrium and strength reduction methods. *Comput. Geotech.* **2007**, *34*, 137–150. [CrossRef]
11. Gokceoglu, C.; Sonmez, H.; Ercanoglu, M. Discontinuity controlled probabilistic slope failure risk maps of the Altindag (settlement) region in Turkey. *Eng. Geol.* **2000**, *55*, 277–296. [CrossRef]
12. Farin, M.; Mangeney, A.; Roche, O. Fundamental changes of granular flow dynamics, deposition, and erosion processes at high slope angles: Insights from laboratory experiments. *J. Geophys. Res. Earth Surf.* **2014**, *119*, 504–532. [CrossRef]
13. Dey, R.; Hawlader, B.; Phillips, R.; Soga, K. Modeling of large-deformation behaviour of marine sensitive clays and its application to submarine slope stability analysis. *Can. Geotech. J.* **2016**, *53*, 1138–1155. [CrossRef]
14. McCulloch, W.S.; Pitts, W. A logical calculus of the ideas immanent in nervous activity. *Bull. Math. Biophys.* **1943**, *5*, 115–133. [CrossRef]
15. Cowan, J.D. Discussion: McCulloch-Pitts and related neural nets from 1943 to 1989. *Bull. Math. Biol.* **1990**, *52*, 73–97. [CrossRef]
16. Anderson, J.A. The BSB Model: A Simple Nonlinear Autoassociative Neural Network. In *Associative Neural Memories: Theory and Implementation*; Oxford University Press: Oxford, UK, 1993; pp. 77–103.
17. Hecht-Nielsen, R. Theory of the Backpropagation Neural Network. In *Neural Networks for Perception*; Elsevier: Amsterdam, The Netherlands, 1992; pp. 65–93.
18. Sadeghi, B. A BP-neural network predictor model for plastic injection molding process. *J. Mater. Processing Technol.* **2000**, *103*, 411–416. [CrossRef]
19. Chen, H.; Li, H.; Goh, Y.M. A review of construction safety climate: Definitions, factors, relationship with safety behavior and research agenda. *Saf. Sci.* **2021**, *142*, 105391. [CrossRef]
20. Boussabaine, A.H. The use of artificial neural networks in construction management: A review. *Constr. Manag. Econ.* **1996**, *14*, 427–436. [CrossRef]
21. Forsythe, D.E. Engineering knowledge: The construction of knowledge in artificial intelligence. *Soc. Stud. Sci.* **1993**, *23*, 445–477. [CrossRef]
22. Hou, C.; Wen, Y.; He, Y.; Liu, X.; Wang, M.; Zhang, Z.; Fu, H. Public stereotypes of recycled water end uses with different human contact: Evidence from event-related potential (ERP). *Resour. Conserv. Recycl.* **2021**, *168*, 105464. [CrossRef]
23. Moselhi, O.; Hegazy, T.; Fazio, P. Potential applications of neural networks in construction. *Can. J. Civ. Eng.* **1992**, *19*, 521–529. [CrossRef]
24. Changwei, Y.; Zonghao, L.; Xueyan, G.; Wenying, Y.; Jing, J.; Liang, Z. Application of BP neural network model in risk evaluation of railway construction. *Complexity* **2019**, *2019*, 2946158. [CrossRef]
25. Shi, G.; Cheng, B.; Li, A. A mathematical model for calculating the "brittleness-ductility" drop coefficient of sandstone in mining zones. *Discret. Dyn. Nat. Soc.* **2020**, *2020*, 2621672. [CrossRef]
26. Zichun, Y. The BP artificial neural network model on expressway construction phase risk. *Syst. Eng. Procedia* **2012**, *4*, 409–415.
27. Shen, T.; Nagai, Y.; Gao, C. Design of building construction safety prediction model based on optimized BP neural network algorithm. *Soft Comput.* **2020**, *24*, 7839–7850. [CrossRef]
28. Wang, H.; Wang, L.; Li, L.; Cheng, B.; Zhang, Y.; Wei, Y. The study on the whole stress–strain curves of coral fly ash-slag alkali-activated concrete under uniaxial compression. *Materials* **2020**, *13*, 4291. [CrossRef] [PubMed]
29. Schultz, A.; Wieland, R. The use of neural networks in agroecological modelling. *Comput. Electron. Agric.* **1997**, *18*, 73–90. [CrossRef]
30. Schultz, A.; Wieland, R.; Lutze, G. Neural networks in agroecological modelling—Stylish application or helpful tool? *Comput. Electron. Agric.* **2000**, *29*, 73–97. [CrossRef]

31. China Railway. *Technical Code for Risk Management of Railway Earthworks*; China Railway Publishing House Co., Ltd.: Beijing, China, 2021.
32. Seibold, E.; Hinz, K. Continental Slope Construction and Destruction, West Africa. In *The Geology of Continental Margins*; Springer: New York, NY, USA, 1974; pp. 179–196.
33. Li, Z.; Huang, H.; Nadim, F.; Xue, Y. Quantitative risk assessment of cut-slope projects under construction. *J. Geotech. Geoenviron. Eng.* **2010**, *136*, 1644–1654. [CrossRef]
34. Cunsen, J.; Shipeng, C.; Zuoxin, H. Reinforcement design and construction quality control for k24 cutting slope of south line of jinan belt highway. *Chin. J. Rock Mech. Eng.* **2004**, 5260–5265. [CrossRef]
35. Jiang, Q.; Yan, F.; Wu, J.; Fan, Q.; Li, S.; Xu, D. Grading opening and shearing deformation of deep outward-dip shear belts inside high slope: A case study. *Eng. Geol.* **2019**, *250*, 113–129. [CrossRef]
36. Wei, L.; Debin, G.; Wankui, N. Research on surface plant protection for highway loess cutting slope. *J. Catastrophology* **2007**, *3*, 45–48.
37. Park, H.; West, T.R.; Woo, I. Probabilistic analysis of rock slope stability and random properties of discontinuity parameters, Interstate Highway 40, Western North Carolina, USA. *Eng. Geol.* **2005**, *79*, 230–250. [CrossRef]
38. Zhou, D.; Xiao, S.; Xia, X. Discussion on rational spacing between adjacent anti-slide piles in some cutting slope projects. *Chin. J. Geotech. Eng.* **2004**, *1*, 132–135.
39. Wyllie, D.C.; Mah, C. *Rock Slope Engineering*; CRC Press: Boca Raton, FL, USA, 2004.
40. Scarpelli, G.; Segato, D.; Sakellariadi, E.; Ruggeri, P.; Fruzzetti, V.M.E.; Vita, A. *Slope Instability Problems in the Jonica Highway Construction, in Landslide Science and Practice*; Springer: New York, NY, USA, 2013; pp. 275–282.
41. Fell, R.; Hartford, D. *Landslide Risk Management, in Landslide Risk Assessment*; Routledge: Oxfordshire, UK, 2018; pp. 51–109.
42. Tiwari, B.; Upadhyaya, S. Effect of Rainfall and Building Construction on a Marginal Slope in Triggering Landslide. In *Landslide Science for a Safer Geoenvironment*; Springer: New York, NY, USA, 2014; pp. 313–318.
43. Zhang, Z.; Wang, T.; Wu, S.; Tang, H.; Liang, C. The role of seismic triggering in a deep-seated mudstone landslide, China: Historical reconstruction and mechanism analysis. *Eng. Geol.* **2017**, *226*, 122–135. [CrossRef]
44. He, Z.; Liu, K.; Fu, H.; Wu, C. Safety risk assessment of high slope blasting construction based on set pair-extension analysis. *J. Cent. South Univ. Sci. Technol.* **2017**, *48*, 2217–2223.
45. Dahal, R.; Hasegawa, S.; Masuda, T.; Yamanaka, M. Roadside Slope Failures in Nepal during Torrential Rainfall and Their Mitigation. In *Disaster Mitigation of Debris Flows, Slope Failures and Landslides*; Universal Academy Press: Irvine, CA, USA, 2006; pp. 503–514.
46. Abdulwahid, W.M.; Pradhan, B. Landslide vulnerability and risk assessment for multi-hazard scenarios using airborne laser scanning data (LiDAR). *Landslides* **2017**, *14*, 1057–1076. [CrossRef]
47. China Railway Eryuan Engineering Group Co. Ltd. *Technical Code for Risk Management of Railway Construction Engineering*; China Railway Publishing House Co., Ltd.: Beijing, China, 2015.
48. Ma, J.; Cai, J.; Lin, G.; Chen, H.; Wang, X.; Wang, X.; Hu, L. Development of LC–MS determination method and back-propagation ANN pharmacokinetic model of corynoxeine in rat. *J. Chromatogr. B* **2014**, *959*, 10–15. [CrossRef]
49. Qu, D.; Cai, X.; Chang, W. Evaluating the effects of steel fibers on mechanical properties of ultra-high performance concrete using artificial neural networks. *Appl. Sci.* **2018**, *8*, 1120. [CrossRef]
50. Shi, E.; Shang, Y.; Li, Y.; Zhang, M. A cumulative-risk assessment method based on an artificial neural network model for the water environment. *Environ. Sci. Pollut. Res.* **2021**, *28*, 46176–46185. [CrossRef]

buildings

MDPI

Article

Risk Coupling Evaluation of Social Stability of Major Engineering Based on N-K Model

Hongyan Yan [1], Zhouwei Zheng [1], Hanjie Huang [2], Xinyi Zhou [3], Yizhi Tang [1] and Ping Hu [1,4,*]

1 Department of Construction Management, Hunan University of Finance and Economics, Changsha 410205, China; yanhongyan@hufe.edu.cn (H.Y.); 202005130214@mails.hufe.edu.cn (Z.Z.); 201805130128@mails.hufe.edu.cn (Y.T.)
2 School of Management, Hangzhou Dianzi University, Hangzhou 310018, China; huanghanjie@hdu.edu.cn
3 School of Economics and Management, Changsha University of Science and Technology, Changsha 410000, China; zhouxy@stu.csust.edu.cn
4 School of Civil Engineering, Central South University, Changsha 410075, China
* Correspondence: huping@hufe.edu.cn

Abstract: In view of the sociality, complexity, and uncertainty of major engineering projects, social stability poses many problems for social contradictions and conflicts in the whole life cycle of the project. This study aimed to investigate the approach of the coupling evaluation method to analyze the coupling influence of social stability risk factors of major projects. First, the potential risk factors of internal and external social stability risk of major projects were abstracted based on literature research and case analysis, and a bow-tie model and a coupling evaluation index system were constructed. Then, a N-K model of social stability risk coupling evaluation of major projects was constructed based on complex network, and the probability and risk value of the coupling of different risk factors were calculated. The studies showed that the coupling ways of social stability risk factors of major projects influence the social stability risk. Multi-factor risk coupling will increase the probability of social stability risk of major projects. The study of this paper provides a theoretical basis for the social stability risk management decision-making of major projects and promotes the sustainable development of major projects.

Keywords: major projects; social stability risk; risk factors; risk coupling evaluation

Citation: Yan, H.; Zheng, Z.; Huang, H.; Zhou, X.; Tang, Y.; Hu, P. Risk Coupling Evaluation of Social Stability of Major Engineering Based on N-K Model. *Buildings* **2022**, *12*, 702. https://doi.org/10.3390/buildings12060702

Academic Editor: Carlos Oliveira Cruz

Received: 1 March 2022
Accepted: 13 May 2022
Published: 24 May 2022

1. Introduction

Major projects are the symbol of human civilization, represent the progress of science and technology of the times, and reflect the degree of economic and social development. In recent years, a number of major infrastructure facilities have been established, designed, constructed and operated in China, such as the Three Gorges Water Conservancy Hub Project, South-to-North Water Transfer Project, West-to-East Gas Transmission Project, Hong Kong-Zhuhai-Macao Bridge, high-speed railroad network, etc. However, due to some special attributes of major engineering projects, such as the social nature, complexity, large scale, and uncertainty, the social risks caused during the construction process will seriously threaten the regional social stability. General Secretary Xi Jinping stressed at the opening ceremony of the seminar on major risks in 2019: improve the ability to prevent major risks and make efforts to resolve them so as to maintain sustainable and healthy economic development and social stability. Therefore, standing at the historical intersection of the "two hundred years" goal, in the macro context of China's social transformation and the vigorous construction of major projects, the construction of major projects should not only ensure a high-quality economic development, but also minimize the risks in order to maintain social stability.

At present, major engineering risks have been widely concerned and studied by the academic community, and a number of research results have been obtained. They mainly focus on the social stability risk factors of major projects (Xiang P C, et.al., 2018 [1]; Munier N, et.al., 2016 [2]; Zhang R, et.al., 2016 [3]; Arukala S R, et.al., 2015 [4]), social stability risk evaluation (Lou X H, et.al., 2018 [5]; Li M, et.al., 2019 [6]; Khameneh A H, et.al., 2016 [7]; Zhou H, et.al., 2015 [8]; Yang SL, et.al., 2014 [9]), social stability risk governance (Zhang W, et.al., 2019 [10]; Cui C, et.al., 2012 [11]; Tan S, et.al., 2015 [12]; Huang Y J, et.al., 2013 [13]) and other aspects. Xiang Pengcheng et.al. [1] proposed that the key risk factors of social instability mainly include poor public opinion expression channels, weak government supervision and illegal project approval procedures. Luo Xiaohui et.al. [5] conducted a social stability risk assessment of major engineering projects under the two situational modes of black box operation and information disclosure for the four stages dynamic game model, and analyzed the impact of the feedback correction mechanism of social stability risk based on the hierarchical Bayesian network model, and they proposed that there were differences in the social stability risk assessment results of major engineering projects under different situations. Zhang Wei et.al. [10] identified 6 categories and 28 factors of social stability risk of major engineering projects, calculated the comprehensive driving force and comprehensive dependence of each risk based on Fuzzy ISM model, and put forward governance priority and governance measures for various risks. The above studies have promoted the development of social stability risk assessment theory for major projects, and provide a rich theoretical framework and knowledge reserve for this paper; however, the above studies all discussed the impact and evaluation of a single risk event on the social stability risk of major projects, ignoring the joint effect of multiple risk factors. The social stability risk of major projects has many influencing factors, the risk factors are strongly correlated, and the occurrence of accidents is often caused by the coupling of multiple factors. These characteristics are in line with the risk coupling analysis theory.

The N-K model [14,15] originated from information theory and is mainly used to analyze the influence of the interaction between the internal elements of the system on the overall adaptability of the system. It is widely used in economic and financial fields, and there is still a lack of research in the field of social stability risks in major projects. Therefore, this paper constructs the social stability risk coupling evaluation model based on the N-K model, and provides a reference for social stability risk management of major projects.

2. Identification of Risk Factors for Social Stability of Major Projects

The identification of risk factors for the social stability of major projects is a prerequisite for risk assessment and governance, and a necessary step before risk analysis and measures are taken. Social stability risk assessment of major projects will face many semi-structural decision-making problems, such as the lack of decision-making information, a large amount of false information, or excess information. However, comprehensively considering the three major factors of economy, society, and natural environment, the principle of triple bottom line (TBL) provides a new idea for the identification of risk factors and a new value standard for the sustainable development of an organization or society. For this reason, through the cases of major projects and the sudden process of group incidents, this paper analyzes the uncertain factors induced by the four-stage game process between both sides of the internal game (the government and the surrounding public), as well as external environment, including economic, natural, social and other exogenous uncertainties. Based on the summary of social stability risks and potential results of major projects, the Bow-tie model is constructed according to the logical sequence of "risk factor analysis-consequence evaluation-model creation".

2.1. *Uncertainty Factors within the Main Players of the Game*

In recent years, the social stability risks of major projects have been on the rise [16], especially in the relevant aspects of emergency decision-making that caused mass incidents. Therefore, on the basis of the principle of protecting the vital interests of the people to the greatest extent possible, it is necessary to resolve the value conflicts of different interest groups, such as the government and the public [17]. Major project construction is a systematic work integrating multi-field management and cooperation. Different stakeholders often face great conflicts of interest due to information asymmetry, benefit unbalance, and relationship cognitive dissonance, which seriously affects the sustainable development of major projects and triggers social stability risks. The main purpose of the game between the government and the public is to maximize their own interests. In this paper, the dynamic game process led by the local government and the public is divided into four stages (as shown in Figure 1 in order to explore the uncertain factors within the main game players of the major project [18].

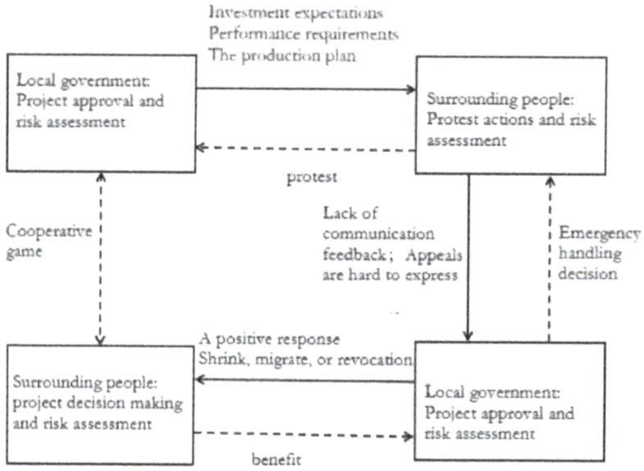

Figure 1. Dynamic game within the main players of the game. (The solid line represents the contradiction generation mechanism, and the dotted line represents the feedback correction system).

In the first stage, as major projects can promote local economic development and meet market demand, the government takes the first action and takes major projects as a decision-making node to realize investment expectations, performance pursuit, production planning, and so on. However, the public in the surrounding areas is worried that chemical products related to major projects will threaten their own safety and health, the quality of the surrounding environment, and the asset chain, and this is taken as the starting point of the game. In the second stage, the surrounding public takes protest actions as the decision-making node according to the first decision-making action of the local government and as a post-actor. When there is no reasonable channel to express their own demands or when they are unable to obtain reasonable and legitimate interest demands in all aspects of major projects, the surrounding public has to stimulate the high spirits of the masses and set off social protests. If the local government actively pays attention to the interests of the public and can communicate with the public about project risks as soon as possible, the public can understand the risk level and prevention and control measures of the project, which will help the public to accept the risk assessment conclusions of the project and form an objective perception of risks, and thus the public may abandon the protests and embrace the project [19]. In the third stage, the government must make scientific and reasonable emergency response decisions, and some governments build an effective interaction mechanism with the public to dispel public doubts; while some local governments take decisions such as reducing, relocating, or even canceling major projects to calm the social situation. In the fourth stage, the surrounding public decides whether to end the protest action so as to achieve the cooperative game according to the emergency response decision of the local government.

2.2. External Environmental Factors

Through the analysis of major engineering accident cases, it is found that there are endless cases of mass incidents caused by external environmental factors such as compensation policies, safety accidents, and environmental pollution. Therefore, external environmental factors are also an important part of the risks that threaten the social stability of major projects. In 1997, John Elkington put forward the representative "triple-bottom-line" theory of social responsibility. He believed that the foundation for an organization to achieve sustainable development is to seek the balance of economic, social, and environmental responsibility on the basis of bottom-line responsibility. Therefore, this paper analyzes the external uncertain factors of the social stability risk of major projects based on the triple bottom line principle of "economy-society-natural environment" (as shown in Figure 2).

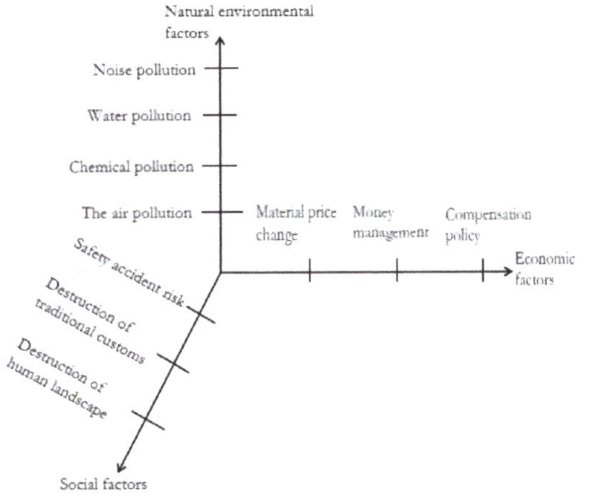

Figure 2. External uncertain factors.

Among the external uncertainties, the first is economic factors [1,16]. The construction period of major projects is long, there are many relevant interest groups, and the construction technology is complex, therefore, there are many unmeasurable risks, and it is easy to be affected by economic factors, including a change in material price, labor supply, fund management, and compensation policy.

The second is the social factors [1,16]. Based on previous literature review, this paper believes that the social factors affecting major projects mainly include the following two aspects: on the one hand, it is the risk of safety accidents brought about by major projects, such as the "five major injuries" of common safety risks; on the other hand, it is the risk caused by the destruction of the surrounding cultural landscape and customs.

Finally, there are natural environmental factors [1,16], through on-the-spot investigation and case analysis, the waste of resources caused by land development, water pollution, chemical pollution, air pollution, waste pollution, and so on, are the triggers of surrounding public protests, parades, and even mass disturbances.

2.3. Building a Bow-Tie Model Based on Internal and External Factors Analysis

The Bow-tie model is a risk management tool that organically combines fault tree analysis and event tree analysis [20]. It is a risk analysis method with strong operability and high visualization. By drawing the bow tie diagram, the potential risk factors of the accident are put on the left as the fault tree part, and the results caused by the accident are put on the right as the event tree part, and lists threats and barriers to reflect the logical development of the event, then build a graphical model. Based on the above analysis, the internal risk factors of social stability of major projects are mainly reflected in the policy risk and public risk induced by the game process between the government and the surrounding public, and the external risk factors are mainly reflected in the economic risk, natural risk and social risk generated by the external environment. Therefore, based on the traditional bow-tie model and the logical idea of "risk source-consequence-evaluation barrier setting", a series of indicators related to the social stability risk of major projects were determined through the forward and reverse push of risk source and consequence, Furthermore, the risk countermeasures of practical value were put forward, and the bow-tie model was constructed (as shown in Figure 3).

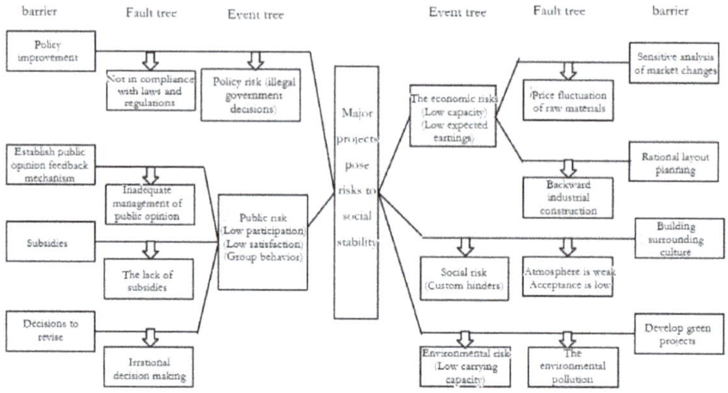

Figure 3. Bow-tie model of social stability risk factors for major projects.

3. Evaluation of Social Stability Risk Coupling for Major Projects

3.1. Construction of Risk Evaluation Index System for Social Stability for Major Projects

Based on the identification of social stability risk factors of major projects and the Bow-tie model, this paper constructs the dimensions of the social stability risk evaluation index system of major projects, including government risk, public risk, economic risk, social risk, and natural environmental risk; comprehensively analyzes the internal uncertain factors of the local government and the surrounding residents, as well as the exogenous uncertain factors such as "economic-social-natural environment" produced by the external environment, dynamically and dialectically considers the risk factors, and summarizes them; and selects 12 program layer indicators according to the principles of comprehensiveness, science, maneuverability, and humanization, as shown in Table 1.

Table 1. Indicators at the program level.

Criterion Layer	Scheme Layer	Indicator Description
Government risk	Decision legitimacy	The government abides by laws and regulations, technical standards, and contract norms in decision-making
	Affiliation institutional perfection	Establish and implement relevant systems at all stages to ensure the smooth implementation of the project and the interests of the relevant masses
Public risk	Social participation	The participation of the public in putting forward reasonable suggestions in major engineering fields
	Social satisfaction	Satisfaction of the public to enjoy environmental subsidies and other preferential policies
	Risk of group behavior	Group events such as group strikes, demonstrations, disturbances, petitions by the masses, etc.
Economic risk	Price changes	The price fluctuation caused by raw materials or supply and demand has an impact on the original profit structure
	Mode of operational management	Failure to coordinate the various management elements, resulting in more labor and material consumption
Social risk	Accident safety risks	Building collapse, occurrence of fire, explosion accidents, etc.
	Traditional customs influence	Migrant masses need to accept the customs of the placement area, culture relearning, etc.

Table 1. *Cont.*

Criterion Layer	Scheme Layer	Indicator Description
Natural environmental risk	Air pollution	When the long-term emission of toxic and harmful gases reaches a certain degree of pollution, it will cause irreversible damage to the atmosphere
	Water pollution	Pollution of water quality caused by the discharge of toxic substances and waste water during construction and production
	Resource occupation	Occupation of surrounding resources by major projects

3.2. Coupling Evaluation of Social Stability Risks of Major Projects Based on N-K Model

The N-K model [14,15] is a general model used to study complex dynamic systems, including two parameters: N is the number of constituent elements of the system, and K is the number of interdependencies in the network. If there are N elements in the system, and each element has n different states, then there are N kinds of possible combinations. The elements of the system are combined in a certain way, that is, a network is formed. The minimum value of K is 0 and the maximum value is N-1.

The steps of using the N-K model to measure the social stability risk coupling of major projects include: major project coupling risk classification, data statistics, and coupling probability calculation.

According to the number of risk factor coupling, the social stability risk coupling of major projects is divided into the following three categories:

(1) Single factor coupling risk: A single risk factor affecting the social stability of major projects will contain multiple risk factors, and each risk factor will interact with each other. Single factor coupling risk includes government (abbreviated G, Code a) factor risk, public (abbreviated P, Code b) factor risk, economic (abbreviated E, Code c) factor risk, social (abbreviated S, Code d) factor risk and natural environmental (abbreviated NE, Code e) factor risk, are recorded as $T_{10}\ (a)$, $T_{11}\ (b)$, $T_{12}\ (c)$, $T_{13}\ (d)$, $T_{14}\ (e)$, respectively, and the total value of coupling risk is recorded as T_1. The single factor coupling risk is shown in Table 2.

Table 2. Single factor coupling risk.

Type	Government Factor Risk	Public Risk	Economic Risk	Social Risk	Natural Environmental Risk
Expression	$T_{10}\ (a)$	$T_{11}\ (b)$	$T_{12}\ (c)$	$T_{13}\ (d)$	$T_{14}\ (e)$

(2) Two-factor coupling risk: Includes 10 types of two-factor coupling risk, and the total value of coupling risk is recorded as T_2. The two-factor coupling risk is shown in Table 3.

Table 3. Two-factor coupling risk.

Type	G-P Coupling Risk	G-E Coupling Risk	G-S Coupling Risk	G-NE Coupling Risk	P-E Coupling Risk
Expression	$T_{20}\ (a,b)$	$T_{21}\ (a,c)$	$T_{22}\ (a,d)$	$T_{23}\ (a,e)$	$T_{24}\ (b,c)$
Type	P-S coupling risk	P-NE coupling risk	E-S coupling risk	E-NE coupling risk	S-NE coupling risk
Expression	$T_{25}\ (b,d)$	$T_{26}\ (b,e)$	$T_{27}\ (c,d)$	$T_{28}\ (c,e)$	$T_{29}\ (d,e)$

(3) Multi-factor coupling risk: Refers to the interaction of three or more risk factors affecting the social stability of major projects, and the total value of coupling risk is recorded as T_3. The multi-factor coupling risk is shown in Table 4.

Table 4. Multi-factor coupling risk.

Type	G-P-E Coupling Risk	G-P-S Coupling Risk	G-P-NE Coupling Risk	G-E-S Coupling Risk	G-E-NE Coupling Risk	G-S-NE Coupling Risk
Expression	T_{30} (a,b,c)	T_{31} (a,b,d)	T_{32} (a,b,e)	T_{33} (a,c,d)	T_{34} (a,c,e)	T_{35} (a,d,e)
Type	P-E-S coupling risk	P-E-NE coupling risk	P-S-NE coupling risk	E-S-NE coupling risk	G-E-S-NE coupling risk	G-P-S-NE coupling risk
Expression	T_{36} (b,c,d)	T_{37} (b,c,e)	T_{38} (b,d,e)	T_{39} (c,d,e)	T_{310} (a,c,d,e)	T_{311} (a,b,d,e)
Type	G-P-E-NE coupling risk	G-P-E-S coupling risk	P-E-S-NE coupling risk	G-P-E-S-NE coupling risk	-	-
Expression	T_{312} (a,b,c,e)	T_{313} (a,b,c,d)	T_{314} (b,c,d,e)	T_4 (a,b,c,d,e)	-	-

In this paper, by calculating the interactive information among five types of social stability risk factors of major projects, the coupling effect is evaluated to form a new risk state. The probability that this type of coupling occurs is measured in terms of the number of times that it occurs more rapidly. The coupling risk magnitude and the accident probability are measured in terms of the coupling value magnitude, i.e., if the resulting value healed with some form of coupling, then the coupling risk healed with the resulting probability healed.

Firstly, the calculation formula of single factor coupling is shown in Formula (1).

$$T(a,b,c,d,e) = \sum_{h=1}^{H}\sum_{i=1}^{I}\sum_{j=1}^{J}\sum_{k=1}^{K}\sum_{l=1}^{L}[P_{hijkl} \times \log_2(\frac{P_{hijkl}}{P_{h....} \times P_{.i...} \times P_{..j..} \times P_{...k.} \times P_{....l}})] \quad (1)$$

where a, b, c, d and e represent five coupling element numbers (where a represents government risk, b represents public risk, c represents economic risk, d represents social risk, and e represents natural environmental insurance); T represents the coupling value, and the larger the coupling value is, the more likely the risk accident caused by this method is; h, i, j, k, l represent the state of the five factors respectively; P_{hijkl} represents the probability of the coupling of the five factors; $P_{h....}$ represents when the government risk factor is in the h state, the single factor coupling probability; $P_{.i...}$ represents the probability of single factor coupling when the public risk factor is in the i state; $P_{..j..}$ represents the probability of single factor coupling when the economic risk factor is in the j state; $P_{...k.}$ represents the probability of single factor coupling when the social risk factor is in k state; $P_{....l}$ represents the probability of single factor coupling when the natural environmental risk factor is in the l state.

Two-factor coupling refers to a form of pairwise coupling among the risk coupling factors of social stability in major projects. The two risk couplings will produce 10 cases; taking T_{20} (a, b) as an example, its calculation formula is shown in Formula (2).

$$T_{21}(a,b) = \sum_{h=1}^{H}\sum_{i=1}^{I}[P_{hi...} \times \log_2(\frac{P_{hi...}}{P_{h....} \times P_{.i...}})] \quad (2)$$

Multi-factor coupling refers to the interaction of more than two factors in the coupling factors that affect the social stability risk of major projects, with a total of 13 cases; taking T_{30} (a, b, c) as an example, its calculation formula is shown in Formula (3).

$$T_{31}(a,b,c) = \sum_{h=1}^{H}\sum_{i=1}^{I}\sum_{j=1}^{J}[P_{hij..} \times \log_2(\frac{P_{hij..}}{P_{h....} \times P_{.i...} \times P_{..j..}})] \quad (3)$$

Finally, according to the order of each coupling value, the conclusion of coupling evaluation is drawn.

4. Example

4.1. Example Statistics of Social Stability Risk Events in Major Projects

This paper collected from website news reports, papers, paper press publications at home and abroad to analyze the cases of stable risk events of major engineering societies at home and abroad, and counted 108 risk events occurring at home and abroad between 2000 and 2020, including 72 risk events at home and 36 risk events abroad; the major engineering social stability risk events are shown in Table 5.

Table 5. Information on social stability risk events of major projects.

Project Name	Government Risk Factors	Public Risk Factors	Economic Risk Factors	Social Risk Factors	Natural Environmental Risk Factors
Three Gorges Project (China)	No	Social satisfaction problem	No	Traditional customs problem	Resource occupation problem
Hong Kong-Zhuhai-Macao Bridge (China)	No	Social participation problem	The problem of the mode of management	No	No
Bird's Nest (China)	No	Social satisfaction problem	The problem of the mode of management	No	No
New Federal Building of San Francisco (United States)	Legitimacy of decision	Other	The problem of the mode of management	No	No
Kemper thermal power plants (United States)	No	No	The problem of the mode of management	No	Air Pollution
Sampoong Department Store(South Korea)	Legitimacy of decision	Social satisfaction problem	Other	Risk of safety accident	No

In the model, 1 and 0 are used to indicate whether each factor is in an unsafe state, 1 indicates occurrence, and 0 indicates that the risk occurrence probability is not present and the coupling value is used to quantify the risk occurrence probability (as shown in Table 6). The coupling factors with a frequency (frequency) of 0 on the way are not marked.

Table 6. Statistics of social stability risk events of major projects under different coupling modes.

Single Factor Coupling			Two-Factor Coupling			Multi-Factor Coupling		
Coupling Factor	Frequency/ Time	Frequency	Coupling Factor	Frequency/ Time	Frequency	Coupling Factor	Frequency/ Time	Frequency
10,000	3	0.028	11,000	7	0.065	11,010	4	0.037
01,000	9	0.083	10,100	2	0.019	10,110	1	0.009
00,100	5	0.046	10,010	4	0.037	10,011	2	0.019
00,010	22	0.204	01,100	1	0.009	01,011	6	0.056
00,001	6	0.055	01,010	30	0.278	-	-	-
-	-	-	01,001	5	0.046	-	-	-
-	-	-	00,011	1	0.009	-	-	-

In Table 3, 00000 of the single factor coupling indicated that none of the five coupling factors had an impact on the social stability risk of major projects, 10,000 said that only government factors had an impact on the social stability risk of major projects, and there were three such accidents with a frequency of 0.028; 11,000 in the two factor coupling indicates that the risk is the result of the coupling of government factors as well as social public factors, and there are seven such accidents with a frequency of 0.065; 11,010 of the multi-factor coupling indicates that the risk is the result of the coupling of government factors, social public factors, and social factors, and there are four such accidents with a frequency of 0.037.

4.2. Risk Coupling Value Calculation

(1) Single factor coupling probability

By analyzing the statistical data of social stability risk accidents of major projects, the probability that government factors do not affect the social stability risk of major projects is as follows: $P_{0\ldots.} = P_{01,000} + P_{00,100} + P_{00,010} + P_{0,0001} + P_{01,100} + P_{01,010} + P_{01,001} + P_{00,110} + P_{00,101} + P_{00,011} + P_{01,110} + P_{01,011} + P_{00,111} + P_{0,1111} = 0.786$. Similarly, $P_{1\ldots.}$, $P_{.0\ldots}$, $P_{.1\ldots}$, $P_{..0..}$, $P_{..1..}$, $P_{\ldots 0.}$, $P_{\ldots 1.}$, $P_{\ldots .0}$, $P_{\ldots .1}$ can be calculated, and the calculated results were tabulated in Table 7.

Table 7. Single factor coupling probability.

Coupling Mode	Probability	Coupling Mode	Probability
$P_{0\ldots.}$	0.786	$P_{..1..}$	0.148
$P_{1\ldots.}$	0.214	$P_{\ldots 0.}$	0.351
$P_{.0\ldots}$	0.426	$P_{\ldots 1.}$	0.649
$P_{.1\ldots}$	0.574	$P_{\ldots .0}$	0.815
$P_{..0..}$	0.852	$P_{\ldots .1}$	0.185

(2) Tow-factor coupling probability

By analyzing the statistical data of social stability risk accidents of major projects, the probability of accidents when government and public factors do not participate in risk coupling is $P_{00\ldots} = P_{00,000} + P_{00,100} + P_{00,010} + P_{00,001} + P_{00,110} + P_{00,101} + P_{0,0011} + P_{00,111} = 0.314$. Similarly, $P_{01\ldots}$, $P_{10\ldots}$, $P_{11\ldots}$ can be calculated, and the calculated results were shown in Table 8.

Table 8. Two-factor coupling probability.

Coupling Mode	Probability	Coupling Mode	Probability	Coupling Mode	Probability	Coupling Mode	Probability
$P_{00\ldots}$	0.314	$P_{1..0.}$	0.112	$P_{.0.0.}$	0.148	$P_{..10.}$	0.074
$P_{01\ldots}$	0.472	$P_{1..1.}$	0.102	$P_{.0.1.}$	0.278	$P_{..11.}$	0.009
$P_{10\ldots}$	0.103	$P_{0\ldots 0}$	0.287	$P_{.1.0.}$	0.203	$P_{..0.0}$	0.741
$P_{11\ldots}$	0.102	$P_{0\ldots 1}$	0.166	$P_{.1.1.}$	0.371	$P_{..0.1}$	0.185
$P_{0.0..}$	0.731	$P_{1\ldots 0}$	0.195	$P_{.0..0}$	0.343	$P_{..1.0}$	0.083
$P_{0.1..}$	0.055	$P_{1\ldots 1}$	0.019	$P_{.0..1}$	0.083	$P_{..1.1}$	0
$P_{1.0..}$	0.186	$P_{.00..}$	0.352	$P_{.1..0}$	0.472	$P_{\ldots 00}$	0.25
$P_{1.1..}$	0.028	$P_{.01..}$	0.074	$P_{.1..2}$	0.102	$P_{\ldots 01}$	0.101
$P_{0..0.}$	0.184	$P_{.10..}$	0.565	$P_{..00.}$	0.277	$P_{\ldots 10}$	0.565
$P_{0..1.}$	0.547	$P_{.11..}$	0.009	$P_{..01.}$	0.64	$P_{\ldots 11}$	0.084

(3) Multi-factor coupling probability

By analyzing the statistical data of social stability risk accidents of major projects, the probability of accidents when the government, the public, and economic factors do not participate in the risk coupling is $P_{000..} = P_{00,000} + P_{00,001} + P_{0,0010} + P_{00,011} = 0.268$. Similarly, $P_{100..}$, $P_{010..}$, $P_{001..}$ can be calculated, in which Tables 9 and 10 present the results.

Table 9. Three-factor coupling probability.

Coupling Mode	Probability	Coupling Mode	Probability	Coupling Mode	Probability	Coupling Mode	Probability
$P_{000..}$	0.268	$P_{1.01.}$	0.056	$P_{.100.}$	0.148	$P_{.1.01}$	0.046
$P_{100..}$	0.084	$P_{1.11.}$	0.009	$P_{.001.}$	0.269	$P_{.0.11}$	0.019
$P_{001..}$	0.046	$P_{0..00}$	0.138	$P_{.110.}$	0.009	$P_{.1.11}$	0.056
$P_{110..}$	0.102	$P_{1..00}$	0.112	$P_{.101.}$	0.371	$P_{..000}$	0.176
$P_{101..}$	0.028	$P_{0..01}$	0.101	$P_{.011.}$	0.009	$P_{..100}$	0.074
$P_{011..}$	0.009	$P_{1..10}$	0.074	$P_{.0.00}$	0.093	$P_{..001}$	0.101
$P_{0.00.}$	0.184	$P_{0..11}$	0.009	$P_{.1.00}$	0.157	$P_{..110}$	0.009
$P_{1.00.}$	0.093	$P_{1..11}$	0.019	$P_{.0.01}$	0.055	$P_{..011}$	0.084
$P_{0.01.}$	0.547	$P_{.000.}$	0.083	$P_{.1.10}$	0.315	$P_{1.10.}$	0.019

Table 10. Four-factor coupling probability.

Coupling Mode	Probability	Coupling Mode	Probability	Coupling Mode	Probability	Coupling Mode	Probability
$P_{0000.}$	0.055	$P_{0.000}$	0.083	$P_{10.00}$	0.047	$P_{01.11}$	0.056
$P_{1000.}$	0.028	$P_{1.000}$	0.093	$P_{01.00}$	0.092	$P_{000.0}$	0.204
$P_{0100.}$	0.129	$P_{0.100}$	0.055	$P_{00.10}$	0.204	$P_{100.0}$	0.065
$P_{0010.}$	0.046	$P_{0.010}$	0.482	$P_{00.01}$	0.055	$P_{010.0}$	0.361
$P_{0001.}$	0.213	$P_{0.001}$	0.101	$P_{11.00}$	0.065	$P_{001.0}$	0.046
$P_{1100.}$	0.065	$P_{1.100}$	0.019	$P_{10.10}$	0.046	$P_{000.1}$	0.064
$P_{1010.}$	0.019	$P_{1.010}$	0.074	$P_{01.10}$	0.278	$P_{110.0}$	0.102
$P_{1001.}$	0.056	$P_{0.011}$	0.065	$P_{01.01}$	0.046	$P_{101.0}$	0.028
$P_{0110.}$	0.009	$P_{1.110}$	0.009	$P_{00.11}$	0.009	$P_{100.1}$	0.019
$P_{0101.}$	0.278	$P_{1.011}$	0.019	$P_{11.10}$	0.037	$P_{011.0}$	0.009

(4) T value calculation

The T value can be obtained according to the Formulas (1)–(3) as shown in Table 11.

Table 11. T values under different coupling regimes.

Coupling Mode	T Values	Coupling Mode	T Values	Coupling Mode	T Values
T_{20}	0.084	T_{29}	0.046	T_{39}	0.140
T_{21}	0.033	T_{31}	0.248	T_{310}	0.252
T_{22}	0.101	T_{32}	0.141	T_{311}	0.320
T_{23}	−0.001	T_{33}	0.188	T_{312}	0.196
T_{24}	0.076	T_{34}	0.099	T_{313}	0.408
T_{25}	0.179	T_{35}	0.173	T_{314}	0.304
T_{26}	0.034	T_{36}	0.202	T_4	0.579
T_{27}	0.099	T_{37}	0.138		
T_{28}	−0.003	T_{38}	0.197		

4.3. Conclusion and Discussion of Example Risk Coupling Evaluation

The following conclusions can be drawn from the present study:

(1) The more kinds of coupling risk factors, the greater the risk of social stability of major projects is. From the calculation results, it can be inferred that the five-factor coupling risk value ($T_4 = 0.579$) is greater than the four-factor coupling risk value (T_{310}–T_{314}), the four-factor coupling risk value (T_{310}–T_{314}) is generally greater than the three-factor coupling risk value (T_{30}–T_{39}), and the three-factor coupling risk value (T_{30}–T_{39}) is generally greater than the two-factor coupling risk value (T_{20}–T_{29}), which is consistent with the actual situation of social stability risks of major projects.

(2) Among the four-factor coupling risks, the coupling value of the government-public-economic-social factor (T_{313} = 0.408) is the largest, and that of the government-public-economic-natural environmental factor is the smallest (T_{312} = 0.196). At the same time, the coupling value of the government-public-social-natural environmental factor (T_{311} = 0.320) is larger than that of the government-economic-social-natural environmental factor (T_{310} = 0.252), and it is between the coupling value of the government-public-economic-social factor and the government-public-economic-natural environmental factor. Among the three-factor coupling risks, the coupling value of government-public-social factors (T_{31} = 0.248) is the largest, while that of government-economic-natural environmental factors (T_{34} = 0.099) is the smallest, which shows that social factors and social public factors play a greater role in major project risks, and the range of social factors is relatively wide, which can affect other factors to a certain extent. Among the social factors, safety accidents not only pose a threat to people's lives and property, but also cause huge economic losses directly or indirectly to society, thus affecting social stability; the public will take the hidden dangers of accidents, policy subsidies, environmental pollution as a fuse to cause social stability risks.

(3) Among the two-factor coupling risks, the coupling value of government-economic (T_{21} = 0.033) < public-natural environmental (T_{26} = 0.034) < social-natural environmental (T_{29} = 0.046) < public-economic (T_{24} = 0.076) < government-public (T_{20} = 0.084) <economic-social (T_{27} = 0.099) < government-social (T_{22} = 0.101) < public-social (T_{25} = 0.179), therefore the value of public-social coupling risk is the largest median risk of two factor coupling, and there is a great coupling between social factors and social public factors. Among the social factors, the destruction of traditional customs caused by major projects has a far-reaching impact on the public, that is, the integration of land expropriation immigrants and residents in resettlement areas. For example, the "Three Gorges Project" involves the migration of nearly two million people. These people may face the risk of losing land, declining living standards, unemployment, marginalization, and broken community relations due to a lack of a sense of security and sense of belonging after resettlement. In addition, immigrants who leave their homes not only need to learn different languages and cultures, but also accept local customs, all of which are social factors that may endanger the stability of the local society.

(4) From the multi-factor and two-factor coupling risk, we can see that the coupling values of politics-society (T_{22} = 0.101) and economy-society (T_{27} = 0.099) are relatively large and similar. Therefore, among the government factors, the legitimacy, rationality, and information transparency of government policy, the change of raw material prices, and the coupling between capital chain management and social risk factors are relatively strong.

4.4. Coupling Risk Countermeasure

(1) In order to solve the risk of social-public coupling, one of the meeting points is the media. On the one hand, the public should gradually cultivate the awareness of finding the media for something. When the risks of major projects infringe upon the vital interests of the public, the public should exercise their power within the scope of the law to seek help from the media or pretend to be the media themselves. Through the official channel the network platform can be used to output information to attract attention, improve the ability of thinking and their own comprehensive quality, and strengthen the ability to screen information. On the other hand, the media should strengthen the networking and digitization of information feedback and define the social responsibility objectives of major projects. Through press conferences, Weibo interviews, large forums, and other information media, the integration of public subsidies, environmental feedback, and other data to analyze the causes of the risk of social stability, to answer questions to the public to form a good interaction.

(2) When making decisions on major projects, the government should, on the premise of abiding by national laws and regulations, technical norms, and industry standards, focus on the disclosure of relevant information in the field of approval and implementation of major projects in an all-round way, and show the information to the society and the public in an open and transparent manner. However, the social public group behavior risk is mostly caused by unreasonable decision-making, therefore it is necessary to construct the concept of overall governance and pursue benign interaction for decision-making revision under the guidance of a people-oriented concept.

(3) An early risk warning mechanism for social stability will be formed. The social stability risk of major projects increases with the increase of risk coupling factors, thus it is necessary to predict and warn of the risk factors before the occurrence of social stability risk events. On the one hand, more serious multi-factor coupling risk events can be avoided through the early warning mechanism; on the other hand, the abnormal indicators in the early warning mechanism can be traced back to the hidden risk factors to obtain more efficient and accurate social stability risk management programs and countermeasures.

5. Conclusions

The complexity of major projects makes them prone to have various conflicts of interest in the construction process, leading to the occurrence of group events and the risk of social stability. Based on the analysis of the risk events related to the social stability risk of major projects, the authors constructed a social stability risk coupling evaluation model to study the impact of the coupling of different factors on the social stability risk of major projects.

The main conclusions of this paper are as follows: (1) Five factors, such as government, social public, economy, society, and natural environment, are the main risk factors that affect the social stability of major projects. (2) There are many ways of coupling among the five factors, and the risk of social stability caused by different coupling modes is different, among the five factors coupling, the social stability risk caused by the coupling of the five factors of "government-public-society-environmental-economy" is the greatest risk. In addition, the "politics-public-economy-society" coupling mode among the four factors coupling, the "politics-public-society" coupling mode among the three factors, and the "public-society" coupling mode among the two factors are most at risk. (3) Overall, there is a positive correlation between coupling factors and risk probability, that is: five-factor coupling risk > four-factor coupling risk > three-factor coupling risk > two-factor coupling risk. From a local perspective, some three-factor coupling risks are higher than four-factor coupling risks, such as "government-public-social" coupling risk > "government-public-economic-environmental" coupling risk. (4) Based on the conclusion of social stability risk coupling evaluation of major projects, some countermeasures were put forward.

The NK analysis method is an extension of traditional analysis methods. Traditional analysis approaches do not consider the interdependence between various risks, such as the analytic hierarchy process [21] and the fuzzy method [22]. It is difficult to quantitatively evaluate the coupling relationship between various risk factors. In this study, the NK model can be used to analyze the risk value of the social stability risk of major projects coupled with different risk factors quantitatively, therefore some coupling laws of social stability risk of major projects were drawn. The research of this paper provides a method for decision-makers to assess the social stability risk of major projects, and provides a theoretical basis for the decision-making of social stability risk management of major projects.

There are some requirements on the integrity of the data and the number of samples for the model that were constructed in this paper. Therefore, in the follow-up research, the integrity of the data can be continuously improved, the sample size can be expanded, and the calculation accuracy can be improved to make the calculation results more in line with the actual situation.

Author Contributions: Conceptualization, H.Y. and P.H.; methodology, Y.T.; validation, Z.Z., Y.T.; formal analysis, H.Y. and X.Z.; investigat.ion, X.Z., Z.Z. and H.H.; data curation, Y.T.; writing—original draft preparation, H.Y.; writing—review and editing, P.H.; supervision, P.H.; project administration, H.Y.; funding acquisition, H.Y. and P.H. All authors have read and agreed to the published version of the manuscript.

Funding: This research was funded by [the Natural Science Foundation of Hunan Province] grant number [2021JJ30748], [Scientific Research Key Project of Hunan Provincial Department of Education] grant number [21A0207], [National University Student Innovation and Entrepreneurship Project] grant number [S202011532017], [National University Student Innovation and Entrepreneurship Project] grant number [S202111532023], [Hunan University student innovation and Entrepreneurship Project] grant number [4059].

Conflicts of Interest: The authors declare no conflict of interest.

References

1. Xiang, P.; Wu, X. Construction of social stability risk network for major engineering projects based on SNA. *Constr. Econ.* **2018**, *6*, 41–47. (In Chinese)
2. Munier, N. *Risk Management for Engineering Projects*; Springer: Cham, Switzerland, 2016.
3. Zhang, R.; Feng, H.; Shi, C. Analysis on impacting factors from joint-operation of Key Works for Expansion of Nenjiang River Water Diversion Project. *Water Resour. Hydropower Eng.* **2016**, *1*, 120–123, 126.
4. Reddy, A.S. Risk management in construction industry—A case study. *Int. J. Innov. Res. Comput. Commun. Eng.* **2015**, *4*, 10058–10067.
5. Luo, X.; Hu, L.; Liu, D. Social stability risk assessment of major engineering project under conditions of black-box operation and information disclosure: Dynamic game analysis based on Hierarchical Bayesian Network. *Technol. Econ.* **2018**, *10*, 117–130. (In Chinese)
6. Ma, L.; Cong, X. Social stability risk assessment of NIMBY major projects by OWA, matter-element, and cloud model. *Intell. Fuzzy Syst.* **2019**, *36*, 2545–2556. [CrossRef]
7. Khameneh, A.H.; Taheri, A.; Ershadi, M. Offering a framework for evaluating the performance of project risk management system. *Procedia Soc. Behav. Sci.* **2016**, *226*, 82–90. [CrossRef]
8. Zhou, H.; Liu, Y. Research on the application of FMEA risk evaluation for metro projects combining FAHP and frequency analysis. *Eng. Manag.* **2015**, *1*, 53–58. (In Chinese)
9. Yang, S.-L. Practical research on risk evaluation methods for power engineering construction projects. *Xinjiang Power Technol.* **2014**, *3*, 111–114.
10. Zhang, W.; Duan, Y.-y. Social stability risk relationship model of major engineering projects based on Fuzzy-ISM. *Civ. Eng. Manag.* **2019**, *5*, 102–108. (In Chinese)
11. Cui, C. A research on how to control the risks of construction engineering project in construction enterprises in China. In Proceedings of the International Conference on Sensor Network & Computer Engineering, Xi'an, China, 8–10 July 2016. (In Chinese).
12. Tan, S. The paradox of social stability risk management of large-scale engineering projects in China: Current situation, causes and countermeasures. *China Soc. Public Saf. Res. Rep.* **2015**, *2*, 31–41.
13. Huang, Y.J. Social risk management: Framework, risk assessment and tool application. *Manag. World* **2013**, *9*, 176–177.
14. Chu, F.; Chu, C. Single item dynamic lot-sizing models with bounded inventory and outsourcing. *IEEE Trans. Syst. Man Cybern. Part A Syst. Hum.* **2008**, *38*, 70–77.
15. Lee, F.; Olav, S. Technology as a complex adaptive system: Evidence from patent data. *Res. Policy* **2001**, *30*, 1019–1039.
16. Duan, Y. Research on Social Stability Risk Evaluation of Major Engineering Projects. Master's Thesis, Xi'an University of Architecture and Technology, Xi'an, China, 2019; p. 6. (In Chinese).
17. Zhong, M. Research on Social Stability Risk Assessment and Index System of Major Engineering Projects. Master's Thesis, China University of Mining and Technology, Beijing, China, 2016. (In Chinese).
18. Deng, L.; Hua, J. Public participation game in decision-making social stability risk assessment of major water conservancy projects. *Water Conserv. Econ.* **2017**, *35*, 12–18. (In Chinese)
19. Hao, Y. Analysis of Social Security Risk Mechanism and Quantitative Model of Major Environmental Sensitive Projects. Ph.D. Thesis, China University of Geosciences, Beijing, China, 2018. (In Chinese).
20. Wang, H.; Zhang, Y.; Li, S. Safety risk analysis of high-speed rail operation based on Bow-tie model. *Liaoning Univ. Eng. Technol. Nat. Sci. Ed.* **2017**, *36*, 767–772. (In Chinese)
21. Guo, M. Research on Social Stability Risk Assessment of Major Engineering Projects. Master's Thesis, Henan University of Technology, Henan, China, 2015. (In Chinese).
22. Miao, J.; Huang, D.; He, Z. Social risk assessment and management for major construction projects in china based on fuzzy integrated analysis. *Complexity* **2019**, *2019*, 2452895.

buildings

MDPI

Review

Behavioral Research in Construction Engineering Management: A Review from a Neuropsychological Perspective

Yuan Li [1,2,*], Jiaqi Liang [1,2], Jingxiong Huang [3], Mengsheng Yang [1,2] and Runyan Li [1,2]

1 School of Architecture and Civil Engineering, Xiamen University, Xiamen 361005, China
2 Fujian Key Laboratory of Sensing and Computing for Smart City, Xiamen 361005, China
3 School of Architecture, Tsinghua University, Beijing 100084, China
* Correspondence: liyuan79@xmu.edu.cn

Abstract: In construction engineering, there are many interactive and decision-making behaviors which could affect the progress and final performance. Based on the people-oriented concept, managing construction engineering should not ignore the understanding of individual behavior, and neuropsychology provides a refined microscopic perspective. This paper employed a bibliometric analysis of 1254 studies from the Web of Science related to behavioral research in construction engineering management using VOSviewer and summarized the neuropsychological mechanisms and research methods of behavior by systematic review. This paper found that: (1) Neuropsychological mechanisms of behavior include basic mechanisms about the brain and function and range from sensory to decision processes. Core factors are the functional ingredients. (2) Behavior research in construction engineering management is turning to neuropsychological experiments. Understanding the complex correlation mechanisms are the research trends in recent years. (3) Construction engineering management studies provide the means and methods to improve the validity and efficiency of management in the construction industry. The results confirm the impact of sensory perception on behavior and managerial performance. (4) The research trend in this field in the future is multidisciplinary. In total, this paper provides a potential effective reference for improving the performance of construction engineering management, developing sustainable construction production and consumption, and building a people-oriented livable city.

Keywords: construction engineering; management; behavior; neuropsychology; bibliometric analysis

Citation: Li, Y.; Liang, J.; Huang, J.; Yang, M.; Li, R. Behavioral Research in Construction Engineering Management: A Review from a Neuropsychological Perspective. *Buildings* **2022**, *12*, 1591. https://doi.org/10.3390/buildings12101591

Academic Editors: Yongjian Ke, Jingxiao Zhang and Simon P. Philbin

Received: 22 August 2022
Accepted: 27 September 2022
Published: 2 October 2022

Publisher's Note: MDPI stays neutral with regard to jurisdictional claims in published maps and institutional affiliations.

1. Introduction

People have been thinking about themselves and understanding their behaviors since ancient times, and this is reflected in ancient Western and Eastern culture. However, limited by science and developing technology, it has mainly manifested in the level of philosophical thinking. Modern science has shown that all human behaviors are inseparable from neural mechanisms. It has become common knowledge that the brain controls human behavior. However, understanding how it works requires thinking about the nature of the mind, the representation of emotions, and the responsibility for behavior [1]. The development of neuroscience and modern technology provided more possibilities for the theoretical exploration and practical application of neuropsychology.

In the relationship between the brain and behavior, the early positions were mainly localizationism and equipotentialism [2]. Localism believes there is a connection between certain areas of the brain and specific behaviors, for example, movement and perception. Meanwhile, other research argues behavior has more of a quantitative role than the type and location of brain tissue. Modern neuroscience rejects the extremes of these two opinions but draws inspiration from both and merges them. Recognizing the crucial role of the brain in behavior, the research of medicine and psychology began to consider using behavioral interventions on brain damage diagnosis, evaluation, and repair [3].

Neuropsychology in behavioral research refers to applying neuroscience theories, techniques, and methods to better understand the process of individual behavioral decision-making and results [4]. Similar behaviors within and among individuals may result from different underlying neuropsychological mechanisms that are imperceptible to many traditional research methods [5]. There is a view that objective neuropsychological data are less susceptible to subjective bias and are therefore more reliable than self-reports [6]. However, interpreting objective data needs to be undertaken with rigorous neuropsychological knowledge, and plenty of reported information can often help. In this sense, the self-report of the subjects can have a better auxiliary effect. Therefore, more researchers preferred the mixed method of combining self-report with experimental measurement.

In construction engineering management (CEM), many interactive and decision-making behaviors exist in actual works; for example, communication among workers, interaction between workers and equipment, spatial relationships between workers and the environment, and behavioral decisions that occur in the construction. All these critical or daily behaviors could affect the progress and the final performance of the construction project. As proof, safety and risk control in the construction progress is of great concern to both academia and industry. In past research and practices of CEM, managers paid more attention to structure, material, and the use of technical tools, and sometimes as well as the construction process control and standardization of workers' operating behavior. Instead, they focused little on the effects of human neurological function and psychological feedback on behavior. Recently, researchers have begun to explore the influence of neuropsychological factors on personnel behavior, especially the factors related to construction engineering, including physical fatigue [7], mental fatigue [8], attention failure [9], and others. However, there is still a lack of neuropsychological considerations in the preliminary design, operation, supervision, and acceptance of construction engineering management.

This review implemented people-oriented development and humanistic care and tries to offer a systemic outline of human behavior research in construction engineering management from the perspective of neuropsychology. Furthermore, we tried to deeply understand the intrinsic neural mechanisms and psychological response of a person participating in construction engineering. The target is to improve work performance and management effectiveness from a micro-scale, and aims at more intelligent, refined, and humanized construction engineering management practices.

The remainder of this paper is organized as follows. Section 2 describes the research design and method employed in this study. In Section 3, we briefly introduce the neuropsychological mechanism of behavior, and the relationship between neuropsychology and CEM. Section 4 describes the neuropsychological methods and their application in CEM. Section 5 discusses the behavioral turn of the experimental method. Moreover, we generalize current findings of behavior research in the CEM field and try to reveal the complex correlation mechanisms of neuropsychology, individual behavior, and CEM. As a result, Section 6 summarizes the review and presents potential future trends, as well as discusses the contributions and limitations.

2. Research Design and Methods

This research is composed of quantitative bibliometric analysis and qualitative systematic literature reviews (Figure 1). On the one hand, we undertook systematic literature reviews to reveal the theory and method of neuropsychology and its implication in the CEM field. The procedures were carried out by the authors manually and the analysis was mostly qualitative description. The results of systematic literature reviews are shown in Sections 3–5, where we generalized the neuropsychological mechanisms and methods, and compared the advantages and disadvantages in detail. Then we reviewed their application in CEM, and discussed the changing methods, hot issues, and trends. Furthermore, a correlation mechanism was generated for future research.

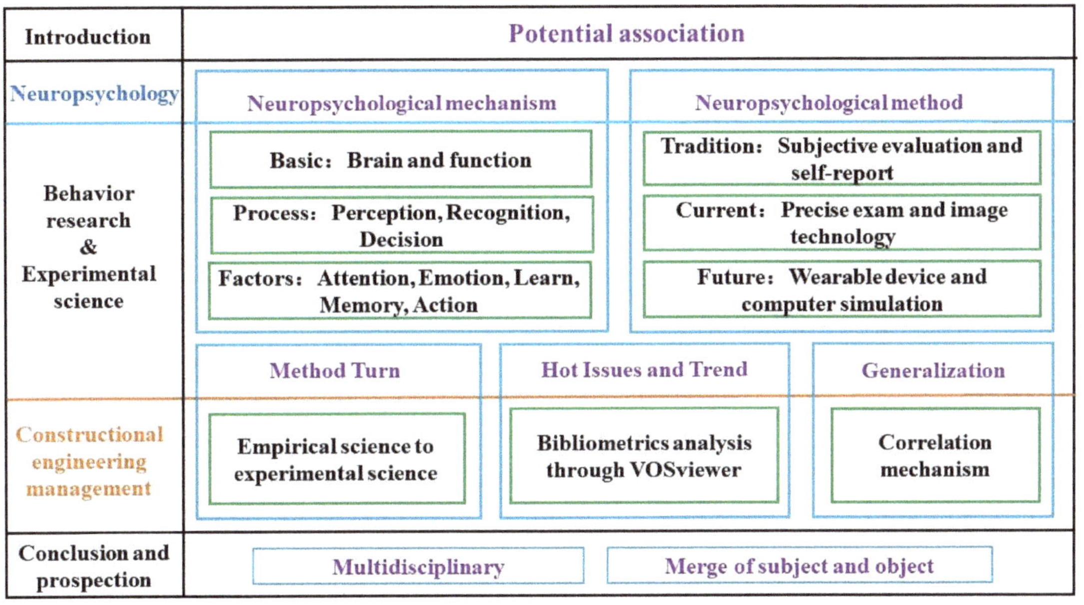

Figure 1. Research framework.

On the other hand, we employed a bibliometric analysis of 1254 studies from Web of Science using VOSviewer. The bibliometric methodology encapsulates the application of quantitative techniques on bibliometric data [10,11]. Bibliometric analysis is useful for deciphering and mapping the cumulative scientific knowledge and evolutionary nuances of well-established fields by making sense of large volumes of unstructured data in rigorous ways [12]. In this research, bibliometric study was used to survey retrospective performance and science of the behavior research in the CEM field from a neuropsychological perspective. The scope for this study was large enough with thousands of papers to warrant bibliometric analysis [12,13]. We chose a bibliographic coupling technique for performance analysis of countries (regions) and journals in the present situation and selected the co-occurrence analysis for notable words for future research directions. To collect the data, the authors consulted the literature and brainstormed to identify relevant combinations of search terms. Finally, terms were combined with construct, build, architect, engineer, manage, behavior, neuro, psychology, brain, experiment, mental, perceptive, perception, cognitive, cognition, neuroscience, and neuroergonomic. Then, the bibliometric analysis was run, and the findings are mainly reported in Section 5.

3. Neuropsychological Mechanism of Behavior

3.1. Basic Mechanism: Brain and Function

The cerebral cortex of the human brain is roughly two cerebral hemispheres in a longitudinal division. When assuming normal right-handedness, the left hemisphere controls the right side of the body, while the right hemisphere controls the left side. In cognitive processing of information, the left hemisphere shows a more sequential fashion, and the right hemisphere appears in a simultaneous manner. Therefore, the left hemisphere is better at dealing with language material, and the right side is good at planning visual space, learning, and adapting to new situations.

The sulcus also divides the brain into parts. The area anterior to the central sulcus is the frontal lobe, and posterior to the central sulcus lies the temporal, parietal, and occipital lobes. The behavioral outputs associated with the frontal lobe include motor programming, abstraction, planning, and self-regulation. The temporal lobes mainly link to the auditory channel. The parietal and occipital lobes correspond to somatosensory and visual stimuli [3,14]. As Figure 2 shows, in electroencephalogram (EEG) experiments and analysis, researchers often divide the brain into 21 electrode positions, which correspond to different function divisions by the international 10–20 electrode lead system [15]. In the EEG electrode system, odd numbers represent left and even numbers represent right. The prefix AF denotes electrodes of the frontal lobe and FC denotes electrodes between the frontal and central lobes; F, T, C, O, and P denote frontal, temporal, central, occipital, and parietal lobes; and the others such as Fp, Fz, Cz, and Pz are marker sites. For example, T8 stands for the electrode of the temporal lobe on the right side. All these channels are used for recording and describing the application locations of scalp electrodes in an EEG test. To better interpret the results, these channels can be divided into four groups: left cluster (AF3, F7, F3), right cluster (AF4, F8, F4), center cluster (FC5, FC6, T7, T8), and back cluster (P7, P8, O1, O2).

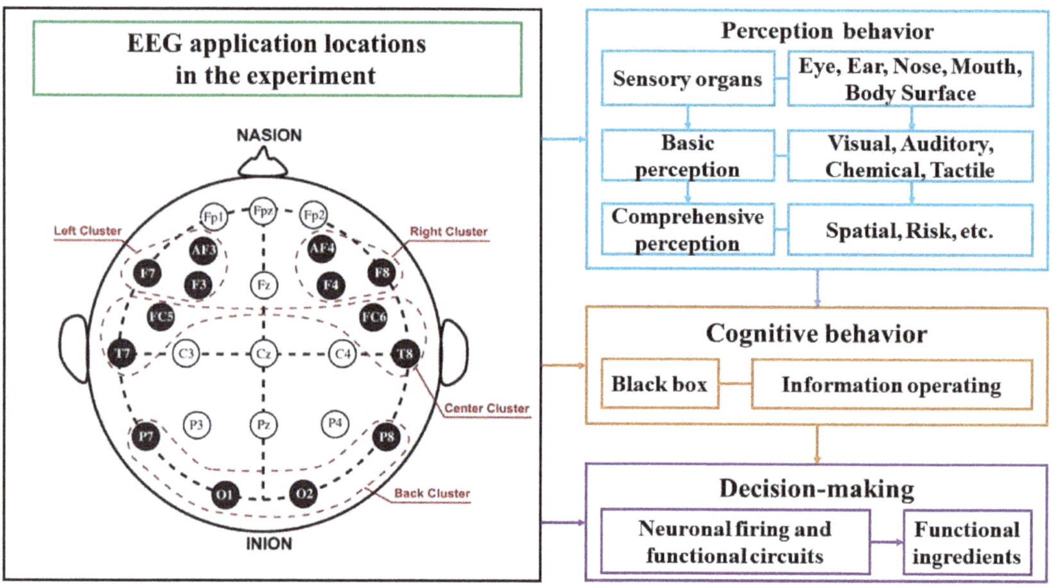

Figure 2. Neuropsychological mechanism of behavior. The electrode lead system on the left is adapted from Wang et al. (2017), and the others are drawn by the authors.

3.2. *Main Process: From Sensory to Decision*

After perceiving external stimuli and cognitively processing the information, the result of individual behavior usually reveals itself as decision-making in various situations and at different times. That is, human behavior usually goes through three main processes: perception, cognition, and decision-making before acting (Figure 2).

Perceptual behavior includes two main stages: sensation and perception. Sensation refers to the stage that organs receive from external stimuli. Perception indicates information understanding and explanation, including the physiological processing of the brain based on scientific information and psychological processing based on experience [16]. We can classify perception as visual perception, auditory perception, chemical perception (mainly olfactory and taste), and tactile perception (including pain). Based on sensation and perception, researchers have proposed more comprehensive perceptual concepts, such as spatial perception [17] and risk perception [18]. More studies pay attention to the connection between self and environment, behavioral individuals and physical objects, independents, and society [19]. In the preliminary design stage of construction, multiple perceptions are involved, such as visual perception when browsing contract content and design drawings, as well as auditory perception of noise, olfactory perception of air quality, and tactile perception of building materials during fieldwork. During the operating stage, equipment operators often must remark on multiple areas at the same time (the building site in front, the control panel in hand, other workers, and equipment in the surrounding areas, etc.). It is necessary to mobilize more complex perception capabilities and maintain them for a long time, such as spatial perception and risk perception [20].

Cognitive behavior also has two views: behaviorist and cognitivism. Behaviorist views take the neuropsychological mechanism as a "black-box", while cognitivism regards neuropsychological mechanisms as the progress of information working [21]. Cognitive behavior research and empirical experience provides many ideas, such as attention and interest stimuli, emotion arousal and valence, mental schema and classification, states of consciousness and memory, mindfulness, flow, and others [22,23]. Cognitive behavior also considers the interrelationships between these ideas; for example, the literature has found a strong connection between attention and memory development [24,25]. In construction engineering, researchers also explored the interrelationship between visual attention and memory experimentally and validated this in building inspections [26].

Decision-making behavior is more common in marketing research. Consumer neuroscience is defined as the application of neuroscience theories and tools to better understand human decision-making and related behaviors [4]. It is considered an interdisciplinary academic branch of marketing and neuroeconomics, including decision neuroscience, and is also supposed to be the intersection of neuroscience and consumer psychology. In fact, marketing processes are also unavoidable for construction management when requiring economic benefits. From this perspective, using relevant theories and methods in neuroscience to understand decision-making processes in construction management is also feasible. Behavior science ties into the design, analysis, evaluation, and choice of proposals, and the visual attention and information acquisition when reading a contract. Moreover, this includes the complex cognition of final discrimination [27,28]. For a more precise task, the operator's mobilization of various sensory functions of the body and the feedback of the brain's neural mechanisms would affect each control action [29].

3.3. Core Factors: Functional Ingredients

Motional behavior is implemented through the interaction of neuronal firing and functional circuits. Simplified brain regions and abstract concepts are often used to organize relevant knowledge, which means using basic functional ingredients to describe neuropsychological progress. Neural circuits of attention, emotion, and memory are ubiquitous in decision-making [30]. Neural control of learning and motion is also produced when a behavior takes place [31,32], which constitutes the main neuropsychological unit of behavior analysis [33].

Attention refers to mental engagement that focuses on specific information. When the information enters consciousness, the mind will learn about the specific items and decides whether to act [34]. Attention is also defined as the individual's choice of previous information from specific stimuli in a social context [35]. As conscious visual evidence, an approximately stable nervous system is behind attention [36]. Similarly, the neuropsychological mechanism behind visual attention is a cognitive-driven information system in construction engineering [26]. Sweany et al. (2016) have analyzed the effects of the format of engineering deliverables on craft performance based on empirical data [37]. They found the 3-dimensional computer aided design (3D CAD) model can improve the accuracy of workers' spatial cognition and the efficiency of completing preset tasks more than 2D. However, because of additional information of architectural textures, colors, and directions increases the cognitive burden of human beings, 3D models or virtual reality (VR) and augmented reality (AR), etc., are not better than traditional 2D drawings in conveying spatial information [38,39].

Learning and memory often occur at the same time because acquiring knowledge from the surrounding environment is a key condition for memory creation. At the end of the 19th century, German experimental psychologist Ebbinghaus used meaningless syllables to evaluate the effect of learning and memory and proposed the famous forgetting curve. Since the 1980s, the memory system has gained more understanding. People had divided it into sensory memory, short-term memory, and long-term memory, and further, divided short-term memory into working memory and reference memory, as well as dividing long-term memory into declarative memory and non-declarative memory (or explicit memory and implicit memory). In the field of construction engineering, limited studies focus on environmental components and spatial cognition application and lack in-depth exploration of memory [40,41].

The feature of emotion is intense sensory arousal associated with specific behavioral responses, which embodies the assessment of specific stimuli related to or irrelevant of individual or group goals [42]. Such assessment is a cognitive process that relies on the environment or proprioception. Additionally, it is taken as an attempt to maintain, establish, disrupt, or terminate connections between the individual and environment that are of great significance [18]. Therefore, Frijda (2009) argued that emotion is a process rather than a state, and that it is a subconscious event triggered by evaluating relevant stimuli in a limited dimension [43]. Moreover, this evaluation determines the strength and quality of behavioral tendencies, physiological responses, sensations, and behaviors. Current construction engineering management research has paid attention to the measurement and evaluation of emotional situations and their impact on work operations; for instance, mental fatigue [7,8] and stress [44]. Therefore, more neuropsychological analysis is still needed.

Motion control involves neural activation in multiple regions of the cerebral cortex and mass signal transmission by ganglions. Specifically, the auxiliary motor area found in the inner side of the brain controls the fingers moving [45]. When a motion needs to be completed under conditions of visual, auditory, or somatosensory feedback, the premotor cortex of the brain will be activated to extract rich representations of sensory information [46].

4. Neuropsychological Methods and Application

Before the neuropsychological technologies and methods led into behavior research, researchers widely used self-reporting methods in psychology (such as interviews and questionnaires) in behavioral research. However, the retrospective reflection of behavioral subjects is often biased [47]. To seek more reliable and accurate measurements, behavior researchers have been attempting biological psychology and neuropsychological technologies and methods, such as electrodermal response experiments in the 1920s and pupillary dilation experiments in the 1960s [48]. Afterwards, wearable devices, such as eye-tracking and heart rate measures, are becoming more acceptable because of their convenience [49]. Recently, researchers used neuroscience techniques, with electroencephalogram (EEG), functional magnetic resonance imaging (fMRI), and functional near-infrared spectroscopy (fNIRS) as the main types, to study the emotional and cognitive responses of behaving individuals. The results promote the behavior research based on neuroscience and neuropsychology [30]. There are also researchers that are pointing out that the paradigm of neuropsychology is shifting to digital neuropsychology and will form a high-dimensional neuropsychological assessment through simulation computing [50].

4.1. Tradition: Subjective Evaluation and Self-Report

Early neuropsychological assessments rely primarily on evaluation and self-report methods. They can often be categorized as qualitatively or quantitatively focused. Qualitative assessments are derived from approaches of behavioral neurology and rely on intuitive insight, which requires considerable knowledge of neurology [2]. Compared to the subjective evaluation of the evaluator, another perspective is the oral retrospective statements, such as in-depth interviews, ladder interviews, and focus interviews or written language statements of open-ended questionnaires based on self-knowledge. Quantitative assessments are attributed to clinical neuropsychology, emphasizing standardized psychometric tests. Quantitative analysis has formed various scales and tests, including the Wechsler Intelligence Scale (WISC), Minnesota Multiphasic Personality Inventory (MMPI), Tactual Performance Test (TPT), Finger-tapping Test (FT), Speech Sounds Perception Test (SSPT), Rhythm Test (RT), Train Making Test (TMT), and sensory perceptual examination, etc.

In the spatial perception and cognitive behavior closely related to architecture and construction, researchers have put forward and applied the cognitive map [51]. In practice, the cognitive map means "maps in the mind" and hypothetical constructions of map metaphors and is the visual expression of space. Researchers have asked people to express spatial form in the mind using painting. Passini (1984) pointed out in more detail that cognitive maps reflect people's cognitive and behavioral abilities in space [52]. This ability is based on three direct manifestations—information processing, decision-making, and task execution—reflecting the solving ability in response to spatial problems. Considering the differences in psychological comfort, spatial anxiety, and behavioral dependence, relevant staff should have higher spatial perception and expression skills [53–55]. Research on escape behavior during fire drills has also shown that people with higher spatial abilities are better able to understand building structures and symbolic representations, which means higher building space design, disaster prevention, and condition controlling capabilities [56,57].

4.2. Current: Precise Exam and Image Technology

There are three major spectrum techniques for precise examinations and images in neuropsychological diagnosis: physical neurological exam (PNE), computed axial tomography scan (CT), and EEG. PNE needs a standard physical examination (usually lasting more than 20 min) and a detailed medical history, so it is usually performed by medical professionals [58]. CT combines a computer and traditional X-ray machine to take brain pictures from multiple locations through X-ray. Then, they are transmitted to a computer for data conversion and analysis, resulting in a visualization of the density of different brain tissues on a cathode-ray screen. However, despite CT having higher cost-effectiveness, the false-negative errors still have a 50% occurrence, and so it needs to be combined with other measurements.

EEG acquires brain wave data by attaching electrodes to specific regions of the frontal, temporal, parietal, and occipital lobes. It can interpret brain activity based on brain waves, and so it is more closely related to the neural activity in the brain [59]. However, it also should be noted that since 15–20% of the normal population will have abnormal EEGs, the results of EEG analysis may have false-positive errors. An EEG experiment found that noise can affect the information machining process of construction workers [60]. Under the influence of high noise, construction workers will have higher self-depletion and show a behavioral tendency to seek rewards. The result means the workers need a longer response time for behavioral decision-making, increasing the risk of hazardous behavior.

4.3. Future: Wearable Devices and Computer Simulation

Neuropsychologists and behaviorists in the digital age have increasing access to emerging technologies. Relying on the technological advancement of portable wearable devices, behavioral research is no longer limited to the laboratory environment. Researchers can conduct behavioral experiments in real scenes and the real world, and it has been used in some fields such as tourism, marketing, transportation, etc. [49].

Eye-tracking can explain the neuropsychological mechanism of visual perception behavior by recording the process of eye movement. Existing research usually visualized tracking results through a heat map and trajectory map to illustrate potential neuropsychological responses functionally and effectively. Eye-tracking research has developed from the initial observation of appearance characteristics to accurate measurement records and formed modern technology to analyze eye movement patterns. It has become an important method in the research of spatial cognition, map cognition, map systems, and other fields [61]. In construction engineering and management, wearable devices are also applied in research of mental fatigue, hazard detection [8,62], fairness perception [27,28], and performance behavior [63] of construction contractors in contract signing.

Wearable and wireless EEG device systems can quantitatively and automatically assess the attention level of construction workers by recording and analyzing signals of the brain. EEG signal characteristics, such as frequency, power spectral density, and spatial distribution, can effectively reflect and quantify the perceived risk level of construction workers. Meanwhile, the lower gamma frequency band and left frontal lobe EEG cluster directly and appreciably show the worker's state of alertness [15]. Compared to fMRI and fNIRS, EEG provides higher temporal resolution but lacks spatial resolution [64–66]. The sampling frequency of fNIRS is relatively low (less than 20 Hz), but the spatial resolution is higher, which is crucial for monitoring brain regions associated with mental loading [67,68]. In empirical research, Shi et al. (2020) built a virtual industrial maintenance scenario to present a human–subject experiment, and to examine the impact of information format on the performance of a pipe maintenance task, as well as the implications of cognitive costs in both working memory development (information encoding) and retrieval (information recalling) [44]. The research aimed to establish an evaluation system for predicting engineering performance and then inspire the design of personalized training systems driven by cognition in construction engineering.

Moreover, computational simulation is considered a high-dimensional technique of neuropsychology in the continual development of VR, AR, and mixed reality (MR). The behavioral and social science initiative of the National Institutes of Health (NIH) highlights the developing scientific and technological potentials (such as new sensors), and to enhance the characterization of neurocognitive, behavioral, emotional, and social understanding. Jolly and Chang (2019) considered that neuropsychological assessments that are two-dimensional may previously have had a flatland fallacy [69]. They suggest that this fallacy can be overcome by formalizing psychological theories into computational models, making it capable for predicting cognition and behavior precisely. Simulating the parallel, reciprocal, and iterative interactions between the environment and neural function of complex behavior can enhance the ability of actual operations [50]. Building Information Modeling (BIM) is a beneficial attempt at visualization and predictability in construction engineering, as well as Heritage Building Information Modeling (HBIM) and BIM+. However, such models are still limited to the simulation and calculation of architecture or building substance, while the behavior simulation and more subtle neural mechanisms are still difficult to achieve.

As Table 1 shows, whether it is an objective measurement dominated by technical tools or a subjective judgment dominated by self-report, each method has its advantages and limitations. There is a growing consensus that researchers using neuroscientific methods and devices must have qualitative knowledge or experience of the phenomenon. Then, the researchers can reasonably interpret neuropsychological indicators and the signals they represent. As Plassman et al. (2015) claim, the fullest explanation of consumer and managerial behavior requires a combination of neuroscientific measurements and subjective explanations [4]. When combining the methods, researchers will draw the most productive and profound conclusions, as the advantages of one method outweigh the disadvantages of the other.

Table 1. Comparison of neuropsychological methods.

Methodology	Specific Method	Advantage	Disadvantage
Self-report	Interview	Richer explanatory information	Incomplete authenticity
	Retrospective statement	Richer explanatory information	Reflection bias
	Cognitive map	Richer explanatory information	Reflection bias
	Questionnaire	Richer explanatory information	Subjectivity of measurements
Biological measurement	Electrodermal response	Objective and accurate	Weak stimulus response
	Pupillary dilation	Objective and accurate	Lack of criteria for expansion
Spectral imaging	PNE	Sufficient	Long duration, high profession
	CT	Comprehensive angle	Cost effectiveness, false-negative errors, invasiveness
	EEG	Non-invasive, high temporal resolution	False-positive errors
	fMRI	Non-invasive, slice image of any angle	Low spatial resolution
	fNIRS	Non-invasive, high spatial resolution	Low sampling frequency
Wearable devices	EEG	Convenience, accuracy, visible	Low spatial resolution
	Eye-tracking	Convenience, accuracy, visible	Limited visual distance, vision requirements
Computational simulation	BIM	Structural modeling, superfine	Difficult to simulate neural procedure behind behavior
	Simulation Platform	Systematic, processed	Difficult to simulate neural procedure behind behavior

5. Behavior Research in Construction Engineering Management

5.1. Behavioral Turns of Method: Neuropsychological Experiment

In addition to focusing on economic issues such as economic benefits and cost control, engineering issues such as construction technology and schedule control must also be given attention. Construction engineering and management also pays attention to management issues such as organizational systems and contract signing. This shows the research content has gradually shifted from focusing on objects such as equipment, materials, and structures, to focusing on humans; that is, the behavioral response of individuals or organizations to system design and management processes. However, single and traditional methods inevitably have shortcomings in data acquisition or measurement, and it is difficult to infer the influencing factors of management and decision-making.

Through scientific design and critical control, behavioral analysis and experimental methods can improve the reliability of data and the interpretability of research results. Figure 3 summarizes the general process of behavioral experiments and neuropsychological experiments. Applicable experimental designs in construction engineering and management include single-station passive observation, contrast experiments, and randomized experiments [70]. In practical application, these experiments should ensure the controllability of the environment, the systematisms of the research content, and the predictability of the research conclusions. Few Chinese researchers also prefer to apply the behavioral experiment to construction engineering and management. For example, Li et al. (2012) took computational experiments as a research method to discuss the multistage group incentive problem when considering the fairness perception of individual contractors [71]. To sum up, research methods applied in construction engineering and management have transitioned from framed qualitative research to model-based quantitative research. However, the application of neuropsychological experimental methods is still in infancy [72].

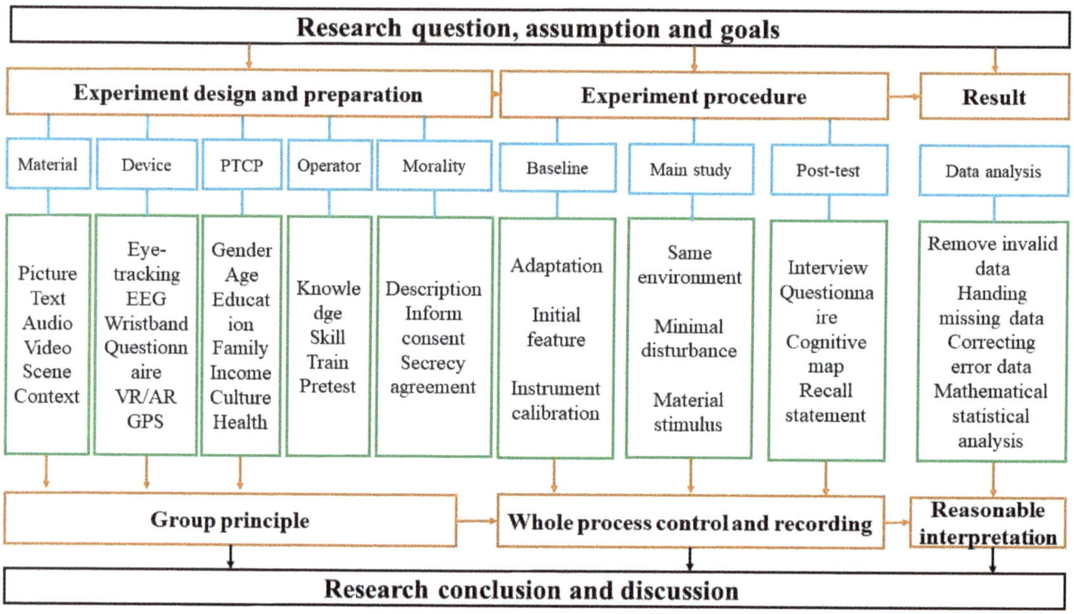

Figure 3. Neuropsychological experiment process. Drawn by the authors.

5.2. Bibliometric Analysis: Countries (Regions) and Hot Issues

5.2.1. Countries (Regions) Cluster and Timing Progress

By choosing bibliographic coupling as the analysis type and countries as the analysis unit, with the minimum number of documents as six, forty countries or regions meet the thresholds. The countries or regions of publications are shown in Figure 4, and the ten countries (regions) with the most publications are shown in Table 2. From the results, researchers from Taiwan, China have begun to pay more attention to the neuropsychological mechanisms of individual behavior in CEM earlier and have published some achievements. Then, American researchers have focused on this field and have published a large number of achievements. After that, researchers from Australia, Canada, England, France, and Spain also began to get involved in neuropsychological research and published some achievements. In these countries (regions), Australia has the most publications. Indian scholars then paid attention to this research topic and carried out some related research. In the most recent time period, Italian researchers also realized the important role of neuropsychology in revealing behavioral mechanisms and carried out lots of research in the CEM field. At the same time, Chinese scholars attended to this field; these researchers tried to introduce neuropsychological theories and methods into construction engineering management and its behavioral research, and thus a large number of achievements have been published.

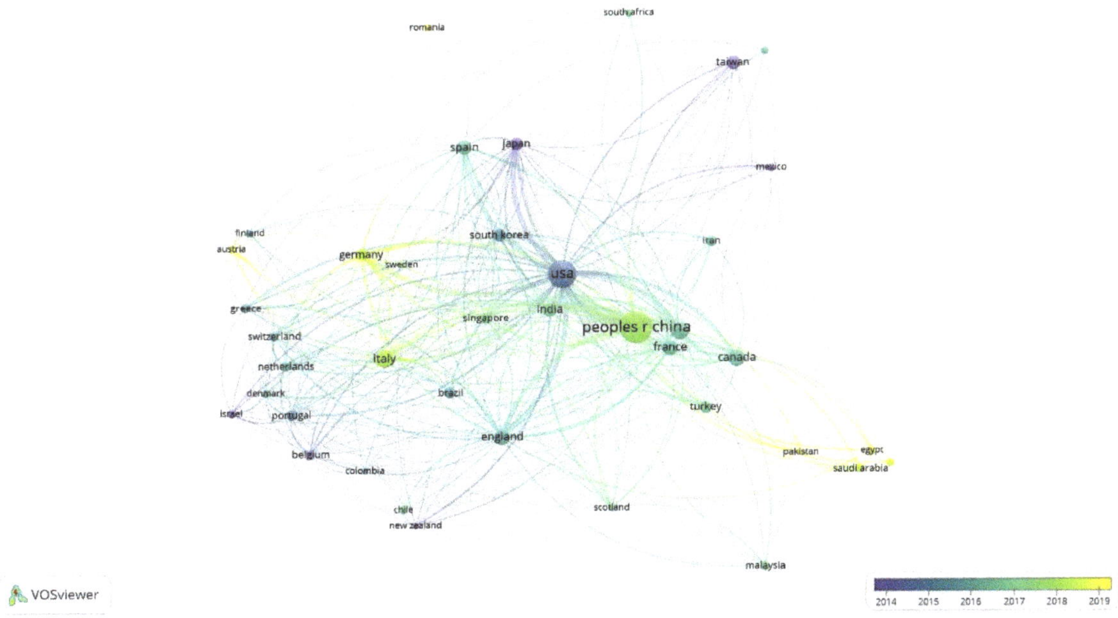

Figure 4. Countries (regions) and timing progress. The "Taiwan" in this figure refers to "Taiwan, China".

Table 2. Distribution and timing progress examples of existing publications.

Countries (Regions)	Average Publication Year	Publication Count
China	2018	370
USA	2015	263
Australia	2016	79
Italy	2018	72
Canada	2016	64
England	2016	56
France	2016	52
India	2017	48
Spain	2016	48
Taiwan, China	2013	38

5.2.2. Hot Issues and Timing Progress

By selecting co-occurrence as an analysis type and keywords as the analysis unit, with the minimum number of occurrences of a keyword as six, 314 of 7443 keywords meet the thresholds. These keywords and their average publication year are shown in Figure 5, and the ten keywords with the most occurrences are shown in Table 3. From the results, "design" is the keyword that occurred earlier at the average publication year of 2015. Then, "model", "construction", and "simulation" occurred frequently near 2016. Scholars were beginning to seek more digital technology and simulation methods in construction research. Then in 2017, "behavior" became the most occurrent keyword, which suggests that behavior research in the CEM field is fresh and hot in recent years. At this period, "performance", "management", "system", "strength", and "impact" are also important keywords, but performance and management occurred much more. It means that researchers have paid more attention to the people-oriented perspective and cared more about the management of the behavior of individuals, such as performance and strength.

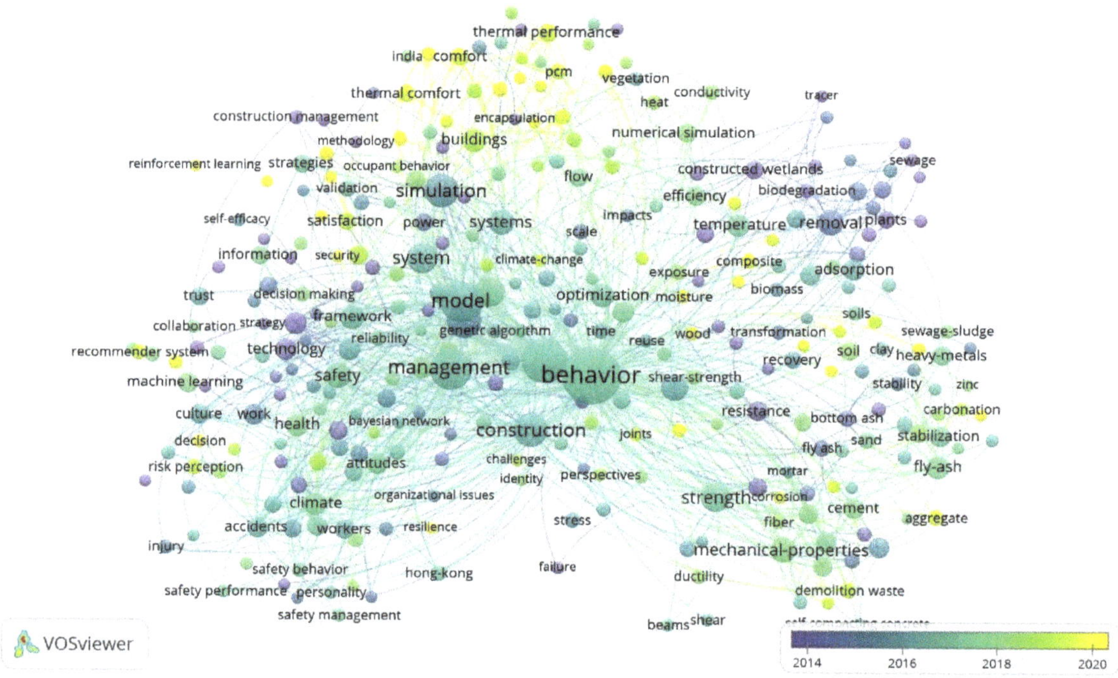

Figure 5. Keywords and timing progress.

Although keywords related to neuropsychology such as risk perception, stress, satisfaction, self-efficacy, stress, attitude, injury, and health are also present in the figure, they are all with little occurrence. This clearly reflects that the application of neuropsychology in construction engineering management and its behavior research is still very insufficient. To sum up, current research on construction engineering management emphasized the analysis of behavior and has shown a tendency to focus on behavioral subjects, but the related theories and methods of neuropsychology are still very lacking in this field. For some research, although neuropsychological theories and methods have been applied, construction engineering management plays a role as a mere research context. However, in interdisciplinary research, the comparative balance between different disciplines is the premise to accomplish complementary advantages and knowledge exchange for interdisciplinary research.

Table 3. Keywords and timing progress examples.

Keyword	Average Publication Year	Count
behavior	2017	261
performance	2017	139
management	2017	106
model	2016	105
construction	2016	68
simulation	2016	67
design	2015	63
system	2017	51
strength	2017	47
impact	2017	45

5.3. Generalization: Complex Correlation Mechanism

Construction engineering management studies provide the means and methods to examine the physical world and organizational management to improve the validity and efficiency of management in the construction industry. Although relevant research is very limited, scholars have explored employing neuropsychology in the behavioral research of construction engineering management. To better understand and reveal behavior mechanisms in CEM, we provide a recap in Table 4 with the examples of CEM behavioral studies and their key methods and findings in the last five years. The main content of CEM behavioral studies includes hazard recognition, construction equipment-related accidents and safety management, mental fatigue, physical fatigue, spatial and work memory, building inspection, engineering information formats, operations and performances, working skills, and others. The research design of CEM behavioral studies is mainly based on experiments with different simulated conditions, such as various task settings and scenarios. The methods are composed of data collection technology and analysis methodology. Neuropsychological methods are usually used for data collection such as NIRS, eye-tracking devices, and wearable EEG systems. Some self-report methods may also be added for subjective data collection, such as questionnaire surveys or cognitive mapping. Information technology and statistical methods are normally taken for data analysis, such as some deep learning or machine learning of information technology, linear discriminant analysis, or MANOVA analysis of statistical methods.

Based on the review of relevant literature findings, this paper tries to establish a behavioral correlation mechanism in CEM from the perspective of neuropsychology. Although this mechanism may be difficult to cover in all research results, it can still intuitively demonstrate how neuropsychological mechanisms affect human behavior and ultimately play a role in CEM. As Figure 6 shows, current neuropsychological research on behavior mainly starts from visual behavior, and the main content of CEM is "risk identification" and "performance evaluation and prediction". For example, Wang et al. (2017) found that EEG signal properties such as frequency, power spectrum density, and spatial distribution can effectively reflect the workers' perceived risk level [15]. Shi et al. (2020) found that stressful training has a strong impact on neural connectivity and gaze movement patterns, which further affect final performance [44]. Some research conclusions clearly point out specific indicators and their positive or negative effects. For example, Shi, Du, & Ragan (2020) found a positive relationship between visual attention (fixation time) and spatial memory [26]. However, many studies only found correlations between certain neuropsychological causes and behavioral outcomes. It is difficult to explain more detailed action paths, degrees, and valence, especially when the complex nervous system of the brain is involved.

Table 4. Current research and main findings.

Authors	Year	Main Content	Design	Method	Findings
Zhou et al. [73]	2021	Neurophysiological mechanics of hazard recognition	Multiple HR tasks experiment in laboratory setting and hemodynamic responses	NIRS; Fisher score; linear discriminant analysis	Left PFC was more engaged in HR
Li et al. [62]	2020	Operator's mental fatigue; construction equipment-related accidents	Simulated excavator operation experiment with wearable eye-tracking devices	TICC method; Supervised learning algorithms; Support Vector Machine	Different levels of mental fatigue had varying effects on the operator's productivity and safety performance
Shi, Du, & Ragan. [26]	2020	Visual attention; spatial memory; building inspection	Human–subject experiment (2D, 3D, VR) with building inspection task	Human–subject experiment; Eye-tracking	There is a positive relationship between visual attention (fixation time) and spatial memory
Shi, Du, & Worthy. [29]	2020	Engineering information formats; construction operations; work memory	Participants reviewed the operational instructions for a pipe maintenance task, performed the task from memory	Human–subject experiment; cognitive load analysis with survey	Larger pupil dilation during encoding, indicative of successful working memory formation, was associated with better subsequent performance
Shi, Zhu, Mehta, & Du. [44]	2020	Industrial shutdown maintenance; training outcome under stress	A virtual reality (VR) system integrated with the eye-tracking function to simulate the operation scenarios	Virtual reality experiment; fNIRS	Stressful training has strong impact on neural connectivity and gaze movement patterns, which further effect final performance
Xing et al. [7]	2020	Physical and mental fatigue of construction workers; safety management	Manual handling tasks for physical fatigue statuses; cognition-required risk identification task for mental fatigue, wearable EEG sensor fatigue detection and measurement	Pilot experimental method	High physical fatigue could accelerate the induction of mental fatigue; more attention sources were required during the intensive manual handling tasks
Hasanzadeh et al. [9]	2017	Construction workers' hazard identification skills	An experiment was designed to track eye movements of construction workers while they searched for hazards in randomly ordered construction scenario images	Eye-tracking experiment; MANOVA analysis	Hazard identification skills significantly impacted workers' visual search strategies; fixation count can discriminate workers with high hazard-identification skills and at-risk workers
Wang et al. [15]	2017	Workers' attention and vigilance in construction activities	On-site experiment to analyze the EEG signal patterns when construction workers avoid different obstacles in their tasks	Wearable EEG system; on-site experiment	EEG signal properties such as frequency, power spectrum density, and spatial distribution can effectively reflect the workers' perceived risk level

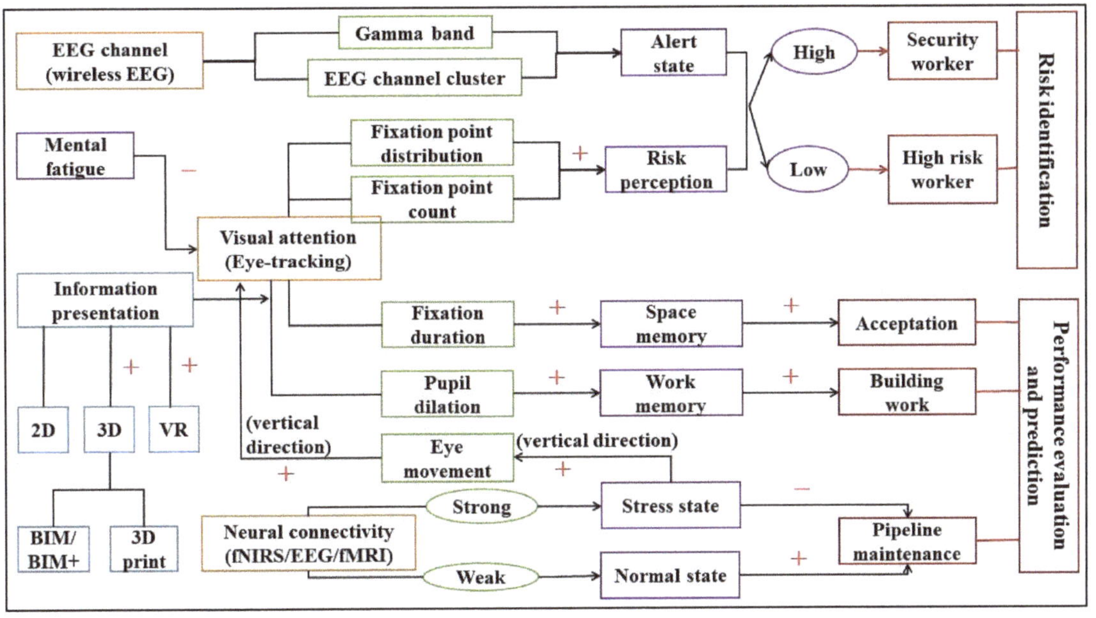

Figure 6. Complex behavioral associations. "+" indicates enhancing or positive promotion, "−" indicates weakening or negative inhibition, "unsigned" paths represent correlation only.

Eye-tracking research on visual behavior has successfully found that specific eye-movement indicators have an indicative effect on certain perceptions and cognitive behaviors, resulting in improved or weakened effects. For example, the distribution and number of fixation points can positively enhance risk perception, to identify whether the participant can keep a high level of safety at work. Moreover, gaze time and pupil dilation can positively promote the participant's memory of space and task content, respectively, and then positively promote the acceptance or building work of construction, and supremely help predict behavioral performance. It is worth noting the cognitive state of the participant may not be the result of neuropsychological effects, but the cause of different neuropsychological manifestations. For instance, mental fatigue will negatively affect workers' visual attention.

In addition, the results in the figure also show the effect of stress on eye movement. When participants are under high stress, their eye movements will improve, and their visual attention will be more concentrated in the vertical direction. It will directly affect the acquiring of visual information. Relevant research shows that in pipeline maintenance work of construction engineering, when the pressure load for the staff is large, the work's precision will reduce. The conclusion in the diagram also shows the influence of information presentation (such as the renderings of construction) on visual attention. As 3D presentations influence the fixation time of visual attention, there is research that found that 3D presentations, such as 3D images, BIM models, 3D printing and virtual reality technologies, are better than 2D planes in rendering effect.

In general, conducting behavioral research in CEM from the perspective of neuropsychology reflects a multidisciplinary research trend. The research in this field involves three complex dimensions: the brain and nervous system, the construction engineering management system, and individual and behavioral research (see Figure 7). For the research field, research between the brain and nervous system and construction engineering, or between the brain and nervous system and individual behavior, is mainly exploring the internal causes of external response. Research between behavior and CEM is mainly focused on the impact on performance. The essence is the result of interaction among the internal neuropsychological mechanisms, the external behavioral mechanisms, and the construction engineering management mechanism. Therefore, the core of this multidisciplinary field is the effect paths among three mechanisms. As for the science of construction engineering management, knowledge from the four disciplines of psychology, neuroscience, management, and architecture should be absorbed. Moreover, the intersection between four disciplines needs to be noticed with an interaction in mind of combining the subjective and objective, the physical and mental, and the micro and macro.

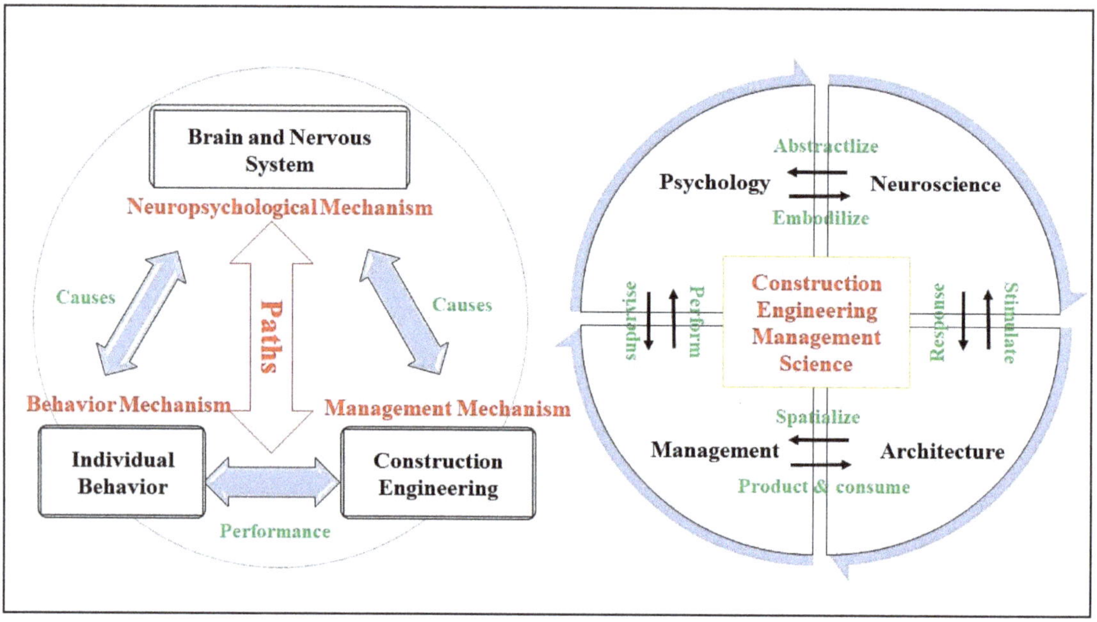

Figure 7. Interdisciplinary interrelationship.

6. Conclusions and Prospection

This paper aimed to reveal the behavioral causes, behavioral performance, and the action paths of the relevant individuals in CEM. Although the research on CEM and its behavior based on neuropsychological theories and methods is still insufficient in volume and incomplete in content, the predictability of its development prospects is substantively existing. The current method has become increasingly rich, including experiments, surveys and observational studies, modeling and simulations, theory building, and case studies and their various subtypes. Moreover, some researchers are often using more than one method, which is both an opportunity and a challenge for construction engineering and management research [70]. Introducing new theories and new technology methods for any scientific research needs to ensure rigorous verification at every step. The derived results need both academics and practice to test, and then such research can be advanced with the help of new knowledge. Finally, related research can filter to daily use, to achieve the goal of providing welfare to society.

In conclusion, this paper attempted to sort out the behavioral research in construction engineering management from the perspective of neuropsychology: on the one hand, to strengthen the understanding of neuropsychological theories, techniques, and methods, and on the other hand, to explore the more refined internal relationship between construction engineering management and behavior from the perspective of micro-individuals. Based on this knowledge, this paper tried to summarize the specific methods and conclusions that can be used for reference in applying neuropsychology in the research and practice of construction engineering management.

The contribution of this paper is the carrying out of a qualitative sorting and quantitative analysis based on the common goal of multidisciplinary crossover and integration. The results can provide effective references for revealing the existing conclusions, available methods, and future trends in this disciplinary field. In detail, this review from the perspective of neuropsychology, focused on the working mechanism of the human brain and nervous system, individual behavior, and construction engineering, is close to the needs of CEM practice. Starting from the human body, it can assist managers to formulate more refined and humanized measures to improve the efficiency and safety of CEM. Meanwhile, a more accurate analysis of the working mechanism between the brain and behavior provided the reference for improving benefits of CEM. To some extent, it can also avoid ineffectiveness and inefficiency in CEM and promotes the development of sustainable construction production and consumption. Lastly, the review of human factors' methods and research also clarifies the human behavioral mechanism, which has considerable reference for designing and building a people-oriented livable city.

Author Contributions: Conceptualization, Y.L. and J.L.; methodology, Y.L.; formal analysis, J.L.; data curation, R.L.; writing—original draft preparation, J.L. and J.H.; writing—review and editing, J.L. and J.H.; visualization, J.L. and M.Y.; funding acquisition, Y.L. All authors have read and agreed to the published version of the manuscript.

Funding: This research was funded by [the Natural Science Foundation of China] grant numbers [42171219] and [the Natural Science Foundation of Fujian Province] grant numbers [2020J01011].

Institutional Review Board Statement: Not applicable.

Informed Consent Statement: Not applicable.

Data Availability Statement: No data, models, or code were generated or used during the study. The electrode lead system on the left of Figure 2 is adapted from Wang et al. (2017), and the other part of Figure 2 and other figures in this study are drawn by the authors.

Conflicts of Interest: The authors declare no conflict of interest.

References

1. Kalat, J.W. *Biological Psychology*, 13th ed.; Cengage Learning: Singapore, 2015.
2. Horton, A.M.J.; Wedding, D.; Phay, A. A current perspective on assessment of and therapy for brain-damaged individuals. In *Applied Techniques in Behavioral Medicine (59–85)*; Golden, C.J., Graber, B., Strider, F., Strider, M.A., Alcaparres, S.S., Eds.; Grune & Stratton: New York, NY, USA, 1981.
3. Horton, J.A.M.; Miller, W.G. Neuropsychology and behavior therapy. *Prog. Behav. Modif.* **1985**, *19*, 1–55. [CrossRef] [PubMed]
4. Plassmann, H.; Venkatraman, V.; Huettel, S.; Yoon, C. Consumer neuroscience: Applications, challenges, and possible solutions. *J. Mark. Res.* **2015**, *52*, 427–435. [CrossRef]
5. Adolphs, R. Conceptual challenges and directions for social neuroscience. *Neuron* **2010**, *65*, 752–767. [CrossRef] [PubMed]
6. Camerer, C.; Loewenstein, G.; Prelec, D. Neuroeconomics: How neuroscience can inform economics. *J. Econ. Lit.* **2005**, *43*, 9–64. [CrossRef]
7. Xing, X.; Zhong, B.; Luo, H.; Rose, T.; Li, J.; Antwi-Afari, M.F. Effects of physical fatigue on the induction of mental fatigue of construction workers: A pilot study based on a neurophysiological approach. *Autom. Constr.* **2020**, *120*, 103381. [CrossRef]
8. Li, J.; Li, H.; Wang, H.; Umer, W.; Fu, H.; Xing, X. Evaluating the impact of mental fatigue on construction equipment operators' ability to detect hazards using wearable eye-tracking technology. *Autom. Constr.* **2019**, *105*, 102835. [CrossRef]
9. Hasanzadeh, S.; Esmaeili, B.; Dodd, M.D. Impact of construction workers' hazard identification skills on their visual attention. *J. Constr. Eng. Manag.* **2017**, *143*, 4017070. [CrossRef]
10. Broadus, R.N. Toward a definition of "bibliometrics". *Scientometrics* **1987**, *12*, 373–379. [CrossRef]
11. Pritchard, A. Statistical bibliography or bibliometrics. *J. Doc.* **1969**, *25*, 348.
12. Donthu, N.; Kumar, S.; Mukherjee, D.; Pandey, N.; Lim, W.M. How to conduct a bibliometric analysis: An overview and guidelines. *J. Bus. Res.* **2021**, *133*, 285–296. [CrossRef]
13. Ramos-Rodríguez, A.-R.; Ruíz-Navarro, J. Changes in the intellectual structure of strategic management research: A bibliometric study of the *Strategic Management Journal*, 1980–2000. *Strateg. Manag. J.* **2004**, *25*, 981–1004. [CrossRef]

14. Horton, A.M., Jr.; Howe, N.R. Behavioral neuropsychology and the traumatic brain injured adult: A case study. In Proceedings of the 5th Annual Post-Graduate Course on the Rehabilitation of the Brain-Injured Adult (18–19); Virginia Commonwealth University: Richmond, VA, USA, 1981.
15. Wang, D.; Chen, J.; Zhao, D.; Dai, F.; Zheng, C.; Wu, X. Monitoring workers' attention and vigilance in construction activities through a wireless and wearable electroencephalography system. *Autom. Constr.* **2017**, *82*, 122–137. [CrossRef]
16. Solso, R.L.; MacLin, M.K.; MacLin, O.H. *Cognitive Psychology*; Pearson Education New Zealand: Auckland, New Zealand, 2005.
17. Foreman, N.; Gillett, R. *A Handbook of Spatial Research Paradigms and Methodologies*; Psychology Press: London, UK, 1997; Volume 2.
18. Perlman, A.; Sacks, R.; Barak, R. Hazard recognition and risk perception in construction. *Saf. Sci.* **2014**, *64*, 22–31. [CrossRef]
19. Devlin, A.S. *Mind and Maze: Spatial Cognition and Environmental Behavior*; Praeger Publishers/Greenwood Publishing Group: Westport, CT, USA, 2001.
20. Wagstaff, A.S.; Lie, J.S. Shift and night work and long working hours-a systematic review of safety implications. *Scand. J. Work. Environ. Health* **2011**, *37*, 173–185. [CrossRef] [PubMed]
21. Skavronskaya, L.; Scott, N.; Moyle, B.; Le, D.; Hadinejad, A.; Zhang, R.; Gardiner, S.; Coghlan, A.; Shakeela, A. Cognitive psychology and tourism research: State of the art. *Tour. Rev.* **2017**, *72*, 221–237. [CrossRef]
22. Langer, E.J. Matters of mind: Mindfulness/mindlessness in perspective. *Conscious. Cogn.* **1992**, *1*, 289–305. [CrossRef]
23. Nevid, J.S. *Essentials of Psychology: Concepts and Applications*; Cengage Learning: Boston, MA, USA, 2021.
24. Awh, E.; Anllo-Vento, L.; Hillyard, S.A. The role of spatial selective attention in working memory for locations: Evidence from event-related potentials. *J. Cogn. Neurosci.* **2000**, *12*, 840–847. [CrossRef]
25. Awh, E.; Vogel, E.K.; Oh, S. Interactions between attention and working memory. *Neuroscience* **2006**, *139*, 201–208. [CrossRef]
26. Shi, Y.; Du, J.; Ragan, E. Review visual attention and spatial memory in building inspection: Toward a cognition-driven information system. *Adv. Eng. Inform.* **2020**, *44*, 101061. [CrossRef]
27. Yan, L.; Zeng, C.; Guo, L.; Li, Z. Formation of the contractor's justice perception based on eye movement experiment. *Sci. Res. Manag.* **2020**, *41*, 219–227. (In Chinese) [CrossRef]
28. Yan, L.; Guo, L.; Zeng, C.; Li, Z. Study on the Multi-dimensional Structure and Content of the Contract Reference Point in Construction Project Situation: Based on Content Analysis and Eye Movement Experiment. *Manag. Rev.* **2020**, *32*, 179–192. (In Chinese)
29. Shi, Y.; Du, J.; Worthy, D.A. The impact of engineering information formats on learning and execution of construction operations: A virtual reality pipe maintenance experiment. *Autom. Constr.* **2020**, *119*, 103367. [CrossRef]
30. Shaw, S.D.; Bagozzi, R.P. The neuropsychology of consumer behavior and marketing. *Consum. Psychol. Rev.* **2018**, *1*, 22–40. [CrossRef]
31. Bliss, T.V.; Lømo, T. Long-lasting potentiation of synaptic transmission in the dentate area of the anaesthetized rabbit following stimulation of the perforant path. *J. Physiol.* **1973**, *232*, 331–356. [CrossRef] [PubMed]
32. Jeannerod, M. *The Cognitive Neuroscience of Action*; Blackwell Publishing: Oxford, UK, 1997.
33. Garrett, B.L. *Brain & Behavior: An Introduction to Biological Psychology*; SAGE Publications: New York, NY, USA, 2010.
34. Terranova, T. Attention, economy and the brain. *Cult. Mach.* **2012**, *13*, 1–19.
35. Gazzaley, A.; Cooney, J.W.; Rissman, J.; D'Esposito, M. Top-down suppression deficit underlies working memory impairment in normal aging. *Nat. Neurosci.* **2005**, *8*, 1298–1300. [CrossRef] [PubMed]
36. Duchowski, A.T.; Duchowski, A.T. *Eye Tracking Methodology: Theory and Practice*; Springer: Berlin, Germany, 2017.
37. Sweany, J.; Goodrum, P.; Miller, J. Analysis of empirical data on the effects of the format of engineering deliverables on craft performance. *Autom. Constr.* **2016**, *69*, 59–67. [CrossRef]
38. Richardson, A.E.; Montello, D.R.; Hegarty, M. Spatial knowledge acquisition from maps and from navigation in real and virtual environments. *Mem. Cogn.* **1999**, *27*, 741–750. [CrossRef]
39. Verghote, A.; Al-Haddad, S.; Goodrum, P.; Van Emelen, S. The effects of information format and spatial cognition on individual wayfinding performance. *Buildings* **2019**, *9*, 29. [CrossRef]
40. Kondyli, V.; Bhatt, M.; Hartmann, T. Precedent based design foundations for parametric design. *Adv. Comput. Des.* **2018**, *3*, 30. [CrossRef]
41. Kondyli, V.; Bhatt, M. Rotational locomotion in large-scale environments: A survey and implications for evidence-based design practice. *Built Environ.* **2018**, *44*, 241–258. [CrossRef]
42. Niedenthal, P.M.; Brauer, M. Social functionality of human emotion. *Annu. Rev. Psychol.* **2012**, *63*, 259–285. [CrossRef] [PubMed]
43. Frijda, N.H. Emotion experience and its varieties. *Emot. Rev.* **2009**, *1*, 264–271. [CrossRef]
44. Shi, Y.; Zhu, Y.; Mehta, R.K.; Du, J. A neurophysiological approach to assess training outcome under stress: A virtual reality experiment of industrial shutdown maintenance using Functional Near-Infrared Spectroscopy (fNIRS). *Adv. Eng. Inform.* **2020**, *46*, 101153. [CrossRef]
45. Chee, B.P.; Lim, A.; Chia, J.C.; Teh, M. Soft tissue chondroma in the finger: A case report and review of the literature. *Ann. Acad. Med. Singap.* **1999**, *28*, 590–592.
46. Goldberg, G. Supplementary motor area structure and function: Review and hypotheses. *Behav. Brain Sci.* **1985**, *8*, 567–588. [CrossRef]
47. Nisbett, R.E.; Wilson, T.D. Telling more than we can know: Verbal reports on mental processes. *Psychol. Rev.* **1977**, *84*, 231. [CrossRef]

48. Bagozzi, R.P. The role of psychophysiology in consumer research. In *Handbook of Consumer Behavior*; Robertson, T.S., Kassarjian, H.H., Eds.; Prentice-Hall: Hoboken, NJ, USA, 1991; pp. 124–161.
49. Wang, Y.J.; Minor, M.S. Validity, reliability, and applicability of psychophysiological techniques in marketing research. *Psychol. Mark.* **2008**, *25*, 197–232. [CrossRef]
50. Parsons, T.; Duffield, T. Paradigm Shift Toward Digital Neuropsychology and High-Dimensional Neuropsychological Assessments. *J. Med. Internet Res.* **2020**, *22*, e23777. [CrossRef]
51. Kuipers, B. The "map in the head" metaphor. *Environ. Behav.* **1982**, *14*, 202–220. [CrossRef]
52. Passini, R. Spatial representations, a wayfinding perspective. *J. Environ. Psychol.* **1984**, *4*, 153–164. [CrossRef]
53. Hund, A.M.; Minarik, J.L. Getting from here to there: Spatial anxiety, wayfinding strategies, direction type, and wayfinding efficiency. *Spat. Cogn. Comput.* **2006**, *6*, 179–201. [CrossRef]
54. Lyons, I.M.; Ramirez, G.; Maloney, E.A.; Rendina, D.N.; Levine, S.C.; Beilock, S.L. Spatial Anxiety: A novel questionnaire with subscales for measuring three aspects of spatial anxiety. *J. Numer. Cogn.* **2018**, *4*, 526–553. [CrossRef]
55. Vidler, A. Warped space: Architectural anxiety in digital culture. In *Impossible Presence: Surface and Screen in the Photogenic Era*; University of Chicago Press: Chicago, IL, USA, 2001; pp. 285–303.
56. Kobes, M.; Oberijé, N.; Groenewegen, T.M.; Morsche, T. *Serious Gaming for Behavioural Assessment and Research in Case of Emergency. An Evaluation of Experiments in Virtual Reality*; Paper Presented at the SimTecT 2009 Simulation Conference and Exhibition; Citeseer: Adelaide, Australia, 2009.
57. Kobes, M.; Helsloot, I.; De Vries, B.; Post, J.G. Building safety and human behaviour in fire: A literature review. *Fire Saf. J.* **2010**, *45*, 1–11. [CrossRef]
58. Golden, C.J. *Diagnosis and Rehabilitation in Clinical Neuropsychology*; Thomas: Springfield, QLD, Australia, 1978.
59. Horton, A.M., Jr. Behavioral neuropsychology in the schools. *Sch. Psychol. Rev.* **1981**, *10*, 367–372. [CrossRef]
60. Ye, G.; Chen, L.; Feng, X.; Yang, J.; Yue, H. Study on inducing mechanism of construction workers' risk-taking behavior by noise. *China Saf. Sci. J.* **2020**, *30*, 16–23. (In Chinese) [CrossRef]
61. Huang, X.; Li, W.; Yan, S. Research on pattern of eye-tracking behavior based on tourism map. *Tour. Trib.* **2018**, *33*, 87–96. (In Chinese)
62. Li, J.; Li, H.; Umer, W.; Wang, H.; Xing, X.; Zhao, S.; Hou, J. Identification and classification of construction equipment operators' mental fatigue using wearable eye-tracking technology. *Autom. Constr.* **2020**, *109*, 103000. [CrossRef]
63. Yan, L.; Deng, J.; Yan, M.; Zhang, X. The experimental investigation method on engineering projects management: Based on the experiment of contractor performance behavior as an example. *J. Eng. Manag.* **2016**, *30*, 87–92. (In Chinese) [CrossRef]
64. Duschek, S.; Schandry, R. Functional transcranial Doppler sonography as a tool in psychophysiological research. *Psychophysiology* **2003**, *40*, 436–454. [CrossRef]
65. Mehta, R.K.; Parasuraman, R. Neuroergonomics: A review of applications to physical and cognitive work. *Front. Hum. Neurosci.* **2013**, *7*, 889. [CrossRef] [PubMed]
66. Pizzagalli, D.A. *Electroencephalography and High-Density Electrophysiological Source Localization*; Cambridge University Press: Cambridge, UK, 2007; pp. 56–84.
67. Nozawa, T.; Kondo, T. *A Comparison of Artifact Reduction Methods for Real-Time Analysis of fNIRS Data*; Paper Presented at the Symposium on Human Interface; Springer: Berlin/Heidelberg, Germany, 2009.
68. Zama, T.; Shimada, S. Simultaneous measurement of electroencephalography and near-infrared spectroscopy during voluntary motor preparation. *Sci. Rep.* **2015**, *5*, 1–9. [CrossRef] [PubMed]
69. Jolly, E.; Chang, L.J. The flatland fallacy: Moving beyond low-dimensional thinking. *Top. Cogn. Sci.* **2019**, *11*, 433–454. [CrossRef] [PubMed]
70. Lucko, G.; Rojas, E.M. Research validation: Challenges and opportunities in the construction domain. *J. Constr. Eng. Manag.* **2010**, *136*, 127–135. [CrossRef]
71. Li, Z.; Meng, Q.; Sheng, Z.; Li, Q. Analysis on performance and evolution of mass stimulation under project quality optimization. *Chin. J. Manag. Sci.* **2012**, *3*, 112–121. (In Chinese) [CrossRef]
72. Bernold, L.E.; Lee, T.S. Experimental research in construction. *J. Constr. Eng. Manag.* **2010**, *136*, 26–35. [CrossRef]
73. Zhou, X.; Hu, Y.; Liao, P.; Zhang, D. Hazard differentiation embedded in the brain: A near-infrared spectroscopy-based study. *Autom. Constr.* **2021**, *122*, 103473. [CrossRef]

buildings

MDPI

Article

How Private Enterprises' Participation Behaviors Evolve with Incentive Modes in PPPs: An Evolutionary Game View

Yunhua Zhang [1], Hongyang Yi [1], Hongtao Xie [2], Junwei Zheng [1] and Yan Wang [3,*]

1 Faculty of Civil Engineering and Mechanics, Kunming University of Science and Technology, Kunming 650500, China; zhangyh@kust.edu.cn (Y.Z.); yihongyang333@163.com (H.Y.); zjw1989@kust.edu.cn (J.Z.)
2 Faculty of Management and Economics, Kunming University of Science and Technology, Kunming 650500, China; xhtkmust@kust.edu.cn
3 Collaborative Innovation Center for Integration of Terrestrial & Marine Economies, Guangxi University of Finance and Economics, Nanning 530003, China
* Correspondence: 2020210002@gxufe.edu.cn; Tel.: +86-137-2386-7575

Abstract: The high risk of financing, building, and operating Public–private partnerships (PPPs) often results from the event that participants can barely obtain expected economic returns, thus inhibiting private enterprises' willingness to participate in PPPs. To increase private enterprises' desire to participate, this study constructed an evolutionary game model of private enterprises' participation in PPPs, focusing on the perspective of the mode of incentive. This model revealed the evolutionary law of private enterprises' participation behavior under different modes of incentive. The results indicate that: First, there is a positive correlation between the intensity of government incentive, the investment return rate, and the probability of private enterprises choosing to participate in PPPs. Specifically, the impact of the investment return rate is more sensitive than the other factors. Second, the cost rate of financing and the risk cost of project uncertainty are negatively correlated with the probability of private enterprises choosing to participate in PPPs, and the impact of the project risk cost is more sensitive than the other factors in this case.

Keywords: public–private partnership; private enterprises; participation behavior; evolutionary game; project governance

Citation: Zhang, Y.; Yi, H.; Xie, H.; Zheng, J.; Wang, Y. How Private Enterprises' Participation Behaviors Evolve with Incentive Modes in PPPs: An Evolutionary Game View. *Buildings* **2022**, *12*, 709. https://doi.org/10.3390/buildings12060709

Academic Editors: Yongjian Ke, Jingxiao Zhang and Simon P. Philbin

Received: 23 April 2022
Accepted: 23 May 2022
Published: 24 May 2022

1. Introduction

Public–private partnerships (PPPs) are a form of cooperation in which public organizations and private organizations combine to provide public goods or services [1]. Compared with traditional procurement, PPPs can allow the public and private sectors' complementary advantages to complement each other, thereby improving infrastructure efficiency and quality [2]. Since 1982 when Britain first adopted PPPs in public infrastructure, PPPs have been developed and applied on a large scale all over the world [3,4]. In 2014, China experienced the beginning of a new round of PPPs. The State Council, the National Development and Reform Commission, and the Ministry of Finance successively issued more than 100 relevant policy documents to encourage and guide private enterprises in participating in PPPs. Since the implementation of the Belt and Road Initiative, the promotion of the Rural Revitalization Strategy, and the acceleration of the construction of "new infrastructure and urbanization initiatives", China's public infrastructure projects operating through PPPs are showing unprecedented levels of construction, and private enterprises are embracing the strong development opportunities [5]. However, after more than seven years of implementation, according to data released by the Ministry of Finance, the participation rate of private enterprises is not high [6], and the phenomenon of "the state-owned enterprises enter into, but the private enterprises exit" can be seen [7]. As of 20 May 2022, the cumulative investment of PPP projects nationwide was 21.1 trillion yuan, and the number of transactions was 12,853. Of these figures, investments by private enterprises accounted for 15.92% of the total, and the number of private enterprises accounted for 32.50% of the total. As it is well-known that the capital needs of PPPs are gigantic, since state-owned enterprises have ample funds and a strong ability to take risks when it comes to cash flow streams, they are more suitable to participating in PPPs. Currently, state-owned enterprises are indeed involved in 70% of PPPs. However, if the number of state-owned enterprises participating in PPPs is too large, this does not help to share the financial pressure of the government. Worse, in Chinese PPPs, state-owned businesses frequently use the benefits of control and relational relationships based on mutual confidence and shared values to implement tunneling behavior, which harms the projects' overall interests and is even detrimental to the long-term development of PPP schemes [8]. Therefore, improving the participation rate of private enterprises is an essential prerequisite for the effective implementation of PPPs.

Scholars who have examined PPPs have found that the low participation rate of private enterprises in PPPs is caused by external environmental risk factors such as the economy, the system, the society, and the project itself [9]. Osei-Kyei and Chan also found that the three critical factors in attracting private companies to participate in PPPs are government support and recognition, positive government attitudes, and political stability [10]. In addition, other critical factors or barriers inhibiting the involvement of the participation of private sectors were also explored, such as project changes [11], government attention and intervention [6], and delivery modes [12]. In fact, the balance of risk and return is among the most important factor for private enterprises participating in PPPs [13]. Generally, governments procure infrastructure PPPs through tendering processes and select potential participants. The successful bidder then creates a special purpose vehicle (SPV), which is in charge of developing the project [13,14]. According to Dixon et al. [15], there is a direct relationship between PPP project risks assumed by the SPV and the cost of finance. From a private sector perspective, a PPP project is an investment opportunity with a long-term contract and some related risks [14]. When the project is granted, a comparison can be drawn between the project's internal rate of return and the price the public sector pays for the risks transferred to the private sector. In the survey conducted by Demirag et al. [16], three-quarters of respondents to the specific question of target returns reported rates between 9 and 15%. However, in China, the average rate of return is only 6.5% in PPPs [17]. There has been relatively little research into the relationship between the expected profit and the participation behavior of private enterprises during the decision stage of PPPs. More importantly, the existing literature have rarely been concerned with the interaction between incentive modes of the public and strategic behaviors of private enterprises. Specifically, private sectors care about the expected returns under specific incentive or compensation modes. For instance, they may compare target returns with potential risks based on the incentive strategies of public organizations. If the expected return is higher than the risk reward, they would then participate in the PPP project. If the expected return is lower than the risk reward, they would not participate in the PPP project. When taking risk management into consideration, the returns are directly linked to incentive patterns. In this process, since the information about returns and risks is incomplete and strategic interaction between the public and the private is dynamic, the process of strategy selection can be perceived as an evolutionary game.

To increase the willingness of private enterprises to participate in the decision stage of PPP projects, this study aimed to quantify the returns under different incentive modes, taking the return as the criterion for whether private enterprises would participate in PPPs. Furthermore, it analyzed how the behaviors of both parties evolved dynamically with the differences. The remainder of this article is organized as follows. First, a comprehensive literature review concerning cooperation and decision-making in PPPs is presented, along with an account of evolutionary game theory and the participation behavior of private enterprises. Next, an evolutionary game model and the related model solution are constructed and conducted. The dynamic evolution of equilibrium positions and the probability of participation through government incentives are then discussed. Subsequently, the factors and their sensitivities are explained through a numerical example. Finally, the article concludes by stating the implications of the research for private enterprises' participation in PPPs.

2. Theoretical Background

2.1. Cooperation and Participation in PPPs

Infrastructure projects can provide a range of societal, environmental, and economic benefits for many entities. In China, PPPs are common ways of handling such projects for two reasons. First, limitations to public funding have led governments to encourage the private sector to engage in various long-term arrangements for capital-intensive projects. Second, a comprehensive approach for the whole area may be more efficient and profitable than piecemeal development via individual owners' interventions [18]. Thus, a key characteristic of PPPs is that the three tasks of financing, building a facility, and subsequently operating it are bundled and delegated to a group of private partners. Such a group associated with a bidding consortium is selected as preferred participants through tendering processes [14]. The group of private partners then create the SPV and are the main equity investors that manage and develop the PPP project [13], and the main investors assume the risks of financing, building, and operating an infrastructure asset in exchange for suitable economic returns [19]. However, due to the uncertain, asset-specific, and complex issues associated with PPPs, private sectors grouped in SPVs are also vulnerable to multiple long-term returns and risks challenges [20]. Thus, diverse strategies, including whether to adopt a specific PPP or the traditional contracting approach for a specific project [21], how the decision of bundling the financing, building, and operating stages affects the incentives in PPP [22], and how to allocate risk among participants involved in PPP projects [23,24], might result in the distinct economic returns of the participants. Additionally, the risk preference, profit sharing, government attention, and delivery modes would affect the involvement of the private sectors. Therefore, the cooperation of the public and private sectors requires them to negotiate conflicting goals, and each participant in the system would adopt strategic interactions due to differences in their excepted returns.

2.2. Evolutionary Game Theory

In general, game theory is used to analyze strategy selection in terms of expected payoffs [25]. From the point of view of economic optimization, this theory provides a mathematical framework to model strategic behavior and demonstrate resulting decisions [26]. Therefore, game theory is the most effective theory of decision-making. It is an economic tool for studying cooperation, conflict, and strategic interaction in the decision-making process [27].

Furthermore, evolutionary game theory (EGT) explains the process by which entities learn, compete, and adapt to each other in biological evolution [28]. Unlike traditional game theory that focuses on perfect rationality, in EGT, participants are assumed to possess bounded rationality and be under the conditions of imperfect information. The theory emphasizes the dynamic process of changing strategies, rather than static equilibrium. The EGT can fully reflect the relationship between strategy change and payoff fluctuation, and this advantage has led the EGT to be applied in different research fields [29]. It has penetrated many areas of research that deal with social and economic systems [30]. The potential is great in economic applications, especially [31].

Economic conflicts and strategic interactions are common in terms of PPP applications, and private enterprises are not always entirely rational. Simultaneously, they attempt to maximize expected payoffs, but information and risks are not complete. This is especially true in complicated market competition, in which decision-makers seek evolutionarily stable strategies (ESS) through constant trial and error, imitation, and learning [32,33]. Therefore, the EGT is suitable for analyzing the cooperation strategies of the leading participants in PPPs. In the next section, we describe the model and its assumptions.

3. Modeling

3.1. Description of the Model

PPPs are a way of financing projects and an innovative way to marketize the supply of public services. PPPs allow all participants to cooperate to form a complementary decision-making mechanism. They break the double puzzles of government failure and market failure in public projects. They promote reforms to the supply side of the social structure in the field of public services. Participants in PPPs have different goals, especially given that those participants can be government departments and private enterprises. They must constantly interact and play games when they participate in PPPs. To maximize their expected payoffs, they may choose cooperation or non-participation strategies, and their behavioral choices will show bounded rationality under uncertain situations. The evolutionary game model can be used to analyze the decision-making mechanisms of governments and private enterprises as they cooperate in PPPs.

3.2. Assumptions and Parameters

The assumptions are hypothesized as follows: Assumption 1. The game environment for public organizations and private enterprises is one of incomplete information. Both parties are bounded rational decision-makers, and each party independently and dynamically chooses their behavior strategy. Both parties make decisions or take actions to maximize their profits based on past experiences or practices [31].

Assumption 2. The set of strategies of the private enterprises is (E_1, E_2) = (Participation, Non-participation), in which the probability of adopting the "participation" strategy is x $(0 \leq x \leq 1)$ and the probability of adopting the "non-participation" strategy is $1 - x$. The set of strategies of the public organizations is (P_1, P_2) = (Incentive, Neutrality), in which the probability of adopting the "incentive" strategy is y $(0 \leq y \leq 1)$, and the probability of adopting the "neutrality" strategy is $1 - y$.

Assumption 3. To encourage private enterprises to participate in PPPs, when a public organization chooses the "incentive" strategy, it will give private enterprises preferential policies such as tax exemptions and reduced interest. If the incentive cost for the public organization is C_I, then the economic incentive for the private company is λC_I, which is the incentive intensity. At the same time, let the public benefit of the PPP be R; that is, under the encouragement of the public sector, private enterprises can provide better public services, and the public benefit at this time is R_H. If the public sector maintains a neutral attitude towards private enterprises participating in PPPs or has potential thresholds such as "glass doors" or "revolving doors", which increase the institutional transaction costs of a private enterprise participating in a PPP, the public benefit is then R_L, $R_L < R_H$, and the public sector will suffer certain losses of reputation.

Assumption 4. Under a given contract, it is reasonable to anticipate that the private sector will adopt a cost-cutting strategy while retaining an acceptable level of participation to avoid high-risk damage [34]. It is assumed that the total investment of private enterprises in PPP projects is V_E, the investment return rate is r_1, the financing cost rate is r_2, the project risk cost is C_R, and the basic income of private enterprises is $(r_1 - r_2)V_E - C_R$. If the public sector chooses the "incentive" strategy, private enterprises will increase their income λC_I. Currently, the private sector's net income is $(r_1 - r_2)V_E - C_R + \lambda C_I$, and the government's net income is $R_H - C_I$. If the public sector chooses the strategy of "neutrality", the institutional transaction costs of private sectors, C_C, will increase. Currently, the net income of private enterprises is $(r_1 - r_2)V_E - C_R - C_C$, and the government's net income is $R_L - S$.

3.3. Basic Model

The payoff matrix can be obtained based on the preceding assumptions and analyses, as shown in Table 1.

Table 1. Payoff matrix between the public and private enterprises.

Game Players	Strategy	Public Sector (P)	
		Incentive y	Neutrality $1-y$
Private enterprise (E)	Participation x	$(r_1-r_2)V_E-C_R+\lambda C_I$; $R_H-V_P-C_I$	$(r_1-r_2)V_E-C_R-C_C$; R_L-V_P-S
	Non-participation $1-x$	$0;-C_I$	$0;0$

1. The expected payoffs of private enterprise (E)

   The payoffs of private enterprise adopting the strategy of "active participation" are:

$$U_{E1}=y[(r_1-r_2)V_E-C_R+\lambda C_I]+(1-y)[(r_1-r_2)V_E-C_R-C_C] \quad (1)$$

The payoffs of private enterprise adopting the "non-participation" strategy are:

$$U_{E2}=0 \quad (2)$$

The expected payoffs of private enterprise (E) adopting a mixed strategy are:

$$\overline{U}_E=xU_{E1}+(1-x)U_{E2} \quad (3)$$

According to the Malthusian dynamic equation principle, the speed of the dynamic change of strategy of a private enterprise can be expressed by a dynamic differential equation, and the replication dynamic equation of the private enterprise can be obtained as follows:

$$\begin{aligned} f_E(x)=\tfrac{dx}{dt} &= x\left(U_{E1}-\overline{U}_E\right)=x(1-x)(U_{E1}-U_{E2}) \\ &= x(1-x)[(\lambda C_I+C_C)y+(r_1-r_2)V_E-C_R-C_C] \end{aligned} \quad (4)$$

2. The expected payoffs for the public organization (P)

   The payoffs of the public organization adopting the "incentive" strategy are:

$$U_{P1}=x[R_H-V_P-C_I]+(1-x)(-C_I) \quad (5)$$

The payoffs of the public organization adopting the "neutral" strategy are:

$$U_{P2}=x(R_L-V_P-S) \quad (6)$$

The expected payoffs of the public organization adopting a mixed strategy are:

$$\overline{U}_P=yU_{P1}+(1-y)U_{P2} \quad (7)$$

The replication dynamic equation of the public (P) is:

$$f_P(y)=y\left(U_{P1}-\overline{U}_P\right)=y(1-y)(U_{P1}-U_{P2})=y(1-y)[(R_H-R_L+S)x-C_I] \quad (8)$$

3.4. Model Solution: Replicator Dynamics

To sum up, a two-dimensional dynamic system can be obtained. By solving two duplicate dynamic equations $\frac{dx}{dt}=0$ and $\frac{dy}{dt}=0$ simultaneously, we learn that (0,0), (0,1), (1,0), (1,1), and $(x*,y*)$ are the five local equilibrium points of this two-dimensional dynamic system, among which:

$$x*=\frac{C_I}{R_H-R_L+S} \quad (9)$$

$$y* = \frac{(r_2 - r_1)V_E + C_R + C_C}{\lambda C_I + C_C} \tag{10}$$

4. Model Analysis

4.1. Dynamic Evolution Analysis of Equilibrium Points

According to the local stability analysis method proposed by Freedman [35], the stability of equilibrium points for an evolutionary system can be calculated using a Jacobian matrix. The specific method of judging is as follows: if the determinant of the Jacobian matrix of the equilibrium point is $Det(J) > 0$ and the trace is $Tr(J) < 0$, it can be judged that the corresponding equilibrium point has asymptotic stability. This is called the ESS point. If $Det(J) > 0$ and $Tr(J) > 0$, it can be assumed that the related equilibrium point is unstable. If $Det(J) < 0$, it can be judged that the corresponding equilibrium point is a saddle point. Therefore, according to the replicated dynamic Equations (4) and (8), the Jacobian matrix of participatory decision-making in PPPs is shown by Equation (11):

$$J(x,y) = \begin{pmatrix} \frac{\partial f_E(x)}{\partial x} & \frac{\partial f_E(x)}{\partial y} \\ \frac{\partial f_P(y)}{\partial x} & \frac{\partial f_P(y)}{\partial y} \end{pmatrix} = \begin{pmatrix} v_1 & v_2 \\ v_3 & v_4 \end{pmatrix} \tag{11}$$

where $v_1 = (1-2x)[(\lambda C_I + C_C)y + (r_1 - r_2)V_E - C_R - C_C]$; $v_2 = x(1-x)[\lambda C_I + C_C]$; $v_3 = y(1-y)[R_H - R_L + S]$; $v_4 = (1-2y)[(R_H - R_L + S)x - C_I]$.

As for Equation (11), its determinant equation is $Det(J) = v_1 v_4 - v_2 v_3$, and the trace of this Jacobian matrix is $Tr(J) = v_1 + v_4$. The determinants and traces of the four equilibrium points (0,0), (0,1), (1,0), and (1,1) are shown in Table 2.

Table 2. Determinants and traces of Jacobian matrix at the equilibrium points of participation behavior decision in PPPs.

Equilibrium Points	*Det*(*J*)	*Tr*(*J*)
$(0,0)$	$[(r_1 - r_2)V_E - C_R - C_C] \times (-C_I)$	$[(r_1 - r_2)V_E - C_R - C_C] + (-C_I)$
$(0,1)$	$[(r_1 - r_2)V_E + \lambda C_I] \times C_I$	$[(r_1 - r_2)V_E + \lambda C_I] + C_I$
$(1,0)$	$-[(r_1 - r_2)V_E - C_R - C_C] \times [R_H - R_L + S - C_I]$	$-[(r_1 - r_2)V_E - C_R - C_C] + [R_H - R_L + S - C_I]$
$(1,1)$	$[\lambda C_I + (r_1 - r_2)V_E] \times [R_H - R_L + S - C_I]$	$-[\lambda C_I + (r_1 - r_2)V_E] - [R_H - R_L + S - C_I]$

Through an analysis of the equilibrium points of private enterprises' participation behavior, the local stability of the equilibrium point can be obtained, as shown in Table 3. The local stability of the equilibrium point is closely related to the initial state of each parameter. In other words, the strategy of private enterprises' participation changes based on changes in the expected income and cost.

Table 3. Local stability analysis of equilibrium points of participation behavior.

Equilibrium Points	Condition	Stability
$(0,0)$	$r_1 V_E < r_2 V_E + C_R + C_C$; $r_1 V_E > r_2 V_E + C_R + C_C$	ESS; Saddle point
$(0,1)$	$r_1 V_E < r_2 V_E + \lambda C_I$; $r_1 V_E > r_2 V_E + \lambda C_I$	Saddle point; Unstable point
$(1,0)$	$r_1 V_E > r_2 V_E + C_R + C_C$ and $R_H + S < R_L + C_I$; $r_1 V_E > r_2 V_E + C_R + C_C$ and $R_H + S > R_L + C_I$	ESS; Saddle point
$(1,1)$	$r_2 V_E < \lambda C_I + r_1 V_E$ and $R_H + S > R_L + C_I$; $r_2 V_E > \lambda C_I + r_1 V_E$ and $R_H + S < R_L + C_I$	ESS; Unstable point

4.2. Discussion of Influencing Factors

When $r_1V_E < r_2V_E + C_R + C_C$ and $R_H + S > R_L + C_I$, the evolutionary system has five equilibrium points, which are (0,0), (0,1), (1,0), (1,1), and $(x*, y*)$. The group replication dynamics between the public and private enterprises are shown in Figure 1, and there are two stable points: A_1 (0,0) and A_4 (1,1). The broken line $A_2A_5A_3$ in Figure 1 divides the plane area $S_{A_1A_2A_4A_3}$ into two parts: the cooperation area $A_2A_4A_3A_5$ (Participation, Incentive) and the non-cooperation area (Non-participation, Neutrality). If the initial state of the public organization P and the private enterprise E falls in the cooperation area $A_2A_4A_3A_5$, the strategic choice of both parties will eventually converge to A_4 (1,1). In contrast, if the initial state of the public organization P and the private enterprise E falls in the non-cooperative zone $A_1A_2A_5A_3$, both parties' strategies will then eventually evolve to A_1 (0,0). The parameters of the initial state directly affect the final convergence results of both strategies. Each influence factor will be analyzed in detail below.

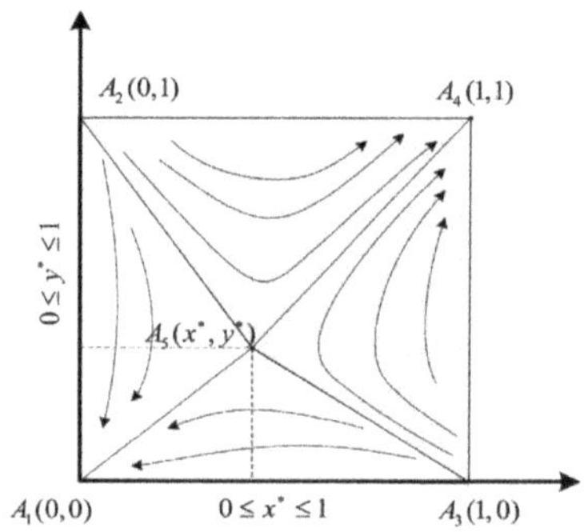

Figure 1. Dynamic of group replication between the public and enterprises.

1. The impact of the incentive intensity of the public sector λ

From the geometric meaning of probability in Figure 1, it can be seen that the probability that the initial state of the private enterprise E and the public organization P after negotiation falls within the (Participation, Incentive) area $A_2A_4A_3A_5$ is equal to its area $S_{A_2A_4A_3A_5}$ (shaded part in Figure 1), namely:

$$P = S_{A_2A_4A_3A_5} \quad \begin{aligned} &= 1 - S_{A_1A_2A_5A_3} = 1 - \left(\tfrac{x*}{2} + \tfrac{y*}{2}\right) \\ &= 1 - \tfrac{C_I}{2(R_H - R_L + S)} - \tfrac{(r_2 - r_1)V_E + C_R + C_C}{2(\lambda C_I + C_C)} \end{aligned} \tag{12}$$

For the $S_{A_2A_4A_3A_5}$ first-order derivative λ, there are:

$$\frac{dS_{A_2A_4A_3A_5}}{d\lambda} = \frac{(r_2 - r_1)V_E + C_R + C_C}{2\lambda^2 C_I} \Rightarrow \begin{cases} > 0, r_1V_E < r_2V_E + C_R + C_C \\ < 0, r_1V_E > r_2V_E + C_R + C_C \end{cases} \tag{13}$$

It can be concluded that the correlation between the incentive intensity λ (through tax reduction or limits on interest) and the region $S_{A_2A_4A_3A_5}$ is related to whether the expected income of private enterprises can cover the expected cost. This is specifically true when other parameter scenarios are fixed, for example, when $r_1V_E < r_2V_E + C_R + C_C$ (the income of private enterprises does not cover the cost), λ is larger, and the probability that the initial state falls within the region $A_2A_4A_3A_5$ is greater. This means that the probability is greater that the public organization P adopts the "incentive" strategy, and the private enterprise E adopts the "participation" strategy. Additionally, when $r_1V_E > r_2V_E + C_R + C_C$ (the income of private enterprises is greater than the cost), λ is larger, and the probability of the initial state falling within the region $A_2A_4A_3A_5$ is smaller. This means that the probability is smaller that the public organization P adopts the "incentive" strategy, and the private enterprise E adopts the "active participation" strategy.

2. The impact of the investment return rate of a private enterprise r_1

For the $S_{A_2A_4A_3A_5}$ first-order derivative r_1, there are:

$$\frac{dS_{A_2A_4A_3A_5}}{dr_1} = \frac{V_E}{2(\lambda C_I + C_C)} > 0 \tag{14}$$

It can be concluded that the investment return rate r_1 is positively correlated with the region $S_{A_2A_4A_3A_5}$. When other parameter scenarios are fixed and r_1 is large, a public organization is more likely to adopt the "incentive" strategy. At the same time, the probability of a private enterprise adopting participation behavior increases.

3. The impact of the financing cost of a private enterprise r_2

For the $S_{A_2A_4A_3A_5}$ first-order derivative r_2, there are:

$$\frac{dS_{A_2A_4A_3A_5}}{dr_2} = \frac{-V_E}{2(\lambda C_I + C_C)} < 0 \tag{15}$$

Therefore, the financing cost of private enterprises r_2 is negatively correlated with the region $S_{A_2A_4A_3A_5}$. When other parameter scenarios are fixed and r_2 is large, a public organization is more inclined to adopt the "incentive" strategy. At the same time, the probability of a private enterprise adopting participation behavior decreases.

4. The impact of the cost of risk for the project C_R

For the $S_{A_2A_4A_3A_5}$ first-order derivative C_R, there are:

$$\frac{dS_{A_2A_4A_3A_5}}{dC_R} = \frac{-1}{2(\lambda C_I + C_C)} < 0 \tag{16}$$

It can be concluded that the risk cost C_R is negatively correlated with the area $S_{A_2A_4A_3A_5}$. When other parameter scenarios are fixed and C_R is large, a public organization is more inclined to adopt the "incentive" strategy. At the same time, the probability of a private enterprise adopting participation behavior decreases.

4.3. Sensitivity Analysis of Influencing Factors

There are many parameters involved in the above model. To better reflect the influencing factors and the effects of private enterprises' participation in PPPs, this study took an infrastructure PPP project in a high-tech part of a city as an example. It analyzed the influence of the mode of incentive, the cooperation costs, and the risk costs. There were three parts to the PPP: municipal road construction, landscape engineering, and an underground utility tunnel. Private enterprises invested capital $V_E = 100$ million; the investment return rate was $r_1 = 0.6$; the financing cost rate was $r_2 = 0.3$; the risk cost was $C_R = 30$ million; the institutional transaction cost was $C_C = 50$ million; the benefits for the public were $R_H = 60$ million and $R_L = 40$ million; the incentive intensity was $\lambda = 1.0$; the incentive cost was $C_I = 20$ million; and the public loss was $S = 20$ million. In this scenario, the initial state of the private enterprises participating in the PPP is shown in Figure 2a. In this scenario, it was assumed that the probability of each strategy chosen by the public and private enterprises was 0.5. The sensitivity of each influencing factor is discussed in turn.

1. The sensitivity of the incentive intensity λ

From the scenario in Figure 2a $r_1V_E < r_2V_E + C_R + C_C$, the incentive intensity was set to $\lambda = 0.3$, $\lambda = 1.5$, and $\lambda = 2.5$, and the other parameters remained unchanged. The dynamic evolution of private enterprises choosing the strategy of "Participation" is shown in Figure 2b. When $r_1V_E < r_2V_E + C_R + C_C$, the incentive intensity was greater, and the private enterprises were more likely to select the strategy of "Participation" more quickly

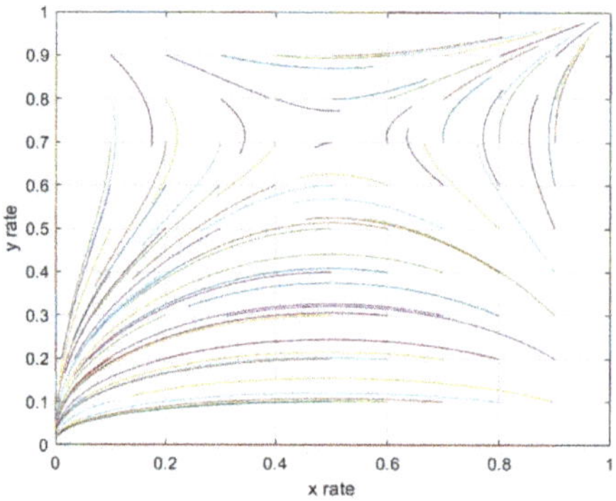

(a) Initial state

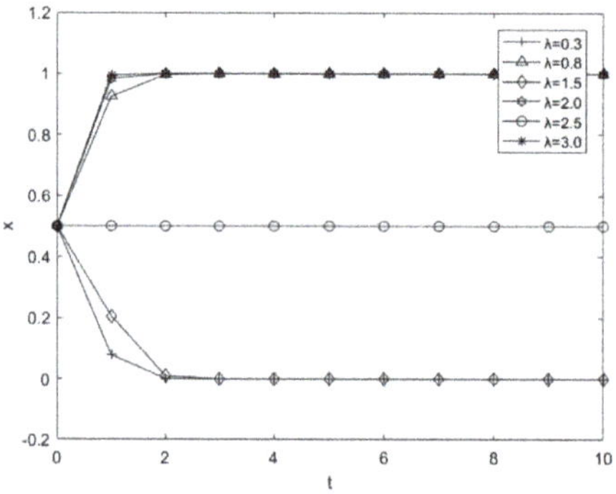

(b) Sensitivity of incentive intensity

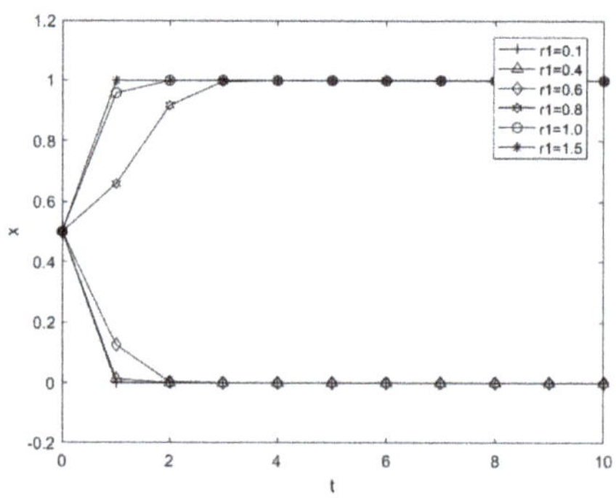

(c) Sensitivity of return rate

Figure 2. Sensitivity of key factors of private enterprises' participation behavior.

2. The sensitivity of the investment return rate r_1

From the scenario in Figure 2a, the investment return rates of private enterprises were set as $r_1 = 0.1, 0.4, 0.6, 0.8, 1.0, 1.5$. The other parameters were unchanged. The dynamic evolution of private enterprises choosing the strategy of "participation" is shown in Figure 2c. It can be seen from Figure 2c that the greater the investment return rate of private enterprises, the greater the probability that private enterprises would choose the strategy of "participation", and the faster the evolution would be. This result further proves that there is a positive correlation between investment return rate and private enterprises' adoption of the "participation" strategy.

3. The sensitivity of the financing cost rate r_2

From the scenario in Figure 2a, the financing cost rates of private enterprises were set as $r_2 = 0.1, 0.3, 0.5, 0.7, 0.9, 1.5$. The other parameters were unchanged. The dynamic evolution of private enterprises choosing to participate is shown in Figure 3a. The higher the financing cost ratio, the lower the probability that private enterprises would choose "participation". The final result of the evolution was that groups chose not to participate in PPPs. This further proves that there is a negative correlation between the financing cost rate and private enterprises' "participation" strategy choices.

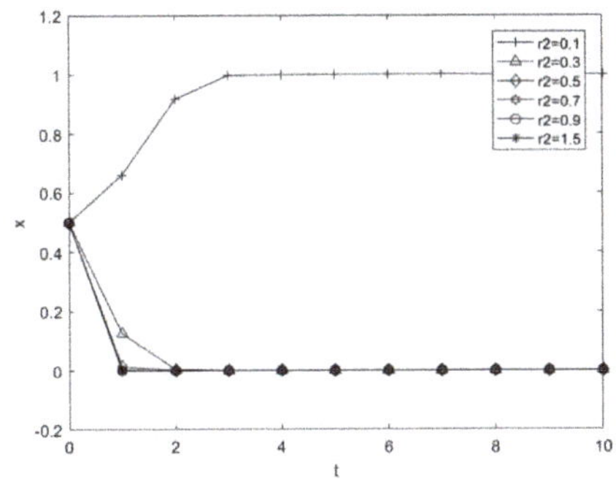

(**a**) Sensitivity of financing cost rate

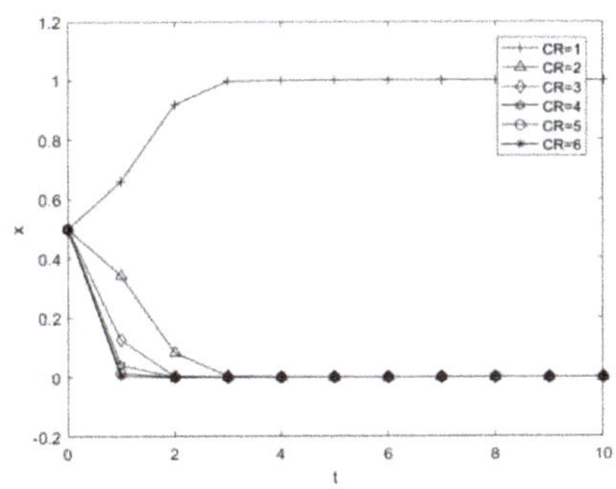

(**b**) Sensitivity of risk cost

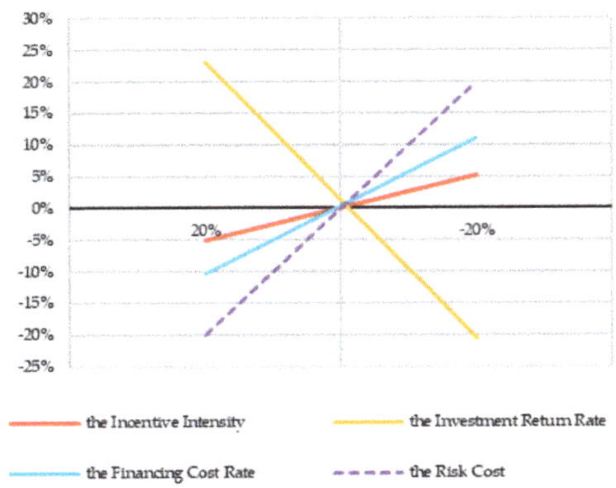

(**c**) Sensitivity comparison of key factors

Figure 3. Sensitivity of key factors of participation behavior.

4. The sensitivity of the risk cost C_R

From the scenario in Figure 2a, the risk costs were set at $C_R = 1, 2, 3, 4, 5$, and 6. The other parameters were unchanged. The dynamic evolution of private enterprises choosing to participate is shown in Figure 3b. This shows that the greater the risk cost, the less likely private enterprises were to choose the "participation" strategy. With the continuous increase C_R, the evolution accelerated and finally converged on groups choosing not to participate in PPPs. The result further proves that the risk cost of a project is negatively correlated with a private enterprise's decision to participate in PPPs.

5. A comparative analysis of the sensitivity of key factors

From the sensitivity chart of each influencing factor (see Figure 3c), it can be seen that although the incentive intensity, the investment return rate of private enterprises, and the participation probability of private enterprises are all positively correlated, the influence effect of investment return rate of private enterprises is the most sensitive. In contrast, the financing cost rate, the risk cost, and the participation probability of private enterprises are negatively correlated, and the effect of the risk cost is the most sensitive.

5. Conclusions and Implications

To encourage private enterprises to actively participate in PPPs, this study constructed a decision-making model of participation behavior for private enterprises in PPPs based on evolutionary game theory. It analyzed the influencing factors and their sensitivity, and it verified the relevant conclusions regarding the theoretical model through simulation. Through the above discussion and analysis, it can be concluded that: First, the incentive intensity and investment return rate are positively correlated with the probability of private enterprises choosing the "participation" strategy, and the influencing effect of the investment return rate of private enterprises is the most sensitive. Second, the financing cost rate, the risk cost, and the probability of private enterprises selecting the "participation" strategy are negatively correlated, and the influencing effect of the risk cost is the most sensitive.

The implications for management that can be taken from the above conclusions is the following:

(1) Government guarantees. Government guarantees involve government incentives, perfecting laws and regulations, standardizing supervision and support mechanisms, and transforming the role of the government. The implementation of PPPs in China is still in the early stages. To allow PPPs to be as effective as possible, it is necessary to improve laws and regulations, establish a superior supervision mechanism, and clarify the scope of the powers, responsibilities, and interests of the government departments and agencies responsible for implementing PPPs.

(2) Ensure a strong rate of return on investment and reduce the financing cost. Private enterprises aim to make profits, and they will only participate in PPPs if they can get a return on their investment. At the same time, the scale of investment in PPPs is large, and there are problems with financing difficulties and expensive financing that make it difficult for private enterprises to participate in PPPs. Compared with state-owned enterprises, the financial strength of private enterprises is relatively weak. The high cost of financing directly dampens the enthusiasm of private enterprises to participate in PPPs. On the one hand, public bodies can increase the profit margins of private enterprises through tax reductions, interest bonuses, financial subsidies, and bundled development. On the other hand, it is necessary to improve the financial market, provide diversified financing tools, and develop low-cost financing channels so that private enterprises will participate in PPPs.

(3) Clarity and controllability of projects are key to attracting private enterprises. The risk of uncertainty is the biggest obstacle for private enterprises when it comes to participating in PPPs. For large-scale, uncertain projects, private enterprises should establish a consortium or cooperative alliance to limit the risks. This can create synergy and allow the companies to provide mutual assistance effectively while still obtaining

the expected benefits. This can help them to increase their market share, leading to a win-win situation.

Author Contributions: Conceptualization and methodology, Y.Z.; formal analysis, H.Y.; writing—original draft preparation, Y.Z. and H.Y.; writing—review and editing, H.X. and J.Z.; supervision and funding acquisition, Y.W. All authors have read and agreed to the published version of the manuscript.

Funding: This research was funded by the Applied Basic Project of Yunnan Province (Grant Nos. 202101AT070088 and 2019FB084), the Youth Program of Humanities and Social Sciences, Ministry of Education (Grant No. 19YJCZH260), the National Natural Science Foundation of China (Grant Nos. 71761021 and 71701083), and Guangxi Philosophy and Social Science Planning Project (Grant No. 20FJL001).

Institutional Review Board Statement: Not applicable.

Informed Consent Statement: Not applicable.

Data Availability Statement: The data presented in this study are available on request from the corresponding author.

Conflicts of Interest: The authors declare no conflict of interest.

References

1. Solheim-Kile, E.; Lædre, O.; Lohne, J. Public-private partnerships: Agency costs in the privatization of social infrastructure financing. *Proj. Manag. J.* **2019**, *50*, 144–160. [CrossRef]
2. Wang, H.; Xiong, W.; Wu, G.; Zhu, D. Public-private partnership in public administration discipline: A literature review. *Public Manag. Rev.* **2017**, *20*, 293–316. [CrossRef]
3. Acerete, J.B.; Shaoul, J.; Stafford, A.; Stapleton, P. The cost of using private finance for roads in Spain and the UK. *Aust. J. Public Adm.* **2010**, *69*, S48–S60. [CrossRef]
4. Hwang, B.G.; Zhao, X.; Gay, M.J.S. Public private partnership projects in Singapore: Factors, critical risks and preferred risk allocation from the perspective of contractors. *Int. J. Proj. Manag.* **2013**, *31*, 424–433. [CrossRef]
5. Zhang, S.; Gao, Y.; Feng, Z.; Sun, W. PPP application in infrastructure development in China: Institutional analysis and implications. *Int. J. Proj. Manag.* **2015**, *33*, 497–509. [CrossRef]
6. Wang, L.; Yan, D.; Xiong, Y.; Zhou, L. A review of the challenges and application of public-private partnership model in Chinese garbage disposal industry. *J. Clean. Prod.* **2019**, *230*, 219–229. [CrossRef]
7. Zeng, L.; Luo, S. Research on the participation of private enterprises and its influencing factors in Chinese PPP Practice—Empirical analysis based on 731 county samples. *Soft Sci.* **2020**, *34*, 33–38. (In Chinese)
8. Gao, R.; Hu, J.; Liu, J. Tunnelling analysis of state-owned enterprises in PPP projects. *Oper. Res. Manag. Sci.* **2021**, 226–231. (In Chinese)
9. Tan, J.; Zhao, J.Z. Explaining the adoption rate of public-private partnerships in Chinese provinces: A transaction cost perspective. *Public Manag. Rev.* **2019**, *23*, 590–609. [CrossRef]
10. Osei-Kyei, R.; Chan, A.P.C. Factors attracting private sector investments in public-private partnerships in developing countries: A survey of international experts. *J. Financ. Manag. Prop. Constr.* **2017**, *22*, 92–111. [CrossRef]
11. Chan, A.P.C.; Lam, P.T.I.; Wen, Y.; Ameyaw, E.E. Cros-sectional analysis of critical risk factors of PPP water projects in China. *J. Infrastruct. Syst.* **2015**, *21*, 04014031. [CrossRef]
12. Xiong, W.; Chen, B.; Wang, H.; Zhu, D. Public-private partnerships as a governance response to sustainable urbanization: Lessons from China. *Habitat Int.* **2020**, *95*, 102095. [CrossRef]
13. Carrillo de Albornoz, V.A.; Molina Millán, J.; Sánchez Soliño, A. Managing a portfolio of public–private partnerships: Concessionaire perspective. *J. Manag. Eng.* **2018**, *34*, 04018044. [CrossRef]
14. Guevara, J.; Herrera, L.; Salazar, J. Interorganizational sponsor networks in road and social infrastructure PPP equity markets. *J. Constr. Eng. Manag.* **2022**, *148*, 04022009. [CrossRef]
15. Dixon, T.; Pottinger, G.; Jordan, A. Lessons from the private finance initiative in the UK: Benefits, problems and critical success factors. *J. Prop. Invest. Financ.* **2005**, *23*, 412–423. [CrossRef]
16. Demirag, I.; Iqbal, S.; Stapleton, P.; Stevenson, C. *Public Private Partnership Financiers' Perceptions of Risks*; The Institue of Chartered Accountants of Scotland: Edinburgh, UK, 2010.
17. Bridata. ROI Trend for the Closed PPP Projects. Available online: http://www.bridata.com/project?type=tzhb&in_cpppc=1. (accessed on 20 May 2022).
18. Glumac, B.; Han, Q.; Schaefer, W.; van der Krabben, E. Negotiation issues in forming public-private partnerships for Brownfield redevelopment: Applying a game theoretical experiment. *Land Use Policy* **2015**, *47*, 66–77. [CrossRef]

19. Nguyen, D.A.; Garvin, M.J.; Gonzalez, E.E. Risk allocation in U.S. public-private partnership highway project contracts. *J. Constr. Eng. Manag.* **2018**, *144*, 04018017. [CrossRef]
20. Burke, R.; Demirag, I. Risk management by SPV partners in toll road public private partnerships. *Public Manag. Rev.* **2018**, *21*, 711–731. [CrossRef]
21. Anastasopoulos, P.C.; Islam, M.B.; Volovski, M.; Powell, J.; Labi, S. Comparative evaluation of public-private partnerships in roadway preservation. *Transp. Res. Rec. J. Transp. Res. Board* **2011**, *2235*, 9–19. [CrossRef]
22. Hoppe, E.I.; Schmitz, P.W. Public-private partnerships versus traditional procurement: Innovation incentives and information gathering. *RAND J. Econ.* **2013**, *44*, 56–74. [CrossRef]
23. Ke, Y.; Wang, S.; Chan, A.P.C.; Lam, P.T.I. Preferred risk allocation in China's Public-Private Partnership (PPP) projects. *Int. J. Proj. Manag.* **2010**, *28*, 482–492. [CrossRef]
24. Shrestha, A.; Chan, T.-K.; Aibinu, A.A.; Chen, C.; Martek, I. Risk allocation inefficiencies in Chinese PPP water projects. *J. Constr. Eng. Manag.* **2018**, *144*, 04018013. [CrossRef]
25. Nash, J. The bargaining problem. *Econometrica* **1950**, *18*, 155–162. [CrossRef]
26. Gao, R.; Liu, J. Selection of government supervision mode of PPP Projects during the operation stage. *Constr. Manag. Econ.* **2019**, *37*, 584–603. [CrossRef]
27. Rosas, A. Evolutionary game theory meets social science: Is there a unifying rule for human cooperation? *J. Theor. Biol.* **2010**, *264*, 450–456. [CrossRef]
28. Taylor, P.D.; Jonker, L.B. Evolutionary stable strategies and game dynamics. *Math. Biosci.* **1978**, *40*, 145–156. [CrossRef]
29. Ji, P.; Ma, X.; Li, G. Developing green purchasing relationships for the manufacturing industry: An evolutionary game theory perspective. *Int. J. Prod. Econ.* **2015**, *166*, 155–162. [CrossRef]
30. Badu, S.; Mohan, U. An integrated approach to evaluating sustainability in supply chains using evolutionary game theory. *Comput. Oper. Res.* **2018**, *89*, 269–283. [CrossRef]
31. Friedman, D. On economic applications of evolutionary game theory. *J. Evol. Econ.* **1998**, *8*, 15–43. [CrossRef]
32. Xiong, W.; Zhao, X.; Wang, H. Information asymmetry in renegotiation of public-private partnership projects. *J. Comput. Civ. Eng.* **2018**, *32*, 04018028. [CrossRef]
33. Ho, S.P. Model for financial renegotiation in public-private partnership projects and its policy implications: Game theoretic view. *J. Constr. Eng. Manag.* **2006**, *132*, 678–688. [CrossRef]
34. Shang, L.; Aziz, A.M.A. Stackelberg game theory-based optimization model for design of payment mechanism in performance-based PPPs. *J. Constr. Eng. Manag.* **2020**, *146*, 04020029. [CrossRef]
35. Freedman, D.A. Statistical models and shoe leather. *Sociol. Methodol.* **1991**, *21*, 291–313. [CrossRef]

buildings MDPI

Article

Organization Synchronization in Response to Complex Project Delays: Network-Based Analysis

Lin Yang [1], Xinran Hu [1] and Xianbo Zhao [2,*]

[1] School of Civil Engineering, Wuhan University, Wuhan 430072, China; yang.lin@whu.edu.cn (L.Y.); xrhu@whu.edu.cn (X.H.)
[2] School of Engineering and Technology, Central Queensland University, Sydney, NSW 2000, Australia
* Correspondence: b.zhao@cqu.edu.au

Abstract: In response to frequent complex project delays, organization synchronization, a set of interactions, is a dynamic behavior that helps to restore the stability of complex projects after delays. However, few studies have figured out how organizations synchronize effectively in order to deal with delay issues. To solve this problem, this study first provides a preliminary list of CDFs and indices of organization interactions are also given. A total of 15 key CDFs and 10 interaction ways were refined according to a questionnaire survey. In addition, the complex network synchronization (CNS) theory was adopted to analyze the synchronizability and importance of nodes by comprehensively using multiple parameters. A complex metro project with 51 project organizations was used as a case study and we found that specific signal organizations synchronized through three effective interaction ways (meetings, discussion and study, and the Internet) to cope with six CDFs (safety accidents, prominent problems of land expropriation, unreasonable timelines by clients, improper construction designs, delayed payments, and high financial risks). This study contributes to defining organization synchronization, providing a feasible research framework for assessing network synchronizability and identifying signal organizations in complex projects, and guiding practitioners to effectively cope with delays by interactions between signal organizations.

Keywords: organizational synchronization; interactions; construction delay factors; complex network synchronization theory; complex projects

Citation: Yang, L.; Hu, X.; Zhao, X. Organization Synchronization in Response to Complex Project Delays: Network-Based Analysis. *Buildings* **2022**, *12*, 662. https://doi.org/10.3390/buildings12050662

Academic Editor: Audrius Banaitis

Received: 4 April 2022
Accepted: 13 May 2022
Published: 16 May 2022

1. Introduction

Complex projects are a focus of research in project management to meet social demands. For instance, infrastructure projects are being pursued at a breakneck pace in developing countries in water and sewage, electricity, and transportation and communications, while developed countries are improving infrastructure systems to solve outstanding problems [1]. Complex projects refer to construction projects that are large in scale and involve large investments, multiple stakeholders, complex interactions, and a dynamic environment. They thus feature a high level of uncertainty, unknown dependencies, and unpredictability [2]. Uncertainty in complex projects leads to unexpected events, such as incorrect on-site exploration and accidents, while organizations handling such projects need to be flexible because of the unknown dependencies [2]. Unexpected situations often lead to construction delays [3,4]. Therefore, project success is closely related to project complexity, such as the scale and technical difficulty [5,6].

Breaking down project complexity is useful to help practitioners identify complex projects more accurately. It is recognized that project complexity has organizational and technical types [7]. Given the long project life cycle and uncertain final project scope, the dynamic and growth-related traits of projects helped researchers define project complexity from the perspectives of time and space [8]. The composition of project complexity also includes information and targets [9], communication [10], interface management [11], cost performance [12], and a dynamic environment [13]. Hence, it is important to summarize existing interaction ways among complex project organizations [14].

Organizations dynamically interact with each other in complex projects and form the behavior of organization synchronization once an accident, such as a delay, occurs [15,16]. Organization interactions are generally made up of different types, such as written, oral, and technical interactions [17,18]. However, it remains unclear in which interaction ways organizations synchronize effectively and which organizations have strong synchronization capabilities. Therefore, this study aimed to resolve the organization synchronization issue by (1) finding the most key construction delay factors (CDFs); (2) establishing an index system of organization interactions; and (3) determining the synchronizability of important organizations and effective interaction ways. The complex network synchronization (CNS) theory was adopted to analyze the process of organization synchronization and a case study was used to validate the application of the CNS theory. Thus, this study contributes to the theory and practice of organizational synchronization and provides a comprehensive way to evaluate synchronizability and the importance of nodes.

2. Literature Review

2.1. Construction Delay Factors

Construction delays in complex projects refer to time extensions that result from multiple causes related to different stakeholders at different project phases. The causes of delays are mainly related to project partners, such as clients, contractors, designers, suppliers, investors, laborers, supervisors, and governments [19–23]. Other researchers indicated that some external factors have a large influence on project performance, such as dangerous environments and terrible weather [21,24], rising prices [20,25,26], and cultural influences [27]. As shown in Table 1, we summarized a total of 27 CDFs gleaned from the literature review and deconstructed the causes of delays from the perspective of stakeholders, including financial institutions, clients, contractors, designers, supervisors, governments, and external factors.

Previous studies indicated that the responsibility for delays should be attributed to contractors and their direct stakeholders, such as suppliers and clients [20,26]. For these organizations, the causes of delays include construction mistakes, site management, delayed payments and other problems with materials, personnel, and equipment [3,20,28]. The coordination of service providers related to project sites and project works was particularly weak [29], and poor communication or coordination problems may lead to rework [3,24,27,30,31]. Therefore, smooth coordination is required in order to successfully remain on schedule to prevent variations in critical activities [26]. However, few previous studies have explained how delay issues are solved by organization synchronization in the context of complex projects, which involve large numbers of organizations and complex organization interactions.

Table 1. Summary of construction delay factors.

Sources	Factors	[3]	[20]	[21]	[22]	[23]	[24]	[25]	[26]	[27]	[31]	[31]	[32]	[33]	[34]	[35]
Clients	Prominent problems of land expropriation								√							
	Slow decisions by clients	√		√	√	√	√	√	√	√		√			√	
	Design alterations by clients	√			√	√	√		√	√		√	√	√		
	Unreasonable timelines by clients	√	√	√	√	√		√	√	√	√	√				√
	Delayed payments	√	√	√	√	√		√	√			√	√		√	√
	Deferred transmission of the construction site					√		√				√				
	Supply problems from clients	√	√		√	√		√		√	√	√	√		√	√
	Improper financing	√		√	√	√	√		√	√		√	√		√	
Contractors	Safety accidents														√	√
	Improper organizational construction design				√	√		√	√	√				√		√
	Limited capability of project managers			√	√	√		√	√	√		√	√	√	√	√
	Lack of labor or unqualified labor	√	√	√	√		√	√	√	√	√	√		√	√	√
	Machinery breakdowns							√	√		√				√	√
	Disharmony with neighbors					√				√						
Designers	Lack of communication between designers and contractors	√	√		√		√	√	√	√		√			√	
	Incorrect design basis															
Supervisors	Incompetent supervision	√	√	√	√	√	√	√	√	√				√	√	√

Table 1. *Cont.*

Sources	Factors	[3]	[20]	[21]	[22]	[23]	[24]	[25]	[26]	[27]	[31]	[31]	[32]	[33]	[34]	[35]
	Conflicts with designers	√	√		√		√	√	√	√		√			√	√
Financial institutions	High financial risk			√				√			√					
Governments	Unreasonable government intervention					√										
	Variations in law and regulations								√	√					√	
	Lack of supervisory strength	√	√	√	√	√		√						√		
External factors	Dangerous environment and terrible weather	√		√	√		√	√	√	√	√				√	√
	Excavation of relics															
	International transportation of materials, machines, and labor															
	Rising prices	√	√					√	√							
	Cultural influences									√						

2.2. Organizational Synchronization in Complex Projects

Synchronization is a ubiquitous process emerging from many dynamically interacting units in many different areas, such as biological ecology, electronic circuits, social relationships, and economic management [36,37]. Examples include numerous fireflies suddenly flashing at the same time and clocks swinging at the same frequency after some imperceptible movements. Importantly, synchronization not only involves multiple interacting units but also the interaction of several signal units that transmit synchronization signals to the other units [38]. Once a complex system has been disturbed by an external or internal accident, the synchronization phenomenon first occurs in the local community that is formed by the signal units and then spreads to the whole system. When achieving a state of synchrony, all units show a high degree of coordination and the complex system reaches a high level of stability.

In other words, the synchronization process helps to rebalance and restore a disturbed complex system. In complex project systems, organizations are the actors who interact to positively solve delay issues and achieve a state of synchrony. Thus, analysis of organization synchronization can contribute to our understanding of delays in complex project management. Similar to the definition of synchronization, we assume that 'organization synchronization' means a dynamic process in which project organizations interact with each other to effectively respond to project accidents and restore the stability of complex project systems. After organizations have been synchronized, it is possible to achieve effective cooperation between organizations and faster information transmission in complex systems [39–41].

Moreover, to better understand organization synchronization, studies need to focus on organization interactions and the early interaction behaviors of signal organizations in complex projects. Generally, organization interactions can be generated in many ways, such as by tasks, documents (e.g., contracts), commitments or trust, information technologies, communications, information sharing, knowledge exchanges, and management procedures [42–48]. However, few studies have established an index system of interaction ways and evaluated the frequency of each interaction way in project organizations. As shown in Table 2, we summarized these interactions and classified them into written interactions, oral interactions, technical interactions, and meetings.

Table 2. Organization interactions in complex projects.

Interactions	Content
Written interactions	Investigation data, survey data, design drawings and instructions, engineering calculations, contracts, rules and regulations, organizational construction designs, situation reports, original records, statistics charts, reports, and letters [49].
Oral interactions	Oral task assignments, instructions, oral reports, inspections of work, introductions, negotiations, suggestions, criticism, discussions, and studies [50,51].
Technical interactions	Internet, telephone, telegraph, computer, TV, video recording, sound recording, and radio [52].
Meetings	Site meetings, supervision meetings, and expert meetings [53].

Organization interactions and related communication issues have been widely studied by using the Social Network Analysis (SNA) theory in the construction field [43,54–57]. For example, Li adopted SNA to study the influence of Building Information Modeling on project organizations and the communication patterns [58]. Admittedly, SNA is a useful method for solving organization interaction problems. However, further theoretical development would be needed if SNA is to be used to deal with synchronization issues.

In contrast, the complex network synchronization (CNS) theory, which was derived from the complex network theory, has specifically been developed to explore the synchronization process. Watts and Strogatz (1998) first introduced the network theory to analyze the synchronization of cricket chirps [59]. According to Li, previous studies on the CNS theory can be divided into four types: synchronization of a chaotic system, synchronization within a network, synchronization between different networks, and synchronization of a multi-layer network [60]. Organization synchronization studies belong to the second type, which refers to the synchronization behavior of internal network nodes. In addition, most of the research concentrates on cumbersome mathematical calculations and theoretical modeling in different areas. Few studies have applied the CNS theory to figure out the organization synchronization process, particularly in complex projects.

Network synchronizability is greatly affected by the structural properties [61,62], which rely on the characteristics of nodes and links in a complex network. In addition, the structural properties of a network are generally evaluated by diverse parameters, many of which have been found to have numeric relationships with network synchronizability. For instance, degree, average path length, heterogeneity in the degree distribution, and node betweenness centrality vary inversely with network synchronizability [63,64], and modularity, which measures the density between different communities, is positively proportional to network synchronizability [38]. However, these studies assessed network synchronizability based on one single parameter. It is important to provide a sound analysis by combining the relationships between network synchronizabilty and as many parameters as possible.

As mentioned above, organization synchronization emerges from a local synchronization initiated by several signal organizations. Some investigations have shown that flows of large amounts of information or intensive interactions between the signal organizations contribute to a faster synchronization process [38,65]. Hence, signal nodes in complex networks have two general traits: (1) a high degree of connection with the other nodes; and (2) initial nodes that perceive disturbance factors in complex networks and transmit synchronization signals. Correspondingly, in complex projects, the first organization to perceive delay factors and its closely related stakeholders may be signal nodes if they can be proved to be highly linked with the other organizations. To find signal organizations, it is critical to evaluate the importance of nodes from the network's perspective. Node importance is generally assessed by two parameters: (1) node degree [66,67]; and (2) node centrality, including degree centrality, closeness centrality, betweenness centrality, and eigenvector centrality [68]. Nevertheless, conflicting results frequently occur because node importance usually ranks differently between the analysis of the five parameters. Hence, we ranked network nodes and found signal organizations by unifying the above-mentioned five node parameters.

3. Methodology

3.1. Research Framework

As shown in Figure 1, a research framework involving five steps was developed to explore organization synchronization in terms of complex networks. First, a literature review was conducted to identify 27 construction delay factors (Table 1) and four kinds of organization interaction ways (Table 2). Then, an expert interview was conducted to discuss the reasonableness of the 27 CDFs, determine the types of organizations, and determine the possible organization interactions in complex projects. Second, a questionnaire survey was used to assess the influence of the 27 CDFs on the project schedule and evaluate four kinds of relationships among different units (inter-organization (IO), organizations and CDFs (OCDFs), organizations and interaction ways (OIWs) before the delay, and organizations and interaction ways after the delay). Third, the software NetMiner 4.0 was utilized to visualize the four corresponding complex networks (the IO network, the OCDF network, the OIW network before the delay, and the OIW network after the delay). Subsequently, several parameters were analyzed to compare the network synchronizability, find the key CDFs, and determine the signal organizations in the early stage of local synchronization.

During the above processes, statistical analysis (Step 2) and network analysis (Step 4) were used to refine the key CDFs.

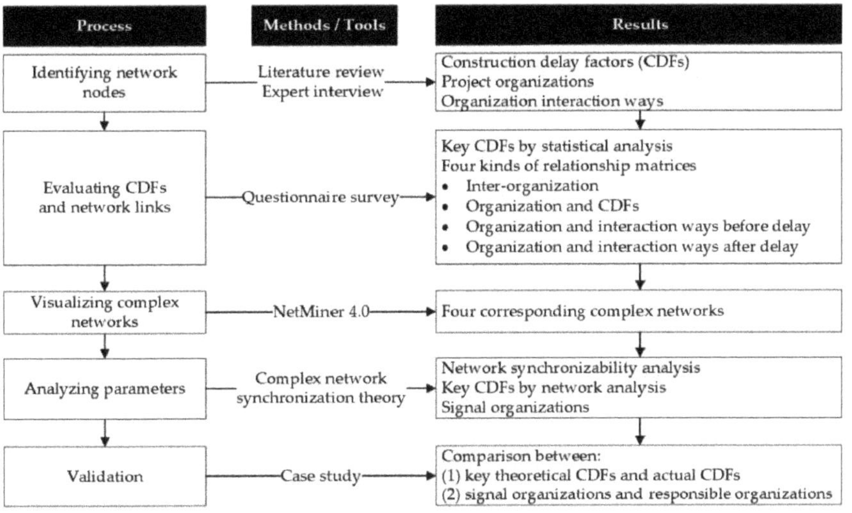

Figure 1. Research framework for studying organization synchronization.

Finally, to validate the effectiveness of the CNS theory's application, a typical complex project in Hangzhou, China was adopted as a case project and studied by implementing the above steps. The complex metro project in the case study was assumed to be a project under construction and we also assumed that no delay had occurred. Thus, the key CDFs and the signal organizations could be generated from the theoretical research. If the theoretical results are highly commensurate with the actual reasons for the delay and the responsible organizations, then the feasibility of the CNS theory's application can be proved.

Therefore, by applying the research framework, researchers can predict key causes of delays and the responsible parties in complex projects where project delays have not happened, determine the network synchronizability of the IO network, the OCDF network, and the OIW network before the delay, and adjust organization interactions to achieve better synchronizability in response to project delays.

3.2. Data Collection

The preliminary literature review yielded a list of CDFs and interaction ways. To improve data accuracy, a further expert interview was conducted with three experienced project managers from contractor and client organizations. All experts had either more than 15 years of experience in construction projects or participated in complex projects. They were required to assess the results of the literature review, summarize 10 types of common organizations in complex projects, and initially discuss general interaction ways between these organizations. According to the interview results, complex project organizations include financial institutions (F), governments (G), including government-funded representatives and departments approving and supervising construction projects, clients (B), contractors (C), supervisors (S), designers (D), operation units (O), the public (P), investors (I), and academic researchers. In addition, some interactions existing only between particular stakeholders and relevant contract relationships were removed according to the suggestions from the expert team. Thus, the 10 refined common interactions were oral task assignments, instructions, oral reports, inspections of work, discussions and studies, meetings, letters, networks, telephone communications, and telegraph communications.

Focusing on the 10 types of organizations, a questionnaire survey was designed to judge the effect of the 27 CDFs on the project period and the four types of relationships mentioned above. The questionnaire consisted of four main sections. The first section was used to obtain a profile of each respondent and collected information such as work experience, geological distribution, and organization types. The second section assessed the influence of the 27 CDFs on complex projects by a five-point Likert scale (1 = very low, 2 = low, 3 = medium, 4 = high, 5 = very high). The third section quantified the above-mentioned four kinds of relationships (inter-organization, organizations and CDFs, organizations and interaction ways before the delay, and organizations and interaction ways after the delay). The last section was used to obtain text solutions to construction delay problems from these experienced respondents.

The specific participants were first selected via the stratified sampling strategy [69]. Then, a total of 175 questionnaires were sent out via paper or electronic files and 173 questionnaires were returned. There were 169 valid questionnaires, meaning a valid response rate of 96.57%. The rate was relatively high compared with other studies and acceptable rates, including 20.13% in [28], 11.1% in [69], 13.02% in [70], and 70% in [71]. The sample size of 169 was adequate for the data analysis and well above the minimum requirement of 30 according to the central limit theorem [35]. Around 59.76% of the respondents had at least 5 years of experience in working with complex projects. A total of 38.46% of the respondents were academic staff from universities, and the other respondents had an even distribution in terms of organizations and geography.

3.3. Statistical Analysis

Cronbach's alpha (α) is widely adopted to prove the inter-reliability of survey data and its value ranges from 0 to 1 [72,73]. As stated in many studies [74–76], an acceptable α score is generally larger than 0.70. As tested by SPSS 25.0, the α of 0.933 within the entire survey sample indicated that the questionnaire survey had a high degree of reliability and the fact that the α of each CDF was below 0.933 proved the strong inner consistency of all scales. As shown in Table 3, the mean scores of the CDFs range from 3.06 to 4.06. To select the critical CDFs, the normalized values of the mean scores were calculated, as recommended by Xu et al. and Zhao et al. [77,78]. A total of 15 CDFs with normalized values above 0.50 were regarded as critical factors.

Figure 2 shows the proportion of organization interaction ways before and after delays. Whether delays occur or not, the most frequent means of interaction was meetings. After project delays, the ratio of utilization of oral task assignments, networks, and instructions decreased by 41.38%, 33.14%, and 28.17%, respectively. This indicates that inefficiencies in organizational cooperation became evident in cases of accidents, leading to a further loss of control and poor outcomes. Attention needs to be paid to effective interaction ways, such as meetings and discussions, in order to enhance the efficiency of collaboration and information sharing.

3.4. Network Parameters

According to the literature review, network synchronizability is related to two types of parameters: (1) global network parameters, including average length path, modularity, and clustering coefficient; and (2) node parameters, including degree and betweenness centrality. The latter type is indirectly related to network synchrony by illustrating signal nodes. Hence, we explain network synchronizability in terms of global network parameters and explore signal organizations and key CDFs in light of node parameters. As shown in Table 4, all involved parameters are explained. Both key CDFs and signal organizations are explained by multiple parameters; thus, conflicting results are prone to occur. Therefore, we summed the rank value of the corresponding parameters to obtain a comprehensive rank.

Table 3. Ranking of CDFs.

CDFs	Rank	Mean	NV [1]	CDFs	Rank	Mean	NV
Prominent problems of land expropriation	1	4.06	1.00	Incorrect design basis	15	3.56	0.50
Safety accidents	2	3.93	0.87	Lack of labor or unqualified labor	16	3.53	0.47
Slow decisions by clients	3	3.92	0.86	Bad weather	17	3.5	0.44
Design alterations by clients	4	3.89	0.83	Excavation of relics	18	3.47	0.41
Unreasonable timelines by clients	5	3.85	0.79	Variations in laws and regulations	19	3.46	0.40
Improper organizational construction design	6	3.77	0.71	Incompetent supervision	20	3.42	0.36
Delayed payments	7	3.75	0.69	International transportation of materials, machines, and labor	21	3.4	0.34
Unreasonable government intervention	8	3.75	0.69	Rising prices	22	3.35	0.29
High financial risk	9	3.71	0.65	Machinery breakdowns	23	3.34	0.28
Deferred transmission of the construction site	10	3.7	0.64	Conflicts with designers	24	3.25	0.19
Limited capability of project managers	11	3.69	0.63	Disharmony with neighbors	25	3.19	0.13
Supply problems from clients	12	3.66	0.60	Lack of supervisory strength	26	3.17	0.11
Lack of communication between designers and contractors	13	3.6	0.54	Cultural influences	27	3.06	0.00
Improper financing	14	3.58	0.52				

[1] Normalized value (NV) = (mean − minimum mean)/(maximum mean − minimum mean).

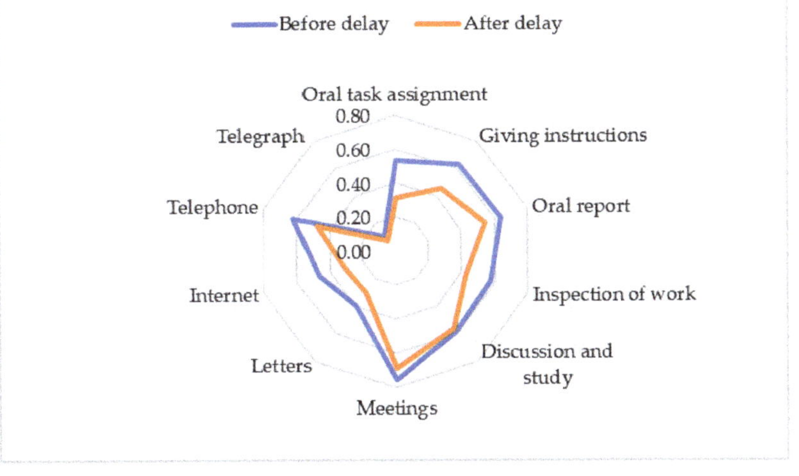

Figure 2. Comparison between organization interaction ways before and after delays.

Table 4. Parameters of complex networks.

Parameters of Complex Network		Explanation
Network overview	Density	Measures the network's cohesiveness.
	Average path length	The average distance between all node pairs.
	Clustering coefficient	Gauges the aggregation of networks. The larger it is, the more connected the network is.
	Modularity	Measures the density between communities.
Node parameters	Eigenvector centrality	Regards nodes around nodes with a high degree of eigenvector centrality as key nodes.
	Betweenness centrality	Measures the extent to which a vertex plays a bridging role [79].
	Closeness centrality	Measures each node's position in the network and means the inverse of the average distance to others in some cases [79].
	Degree centrality	Counts the total number of connections linked to a node [79].
	Degree	The weighted sum of edges for a node [80].

4. Case Study

Phase I of the Hangzhou Metro Line 4 project was selected as a case study for the following reasons. First, it is a typical complex project considering the interacting project complexities due to the organizational structure, technologies, etc. Second, seven delays indeed occurred in the case project and led to a huge final cost of as much as 14.35 billion RMB. It was supposed to be operational on 29 December 2017, but the operational trial was postponed by six months to 6 June 2018. The actual causes of the delay reported by the news included a huge financial risk, prominent problems of land expropriation, an unreasonable timeline for clients, an improper organizational construction design, safety accidents, disharmony with neighbors, and the excavation of aerial bombs. Third, as a completed complex project, ample data could be collected to determine the project organizations. Three official websites (Zhejiang Provincial Development and Reform Commission, Hangzhou City Development and Reform Commission, and Hangzhou Metro) were the major sources of data on project organizations and publish project information such as progress and participants. By a full-text search and the use of Web crawlers on the websites in chronological order, a total of 357 texts were obtained and split into substantial quantities of words. By selecting words with the organization property, more than one hundred organizations were found to have participated in the completion of this project.

4.1. Network Nodes and Links

According to the research framework, there are three types of network nodes in a theoretical complex project system, including 15 CDFs (Table 3), 10 types of project organizations, and 10 common organization interaction ways. Before the evaluation of network links, a focus group was formed to discuss the suitability of the theoretical nodes in the case study. Seven experts participated in this project and came from various organizations, including the client, two construction units, two supervision units, one material supplier, and the investor. They consistently agreed that the initial selection of 15 CDFs and 10 interaction ways could be applied to the case study. Importantly, the expert team selected 51 major project organizations according to their experience in the case study. As shown in Table 5, the project organizations in the case study consisted of the financial institution, the client, 14 contractors, 10 supervision units, 3 design units, 13 material suppliers, the government, the investor, the operation unit, and 7 public units. The rule used when coding the nodes was that the first number represents the organizational type and the second number represents the numerical reference method.

Table 5. Node coding.

Organizations	Coding	CDFs	Coding	Interactions	Coding
Financial institutions	F	Prominent problems of land expropriation	Delay 1	Oral task assignment	Inf1
Clients	B	Safety accidents	Delay 2	Giving instructions	Inf2
Contractors	C (1~14)	Slow decisions by clients	Delay 3	Oral reports	Inf3
Designers	D (1~3)	Design alterations by clients	Delay 4	Inspections of work	Inf4
Supervisors	S (1~10)	Unreasonable timelines by clients	Delay 5	Discussions and studies	Inf5
Material suppliers	M (1~13)	Improper construction organizational designs	Delay 6	Meetings	Inf6
Governments	G	Delayed payments	Delay 7	Letters	Inf7
Operation units	O	Unreasonable government intervention	Delay 8	The Internet	Inf8
Investors	I(G)	Huge financial risk	Delay 9	Telephone	Inf9
Public	P (1~7)	Deferred transmission of the construction site	Delay 10	Telegraph	Inf10
		Limited capability of project managers	Delay 11		
		Supply problems from clients	Delay 12		
		Lack of communication between designers and contractors	Delay 13		
		Improper financing	Delay 14		
		Incorrect design basis	Delay 15		

Network links refer to the four kinds of relationships between the 15 CDFs, the 51 project organizations, and the 10 interaction ways. The weights of network links were assessed by a questionnaire survey similar to the one described in Section 3.2 (Data Collection), and four corresponding matrices were formed for network visualization. As shown in Table 6, a sample matrix consists of i row nodes ($N_{r1} \sim N_{ri}$) and j column nodes ($N_{c1} \sim N_{cj}$). The weight of the link between N_{ri} and N_{cj} is defined as w_{ij}. Note that diagonal values such as w_{11} are 0 in complex network analysis. For example, the 51 project organizations are both row and column nodes in the IO matrix, and 782 links were found and weighted by the questionnaire survey; in the OCDF matrix, the 51 project organizations are row nodes, the 15 CDFs are column nodes, and 66 links exist between these nodes.

Table 6. Sample matrix of complex networks.

Nodes	N_{c1}	N_{c2}	...	N_{cj}
N_{r1}	w_{11}	...	...	w_{1j}
N_{r2}	w_{21}	w_{22}	...	w_{2j}
...	...	...	...	
N_{ri}	w_{i1}	w_{i2}	...	w_{ij}

The tools available for analyzing complex networks include GEPHI, MATLAB [67], NETWORKX, IGRAPH, Python, UCINET [81], NetMiner, and Pajek [82,83]. Among these tools, NetMiner is user-friendly and provides a sufficient parameter analysis. Therefore, the four matrices were input into NetMiner 4.0 and nodes were styled using different colors and types. For instance, Figure 3 shows the inter-organization network visualized by NetMiner 4.0, where nodes are styled as follows: public, red heart; investors/the government, yellow crisscross; operation units, orange star; financial institution, pink pentagon; the client, blue triangles; other construction organizations, rectangles in different degrees of blue. CDFs and interaction ways are styled in purple diamonds and green circles, respectively, in other networks.

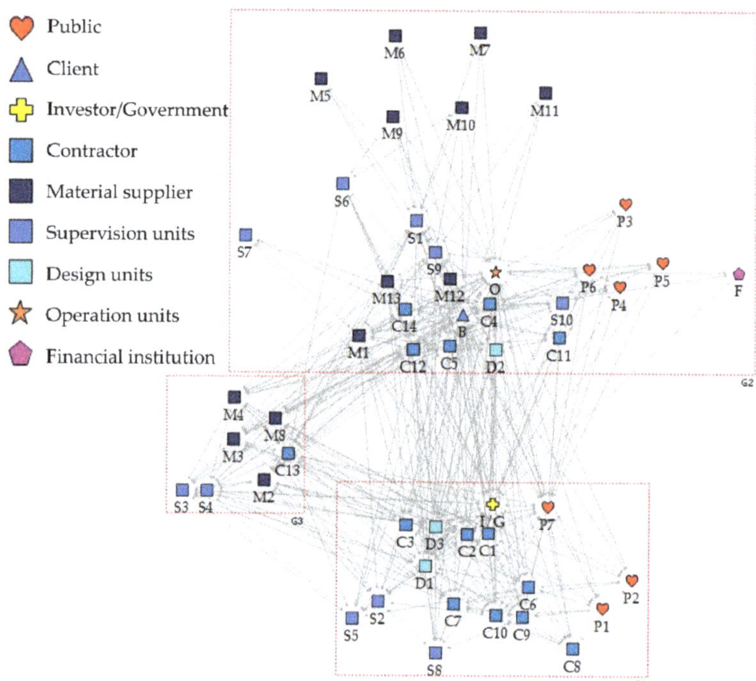

Figure 3. Inter-organization network.

4.2. Parameter Analysis

4.2.1. Network Synchronizability Analysis

Table 7 summarizes the global features of complex networks based on the four parameters. Network synchronizability is reflected from two perspectives: average path length and modularity. Apart from the IO network, the OCDF network and the OIW network before and after the delay have the same average path length of 1. Thus, the relative synchronizability of the three networks depends on their modularity values. The higher the modularity value is, the better synchronizability the complex network has. As can be seen in Table 6, OCDF ranks first in terms of modularity. Compared with the IO network, the OCDF network has a shorter average path length, showing greater synchronizability. In addition, the organizations interacted more frequently and effectively after the delay because the modularity increased by 47.9%.

Table 7. Global features of complex networks.

Network Model	Clustering Coefficient	Density	Average Path Length	Modularity
IO network	0.566	0.307	1.693	0.164
OCDF network	0	0.045	1	0.284
OIW network before the delay	0	0.128	1	0.071
OIW network after the delay	0	0.123	1	0.105

The IO network has the highest density and clustering coefficient. However, the network performs the best when the values of both parameters are close to 1. This indicates that, in the case study project, organizational relationships were supposed to be enhanced for rapid information flow and effective cooperation.

4.2.2. Key Construction Delay Factors

Table 8 presents a comprehensive ranking of the 15 CDFs based on the analysis of the five parameters. Delay 2 (safety accidents) was recognized as the most influential factor in the case study project. This result is highly consistent with the case study because a total of seven safety accidents occurred from April 2014 to July 2016 and resulted in a project delay as well as four deaths and one injury. The second-ranked CDF is Delay 1 (Prominent problems of land expropriation). It also postponed the project because neighbors around the construction site (Lianzhuang Station) were afraid of damage to the environment. Another four key CDFs include Delay 5 (Unreasonable timelines by clients), Delay 6 (Improper construction design), Delay 7 (Delayed payments), and Delay 9 (High financial risk). After comparing the theoretical key CDFs and the actual reasons for delays, we found five 'overlapping' factors among the seven actual causes of delay, proving the applicability of the CNS theory to complex projects.

Table 8. Node parameters of construction delay factors.

CDFs	Rank	Betweenness Centrality	Closeness Centrality	Degree Centrality	Eigenvector Centrality	Degree
Delay2	1	0.175	0.714	0.51	0.464	26
Delay1	2	0.082	0.604	0.373	0.365	19
Delay5	3	0.036	0.567	0.353	0.399	18
Delay6	4	0.076	0.579	0.333	0.321	13
Delay7	5	0.075	0.534	0.294	0.309	17
Delay9	6	0.035	0.556	0.294	0.309	4
Delay12	6	0.015	0.514	0.255	0.321	13
Delay13	8	0.015	0.514	0.255	0.309	13
Delay11	9	0.005	0	0.078	0.321	15
Delay14	10	0	0.426	0.294	0.303	15
Delay10	11	0.003	0.426	0.078	0.044	4
Delay3	12	0	0	0	0	0
Delay4	12	0	0	0	0	0
Delay8	12	0	0	0	0	0
Delay15	12	0	0	0	0	0

4.2.3. Signal Organizations in Synchronization

Signal organizations are closely connected to other units and are usually the first ones to identify delays. In this study, five parameters (degree, degree centrality, betweenness centrality, closeness centrality, and eigenvector centrality) were analyzed to find critical organizations in the IO network. Figure 4a, Figure 4b, and Figure 4c represent the degree, degree centrality, and betweenness centrality in the IO network, respectively. The other two-parameter analysis yielded consistent results and had similar graphs to those shown in Figure 4. As shown in Figure 4, the signal organizations consist of the client, investors/governments, the operation unit, two design units (D1 and D3), one public unit (P7), one supervision unit (S1), and one contractor (C12).

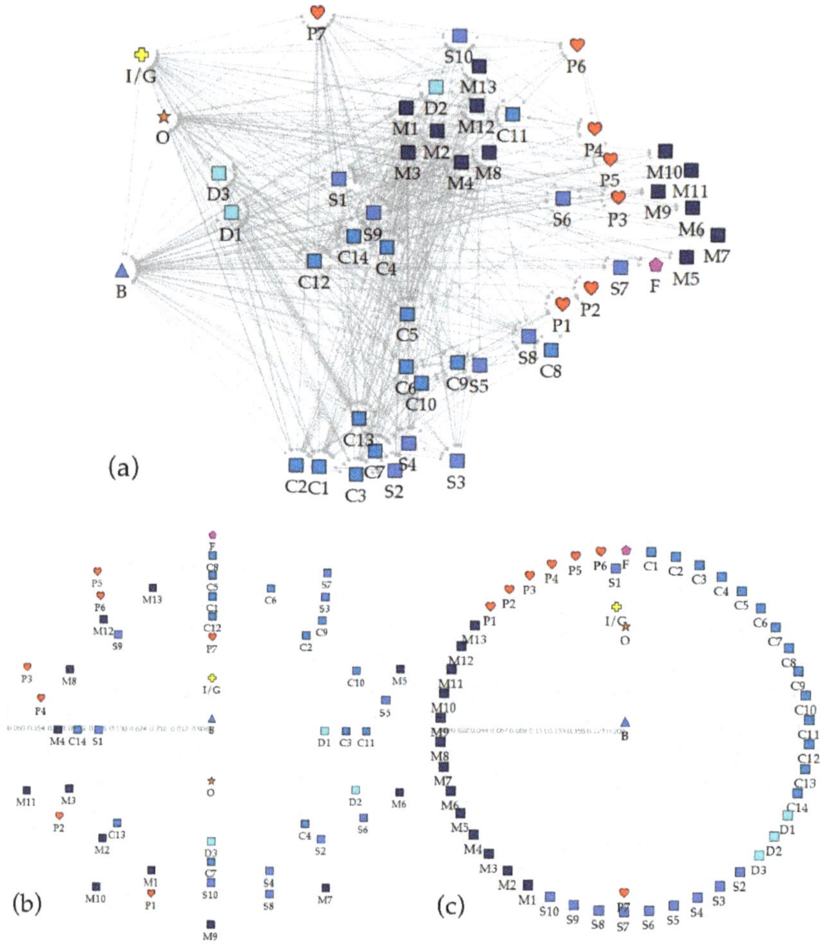

Figure 4. IO network: (**a**) degree; (**b**) degree centrality; (**c**) betweenness centrality.

Figure 5 presents the visualization of the OCDF network, revealing the relationships between organizations and CDFs. One can easily observe that those signal organizations are close to the seven key CDFs. S1 and C12 are the supervisors and contractors, respectively, in the southern project section who were the parties responsible for safety accidents that occurred in this section. P7 represents the community protesting the construction of Lianzhuang Station; the construction design supplied by D1 and D3 had several mistakes and led to rework. In addition, the government and the client failed to solve the problems of land expropriation and proposed an unreasonable project period due to an underestimation of the construction project's difficulty. The consistency of signal organizations and responsible units also proves the feasibility of the CNS theory's application. Thus, it is possible to predict the specific organizations responsible for delays in complex projects.

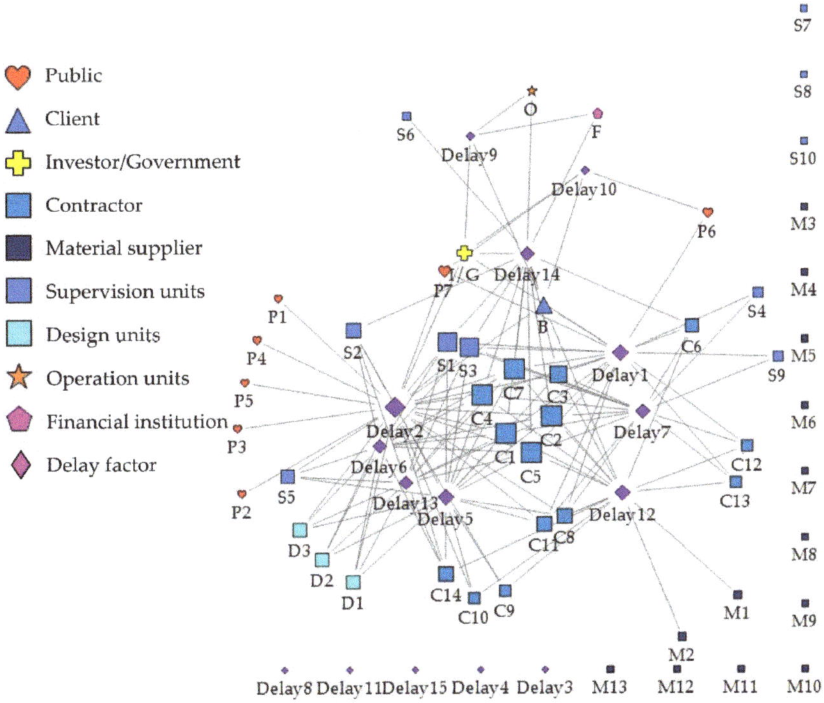

Figure 5. OCDF network.

4.2.4. Effective Interactions in Organization Synchronization

The above-described research results show that the network synchronizability after the delay increased, and seven key CDFs and related signal organizations were accurately identified. In other words, when faced with the key CDFs, signal organizations interacted with each other to achieve better cooperation. Hence, we determined the variations in organization interaction ways. Figure 6 illustrates the degree distribution of the OIW network before and after the delay. As can be intuitively seen in Figure 6, the values for Inf 6 (Meetings), Inf 3 (Oral reports), Inf 5 (Discussions and studies), and Inf 2 (Giving instructions) are large; thus, they possibly play leading roles in organization synchronization.

Further comprehensive studies were conducted to determine the importance of interaction ways in organizations and make a comparison before and after the delay. Tables 9 and 10 present the parameter values and rankings of the 10 interaction ways. We found that discussions and studies, meetings, and the Internet were used more frequently after the delay, indicating the effectiveness of these three interaction ways in organization synchronization.

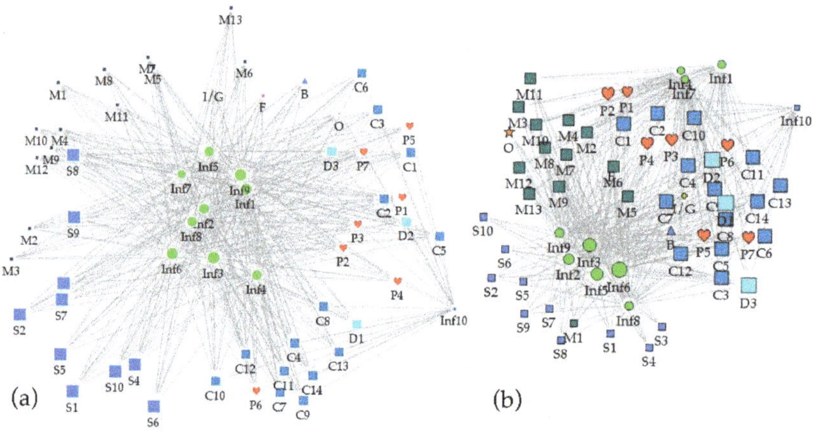

Figure 6. Degree distribution: (**a**) OIW network before the delay; (**b**) OIW network after the delay.

Table 9. Ranking of interaction ways before the delay.

Nodes	Rank	Degree	Betweenness Centrality	Closeness Centrality	Degree Centrality	Eigenvector Centrality
Inf1	4	32.12	0.086	1	1	0.332
Inf2	5	30	0.086	1	1	0.322
Inf3	1	36.41	0.086	1	1	0.392
Inf4	8	29.34	0.044	0.726	0.745	0.336
Inf5	6	28.81	0.086	1	1	0.309
Inf6	3	34.75	0.086	1	1	0.358
Inf7	7	23.92	0.086	1	1	0.271
Inf8	9	28.76	0.082	0.972	0.980	0.294
Inf9	2	35.41	0.086	1	1	0.367
Inf10	10	3.26	0.017	0.570	0.490	0.035

Table 10. Ranking of interaction ways after the delay.

Nodes	Rank	Degree	Eigenvector Centrality	Betweenness Centrality	Degree Centrality	Closeness Centrality
Inf1	7	17.15	0.211	0.049	0.784	0.758
Inf2	4	26.97	0.334	0.1	1.0	1.0
Inf3	2	36.17	0.447	0.1	1.0	1.0
Inf4	9	13.89	0.167	0.049	0.784	0.758
Inf5	3	32.36	0.402	0.1	1.0	1.0
Inf6	1	44.21	0.549	0.1	1.0	1.0
Inf7	8	15.19	0.187	0.049	0.784	0.758
Inf8	6	18.39	0.226	0.096	0.980	0.972
Inf9	5	20.99	0.255	0.1	1.0	1.0
Inf10	10	3.73	0.046	0.016	0.471	0.561

5. Discussion

The application of the CNS theory lends credence to the study of organization synchronization in complex projects. To achieve our research objectives, this study involved network synchronizability analysis, six key CDFs, specific signal organizations, and effective interaction ways in the synchronization process.

Network synchronizability was tested in four kinds of networks by two global parameters (average path length and modularity). Results show that the network synchronizability was improved because the modularity increased by 47.9%, while the average path length remained the same after the project delay. However, it is not clear whether the synchronizability after the delay reached a level that is sufficient for complex projects since there is little similar research. Thus, to discuss the reasonableness of the results, a further review of the literature was used to check whether the case study project had met the boundary conditions. This study was based on a rule that states that network synchronizability is inversely proportional to average path length and positively proportional to modularity [38]. However, studies have proved that the rule related to average path length is useful when the number of all nodes and new nodes remains the same during the project period [84]. The case study involved a static network analysis focusing on the entire project life cycle, meaning that there was no variation in the number of nodes.

Six key CDFs were identified by statistical and network analysis (Delay 2 (Safety accidents), Delay 1 (Prominent problems of land expropriation), Delay 5 (Unreasonable timelines by clients), Delay 6 (Improper construction designs), Delay 7 (Delayed payments), and Delay 9 (High financial risk)). Previous studies also recognized the effect of these factors, and they generally pointed out that delays are most related to clients, contractors, and designers [20,25,32,85,86]. In green buildings, the delivery of materials by suppliers was found to play an important role in the construction process [35]. In contrast, we developed an OCDF network to identify the critical organizations that perceived delays the earliest and transmitted signals the fastest.

Specific signal organizations were found in the case study, including the client, investors/governments, the operation unit, two design units (D1 and D3), one public unit (P7), one supervision unit (S1), and one contractor (C12). Compared with previous studies, we achieved a relatively accurate identification of organizations who are responsible for coping with project delays. It is worth noting that supervisors could play a mediating role in complex projects and contribute to the safety of workers [87,88]. Regarding the frequent safety accidents in complex projects, both contractors and supervisors should be cared for, particularly in terms of psychological needs [89].

Researchers also suggest that organizations are supposed to enhance communication and cooperation, but few have proposed specific strategies. As reviewed above, organization interactions had a positive influence on project performance [90]. Therefore, in the case study, effective interactions, such as discussions and studies, meetings, and the Internet, were recognized as ways to enhance organization synchronization after a delay. In addition to the interaction ways presented in Table 2, invisible ways such as organizational culture and national culture affect decision-making performance and the quality of organization interactions [91,92]. Therefore, it is necessary to perfect the index system of interactions, advance methods for measuring the indices, and understand the mechanisms underlying organization interactions and organization synchronization.

Regarding future research, we recommend that further pilot studies be conducted on diverse complex projects, such as road and bridge construction projects and hydraulic projects. Cross-sectional analysis of these typical projects would contribute to perfecting the index system of interactions and determining the range of good synchronizability levels. In addition, the dimensions of interactions can be broadened to social effects such as communication skills and the degree of truth between two organizations. With a combination of information transmission and social effects, a more comprehensive understanding of organization synchronization can be obtained and implemented to cope with delays or other accidents in complex projects.

6. Conclusions

Organization synchronization is the dynamic process of recovering a complex project system from a disturbed state to an efficient state and can be used to deal with delays in complex projects. In this study, we adopted the CNS theory to break down the synchronization process by assessing network synchronizability, identifying key CDFs, and finding signal organizations and productive interactions after delays. To address these points, we established a research framework involving multiple methods and validated the feasibility of applying the CNS theory through a case study.

Our research results can be summarized as follows. First, the network synchronizability was enhanced after the delay in the case study. Second, the six key CDFs in complex projects are Delay 2 (Safety accidents), Delay 1 (Prominent problems of land expropriation), Delay 5 (Unreasonable timelines by clients), Delay 6 (Improper construction designs), Delay 7 (Delayed payments), and Delay 9 (High financial risk). The theoretical CDFs were also found to be commensurate with the actual causes of delay in the case study project. Third, a broad range of signal organizations were accurately identified in the complex project and effective interaction ways (meetings, discussions and studies, and the Internet) can contribute to organization synchronization.

Therefore, this study contributes to both the theory and practice of organization synchronization in complex projects. First, it provides an innovative application of the CNS theory to the field of complex project management. Second, this study offers a comprehensive way to assess network synchronziability and node importance by considering multiple parameters simultaneously. Third, the case study based on a complex project may help researchers implement the research framework and provide useful strategies and practical guidance.

This study has some limitations. There was only one case project in China that was suitable for use in the application of the CNS theory. This brought about the limitation on generalization, which is a common problem when a case study method is used. Nonetheless, Yin argues that the aim of multiple case studies is analytical generalization using the theoretical framework of a study to establish a logic that might apply to other situations rather than statistical generalization as in surveys [93].

Author Contributions: Conceptualization, L.Y. and X.Z.; methodology, L.Y. and X.H.; software, X.H.; validation, L.Y. and X.H.; formal analysis, X.H.; investigation, L.Y. and X.H.; resources, L.Y. and X.H.; data curation, L.Y. and X.H.; writing—original draft preparation, L.Y. and X.H.; writing—review and editing, L.Y., X.H. and X.Z.; visualization, L.Y. and X.H.; supervision, L.Y. and X.Z.; project administration, L.Y.; funding acquisition, L.Y. All authors have read and agreed to the published version of the manuscript.

Funding: This research was funded by the National Science Foundation of China, grant number 71702136.

Informed Consent Statement: Not applicable.

Data Availability Statement: All the data used in this study are presented in Sections 3 and 4.

Conflicts of Interest: The authors declare no conflict of interest.

References

1. Hu, Y.; Chan, A.P.C.; Le, Y.; Jin, R.-Z. From Construction Megaproject Management to Complex Project Management: Bibliographic Analysis. *J. Manag. Eng.* **2015**, *31*, 04014052. [CrossRef]
2. Complex Project Management—What It Is, and What Success Looks Like-EXEPRON-CCPM Software. Available online: https://exepron.com/complex-project-management-what-it-is-and-what-success-looks-like/ (accessed on 11 April 2020).
3. Hwang, B.-G.; Zhao, X.; Ng, S.Y. Identifying the critical factors affecting schedule performance of public housing projects. *Habitat Int.* **2013**, *38*, 214–221. [CrossRef]
4. Zhao, X.; Hwang, B.-G.; Phng, W. Construction project risk management in Singapore: Resources, effectiveness, impact, and understanding. *KSCE J. Civ. Eng.* **2014**, *18*, 27–36. [CrossRef]
5. Luo, L.; He, Q.; Xie, J.; Yang, D.; Wu, G. Investigating the Relationship between Project Complexity and Success in Complex Construction Projects. *J. Manag. Eng.* **2017**, *33*, 04016036. [CrossRef]
6. Flyvbjerg, B. Survival of the unfittest: Why the worst infrastructure gets built—and what we can do about it. *Oxf. Rev. Econ. Policy* **2009**, *25*, 344–367. [CrossRef]
7. Baccarini, D. The concept of project complexity—A review. *Int. J. Proj. Manag.* **1996**, *14*, 201–204. [CrossRef]
8. Shane, J.; Strong, K.; Gransberg, D.; Strategic Highway Research Program; Strategic Highway Research Program Renewal Focus Area; Transportation Research Board; National Academies of Sciences, Engineering, and Medicine. *Guide to Project Management Strategies for Complex Projects*; Transportation Research Board: Washington, DC, USA, 2013.
9. Hui, L. Research on the Structure of the Complexity of Complex Project System. *Soft Sci.* **2009**, *23*, 75–79.

10. Senescu, R.R.; Aranda-Mena, G.; Haymaker, J.R. Relationships between Project Complexity and Communication. *J. Manag. Eng.* **2013**, *29*, 183–197. [CrossRef]
11. Ahn, S.; Shokri, S.; Lee, S.; Haas, C.T.; Haas, R.C.G. Exploratory Study on the Effectiveness of Interface-Management Practices in Dealing with Project Complexity in Large-Scale Engineering and Construction Projects. *J. Manag. Eng.* **2017**, *33*, 04016039. [CrossRef]
12. Safapour, E.; Kermanshachi, S.; Habibi, M.; Shane, J. Resource-Based Exploratory Analysis of Project Complexity Impact on Phase-Based Cost Performance Behavior. In Proceedings of the Construction Research Congress 2018, New Orleans, LA, USA, 2–4 April 2018; pp. 439–448.
13. Trinh, M.T.; Feng, Y. Impact of Project Complexity on Construction Safety Performance: Moderating Role of Resilient Safety Culture. *J. Constr. Eng. Manag.* **2020**, *146*, 04019103. [CrossRef]
14. Ahern, T.; Leavy, B.; Byrne, P.J. Complex project management as complex problem solving: A distributed knowledge management perspective. *Int. J. Proj. Manag.* **2014**, *32*, 1371–1381. [CrossRef]
15. Bygballe, L.E.; Swärd, A.R.; Vaagaasar, A.L. Coordinating in construction projects and the emergence of synchronized readiness. *Int. J. Proj. Manag.* **2016**, *34*, 1479–1492. [CrossRef]
16. Jacobsson, M. On the importance of liaisons for coordination of projects. *Int. J. Manag. Proj. Bus.* **2011**, *4*, 64–81. [CrossRef]
17. Wu, G.; Zhao, X.; Zuo, J.; Zillante, G. Effects of contractual flexibility on conflict and project success in megaprojects. *Int. J. Confl. Manag.* **2018**, *29*, 253–278. [CrossRef]
18. Wu, G.; Zhao, X.; Zuo, J.; Zillante, G. Effects of team diversity on project performance in construction projects. *Eng. Constr. Arch. Manag.* **2019**, *26*, 408–423. [CrossRef]
19. Assaf, S.A.; Al-Hejji, S. Causes of delay in large construction projects. *Int. J. Proj. Manag.* **2006**, *24*, 349–357. [CrossRef]
20. Gomarn, P.; Pongpeng, J. Causes of construction delay from contractors and suppliers in Thailand's oil and gas platform projects. *MATEC Web Conf.* **2018**, *192*, 02008. [CrossRef]
21. Zarei, B.; Sharifi, H.; Chaghouee, Y. Delay causes analysis in complex construction projects: A Semantic Network Analysis approach. *Prod. Plan. Control* **2017**, *29*, 29–40. [CrossRef]
22. Abbasi, O.; Noorzai, E.; Jafari, K.G.; Golabchi, M. Exploring the Causes of Delays in Construction Industry Using a Cause-and-Effect Diagram: Case Study for Iran. *J. Arch. Eng.* **2020**, *26*, 05020008. [CrossRef]
23. Chen, L.; Lu, Q.; Zhao, X. Rethinking the Construction Schedule Risk of Infrastructure Projects Based on Dialectical Systems and Network Theory. *J. Manag. Eng.* **2020**, *36*, 04020066. [CrossRef]
24. Larsen, J.K.; Shen, G.Q.; Lindhard, S.M.; Brunoe, T.D. Factors Affecting Schedule Delay, Cost Overrun, and Quality Level in Public Construction Projects. *J. Manag. Eng.* **2016**, *32*, 04015032. [CrossRef]
25. Gebrehiwet, T.; Luo, H. Analysis of Delay Impact on Construction Project Based on RII and Correlation Coefficient: Empirical Study. *Procedia Eng.* **2017**, *196*, 366–374. [CrossRef]
26. Choong Kog, Y. Major Construction Delay Factors in Portugal, the UK, and the US. *Pract. Period. Struct. Des. Constr.* **2018**, *23*, 04018024. [CrossRef]
27. Ruqaishi, M.; Bashir, H.A. Causes of Delay in Construction Projects in the Oil and Gas Industry in the Gulf Cooperation Council Countries: A Case Study. *J. Manag. Eng.* **2015**, *31*, 05014017. [CrossRef]
28. Zhang, X.; Tariq, S. Failure Mechanisms in International Water PPP Projects: A Public Sector Perspective. *J. Constr. Eng. Manag.* **2020**, *146*, 04020055. [CrossRef]
29. Alsuliman, J.A. Causes of delay in Saudi public construction projects. *Alex. Eng. J.* **2019**, *58*, 801–808. [CrossRef]
30. Hwang, B.-G.; Zhao, X.; van Do, T.H. Influence of Trade-Level Coordination Problems on Project Productivity. *Proj. Manag. J.* **2014**, *45*, 5–14. [CrossRef]
31. Tafazzoli, M.; Shrestha, P. Factor Analysis of Construction Delays in the U.S. Construction Industry. In Proceedings of the International Conference on Sustainable Infrastructure 2017, New York, NY, USA, 26–28 October 2017; pp. 111–122.
32. Gamil, Y.; Rahman, I.A. Assessment of critical factors contributing to construction failure in Yemen. *Int. J. Constr. Manag.* **2018**, *20*, 429–436. [CrossRef]
33. Bashir, H.; Ojiako, U.; Mota, C. Modeling and Analyzing Factors Affecting Project Delays Using an Integrated Social Network-Fuzzy MICMAC Approach. *Eng. Manag. J.* **2019**, *32*, 26–36. [CrossRef]
34. Hwang, B.G.; Leong, L.P. Comparison of schedule delay and causal factors between traditional and green construction projects. *Technol. Econ. Dev. Econ.* **2013**, *19*, 310–330. [CrossRef]
35. Hwang, B.-G.; Zhao, X.; Tan, L.L.G. Green building projects: Schedule performance, influential factors and solutions. *Eng. Constr. Archit. Manag.* **2015**, *22*, 327–346. [CrossRef]
36. Lehnert, J. (Ed.) Synchronization in Complex Networks. In *Controlling Synchronization Patterns in Complex Networks*; Springer International Publishing: Cham, Switzerland, 2016; pp. 33–41.
37. Li, C.; Chen, G. Synchronization in general complex dynamical networks with coupling delays. *Phys. A Stat. Mech. Its Appl.* **2004**, *343*, 263–278. [CrossRef]
38. Arenas, A.; Diaz-Guilera, A.; Kurths, J.; Moreno, Y.; Zhou, C. Synchronization in complex networks. *Phys. Rep.* **2008**, *469*, 93–153. [CrossRef]
39. Mogan, R.; Fischer, R.; Bulbulia, J.A. To be in synchrony or not? A meta-analysis of synchrony's effects on behavior, perception, cognition and affect. *J. Exp. Soc. Psychol.* **2017**, *72*, 13–20. [CrossRef]

40. Macpherson, M.C.; Fay, N.; Miles, L.K. Seeing synchrony: A replication of the effects of task-irrelevant social information on perceptions of interpersonal coordination. *Acta Psychol.* **2020**, *209*, 103140. [CrossRef]
41. Li, L.; Xiao, J.; Peng, H.; Yang, Y.; Chen, Y. Improving synchronous ability between complex networks. *Nonlinear Dyn.* **2012**, *69*, 1105–1110. [CrossRef]
42. Lu, P.; Yuan, S.; Wu, J. The interaction effect between intra-organizational and inter-organizational control on the project performance of new product development in open innovation. *Int. J. Proj. Manag.* **2017**, *35*, 1627–1638. [CrossRef]
43. Herrera, R.F.; Mourgues, C.; Alarcón, L.F.; Pellicer, E. Understanding Interactions between Design Team Members of Construction Projects Using Social Network Analysis. *J. Constr. Eng. Manag.* **2020**, *146*, 04020053. [CrossRef]
44. Melander, L.; Lakemond, N. Governance of supplier collaboration in technologically uncertain NPD projects. *Ind. Mark. Manag.* **2015**, *49*, 116–127. [CrossRef]
45. Chinowsky, P.; Taylor, J.E.; Di Marco, M. Project Network Interdependency Alignment: New Approach to Assessing Project Effectiveness. *J. Manag. Eng.* **2011**, *27*, 170–178. [CrossRef]
46. Khouja, A.; Lehoux, N.; Cimon, Y.; Cloutier, C. Collaborative Interorganizational Relationships in a Project-Based Industry. *Buildings* **2021**, *11*, 502. [CrossRef]
47. Solis, F.; Sinfield, J.V.; Abraham, D.M. Hybrid Approach to the Study of Inter-Organization High Performance Teams. *J. Constr. Eng. Manag.* **2013**, *139*, 379–392. [CrossRef]
48. Sun, J.; Lei, K.; Cao, L.; Zhong, B.; Wei, Y.; Li, J.; Yang, Z. Text visualization for construction document information management. *Autom. Constr.* **2019**, *111*, 103048. [CrossRef]
49. Choi, Y.-K. *Principles of Applied Civil Engineering Design: Producing Drawings, Specifications, and Cost Estimates for Heavy Civil Projects*, 2nd ed.; American Society of Civil Engineers: Reston, VA, USA, 2017.
50. Shohet, I.M.; Frydman, S. Communication Patterns in Construction at Construction Manager Level. *J. Constr. Eng. Manag.* **2003**, *129*, 570–577. [CrossRef]
51. Ellis, R.; Li, S.; Zhu, Y. The effects of pre-task explicit instruction on the performance of a focused task. *System* **2019**, *80*, 38–47. [CrossRef]
52. Forcada, N.; Serrat, C.; Rodríguez, S.; Bortolini, R. Communication Key Performance Indicators for Selecting Construction Project Bidders. *J. Manag. Eng.* **2017**, *33*, 04017033. [CrossRef]
53. Loosemore, M. *Crisis Management in Construction Projects*; American Society of Civil Engineers: Reston, VA, USA, 2000.
54. Castillo, T.; Alarcón, L.F.; Pellicer, E. Influence of Organizational Characteristics on Construction Project Performance Using Corporate Social Networks. *J. Manag. Eng.* **2018**, *34*, 04018013. [CrossRef]
55. Lee, C.Y.; Chong, H.-Y.; Liao, P.-C.; Wang, X. Critical Review of Social Network Analysis Applications in Complex Project Management. *J. Manag. Eng.* **2018**, *34*, 04017061. [CrossRef]
56. Yang, M.; Chen, H.; Xu, Y. Stakeholder-Associated Risks and Their Interactions in PPP Projects: Social Network Analysis of a Water Purification and Sewage Treatment Project in China. *Adv. Civ. Eng.* **2020**, *2020*, 8897196. [CrossRef]
57. Malisiovas, A.; Song, X. Social Network Analysis (SNA) for Construction Projects' Team Communication Structure Opti-mization. In Proceedings of the Construction Research Congress 2014, Atlanta, GA, USA, 19–24 May 2014; pp. 2032–2042.
58. Lu, W.; Xu, J.; Söderlund, J. Exploring the Effects of Building Information Modeling on Projects: Longitudinal Social Network Analysis. *J. Constr. Eng. Manag.* **2020**, *146*, 04020037. [CrossRef]
59. Watts, D.J.; Strogatz, S.H. Collective dynamics of 'small-world' networks. *Nature* **1998**, *393*, 440–442. [CrossRef] [PubMed]
60. Kuan, L.; Qixin, S. Research progress of complex network synchronization control. In Proceedings of the 2020 International Conference on Intelligent Design (ICID), Xi'an, China, 11–13 December 2020; pp. 23–28.
61. Lu, J. Synchronization of complex networks: Theories, approaches, applications and prospects. *Adv. Appl. Mech.* **2008**, *38*, 713–722.
62. Xing, W.; Shi, P.; Agarwal, R.K.; Zhao, Y. A survey on global pinning synchronization of complex networks. *J. Frankl. Inst.* **2019**, *356*, 3590–3611. [CrossRef]
63. Hong, H.; Kim, B.J.; Choi, M.; Park, H. Factors that predict better synchronizability on complex networks. *Phys. Rev. E* **2004**, *69*, 067105. [CrossRef]
64. Zhou, C.; Kurths, J. Dynamical Weights and Enhanced Synchronization in Adaptive Complex Networks. *Phys. Rev. Lett.* **2006**, *96*, 164102. [CrossRef]
65. Berec, V. Complexity and dynamics of topological and community structure in complex networks. *Eur. Phys. J. Spec. Top.* **2017**, *226*, 2205–2218. [CrossRef]
66. Zhao, M.; Qu, H.; Li, G. Vulnerable Stations Identification of Urban Rail Transit Network: A Case Study of the Shenzhen Metro. In Proceedings of the ICTE 2019, Chengdu, China, 20–22 September 2019; pp. 549–559.
67. Jia, H.-F.; Han, J.-Q.; Li, Y.-X. Reliability Analysis of a Typical Road Network Based on the Complex Network Theory. In Proceedings of the CICTP 2016, Shanghai, China, 6–9 July 2016; pp. 2187–2200.
68. Yang, L.; Lou, J.; Zhao, X. Risk Response of Complex Projects: Risk Association Network Method. *J. Manag. Eng.* **2021**, *37*, 05021004. [CrossRef]
69. Pan, M.; Pan, W. Stakeholder Perceptions of the Future Application of Construction Robots for Buildings in a Dialectical System Framework. *J. Manag. Eng.* **2020**, *36*, 04020080. [CrossRef]

70. Yuan, J.; Wang, C.; Skibniewski, M.J.; Li, Q. Developing Key Performance Indicators for Public-Private Partnership Projects: Questionnaire Survey and Analysis. *J. Manag. Eng.* **2012**, *28*, 252–264. [CrossRef]
71. Shen, J.; Yang, G. Integrated Empirical Analysis of the Effect of Variable Message Sign on Driver Route Choice Behavior. *J. Transp. Eng. Part A Syst.* **2020**, *146*, 04019063. [CrossRef]
72. Collins, L.M. Research Design and Methods. In *Encyclopedia of Gerontology*, 2nd ed.; Birren, J.E., Ed.; Elsevier: New York, NY, USA, 2007; pp. 433–442.
73. DeVellis, R.F. Inter-Rater Reliability. In *Encyclopedia of Social Measurement*; Kempf-Leonard, K., Ed.; Elsevier: New York, NY, USA, 2005; pp. 317–322.
74. Smith-Colin, J.; Amekudzi-Kennedy, A.; Kingsley, G. Role of Inputs, Processes, and Relations in Transportation System Performance Management: Case Study of Regional Transportation Collaborations. *J. Manag. Eng.* **2021**, *37*, 04020093. [CrossRef]
75. Ahmed, H.; Edwards, D.J.; Lai, J.H.K.; Roberts, C.; Debrah, C.; Owusu-Manu, D.-G.; Thwala, W.D. Post Occupancy Evaluation of School Refurbishment Projects: Multiple Case Study in the UK. *Buildings* **2021**, *11*, 169. [CrossRef]
76. Debrah, C.; Owusu-Manu, D.-G.; Kissi, E.; Oduro-Ofori, E.; Edwards, D.J. Barriers to green cities development in developing countries: Evidence from Ghana. *Smart Sustain. Built Environ.* **2020**. [CrossRef]
77. Xu, Y.; Chan, A.P.C.; Yeung, J.F.Y. Developing a Fuzzy Risk Allocation Model for PPP Projects in China. *J. Constr. Eng. Manag.* **2010**, *136*, 894–903. [CrossRef]
78. Zhao, X.; Hwang, B.-G.; Low, S.P.; Wu, P. Reducing Hindrances to Enterprise Risk Management Implementation in Construction Firms. *J. Constr. Eng. Manag.* **2015**, *141*, 04014083. [CrossRef]
79. Hansen, D.L.; Shneiderman, B.; Smith, M.A.; Himelboim, I. (Eds.) Chapter 3-Social network analysis: Measuring, mapping, and modeling collections of connections. In *Analyzing Social Media Networks with NodeXL*, 2nd ed.; Morgan Kaufmann: Boston, MA, USA, 2020; pp. 31–51.
80. Golbeck, J. (Ed.) Chapter 3-Network Structure and Measures. In *Analyzing the Social Web*; Morgan Kaufmann: Boston, MA, USA, 2013; pp. 25–44.
81. Li, C.; Wang, Z. Analysis of Urban Slow-Moving Traffic Network Structure Characteristics Based on Complex Network Theory. In Proceedings of the ICCTP 2011, Nanjing, China, 14–17 August 2011; pp. 497–505.
82. Gong, Y.; Tang, L.; Yi, H. Selection of Introduction Schemes for Guanzhong Intercity Railway Network Based on Complex Network. In Proceedings of the ICTE 2019, Chengdu, China, 20–22 September 2020; pp. 130–140.
83. Feng, L.; Hu, X. Construction Rules of Urban Rail Transit Network Based on Complex Network Eigenvalue. In Proceedings of the ICTE 2019, Chengdu, China, 20–22 September 2020; pp. 540–548.
84. Wang, L.; Yan, B.; Li, G.; Ma, Y.; Yang, L. Synchronization in collaboration network. *Expert Syst. Appl.* **2021**, *170*, 114550. [CrossRef]
85. Zhang, D.; Zhang, H.; Cheng, T. Causes of Delay in the Construction Projects of Subway Tunnel. *Adv. Civ. Eng.* **2020**, *2020*, 8883683. [CrossRef]
86. Barqawi, M.; Chong, H.-Y.; Jonescu, E. A Review of Employer-Caused Delay Factors in Traditional and Building Information Modeling (BIM)-Enabled Projects: Research Framework. *Adv. Civ. Eng.* **2021**, *2021*, 6696203. [CrossRef]
87. Zhang, P.; Li, N.; Fang, D.; Wu, H. Supervisor-Focused Behavior-Based Safety Method for the Construction Industry: Case Study in Hong Kong. *J. Constr. Eng. Manag.* **2017**, *143*, 05017009. [CrossRef]
88. Zhang, R.P.; Lingard, H.; Oswald, D. Impact of Supervisory Safety Communication on Safety Climate and Behavior in Construction Workgroups. *J. Constr. Eng. Manag.* **2020**, *146*, 04020089. [CrossRef]
89. Newaz, M.T.; Davis, P.; Jefferies, M.; Pillay, M. Examining the Psychological Contract as Mediator between the Safety Behavior of Supervisors and Workers on Construction Sites. *J. Constr. Eng. Manag.* **2020**, *146*, 04019094. [CrossRef]
90. Stewart, R.A. IT enhanced project information management in construction: Pathways to improved performance and strategic competitiveness. *Autom. Constr.* **2007**, *16*, 511–517. [CrossRef]
91. Arditi, D.; Nayak, S.; Damci, A. Effect of organizational culture on delay in construction. *Int. J. Proj. Manag.* **2017**, *35*, 136–147. [CrossRef]
92. Naoum, S.G.; Alyousif, A.-R.T.; Atkinson, A.R. Impact of National Culture on the Management Practices of Construction Projects in the United Arab Emirates. *J. Manag. Eng.* **2015**, *31*, 04014057. [CrossRef]
93. Yin, R.K. *Case Study Research: Design and Methods*, 4th ed.; Sage: Thousand Oaks, CA, USA, 2009.

buildings

MDPI

Article

Regret Theory and Fuzzy-DEMATEL-Based Model for Construction Program Manager Selection in China

Hongyan Yan [1], Yuxuan Yang [2], Xi Lei [1], Qing Ye [1], Wenzhen Huang [1] and Ce Gao [3,4,*]

1 School of Construction Management, Hunan University of Finance and Economics, Changsha 510000, China
2 SWJTU-LEEDS Joint School, Southwest Jiaotong University, Chengdu 610000, China
3 School of Civil Engineering, Guangzhou University, Guangzhou 510000, China
4 School of Civil Engineering and Architecture, Guangzhou City Construction College, Guangzhou 510000, China
* Correspondence: gaoce89@gzhu.edu.cn

Abstract: During the drastic changing process of the construction industry in China, construction program management has been given significant attention. Due to the complexity of construction programs, selecting competent managers is crucially important to its success. Therefore, based on a comprehensive literature review, this paper combines regret theory and the Fuzzy-DEMATEL method to develop a multi-attribute model for construction program manager selection. Firstly, six competence elements are extracted, then the manager selection and evaluation index system are constructed. Secondly, the regret theory is used to simulate the psychological characteristics of the decision makers, combined with Fuzzy-DEMATEL, the comprehensive weights for each element are calculated. Lastly, all alternatives for the selection are sorted and the competent ones are selected. A case study is provided to exam the effectiveness of the developed model. Results shows that the proposed model adopted multi-attribute evaluation and group decision making and took into account the psychological behavior of decision makers as well as influences from the relationships between different attributes. Such results indicate that the proposed model is able to provide more comprehensive and scientific construction program manager selections, which can further improve the management of construction programs.

Keywords: construction program manager selection; group decision-making; multi-attribute evaluation; regret theory; Fuzzy-DEMATEL

Citation: Yan, H.; Yang, Y.; Lei, X.; Ye, Q.; Huang, W.; Gao, C. Regret Theory and Fuzzy-DEMATEL-Based Model for Construction Program Manager Selection in China. *Buildings* **2023**, *13*, 838. https://doi.org/10.3390/buildings13040838

Academic Editor: Maxim A. Dulebenets

Received: 7 February 2023
Revised: 19 March 2023
Accepted: 20 March 2023
Published: 23 March 2023

1. Introduction

In the new journey of the construction industry's development, construction program management, which manages construction projects in groups has gradually become a trend in China [1]. Construction program is more presented in the form of large-scale, complex and groups of "giant projects", such as the West-East Gas Transmission Project, the Yangtze River Three Gorges Project, the Beijing Olympic venues construction project, the South China Sea Petrochemical Project, etc. Construction program management is conducive to enhancing the core competitiveness of enterprises, expanding the market of construction enterprises and realizing enterprises' organization strategy [2]. However, the complexity and uncertainty of different construction projects, especially large-scale projects, has brought great challenges to construction program management, which requires the construction program managers being able to handle rapidly changing programs. As the core personnel in the management process, the construction program manager has a huge impact on the success of the project [3]. Therefore, selecting a competent construction program manager is one of the key factors of the construction program.

However, the current process of selecting and hiring project managers usually uses simple factors in the decision-making process. Such process has significant cognitive limitations while psychological factors and preferences of the decision makers will affect the decision results, thus potentially causing regrets. In such cases, the decision made based on expectation theory is often unable to explain the actual decision-making behavior, and most of the decision makers evaluate alternative construction project managers not in precise numbers, but in fuzzy language, and there are defects such as information loss and poor accuracy of results in the quantification process.

Thus, based on the limitations of currently used construction project manager selection process, this paper combined regret theory, Fuzzy-DEMATEL method, disparity minimization, and multi-attribute group decision making method to quantify, rank, and selects the optimal construction program manager. The proposed model can further improve the scientific nature of the selection of construction program managers for construction enterprises, thus, to improve the overall management effectiveness and efficiency of the programs.

The rest of the paper is organized as follows: (1) a literature review regarding related topics is provided; (2) the data collection process of this study is explained; (3) the competency model of construction program manager is explained; (4) the construction program manager selection model's development process is provided; (5) a case study is provided to prove the feasibility of the proposed model, and (6) the results of the case study as well as main conclusions and future works are presented.

2. Literature Review

The literature review of this paper mainly covered literatures related to methods for construction program (project) manager selection, construction project manager's evaluation indices, language assessment methods, and regret theory.

2.1. Construction Program (Project) Manager Selection

After a comprehensive literature review, only a few studies were found that focused on the selection of construction program managers while most of the documents were aimed at the selection of single project managers. The traditional methods for selecting construction managers in China are mainly achieved through recruitment and open competition [4]. Although organized recruitment method is easy to apply, it is not conducive to innovation and may lead to lack of vitality [5]. Open competition, on the other hand not only selects the best option, but also enhances the competition awareness and the sense of responsibility of the program manager. For open competition, most scholars mainly study the competency of construction project managers [6–8]. In the selection of construction project managers, binary semantic analysis [9], vector angle cosine method [10], analytic hierarchy process [11], support vector basis method [12] are used for selection.

Related research overseas is more in-depth than that in China. Overseas, the selection often requires the application of multi-criteria decision-making (MCDM) methods for robust recruitment [13]. Most foreign scholars also studied the competency of construction managers [14,15], using comprehensive mathematics method [16], analytic hierarchy process [17], Delphi and fuzzy language evaluation method, target programming and topsis [18,19], and other methods. Their main objective was to establish the selection model index system, ranking the alternative construction managers, then selecting the optimal construction managers.

2.2. Construction Project Manager'S Evaluation Indices

For construction project manager's evaluation indices, Skulmoski and Hartman considered that excellent personality charm is one of the strengths of construction project managers to ensure the trust and support of project team members and smooth cooperation with other stakeholders [20]; Krchová's study showed that as the leader of a construction project, the manager must have a convincing personality charm, which plays an important role in mobilizing the team's motivation, uniting the team's fighting force, and successfully implementing the project. Therefore, developing and demonstrating personal charm has become an important challenge for project team managers. Management skills are considered to be a direct reflection of the high level of working ability of project team managers and one of the key hard skills they must master [21]. Considering the characteristics of construction progam managers in practice, most of the time is spent on communication and emotional connection between the program's stakeholders. Therefore, effective communication can help with reducing conflicts between the construction program organization and stakeholders and support the manager to make the right decisions [22]. As a construction program manager, having excellent professional skills is necessary to ensure the program's success [21]. The managers should have the ability to control the risks and discover potential risks of the construction program, to propose effective solutions in a timely manner, and to adopt effective ways to achieve the management goal of the construction program [23]. Furthermore, strategic vision and stress resistance ability are also emphasized as key characteristics of the managers considering the uncertainty and high stress environment of construction program management [22,24,25].

2.3. Language Assessment Methods

Language assessment scale is a prerequisite for group decision making that generally required to select appropriate assessment scale. Levrat and Bordogna et al. [26] used linguistic terms such as "very high", "high", "general", "low", and "very low" to evaluate the scheme. Herrera and Martinez (2000) transformed the evaluation terms into binary groups formed by the values in [−0.5, 0.5]. Dai et al. [27] conducted comparative analysis on the commonly used uniform scale and non-uniform scale. The research shows that the non-uniform language scale is more consistent with people's expression habits as well as the conclusions.

2.4. Regret Theory

For regret theory, it was firstly applied to the choice between two schemes, and Quiggin (1994) extended it to the field of multiple schemes. Bleiehrodt et al. [28] constructed regret-happy function based on exact number and provided its calculation method. On this basis, Zhang, Fan, and Chen [29] used interval number to express the uncertainty of information, while Zhang, Zhu, and Liu [30] used fuzzy number concept to describe it. However, in practice, decision makers tend to use language to evaluate a scheme. Therefore, Zhang and Wang [31] designed a regret-happy function based on language information which shows advantages in language identification.

3. Construction Program Manager Selection Evaluation Index System Based on Grounded Theory

Grounded Theory is a bottom-up qualitative research method based on research questions from field observations by collecting and analyzing data to refine concepts and categories, thus rising to the theoretical level [32]. Grounded theory is able to resolve the differences between traditional quantitative research and qualitative research. At present, academic research on the selection and evaluation index of construction program manager is still in its infancy. Therefore, this paper chooses grounded theory method to construct the selection and evaluation index system of construction program manager selection.

3.1. Data Collection

Considering the number of interviewees available and the purpose of this research, purposive sample is used to select proper interviewees in the data collection process. Purposive sampling, as known as judgment sampling, is typically used to identify and select the information-rich cases for the most proper utilization of available resources [33]. So, managers who have rich experience in construction project/program management and have high credibility in constructing the selection evaluation index system for construction program managers are selected. Based on the literature analysis, this paper prepared a preliminary interview outline. To ensure that the interviewees could accurately understand the purpose of the interview and that the recovered interview data could fully support the study, experts were invited to make preliminary modifications to the interview outline.

On this basis, the first two interviewees were interviewed using the preliminary revised interview outline, and the outline was revised again based on the suggestions of the interviewees to obtain a final interview outline that met the requirements of the study. The interviews were conducted by telephone from January 2021 to March 2021 with 15 interviewees and nearly 20,000 pieces of information were collected, as shown in Figure 1.

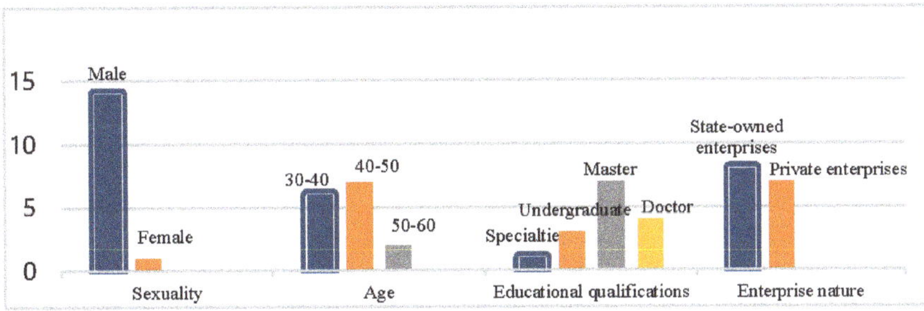

Figure 1. Basic information of interviewees.

Most of the interviewees were males aged between 30 and 50 with considerable experiences. In terms of academic qualifications, the majority of the interviewees had master's degrees or PhDs, reflecting that the interviewees had high level education and social perspective, and could understand the interview questions to give credible responses. In terms of gender, males were significantly more numerous than females in this study's sample. This is mainly caused by the fact that, on the whole, males are more frequently hired in construction project manager positions due to the working condition and intensities. For the employers of the interviewees, the number of state-owned enterprises and private enterprises was equal, reflecting that the work environment was selected in a more balanced manner in this study.

3.2. Development of Construction Program Manager Selection Evaluation Index System Based on Grounded Theory

In this paper, the interview data of 13 construction program manager are used for theoretical modeling, and the rest are used to validate the model. Firstly, the collected raw data are imported into qualitative analysis software, Nvivo11, for spindle and selective coding to identify the characteristic indices and structural dimensions of the managers. Based on the software analysis results and the conducted literature review presented in the previous section, the construction program manager competency model use in this study is shown in Table 1.

Table 1. Competency model of construction program manager.

Serial Number	Index	Secondary Index	Index Connotation
C_1	Personality charm	B_1 Boldness, decisive	Personality charm refers to the manager's ability to attract people in terms of temperament, personality, ideology and morality. In specific work, managers can make decisive decisions and seize key opportunities; managers can communicate with others sincerely and consider problems from the other side's standpoint; managers can solve problems impartially and deal with matters fairly; managers can take responsibility courageously.
		B_2 Sincerity	
		B_3 Responsibility	
		B_4 Transpositional consideration	
		B_5 Act fairly	

Table 1. *Cont.*

Serial Number	Index	Secondary Index	Index Connotation
C_2	Management ability	B_6 Whole process management B_8 Team management B_{20} Experience in management B_{21} Establishing management system	Management ability refers to the ability of managers to formulate a good plan in advance and execute it according to the plan. Managers can tap the maximum potential of members in team management, build a team culture actively, have rich management experience, and establish a suitable management system to manage the whole process.
C_3	Communication and coordination	B_7 Arouse the enthusiasm B_9 Path unification B_{18} Flexible adaptation B_{19} Participant management	Communication and coordination means that managers can coordinate and communicate well with all participants in the construction program, and reach a unified strategy, purpose, and path; managers can actively use their own subjective initiative to coordinate well the relationship between the participants in the construction program.
C_4	Professional skills	B_1 Professional knowledge B_{17} Information technology	Vocational skills refer to managers who have a certain level of professional skills, and have sufficient depth and breadth in professional knowledge, so as to be able to guide the team directionally.
C_5	Risk control	B_{13} Policy risk control B_{14} Market risk control B_{15} Employee turnover risk control	Risk control means that managers can predict and control possible risks and establish risk prevention measures.
C_6	Strategic vision	B_{10} Strategic objective B_{11} Prospective B_{12} Systematic thinking	Strategic perspective means that managers have forward-looking strategic thinking based on system thinking and starting from long-term interests. The program manager needs to maintain the sensitivity and attention to the organizational strategy at all times to serve the realization of the long-term organizational strategic goals.

Based on grounded theory, this paper constructs the selection and evaluation indices of construction program managers, namely C = {C_1 = Personality charm, C_2 = Management ability, C_3 = Communication and coordination, C_4 = Professional skills, C_5 = Risk control, C_6 = Strategic vision}.

4. Construction Program Manager Selection Model

4.1. Method Selection

Due to the complexity of construction programs, the competent construction program manager should have multi-factor competencies. In reality, due to the limitation of personal knowledge and experiences, it is often difficult to achieve optimal decisions only judging personal abilities. Therefore, the selection of construction program managers should be a multi-attribute group decision-making process.

In this process, due to the "bounded rationality" characteristics of the decision makers, the psychological factors and preferences will affect the results. When decision makers have the choice to compare the selected results with other alternatives, it may result in regret, which most decision makers prefer to avoid. Considering the cognitive limitations and subjective psychological preferences of the decision makers, "complete rational" decision-making based on expectation theory is often unable to explain the actual decision-making behavior. To solve such problems, Kahneman and Tversk [34], Bell [35], and Loomes and Sugden [36] proposed prospect theory and regret theory, respectively. Prospect theory considers a series of factors such as reference point dependence, loss dependence and subjective probability of the decision makers, while regret theory focuses on the influence of "regret" of decision makers on decision effectiveness. Compared with prospect theory, regret theory has fewer assumptions which can better describe and explain the paradoxes such as Alai paradox and preference reversal effect in actual decision-making behavior, so it is more widely used in decision-making problems. Therefore, this paper chooses regret theory to take the psychological characteristics of decision makers into consideration and obtains the multi-attribute evaluation matrix to calculate the perceived utility value of each alternative construction program manager.

In the multi-attribute group decision-making of construction project group manager selection, determining the weight of the evaluation attributes is the basis for the ranking and selection process. The current methods for attribute weight determination include MCDM (Multi-criteria decision-making) proposed by Linett Montano Guzman [37], and DEMATEL proposed by Gabus and Fontela [38,39] of Geneva Research Center. MCDM refers to an analytical method for making decisions under multiple decision criteria or objectives, and commonly used methods include hierarchical analysis (AHP), fuzzy comprehensive evaluation (FCE), grey correlation analysis (GRA), and entropy weighting. These methods mostly try to obtain the best decision result by assigning weights to decision factors and evaluating them comprehensively. DEMATEL was originally used to solve complex social events, based on graph theory and matrix to construct structural models through analysis to study the causal relationships between attributes of complex events and identify key attributes for better analysis of events. Therefore, MCDM and DEMATEL are slightly different in their scope of application. MCDM is more suitable for scenarios that require selection from multiple decision criteria or objectives, while DEMATEL is more suitable for scenarios that require understanding the interactions between factors in complex systems [40–42]. With the advancement of related techniques, multi-criteria decision analysis method like Ordinal Priority Approach (OPA) have also emerged. The core idea of OPA is to compare the relative importance of different attributes to determine the best decision. The method usually involves assigning different attributes to different importance levels, such as high, medium and low, and scoring and weighting each attribute to calculate a composite score [43]. Compared to DEMATEL, OPA places more emphasis on the comparison and ranking between attributes, instead of the interdependence between the attributes. For the above considerations, DEMATEL is chosen in this paper.

The key to the application of group decision making is the aggregation of decision makers' judgment information. Due to the differences in professional background, knowledge and ability, the evaluation quality and level of each decision maker, the application in this study should be a heterogeneous decision group. The closer the attribute evaluation result of a decision-maker is to the group attribute evaluation result, the more credible the decision-maker is, and the greater its weight. Therefore, this paper uses deviation minimization and attribute weights to obtain the weights of the decision makers. By comparing the attribute weights determined by each decision maker based on Fuzzy-DEMATEL, smaller weights were given to decision makers with larger difference values. Therefore, this paper selects the regret theory to consider the psychological characteristics of decision makers to obtain the perceived utility matrix. The algorithm flow of linguistic multi-attribute selection of construction program managers based on regret theory and Fuzzy-DEMATEL is shown in Figure 2.

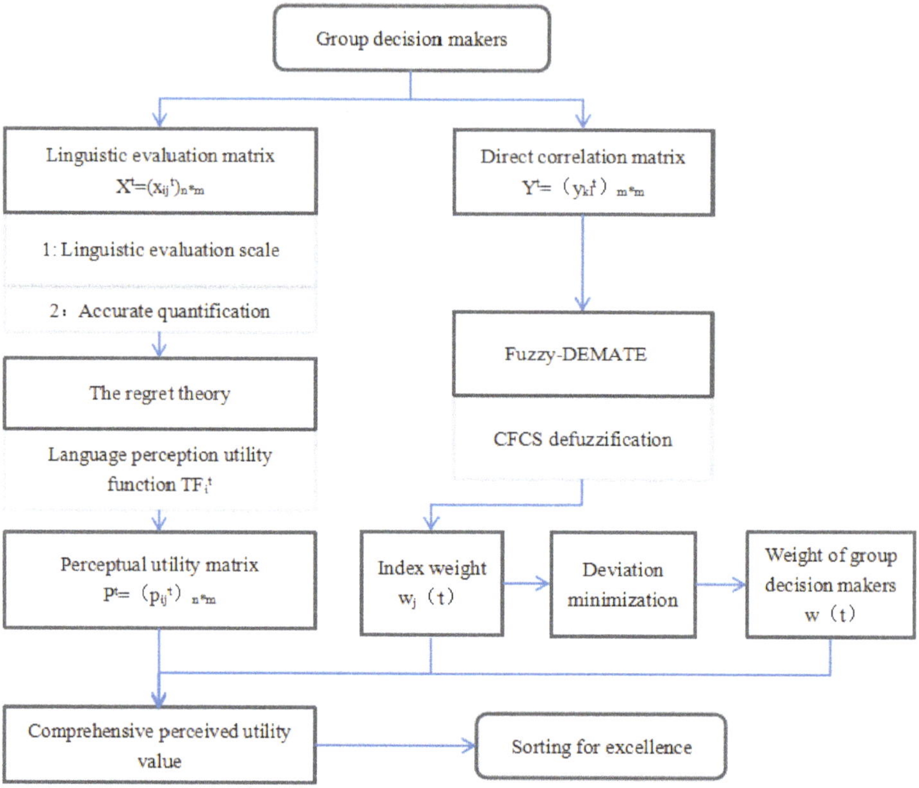

Figure 2. Linguistic Multi-attribute selection of construction program managers based on Regret Theory and Fuzzy-DEMATEL.

4.2. Construction Program Manager Selection

4.2.1. Model Preparation

Assuming the alternative construction program manager set $A = \{A_1, A_2, A_3, \ldots, A_n\}$, where A_i denotes the ith program manager, $i \in N$, $N = \{1, 2, 3, \ldots, n\}$. Property (index) set $C = \{C_1, C_2, C_3, \ldots, C_m\}$, where C_j denotes the j property, $j \in M$, $M = \{1, 2, 3, \ldots, m\}$. The decision maker (expert) set $D = \{D_1, D_2, D_3, \ldots, D_p\}$, where D_t denotes the t_{th} decision maker, $t \in P$, $P = \{1, 2, 3, \ldots, p\}$. The linguistic multi-attribute evaluation matrix $X^t = (x_{ij}^t)_{n\times m}$, where x_{ij}^t denotes the linguistic evaluation value of decision maker D_t on the selection attribute C_j of construction program manager A_i.

In order to make better use of the knowledge and personal experience of the decision makers, let the decision makers to evaluate the degree of the interaction between the attributes, and then construct an attribute correlation matrix, so the direct correlation matrix $Y^t = (y_{kl}^t)_{m\times m}$, where y_{kl}^t represents the impact assessment value of the t decision maker on the k_{th} attribute and the l_{th} attribute.

4.2.2. Evaluation Language Based on Regret Theory

(1) Basic definition

This paper uses the non-uniform language scale and are defined as follows:

The linguistic term set of *a* on the right side of numerical zero is

$$S^+ = \left\{ S_a \middle| a = \frac{2(i-1)}{\sigma+2-i}, \ i = 2, \cdots, \sigma-1, \sigma \right\} \tag{1}$$

The linguistic term set of a on the left side of numerical zero is

$$S^{-} = \left\{ S_a \middle| a = \frac{2(i-1)}{\sigma + 2 - i}, i = \sigma, \sigma - 1, \cdots, 2 \right\} \quad (2)$$

Therefore, the language assessment scale is

$$S = \left\{ S_a \middle| \alpha = -(\sigma - 1), -\frac{2(\sigma - 2)}{3}, \cdots, 0, \cdots, \frac{2(\sigma - 2)}{3}, (\sigma - 1) \right\} \quad (3)$$

In particular, $S_{-(\sigma-1)}$ and $S_{(\sigma-1)}$ denote the lower and upper limits of the linguistic terms actually used by decision makers, σ is a positive integer, and the number of linguistic terms $(2\sigma - 1)$ is called the granularity of the term set, and S_a satisfies the following properties: If $\alpha > \beta$, then $S_\alpha > S_\beta$; there exists a negative operator $neg(S_a) = S_{-a}$.

For example, when σ is 5, the granularity of the linguistic term set is 9, and then

S = {S_{-4} = extremely poor, S_{-2} = very poor, S_{-1} = poor, $S_{-0.4}$ = slightly poor, S_0 = general, $S_{0.4}$ = slightly good, S_1 = good, S_2 = very good, S_4 = extremely good}.

(2) Utility perception value based on language information

The language identification process used in this study is as follows:

Definition 1. *Let $S_a \in S$ be a term for a language evaluation set, then, a subscript conversion function that converts any language assessment into an exact number is used as shown in function (4).*

$$H(S_a) = a \quad (4)$$

Definition 2. *Let $V(X)$ be a classical utility function, which is a monotone increasing concave function, namely $V'(X) > 0$, $V''(X) < 0$, indicating that the decision makers are risk aversion, then, a language utility function is formed as shown in function (5), in which, $\sigma - 1$ is called the cardinality and satisfies $0 \leq (a + \sigma - 1)/(2(\sigma - 1)) \leq 1$.*

$$UV(S_a) = V\left(\frac{H(S_a) + \sigma - 1}{2(\sigma - 1)}\right) = V\left(\frac{a + \sigma - 1}{2(\sigma - 1)}\right) \quad (5)$$

From Definition 2, when S_a takes the maximum value, the language utility function is also the largest. When S_a takes the minimum value, the language utility function is the smallest. Therefore, it will ensure the accuracy of the results without information loss.

Definition 3. *Let S_a, $S_a{}^*$, $S_a{}^-$, $S_a{}^+$ be the current selected construction program manager, the ideal construction program manager, the negative ideal construction program manager and the positive ideal construction program manager, respectively. $R(y)$ is a classical regret-happy function, which is also a monotone increasing concave function, where $R'(y) > 0$, $R''(y) < 0$, and to meet the intuitive judgment result $R(0) = 0$, then, there is function (6) that helps the decision maker to chooses the current project construction program manager A_i and abandons the ideal construction program manager A^*.*

$$TR_i = R\left(V\left(\frac{a + \sigma - 1}{2(\sigma - 1)}\right) - V\left(\frac{a^* + \sigma - 1}{2(\sigma - 1)}\right)\right) \quad (6)$$

When $S_a{}^* = S_a{}^-$, expressed as the negative ideal construction program manager language evaluation value, namely

$$TR_i{}^- = R\left(V\left(\frac{a + \sigma - 1}{2(\sigma - 1)}\right) - V\left(\frac{a^- + \sigma - 1}{2(\sigma - 1)}\right)\right) \quad (7)$$

When $S_a{}^* = S_a{}^+$, expressed as the positive ideal construction program manager language evaluation value, namely

$$\mathrm{TR_i}^+ = \mathrm{R}\left(\mathrm{V}\left(\frac{\mathrm{a}+\sigma-1}{2(\sigma-1)}\right) - \mathrm{V}\left(\frac{\mathrm{a}^+ + \sigma - 1}{2(\sigma-1)}\right)\right) \tag{8}$$

Definition 4. *Suppose that the language utility function of decision maker Dt for the evaluation value S_a of A_i of the selected project construction group manager is $UV(S_a)$, and the language regret-happiness function is TR i, then, decision maker D_t chooses the language perception utility function of the construction program manager A_i.*

$$\mathrm{TF_i^t} = \mathrm{UV(S_a)} + \mathrm{TR_i} = \mathrm{UV(S_a)} + \mathrm{TR_i}^- + \mathrm{TR_i}^+ = \mathrm{V}\left(\frac{\mathrm{a}+\sigma-1}{2(\sigma-1)}\right) + \mathrm{R}\left(\mathrm{V}\left(\frac{\mathrm{a}+\sigma-1}{2(\sigma-1)}\right) - \mathrm{V}\left(\frac{\mathrm{a}^- + \sigma-1}{2(\sigma-1)}\right)\right) + \mathrm{R}\left(\mathrm{V}\left(\frac{\mathrm{a}+\sigma-1}{2(\sigma-1)}\right) - \mathrm{V}\left(\frac{\mathrm{a}^+ + \sigma-1}{2(\sigma-1)}\right)\right) \tag{9}$$

Through the function TF_i^t, the language perception utility value of any construction program manager of any decision maker can be obtained.

In this paper, we take the function $R(y) = 1 - exp(\delta \cdot y)$, where the parameter $\delta \in [0, +\infty]$ is the regret aversion coefficient of the decision makers. The greater the parameter δ, the greater the regret aversion degree of the decision makers, and vice versa. In addition, the power function $V(x) = X^{\mu}(0 < \mu < 1)$ is used as the utility function, where μ denotes the risk aversion coefficient of decision makers, and the greater the μ, the smaller the degree of risk aversion of decision makers.

4.2.3. Determination of Index Weights Based on Fuzzy-DEMATEL

The key to traditional DEMATEL method is to invite experts to evaluate the mutual influence of each attribute based on their knowledge and experience to form a direct correlation matrix. Due to the uncertainty of practical problems, the complexity of evaluation and the differences between invited experts, most of the evaluations given by experts are not accurate but are similar to fuzzy semantic expressions such as "important" or "satisfied". Therefore, this paper introduces the triangular fuzzy number method to process the initial matrix to improve the accuracy and the steps are as follows [44]:

Step 1: The construction program manager influencing factors system is constructed, denoted as $F_1, F_2, \ldots, F_6$. The semantic scale assessed by experts is designed and divided into five levels according to the degree of influence, namely: no influence "0", very weak influence "1", weak influence "2", strong influence "3" and very strong influence "4", as shown in Table 2.

Table 2. Semantic transformation table.

Semantic Variable	Numerical Value	Corresponding Triangular Fuzzy Number
No influence	0	(0, 0, 0.2)
Very weak influence	1	(0, 0.2, 0.4)
Weak influence	2	(0.2, 0.4, 0.6)
Strong influence	3	(0.4, 0.6, 0.8)
Very strong influence	4	(0.8, 1, 1)

Step 2: Invite experts to use language operators to evaluate the influencing factors of the construction program manager, and convert the evaluation semantics into the corresponding triangular fuzzy number $W_{ij}{}^t = (\beta_{1ij}{}^t, \beta_{2ij}{}^t, \beta_{3ij}{}^t)$ according to the semantic transformation table, which means that the t experts believe that the factor i has an effect on the factor j, where $\beta_{1ij}{}^t$ is a conservative value, $\beta_{2ij}{}^t$ is the closest to the reality value, and $\beta_{3ij}{}^t$ is an optimistic value.

Step 3: Using the CFCS method to de-fuzzify the triangular fuzzy number, the direct influence matrix Z is obtained. Z reflects the direct effect between factors, and the steps are the following three steps.

(1) Standardization of triangular fuzzy numbers:

$$x\beta_{1ij}^{t} = \left(\beta_{1ij}^{t} - \min\beta_{1ij}^{t}\right)/\Delta_{\max}^{\min} \tag{10}$$

$$x\beta_{2ij}^{t} = \left(\beta_{2ij}^{t} - \min\beta_{1ij}^{t}\right)/\Delta_{\max}^{\min} \tag{11}$$

$$x\beta_{3ij}^{t} = \left(\beta_{3ij}^{t} - \min\beta_{1ij}^{t}\right)/\Delta_{\max}^{\min} \tag{12}$$

where $\Delta_{\max}^{\min} = \max\beta_{3ij}{}^{t} - \min\beta_{1ij}{}^{t}$, and $x\beta_{1ij}{}^{t}$, $x\beta_{2ij}{}^{t}$, $x\beta_{3ij}{}^{t}$ are calculated in turn.

(2) Standardize left (ls) and right (rs) values:

$$xls_{ij}^{t} = x\beta_{2ij}^{t}/\left(1 + x\beta_{2ij}^{t} - x\beta_{1ij}^{t}\right) \tag{13}$$

$$xrs_{ij}^{t} = x\beta_{3ij}^{t}/\left(1 + x\beta_{3ij}^{t} - x\beta_{1ij}^{t}\right) \tag{14}$$

(3) Calculate the clarity value after defuzzification:

$$x_{ij}^{t} = \left[xls_{ij}^{t}\left(1 - xls_{ij}^{t}\right) + xrs_{ij}^{t} \times xrs_{ij}^{t}\right]/\left[1 - xls_{ij}^{t} + xrs_{ij}^{t}\right] \tag{15}$$

$$Z_{ij}^{t} = \min\beta_{1ij}^{t} + x_{ij}^{t} \times \Delta_{\max}^{\min} \tag{16}$$

Step 4: Standardize the direct impact matrix Z^t to get the standardized direct impact matrix $G^t = (g_{kl})_{m\times m}$, where

$$g_{kl}^{t} = z_{kl}/\max_{1\le i\le m}\sum_{j=1}^{m} z_{kl} \tag{17}$$

Step 5: Measure the comprehensive influence matrix T^t, namely

$$\begin{aligned} T^{t} &= \lim_{m\to\infty}\left(G^{1} + G^{2} + \cdots + G^{m}\right) \\ &= Z(E - Z)^{-1} \end{aligned} \tag{18}$$

where E is the unit matrix, when $m \to \infty$, $G^m = 0$ is satisfied.

Step 6: Calculate the importance of influencing factors ε_j^t.

Note that each row of the elements in T^t is added to the influence degree $r_k{}^t$, indicating the combined influence value. The addition of each column element in T^t is the affected degree $d_i{}^t$, indicating the comprehensive influence value of this element by other elements. Let k = l = j, then the importance of the influencing factor of ε_j^t is

$$\varepsilon_j^{t} = \sqrt{\left(r_k{}^{t} + d_i{}^{t}\right)^2 + \left(r_k{}^{t} - d_i{}^{t}\right)^2} \tag{19}$$

Step 7: Determine the attribute weight $w_j(t)$. Normalize the importance of the influencing factor ε_j^t in step 6 to obtain the index weight as

$$w_j(t) = \varepsilon_j^{t}/\sum_{j=1}^{m}\varepsilon_j^{t} \tag{20}$$

In the formula, $0 < w_j(t) < 1$, and satisfy $\sum_{j=1}^{m} w_j(t) = 1$, for generality, let $w_j(t) = (w_1(t), w_2(t), \cdots, w_m(t))^T$, $w_j(t)$ represents the attribute weight of the t-th decision maker based on Fuzzy-DEMATEL on the j-th attribute.

4.2.4. Weight Determination of Decision Makers Based on Deviation Minimization

The attribute weights are obtained by Fuzzy-DEMATEL based on the attribute evaluation of each decision maker. Based on the idea that smaller differences mean larger weights, this paper uses the deviation minimization to obtain the weight of decision makers [45].

For attribute weight $w_j(t)$, the difference value of attribute weight between decision maker D_t and other decision makers is $E_j(t)$:

$$E_j(t) = \sum_{t'=1}^{P} \{w_j(t) - w_j(t')\}^2 \ (t' = 1, 2, \cdots, P) \tag{21}$$

Then define the attribute weight difference $E(t)$ of the decision maker D_t with respect to all attributes compared to other decision makers as:

$$E(t) = \sum_{j=1}^{m} \sum_{t'=1}^{p} \{w_j t - w_j(t')\}^2 \ (t' = 1, 2, \cdots, P) \tag{22}$$

The selection of decision maker's weighting vector $\varphi(t)$ should minimize the difference value of the total attribute weight of all decision makers for all attributes. Therefore, an objective weighting model for decision makers is constructed:

$$\min E = \sum_{t=1}^{p} \varphi(t)^2 E(t) \tag{23}$$

$$s.t \sum_{t=1}^{p} \varphi(t) = 1, \ \varphi(t) > 0, t = 0, 1, \cdots, p$$

Introduce the Lagrange function to solve the above model:

$$L(\varphi(t), \theta) = \sum_{t=1}^{p} \varphi(t)^2 E(t) + 2\theta \left[\sum_{t=1}^{p} \varphi(t) - 1\right] \tag{24}$$

The derivations of $\varphi(t)$ and θ are obtained:

$$\begin{cases} \frac{\partial L}{\partial \varphi(t)} = 2\varphi(t)E(t) + 2\theta = 0 \\ \sum_{t=1}^{P} \varphi(t) = 1 \end{cases} \tag{25}$$

Thus:

$$\varphi(t) = \frac{1}{E(t)} \frac{1}{\sum_{t=1}^{p} \frac{1}{E(t)}} \tag{26}$$

The decision maker weight vector $\varphi(t)$ is normalized to get the decision maker weight:

$$W(t) = \varphi(t) / \sum_{t=1}^{p} \varphi(t) \tag{27}$$

In the formula, $0 < w(t) < 1$, and satisfies $\sum_{t=1}^{p} w(t) = 1$, for generality, let $w(t) = (w(1), w(2), \cdots, w(p))^T$, $w(t)$ represents decision maker D_t is based on the weight of decision maker with minimum deviation.

4.2.5. Comprehensive Perceived Utility Value Calculation and Decision-Making

Let that $p_{ij}^{\ t}$ ($i \in N, j \in M, t \in P$) is the perceptual utility value calculated by decision maker D_t for the linguistic assessment value x_{ij} of alternative construction program manager A_i for attribute C_j, $w_j(t)$ is the attribute weight of decision maker D_t to attribute C_j based on Fuzzy-DEMATEL, $w_j(t)$ is the decision maker D_t's decision maker weight based on deviation minimization, then the comprehensive perceptual utility value of alternative construction program manager A_i is:

$$p_i^* = \sum_{t=1}^{p} \sum_{j=1}^{m} [\tau w_j(t) + (1 - \tau) w(t)] p_{ij}(t) \tag{28}$$

In the formula, the parameter $\tau \in [0, 1]$ is the weight preference adjustment coefficient, the larger the value of τ, indicating that the group decision makers pay more attention to the attribute weight based on Fuzzy-DEMATEL.

5. Case Study

5.1. Background

In this case of selecting construction program manager, there are five candidates $A = \{A_1, A_2, A_3, A_4, A_5\}$, and three decision makers $D = \{D_1, D_2, D_3\}$. According to the selection attribute system in this paper, $C = \{C_1$ = Personality charm, C_2 = Management ability, C_3 = Communication and coordination, C_4 = Professional skills, C_5 = Risk control, C_6 = Strategic vision$\}$. Using the language assessment scale $S = \{S_{-4}$ = Extreme poor, S_{-2} = Very poor, S_{-1} = Poor, $S_{-0.4}$ = Slightly poor, S_0 = General, $S_{0.4}$ = Slightly good, S_1 = Good, S_2 = Very good, S_4 = Extreme good$\}$ to evaluate the five candidates, the linguistic evaluation matrix for the group decision makers can be obtained as shown in Tables 3–5. The significant impact between different attributes, such as personality charm, having a significant impact on communication and coordination, has also been taken into consideration. Then, the three decision makers used {no influence "0", very weak influence "1", weak influence "2", strong influence "3", very strong influence "4"} to analyze the influence relationship between the attributes and use the language assessment scale to get the direct correlation matrix between attributes as shown in Tables 6–8.

Table 3. Linguistic Multi-attribute Evaluation Matrix X_1 of decision maker D_1.

Alternative Construction Program Manager	Evaluation Index					
	C_1	C_2	C_3	C_4	C_5	C_6
A_1	S_4	S_1	S_4	S_2	$S_{0.4}$	S_2
A_2	S_2	S_1	$S_{0.4}$	S_4	S_4	S_1
A_3	S_1	S_4	S_4	$S_{0.4}$	$S_{0.4}$	S_4
A_4	S_2	$S_{0.4}$	S_4	S_2	$S_{0.4}$	S_4
A_5	S_4	S_4	S_1	S_1	S_4	S_0

Table 4. Linguistic Multi-attribute Evaluation Matrix X_2 of decision maker D_2.

Alternative Construction Program Manager	Evaluation Index					
	C_1	C_2	C_3	C_4	C_5	C_6
A_1	S_2	S_4	$S_{0.4}$	S_1	$S_{0.4}$	S_2
A_2	S_4	S_2	S_4	$S_{0.4}$	S_4	S_1
A_3	S_4	S_1	$S_{0.4}$	S_4	$S_{0.4}$	S_4
A_4	S_1	S_4	S_0	S_4	$S_{0.4}$	S_4
A_5	$S_{0.4}$	S_2	S_2	S_1	S_4	S_0

Table 5. Linguistic Multi-attribute Evaluation Matrix X_3 of decision maker D_3.

Alternative Construction Program Manager	Evaluation Index					
	C_1	C_2	C_3	C_4	C_5	C_6
A_1	$S_{0.4}$	S_1	S_4	S_2	S_4	S_4
A_2	S_2	$S_{0.4}$	S_4	S_1	S_4	S_1
A_3	S_1	S_1	S_4	S_4	S_1	S_4
A_4	$S_{0.4}$	S_4	S_4	S_2	S_2	S_1
A_5	S_2	S_4	$S_{0.4}$	S_4	S_1	S_2

Table 6. Direct correlation matrix Y_1 between indicators of decision maker D_1.

Evaluation Index	C_1	C_2	C_3	C_4	C_5	C_6
C_1	0	2	3	0	1	2
C_2	2	0	4	2	3	2
C_3	3	3	0	1	2	1
C_4	1	2	1	0	1	0
C_5	1	3	1	1	0	2
C_6	3	1	1	1	2	0

Table 7. Direct correlation matrix Y_2 between indicators of decision maker D_2.

Evaluation Index	C_1	C_2	C_3	C_4	C_5	C_6
C_1	0	3	2	0	1	1
C_2	2	0	3	2	4	2
C_3	4	3	0	1	2	1
C_4	2	2	1	0	1	1
C_5	1	3	1	0	0	2
C_6	4	2	1	1	2	0

Table 8. Direct correlation matrix Y_3 between indicators of decision maker D_3.

Evaluation Index	C_1	C_2	C_3	C_4	C_5	C_6
C_1	0	3	3	0	1	2
C_2	3	0	4	2	4	3
C_3	4	4	0	0	3	1
C_4	1	3	2	0	1	0
C_5	1	4	2	0	0	3
C_6	3	2	1	0	3	0

5.2. Decision-Making Steps

(1) Given the parameters $\mu = 0.88$ and $\delta = 0.3$ [17,18], Formulas (4)–(9) are used to process the linguistic multi-attribute evaluation matrix X given by each decision maker to obtain the perceived utility function $p_{ij}{}^{t}$, as shown in Tables 9–11.

Table 9. Perceived utility value of decision maker D_1.

Alternative Construction Program Manager	Evaluation Index					
	C_1	C_2	C_3	C_4	C_5	C_6
A_1	1.0966	0.5752	1.1155	0.7450	0.4603	0.7644
A_2	0.7409	0.5752	0.4603	1.1155	1.1155	0.5891
A_3	0.5543	1.1155	1.1155	0.4603	0.4603	1.1280
A_4	0.7409	0.4603	1.1155	0.7450	0.4603	1.1280
A_5	1.0966	1.1155	0.5752	0.5752	1.1155	0.3966

Table 10. Perceived utility value of decision maker D_2.

Alternative Construction Program Manager	Evaluation Index					
	C_1	C_2	C_3	C_4	C_5	C_6
A_1	0.7450	1.0966	0.4744	0.5752	1.0966	0.5752
A_2	1.1155	0.7409	1.1280	0.4603	0.7409	0.4603
A_3	1.1155	0.5543	0.4744	1.1155	1.0966	0.4603
A_4	0.5752	1.0966	0.3966	1.1155	0.5543	1.1155
A_5	0.4603	0.7409	0.47644	0.5752	0.7409	1.1155

Table 11. Perceived utility value of decision maker D_3.

Alternative Construction Program Manager	Evaluation Index					
	C_1	C_2	C_3	C_4	C_5	C_6
A_1	0.5369	0.5752	1.1155	0.7409	1.0966	1.0966
A_2	0.8176	0.4603	1.1155	0.5543	1.0966	0.5543
A_3	0.6502	0.5752	1.1155	1.0966	0.5543	1.0966
A_4	0.5369	1.1155	1.1155	0.7409	0.7409	0.5543
A_5	0.8176	1.1155	0.4603	1.0966	0.5543	0.7409

(2) The direct correlation matrix Y between indicators is processed according to Formulas (10)–(20), and the attribute weights are obtained as follows.

$$w_j(1) = (0.1737, 0.2083, 0.1950, 0.1118, 0.1647, 0.1464)^T$$

$$w_j(2) = (0.1778, 0.2076, 0.1751, 0.1108, 0.1727, 0.1559)^T$$

$$w_j(3) = (0.1689, 0.2112, 0.1886, 0.0970, 0.1823, 0.1519)^T$$

(3) According to Formulas (21)–(27), the attribute weight $w_j(t)$ is processed, and the weight of decision makers is obtained as follows.

$$W(t) = (0.3287, 0.3448, 0.3265)$$

(4) According to Formula (28), the preference coefficient $\tau = 0$, which means the process only focus on the weight of decision makers based on the minimization of deviation. So, the perceived utility value of each alternative construction program manager is $p_i^* = (5.3858, 5.1525, 5.4812, 5.3175, 5.2502)^T$. Thus, the order of construction program managers is $A_3 > A_1 > A_4 > A_5 > A_2$, that is, the construction unit chooses A_3 as the optimal construction program manager.

5.3. Parameter Sensitivity Analysis

When calculating the comprehensive weight, there is a preference adjustment parameter τ. Therefore, in the process of ranking the candidate construction program managers, the comprehensive perceived utility value p_i^* of the construction program managers will be different. This will ultimately affect the ranking of the candidates. The following change parameter τ values $\tau = 0$, $\tau = 0.2$, $\tau = 0.4$, $\tau = 0.6$, $\tau = 0.8$, $\tau = 1$, resulted in different alternative rankings as shown in Figure 3.

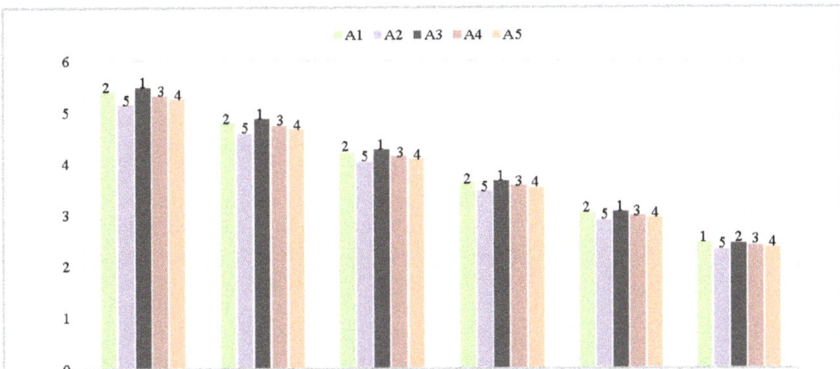

	τ=0	τ=0.2	τ=0.4	τ=0.6	τ=0.8	τ=1
A1	5.3858	4.7976	4.2095	3.6217	3.0337	2.4457
A2	5.1525	4.5853	4.0181	3.4511	2.8840	2.3169
A3	5.4812	4.8719	4.2627	3.6537	3.0446	2.4355
A4	5.3175	4.7311	4.145	3.5591	2.9729	2.3867
A5	5.2502	4.6714	4.0926	3.514	2.9354	2.3566

Figure 3. Ranking of alternative construction program managers.

It can be seen from the sensitivity analysis results that large preference adjustment parameter gives larger weight $w_j(t)$ based on Fuzzy-DEMATEL.

When $\tau = 1$, only attribute weights are considered, the first and second rankings in the ranking result will be switched. When $\tau < 1$, it can be found from the figure that as τ decreases to zero, the ranking results of the candidate construction program managers are consistent and do not change. Based on the above results, on the one hand, when $\tau = 1$, due to ignoring the weight of decision makers, the ranking results change greatly, this indicates that the model may have some instability when only attribute weights are considered, suggesting that decision maker weights should be considered in a comprehensive manner in practical applications. For example, in a construction program, an optimal manager candidate needs to be selected from several alternatives. However, different decision makers may have different views on the importance of the characteristics, so the decision maker weights need to be considered together when ranking. If we only consider the attribute weights and ignore the decision maker weights, it may lead to instability in the ranking results and thus bring negative impact to the selection process. Therefore, in practical application, we need to consider both attribute weights and decision maker weights according to the context in order to better stabilize the ranking results and provide a more credible basis for the selection of construction program managers.

On the other hand, the ranking results did not change when $\tau < 1$. Although, theoretically the different values of the parameter may have an impact on the ranking of the alternative construction program managers. However, if sensitivity analysis of the parameters reveals that the ranking results of the alternatives did not change significantly, this indicates that the model is stable when the parameter changes. This stability may enable decision makers to use the model with more confidence. It is important to note that although the model performs well in such cases, in practical applications, decision makers still need to choose the parameters carefully to ensure reliable results.

6. Conclusions and Contributions

Construction program managers are the core personnel in the management process. They need to respond and handle various events flexibly to promote the success of the program. This paper proposed a multi-attribute manager evaluation and selection model based on grounded theory, regret theory and Fuzzy-DEMATEL methods. During the course of completing this study, the following conclusions and contributions were made:

(1) By using grounded theory and semi-structured interview, this paper proposed a construction program managers selection and evaluation index system that is able to take various attribute into consideration, including personality charm, management ability, communication and coordination, professional skills, risk control, and strategic vision, which comprehensively reflects the competency requirements for construction program managers.

(2) In the process of decision-making, this study identified that the decision makers have "bounded rationality", and their psychological factors and preferences will have an impact on the decision results. Therefore, this paper introduces the idea of "regret theory" to make the decisions more practical.

(3) The method based on Fuzzy-DEMATEL attribute weight determination method used in this paper is found more efficient for complex situations as it considered both fuzzy language evaluation and the mutual influence between attributes.

(4) Based on the analysis of attribute weights, the decision makers' weights are analyzed using deviation minimization method. Their weights and attribute weights are effectively combined to propose more comprehensive and reasonable weights compared to traditional single weight methods.

(5) The case study shows that the construction program manager selection model proposed in this paper takes into account the psychological behavior of decision makers and group rationality and considers the mutual influence relationship between attributes. This provides a new effective way to solve the problem of construction program manager selection.

(6) Compared to past studies, the proposed study took into account the "limited rationality" of decision makers, which is different from the traditional assumption of "perfect rationality". In order to reflect the actual decision-making situation more accurately, this paper introduces "regret theory" to consider the risk attitude and decision pref-

erences of decision makers. This approach makes up for the shortcomings of past studies, and also helps to improve the accuracy and practicality of the selection model of construction program managers.

7. Future Works

This paper used fuzzy numbers to construct the construction program manager evaluation and selection model. In the future, the use of other data forms such as interval numbers and incomplete information can be attempted. In addition, hybrid approaches have been used frequently in personnel, supplier, and key factor selection problems in different industries, so combining different methods such as ANP, TOPSIS, MCDM, and MEMATEL can also be attempted in the future.

Author Contributions: Conceptualization, H.Y.; Software, W.H.; Investigation, Y.Y.; Data curation, C.G.; Writing—original draft, X.L. and Q.Y.; Writing—review & editing, H.Y.; Supervision, H.Y. and C.G. All authors have read and agreed to the published version of the manuscript.

Funding: The present study was supported by the Social Sciences Fund of Hunan Province (22ZDB089, 22JD076) and the National University Student Innovation and Entrepreneurship Project (202111532010).

Data Availability Statement: Some or all data, models, or code that support the findings of this study are available from the corresponding author upon reasonable request.

Conflicts of Interest: The authors declare no conflict of interest.

References

1. Bai, L.; Zhen, K.; Shi, H.; Guo, W.; Du, Q. Construction of collaborative management organization mode of enterprise project group. *J. Eng. Manag.* **2019**, *33*, 91–96. (In Chinese)
2. Jia, G.; Chen, Y.; Xue, X.; Chen, J.; Cao, J.; Tang, K. Program management organization maturity integrated model for mega construction programs in China. *Int. J. Proj. Manag.* **2011**, *29*, 834–845. [CrossRef]
3. Guo, F.; Shi, B.; Chen, Y. Research on the coupling interaction structure between project group construction and enterprise growth of large construction enterprises. *J. Railw. Sci. Eng.* **2015**, *12*, 449–454. (In Chinese)
4. Hong, Y.; Xu, Y. Analysis of project manager selection based on competency model. *China Bus.* **2013**, *18*, 35–37. (In Chinese)
5. Jian, S. Analysis of enterprise youth management personnel training mechanism from the perspective of governance modernization. *China Youth Res.* **2019**, *5*, 37–41. (In Chinese)
6. Liu, H.; Cao, Q. Research on the competency of large-scale project managers. *Sci. Technol. Dev.* **2017**, *13*, 540–546. (In Chinese)
7. Shui, Z.; Fei, K. Analysis on the difference of project manager's competency under the construction mode of DBB and DB/EPC. *J. Civ. Eng.* **2014**, *47*, 129–135. (In Chinese)
8. Sun, C.; Song, H.; Zhai, X.; Zhang, H. Construction of professional ability index system for construction project managers. *Prediction* **2018**, *32*, 72–76. (In Chinese)
9. Li, J.; Zhang, F.; Zhang, W.; Wang, J. Selection method of construction project manager based on capability characteristic analysis. *J. Railw. Eng.* **2019**, *36*, 94–100. (In Chinese)
10. Chen, W.; Wang, H.; Yan, H.; Li, M. Competency evaluation of construction project manager based on cosine of vector angle. *J. Civ. Eng. Manag.* **2018**, *35*, 32–38+84. (In Chinese)
11. Dong, X. Project manager selection system model based on analytic hierarchy process. *J. Pu'er Univ.* **2016**, *32*, 53–55. (In Chinese)
12. Shui, Z.; Fei, K.; Xiang, L. Competency evaluation of construction project manager based on support vector machine. *China Soft Sci.* **2013**, *11*, 83–90. (In Chinese)
13. Ceran, T.; Dorman, A.A. The Complete Project Manager. *J. Archit. Eng.* **1995**, *2*, 67–72. [CrossRef]
14. Sina, M.; Kalle, K.; Jonny, K.O.; Kirsi, A. A Competency Model for the Selection and Performance Improvement of Project Managers in Collaborative Construction Projects: Behavioral Studies in Norway and Finland. *Buildings* **2020**, *11*, 4. [CrossRef]
15. Alvarenga, J.C.; Branco, R.R.; Guedes, A.L.A.; da Silveira e Silva, W. The project manager core competencies to project success. *Int. J. Manag. Proj. Bus.* **2020**, *2*, 277–292. [CrossRef]
16. Hanna, A.S.; Ibrahim, M.W.; Karim, A.W.L. Modeling Project Manager Competency: An Integrated Mathematical Approach. *J. Constr. Eng. Manag.* **2016**, *142*, 04016029. [CrossRef]
17. Kumar, S.K.; Anup, K. Facilitating quality project manager selection for Indian business environment using analytical hierarchy process. *Int. J. Qual. Reliab. Manag.* **2018**, *6*, 1177–1194.
18. Afshari, A.R. Selection of construction project manager by using Delphi and fuzzy linguistic decision making. *J. Intell. Fuzzy Syst.* **2015**, *6*, 2827–2838. [CrossRef]
19. Afshari, A.R. Methods for Selection of Construction Project Manager: Case Study. *J. Constr. Eng. Manag.* **2017**, *143*, 06017003. [CrossRef]

20. Skulmoski, G.J.; Hartman, F.T. Information systems project manager soft competencies: A project-phase investigation. *Proj. Manag. J.* **2010**, *41*, 61–80. [CrossRef]
21. Alshammari, F.; Yahya, K.; Haron, Z.B. Project Manager's Skills for improving the performance of complex projects in Kuwait Construction Industry: A Review. *IOP Conf. Ser. Mater. Sci. Eng.* **2020**, *713*, 012041. [CrossRef]
22. Wu, G.; Hu, Z.; Zheng, J. Role stress, job burnout, and job performance in construction project managers: The moderating role of career calling. *Int. J. Environ. Res. Public Health* **2019**, *16*, 2394. [CrossRef] [PubMed]
23. Wang, C.M.; Xu, B.B.; Zhang, S.J.; Chen, Y.Q. Influence of personality and risk propensity on risk perception of Chinese construction project managers. *Int. J. Proj. Manag.* **2016**, *34*, 1294–1304. [CrossRef]
24. Ghorbani, A. A Review of Successful Construction Project Managers' Competencies and Leadership Profile. *J. Rehabil. Civ. Eng.* **2023**, *11*, 76–95.
25. Shehu, Z.; Egbu, C. The skills and competencies of programme managers. In Proceedings of the RICS-COBRA 2007 Annual Conference, Atlanta, GA, USA, 6–7 September 2007; pp. 6–7.
26. Levrat, E.; Voisin, A.; Bombardier, S.; Brémont, J. Subjective evaluation of car seat comfort with fuzzy set techniques. *Int. J. Intell. Syst.* **1997**, *11–12*, 891–913. [CrossRef]
27. Dai, Y.; Xu, Z.; Li, Y.; Da, Q. New scale of language information assessment and its application. *China Manag. Sci.* **2008**, *2*, 145–149. (In Chinese)
28. Bleichrodt, H.; Cillo, A.; Diecidue, E. A quantitative measurement of regret theory. *Manag. Sci.* **2010**, *56*, 161–175. [CrossRef]
29. Zhang, X.; Fan, Z.; Chen, F. Risk based multiple attribute decision making method based on Regret Theory. *Syst. Eng. Theory Pract.* **2013**, *33*, 2313–2320. (In Chinese)
30. Shi, Z.; Jian, Z.; Xiao, L. Group decision making method based on Regret Theory under multi-dimensional preference information of scheme pair. *China Manag. Sci.* **2014**, *22*, 33–41. (In Chinese)
31. Fa, Z.; Wei, W. Linguistic multiple attribute decision making method based on Regret Theory and DEMATEL. *China Manag. Sci.* **2020**, *28*, 201–210. (In Chinese)
32. Glaser, B.G.; Strauss, A.L.; Strutzel, E. The discovery of Grounded Theory; strategies for qualitative research. *Nurs. Res.* **1968**, *17*, 377–380. [CrossRef]
33. Etikan, I.; Musa, S.A.; Alkassim, R.S. Comparison of convenience sampling and purposive sampling. *Am. J. Theor. Appl. Stat.* **2016**, *5*, 1–4. [CrossRef]
34. Kahneman, D.; Tversky, A. Prospect Theory: An Analysis of Decision under Risk. *Econometrica* **1979**, *2*, 263–291. [CrossRef]
35. Bell, D.E. Regret in Decision Making under Uncertainty. *Oper. Res.* **1982**, *30*, 961–981. [CrossRef]
36. Loomes, G.; Sugden, R. Regret Theory: An Alternative Theory of Rational Choice Under Uncertainty. *Econ. J.* **1982**, *92*, 805–824. [CrossRef]
37. Guzman, L.M. *Multi-Criteria Decision Making Methods: A Comparative Study. Applied optimization. Evangelos Triantaphyllou*; Kluwer Academic Publishers: Alphen am Rhein, The Netherlands, 2001; p. 288.
38. Gabus, A.; Fontela, E. *World Problems, an Invitation to Further thought within the Framework of DEMATEL*; Working Paper; Battelle Geneva Research Centre: Geneva, Switzerland, 1972.
39. Fontela, E.; Gabus, A. DEMATEL: Progress achieved. *Futures* **1974**, *6*, 361–363. [CrossRef]
40. Baykasoglu, A.; Durmusoglu, Z.D. A hybrid MCDM for private primary school assessment using DEMATEL based on ANP and fuzzy cognitive map. *Int. J. Comput. Intell. Syst.* **2014**, *7*, 615–635. [CrossRef]
41. Chen, Y.S.; Chuang, H.M.; Sangaiah, A.K.; Lin, C.K.; Huang, W.B. A study for project risk management using an advanced MCDM-based DEMATEL-ANP approach. *J. Ambient. Intell. Humaniz. Comput.* **2019**, *10*, 2669–2681. [CrossRef]
42. Göncü, K.K.; Çetin, O. A Decision Model for Supplier Selection Criteria in Healthcare Enterprises with Dematel ANP Method. *Sustainability* **2022**, *14*, 13912. [CrossRef]
43. Ataei, Y.; Mahmoudi, A.; Feylizadeh, M.R.; Li, D.-F. Ordinal priority approach (OPA) in multiple attribute decision-making. *Appl. Soft Comput.* **2020**, *86*, 105893. [CrossRef]
44. Li, R.-J. Fuzzy method in group decision making. *Comput. Math. Appl.* **1999**, *38*, 91–101. [CrossRef]
45. Hong, Y.; Fei, Z.; Wen, T.; Yong, D. ANP group decision making model for project group selection of construction enterprises. *Sci. Technol. Prog.* **2011**, *28*, 38–42. (In Chinese)

buildings

MDPI

Article

Current and Future Trends of Resource Misallocation in the Construction Industry: A Bibliometric Review with Grounded Theory

Jingxiao Zhang [1,*], Fangyu Dong [1], Pablo Ballesteros-Pérez [2], Hui Li [3,*] and Martin Skitmore [4]

[1] School of Economics and Management, Chang'an University, Xi'an 710061, China
[2] Centro de Investigación en Dirección de Proyectos, Innovación y Sostenibilidad (PRINS), Departamento de Proyectos Ingeniería, Universitat Politècnica de València, Camino Vera s/n, 46022 Valencia, Spain
[3] School of Civil Engineering, Chang'an University, Xi'an 710061, China
[4] Faculty of Society & Design, Bond University, Robina, QLD 4226, Australia
* Correspondence: zhangjingxiao964@126.com (J.Z.); lihui9922@chd.edu.cn (H.L.); Tel.: +86-159-2973-9877 (J.Z.)

Abstract: Resource misallocation (RM) refers to the existence of marginal output inequalities between different industries or companies in an economy. Prior studies of RM have mostly focused on effect analysis, construction industry structure upgrades, and organization management. However, these studies have been fragmented and unrelated. This paper analyzes the status quo, consequences, and emerging trends of RM research at the macroscopic level based on current problems and with the aim of exploring potential solutions. Drawing on grounded theory, a qualitative analysis using text-mining is used to analyze the characteristics of 124 RM-related papers. The results more comprehensively and systematically reveal that current RM research encompasses four major dimensions of sources and concepts, misallocation degree measurement and characterization, focused issues (field), and RM research deficiencies. Methods for measuring RM have also been developed from the simple proportional method to current mainstream methods (e.g., growth rate decomposition and variant substitution). We conclude that, in order for this discipline to thrive and effectively reduce RM, future research into RM should focus on core categories, especially the reform of market-oriented factors, transformation of government functions, construction industrial structure adjustment, and methods of income distribution. This systematic review provides a discipline oversight and uncovers necessary and potential research directions.

Keywords: resource misallocation; grounded theory; market distortion; total factor productivity

Citation: Zhang, J.; Dong, F.; Ballesteros-Pérez, P.; Li, H.; Skitmore, M. Current and Future Trends of Resource Misallocation in the Construction Industry: A Bibliometric Review with Grounded Theory. *Buildings* **2022**, *12*, 1731. https://doi.org/10.3390/buildings12101731

Academic Editor: Ahmed Senouci

Received: 14 September 2022
Accepted: 13 October 2022
Published: 19 October 2022

1. Introduction

The global economy has developed rapidly in recent years and attempts to achieve sustainable development is gaining momentum in many countries. Resource misallocation (RM) refers to the situation where resources cannot be optimally allocated between industries or companies because of the interference of existing systems or policies [1]. This suboptimal situation greatly affects a country's total factor productivity (TFP) [2]. TFP refers to the increase in output caused by progress in technology and ability other than the input of elements, such as capital and labor [3]. Some studies suggest that RM is an important reason for the huge wealth gap between countries [2]. Therefore, the allocation of resources has an extremely important impact on the success (or failure) of a country's economic development. The adverse effects of RM also make it harder to achieve sustainable development.

China is a representative example of many contemporary countries. Its current and past situation can be used as a reference to extrapolate results to other developing countries. There are serious RM problems among regions, industries, and companies in China that significantly reduce its productivity [4]. These largely explain the difference in productivity between China and other more developed countries [5]. Additionally, China is facing a significant transformation of its industrial economic structure, while deepening the reform of its economic system [6]. RM is an important factor that affects TFP and an important channel to improve China's economic growth. Alleviating RM within and between regions can unlock rapid and coordinated development at both country and regional levels [7].

Previous research into RM has partially clarified the mechanism underlying the effect of resource allocation on output efficiency, improved RM's theoretical analysis, and helped understand how government policies conceive resource allocation. However, still under the COVID epidemic impact, the pressure on China's economy is relatively high. China is facing a very unbalanced state of productivity growth in the economy, reflecting a series of resource allocation problems, such as slow productivity gains in many construction market sectors, while sectors with faster productivity gains in the economy as a whole, or market employment share proportion, are becoming increasingly smaller. The market economy of China's construction industry is currently in a relatively long-term adjustment process. The process of deleveraging is far from over and the industrial structure adjustment process is equally so. With the progress of theoretical knowledge and science technology, improving the quality and construction industry market resource allocation efficiency, and solving the bottleneck of productivity are key issues. Some examples are the application of blockchain technology and digital technology resources including from BIM to extended reality (VR, AR, MR, Digital Twins, data capturing technologies, etc.) in AEC industry [8]. This could also become the key to promoting regional integration and high-quality economic development in China.

At present, the existing research has been mainly focused on effect analysis, industry policies, and organization management, but remains highly fragmented. Moreover, qualitative research into the problems and effects of RM is lacking, and there are no macroscopic analyses explaining the current situation, effects, or trends based on wide empirical evidence. Hence, a systematic literature review will help promote further and more coherent RM research in the industry.

This study provides a comprehensive review of the RM-related literature using a qualitative approach combined with grounded theory in order to provide a comprehensive and systematic literature review, and promote further and more coherent RM research in the construction industry from a global perspective. Grounded theory is convenient here as it focuses more on the systematic processing of concepts compared to traditional literature reviews. This study determines the main dimensions and hierarchical structure of the problems and effects of construction industry RM. It also explores the different interaction relationship dimensions to effectively reduce the misallocations involved. The research results will provide a stronger theoretical support for the further study of construction industry RM, as well as promote the transformation and development of industries such as those in China by improving their TFP.

2. Research Methods

This study provides a more comprehensive review of the RM-related literature using a qualitative approach combined with grounded theory. Grounded theory is convenient here as it focuses more on the systematic processing of concepts compared to traditional literature reviews.

2.1. Data Collection

In the context of this research, Resources refers to a source of supply or support, such as materials, money, or human capital owned by a country or region. They can be divided into natural and social resources. Natural resources include sunlight, air, water, land, forests, grasslands, animals, and minerals; social resources include human resources and information, together with wealth created through labor. Scarcity of resources is a basic problem in economics research, i.e., how to use limited resources to maximize social welfare is a common concern for economists. For the construction industry, this includes life cycle model material resources [9], human resources [10], market investment resources [11], and developing new models [12].

We searched for scientific RM-related papers in the Web of Science, Science Direct, Scopus, and Emerald databases. Both Chinese and English languages were used when looking for potential matches. The main keywords for the literature search were 'resource allocation' or 'resource misallocation' + 'total factor productivity', 'upgrading of construction industry structure', and 'corporate resources'. Chinese papers from 2010 to 2020 and English papers between 2000 and 2020 were selected. Finally, we focused on core journals with high impact factors and/or recognized authors or institutions. As a result, a total of 124 papers were selected with a high relevance to RM. The most recent was published in December 2020 (the month when the literature retrieval stage was concluded). Figure 1 illustrates the literature retrieval and screening process used.

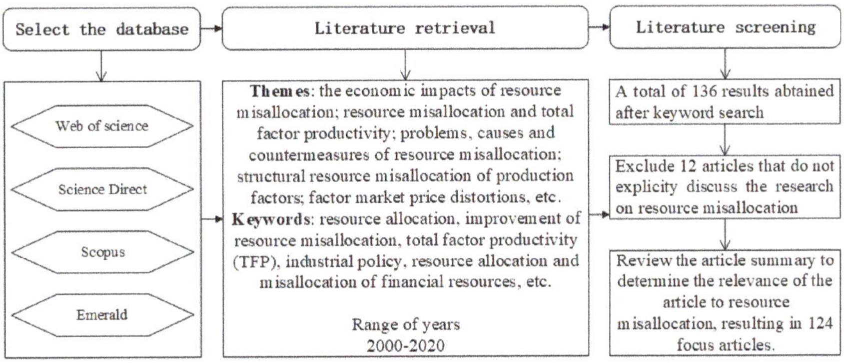

Figure 1. Literature retrieval and screening process.

The systematic review methodology excludes highly influential books, which is recognized as one of the method's potential limitations. However, the exclusion of books is common practice, given these sources are frequently classified as gray literature [13] due to their lack of rigorous peer-review. Many editorial outputs are also not included in scientific databases. Hence, despite some pioneering RM ideas being firstly developed in books -for example, the concept of 'industries assistance' and RM-, they have all been developed in later scientific papers. Consequently, the literature review is expected to be almost as representative as having books included.

2.2. Description of Methods Used

Qualitative research is a general term comprising various research methods, such as ethnography, nature exploration, fragment analysis, case studies, and ecological sample record analysis. Of these, grounded theory is an effective bottom-up qualitative research method that is frequently applied in the social sciences. With grounded theory, information is processed by comparing, analyzing, and transforming data into core concepts to establish theories. Grounded theory's data analysis requirements are quite strict [14]. As Strauss and Corbin (1997) describe, it involves the process of coding the flow of operations that decompose previously collected or translated textual data and identifying phenomena to conceptualize them so that later we can re-abstract, upgrade, and synthesize concepts into new categories. Hence, the aim of grounded theory is to describe the essence and significance of phenomena at the theoretical level but generate it from the analysis of what has been written in the literature. A relatively standardized grounded theory research process is shown in Figure 2 [15].

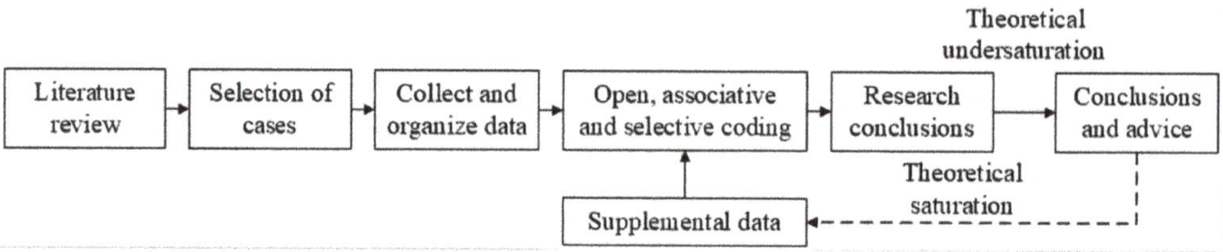

Figure 2. General grounded theory application process.

One grounded theory requirement is that theories are explored and developed from the data in no particular form [16]. The process of data analysis can be divided into three steps, frequently named 'triple coding'. Triple coding comprises open coding, associative coding, and selective coding [17]. Although triple coding is formally presented in three sequential stages, they may contain loops and/or force the researcher to go back while grouping concepts and categories until all the data and categories are consistent [14].

First, open coding focuses on reading relevant texts and discovering a variety of topics. This method is often used in the initial analysis of the literature. It can conceptualize and categorize (downsize) references. Usually, a large number of references is progressively reduced into concepts and categories that correctly reflect the documents' contents. In this process, the original references in the literature, as well as their abstracts and content, are "broken" or "crumbled", then re-instated. The purpose of open coding is to deal with the problem of convergence by identifying recurring phenomena, concepts, and categories.

Associative coding is the second-level coding process in grounded theory. After completing the previous coding task, the categories with high frequency and importance are screened out through the inductive process of "causality → phenomenon → situation → intermediary condition, action/interaction strategy → result". Then, the potential organic context and internal relationships between categories are explored with the use of synonyms, attribute levels, and even personal opinions. Next, we select a certain "axis" category as the core and reintegrate the previously decomposed data to identify any relationships between that main category with other secondary categories and/or problems.

Finally, selective coding involves the continuous comparison of raw materials, concepts, categories, and (especially) category relationships. Selective coding refers to the systematic analysis of the discovered conceptual genera and the selection of a "core genera" [18]. Selective coding is high-level coding in qualitative research. It aims to construct a qualitative theoretical model that describes the research topic based on the conceptual categories and relationships defined in previous relational coding stages. Therefore, selective coding is also called "theoretical coding". Its output involves dividing the research topic into core and supported categories. The core genera are an abstract category with more secondary genera (subgenera) and more coding reference points. This code genera have a large proportion and strong explanatory power in the model. A class outside the core class (but at the same level) is the supported class.

Grounded theory is suited to areas of research where theoretical systems are not yet explanatory and where it is difficult to explain some practical phenomena. It is also suitable when there are significant theoretical gaps and/or new phenomena that keep appearing [14]. As current research into RM is incomplete and its theoretical system is not fully fledged, grounded theory should be particularly useful. Figure 3 provides a flowchart summary of the research method and its major components introduced above.

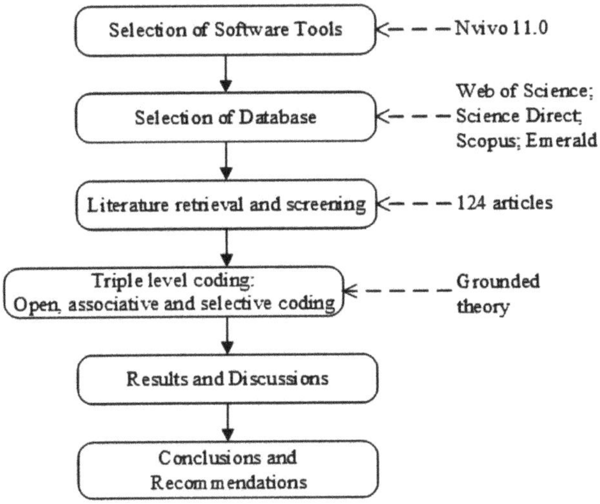

Figure 3. Research method flowchart.

3. Results

Following the grounded theory process outlined above, a system was formed involving 279 coding reference points, 19 basic categories, and 4 core categories as the explanatory model's basis. In this model, the nodes at all levels from the bottom to the top have subordinate relationships, and were formed through a bottom-up induction process.

3.1. Open Coding—Primary Related Factors of RM

The concept genera, or set of primary factors related to RM, was derived from the 124 selected papers, and named from the original texts based on grounded theory's "localization" principle. 19 three-level nodes were obtained by combining and coding the papers' contents. From this, the following direct factors were identified:

- Organizational operation efficiency and regulation, whether the enterprise can overcome financial constraints and the degree of impact of resource mismatching output, control the entry of new enterprises, support or suppress construction companies, and various other government policies;
- Insufficient functional mechanisms and analytical frameworks, research reasons, such as allocating resources through state-owned enterprises instead of the market, economic restructuring costs, trade costs, the informal sectors presence, and resource allocation based on human talent;
- The beginning of the research and the theoretical origin;
- The methods of calculating the RM, such as growth rate decomposition, simple proportional, and variant substitution at the bottom of the subordinate relationship.

3.2. Associative Coding—A Structural Model for RM Research

Associative coding was conducted to further discover and establish the relationships between conceptual genera and their organic correlations. This provided an initial set of 11 s-level nodes, some of which were inferred from the induction and integration of qualitative analysis based on third-level nodes. They represent intermediate elements in the RM analysis. Further qualitative analysis of the 11 secondary nodes revealed four primary nodes located at the top of the subordinate relationship, and embodied macroscopic elements for RM study. Figure 4 shows the structural model.

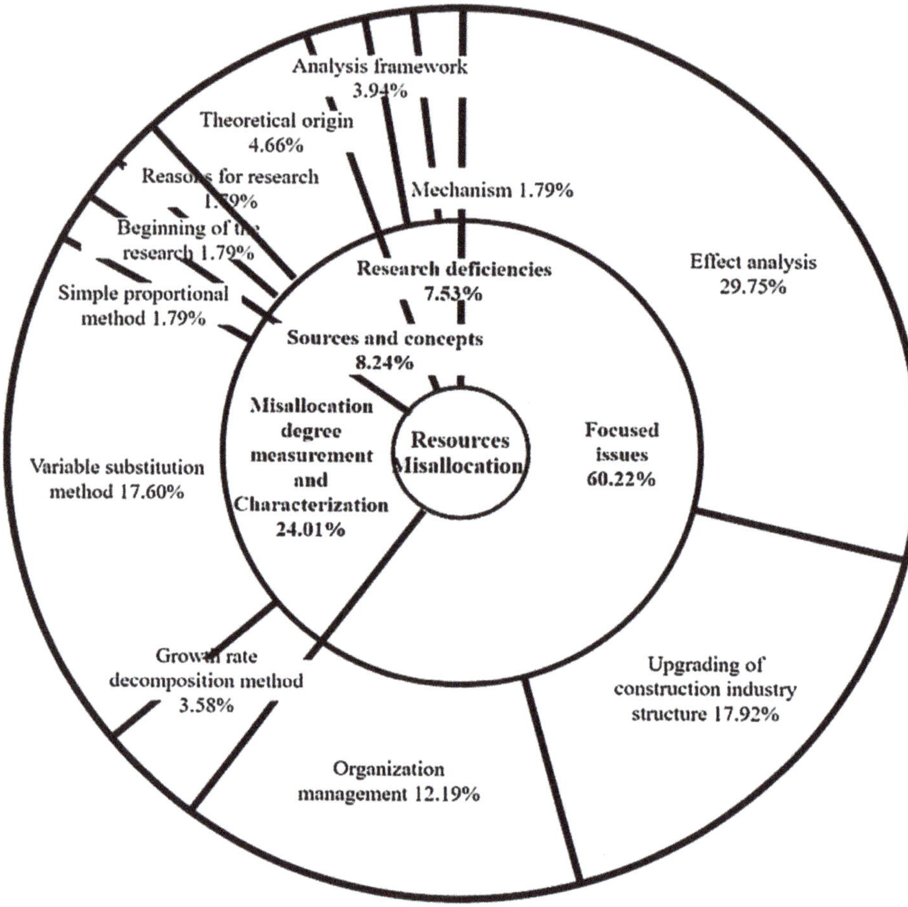

Figure 4. Structural model for RM analysis. Second-level nodes are outside. First-level nodes are inside.

The center of Figure 4 contains the model theme, namely RM. The specific hierarchical structures of the RM-related research elements are represented by two-layer circles. The wedges within each circle reflect the dimensional categories (nodes). As Figure 4 illustrates, current RM studies consist of 4 first-level nodes and 11 s-level nodes. Third-level nodes are not represented due to their large number. The size of each ring sector is determined by the number of coded reference points included. This represents the "volume" (importance) of papers supporting each node and indirectly reflects their influence.

As Figure 4 shows, the *Focused issues* node accounts for most reference points among all first-level nodes. In this category, RM is a popular topic for different effect analyses. Indeed, the analysis of misallocation based on the construction market is also supported by the majority of reference points among all second-level nodes. This indicates that this market environment is still a key factor for improving current RM situations. In particular, the market economy of China's construction industry is currently in the process of relatively long-term adjustment of the industrial structure and needs to be explored further in subsequent studies. In contrast, the *research deficiencies* node contains the least number of reference points and also has the least influence.

3.3. *Selective Coding—The RM Research Core*

In the final stage, focused issues and misallocation degree measurement and characterization were determined as core categories. These two core genera include 7 subgenera, involving 235 coding reference points and account for 60.2% and 24.0% of the total analyzed, respectively. These two core genera occupy the main position among all genera and represent important links in RM studies. Figure 5 summarizes the final data structure of our selective coding research.

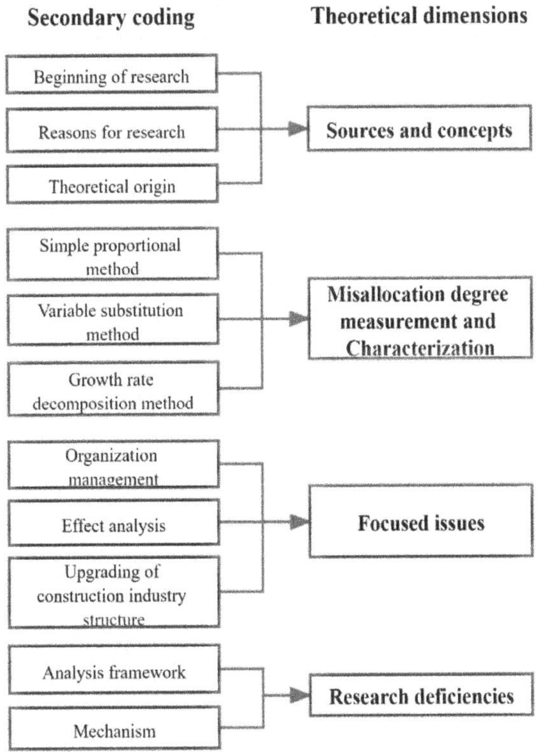

Figure 5. Selective coding data structure.

Focused issues include three specific indicators of upgrading of construction industry: structure, organization management, and effect analysis. Of these, there are a total of 83 effect analysis nodes with the largest number of reference points (49.4%). This again shows that research into the effect analysis system's direction the plays a key role in guiding the effective allocation of resources. Improving the effect analysis system will further improve (optimize) the allocation of corporate resources. Moreover, the *measurement method* contains three specific index points: the simple ratio method, variable substitution method, and growth rate decomposition method. Of these, there are a total of 52 variable substitution method nodes with the largest number of reference points (77.6%) of the effect analysis node. This shows that the measurement method using the variable substitution method for robustness tests reflects the RM level being measured in many practical settings. Conversely, *sources and concepts* and *research deficiencies* are relatively minor.

Moving forward, we focus on the core and supporting *genera*'s high-frequency nodes of *sources and concepts*, *focused issues*, and *misallocation degree measurement and characterization* for the model's explanation. *Research deficiencies* represents a very low-frequency node and is not developed further.

(1) Sources and concepts

As Table 1 shows, *source and concepts* can be considered the entry point in current research.

Table 1. Key literature reviews of sources and concepts.

Research Direction	Research Connotation	Key Literature
TFP	TFP as a measure of technological progress to explain the efficient functioning of an economy and ways to improve its productivity.	Solow [4]
	Assess the scope of input misallocation in Italy and its impact on aggregate output and TFP.	Lenzu [19]
Shortage of service demand	Differentiate between the concepts of "need" and "demand" for medical services and provide a concept of "shortage" in this context.	Jeffers [20]
TFP growth decomposition	Decompose TFP growth into the industry's own TFP growth and the allocation effect of factors among industries.	Syrquin [21]
Structural adjustment policies	Any anti-diversification policy directly limiting the realm or range of business activities without correcting the business environment may not be effective and will create serious RM.	Jwa [22]
Misallocation in various fields	Resource prosperity increases the value of power by providing more resources for politicians and exacerbating misallocation in various fields.	James A [23]
Concept formation	Provide a new analytical perspective for TFP growth.	Hsieh [24]
Enterprise productivity	Imperfect market mechanisms exacerbate the distortion of allocation of and enterprise factors, and restrict improvement in enterprise production efficiency.	Zhigang Yuan [25] Yongwei Chen [26] Deming Luo [27] Huihua Nie [28] Zhen Yang [29]
Changes in ecological efficiency	Environmental regulation and RM factors are introduced to identify the key factors influencing ecological efficiency.	Wang [30]
Outward foreign direct investment (OFDI)	RM, as an essential characteristic of China's "progressive reform", has become a significant factor restricting high-quality OFDI.	Kong [31]

The sources and concepts node includes three secondary nodes: theoretical origin, research beginning, and research reasons, with a total of 23 reference points. Of these, there are 13 theoretical origin nodes, which contain most reference points and occupy a relatively important position. At the beginning of the research, there are 5 reference points and the reason for the research occupies 5 reference points. The number of points included is small and its impact is small too. The number of reference points for each secondary node is equally small, as is their gap, indicating that the source of RM and the concept's dimension have received little attention (apart from the theoretical origin).

In general, the allocation of resources at both macro and micro levels will be continuously optimized and updated due to the continuous development and change of external market environments. Therefore, its sources and concepts are often ignored or not widely mentioned. Table 2 shows the number of nodes at all levels and coding reference points included in sources and concepts node.

Table 2. Number of nodes at all levels and coding reference points included in the sources and concepts node.

First Level Node	Secondary Nodes	Three-Level Nodes	Number of Coding Reference Points
Sources and concepts (3)	Beginning of research		5
	Reasons for research	Allocate resources through state-owned enterprises instead of the market	5
		Economic restructuring costs	
		Trade costs	
		Presence of the informal sector	
		Cannot allocate resources based on human talents	
	Theoretical origin		13

(2) Focused issues

Whether it is the (1) factor allocation efficiency of micro-enterprises; (2) construction industrial organization efficiency at the meso-level; or (3) changes in the macro-level demand structure, technological progress, and the quality of economic growth. RM has caused a decline in economic growth. The RM problem's research and exploration can be divided into such different aspects as the effect analysis and organization management. Table 3 shows the result created from analyzing the data relating to *focused issues*.

Table 3. Key literature relating to *focused issues*.

Research Direction			Research Connotation	Key Literature
Effect analysis	Overcoming the impact of	Financial friction	Financial market friction causes the misallocation of resources in the initial stage, which greatly hinders the convergence of high-speed growth to a steady-state equilibrium path.	Francisco J Buera [32]
			Enterprises can accumulate enough capital and self-correct the "first type of misallocation" caused by financial market friction.	Abhijit V. Banerjee [33]
		Corporate debt	Reveal the truth of the decline of investment efficiency and the misallocation of financial resources behind the deviation between the asset-liability ratio and the macro-debt ratio of Chinese enterprises.	Pan [34]
	Impact on outputs	TFP dispersion and impact	The dynamic process of controlling productivity shocks determines misallocation and income disparities.	Asker [35]
		Productivity gap among enterprises	RM among enterprises is an important reason for a low TFP.	Lewandowskakalina [36]
		TFP differences among countries	The redistribution of factors between heterogeneous production units is an important source for measuring TFP differences between countries.	Restuccia [37]
		Impact on productivity	The misallocation caused by financial market distortions can explain 10% of the total misallocation at most, and its impact is minimal.	Abhijit V. Banerjee [33] Virgiliu Midrigan [38]
	Market distortions	Operating effectiveness	The effectiveness of market operations has an impact on the cost of capital acquisition, allocation efficiency, and entrepreneurial behavior by enterprises.	Buera [32]
		Market malfunction	Banking system malfunctions will increase RM.	Ziebarth [39]

Table 3. *Cont.*

Research Direction		Research Connotation	Key Literature
Upgrading of construction industry structure	Scale control	Policies restricting large enterprises and supporting SMEs have depressed the scale of enterprises in the economy and caused a misallocation of resources in low-productivity SMEs.	Guner [40]
	Construction organizations	RM of manufacturing policy will reduce TFP.	Hsieh [24]
	Entry regulation	The countries with loose entry regulations have higher TFP levels and productivity than strictly regulated countries.	Barseghyan [41]
	Inappropriate organizational policy	Inappropriate industrial organization policies distort resource allocation and cause serious misallocation and loss of TFP and output.	Repetto [42]
	Development zone evaluation	Improved system efficiency can alleviate the policy impact of development zones on the misallocation of regional resources.	Bai [43]
	Government intervention and special rules	The level reached by social planners that allocate the current aggregate output across mines, so as to minimize emissions, conditional on some well-defined extraction rules.	Jac [44]
Organization management	Market distortion	Assesses the degree and impact of distortions in the construction market.	Chang-Tai Hsieh [24]
	Job supply and demand	A serious mismatch between job supply and demand will cause a large number of laborers to be "misallocated" and become unemployed, thereby raising the labor cost of construction enterprises.	Sahin [45]
	Entrepreneur or worker decisions	Fluctuations in input prices lead to the misallocation of talent between entrepreneurs and the salaried sector.	Sargent [46]
	Labor allocation in the construction industrial sector	Increased labor-misallocation of resources significantly hampers low-carbon productivity gains.	Xiaolan Deng [47]
	Internal influence mechanisms of labor allocation	Guiding the diversion and transfer of labor resources according to local conditions and improving allocation efficiency will provide human and capital support for the optimization and upgrade of construction industrial structure.	Ai Ma [48]

The focused issues node has a total of 168 reference points, which is the largest of all the first-level nodes, and also the most important in the entire research field. The effect analysis node is also the main node with more related publications. It has the most reference points of the three second-level nodes included in focused issues (a total of 83). Financial constraints and output efficiency also have a particularly important impact of RM from the three-level nodes included. Without fair, open, and transparent market rules and strict compliance supervision, the party with less information is very likely to be seriously disadvantaged. In this situation, the market mechanism can hardly allocate funds effectively, as effect analyses are markets with extremely asymmetric information. Moreover, the development of effect analysis can lead to more investment in growing industries and the withdrawal of capital from declining industries to improve capital allocation efficiency [34]. Therefore, more research into the misallocation of resources in effect analysis is also needed.

Furthermore, there are a total of 50 upgrading of construction industry structure nodes, which contain more reference points. Since upgrading construction industry structure can identify the industries in need of the Chinese government's intervention currently [49], it also plays an important role in reducing enterprises' RM and improving production efficiency. The organization management node is a minor node with a small number of publications, but still contains 34 reference points. The analysis shows that, in many developing countries, the influence of RM caused by construction market distortions, such as the unemployment rate or enterprise labor costs, is still relatively large [45]. Table 4 shows the focused issues node at all levels and coding reference points.

Table 4. Number of nodes and coding reference points in the focused issues node.

First Level Node	Secondary Nodes	Three-Level Nodes	Coding Reference Points
Focused issues (3)	Effect analysis (main node)	Overcoming financial constraints	83
		Impact on outputs	
	Upgrading construction industry structure	Control the entry of new businesses	50
		Support or suppress construction companies	
		Other government policies	
	Organization management	Organizational efficiency	34
		Corporate regulation	

Table 5, Figures 6 and 7 show the results of a grouping word frequency query applied to the 124 selected publications, identifying the most popular research foci in different periods. Figure 8 also shows the keyword cluster analysis results obtained by VOS viewer software based on keyword classification.

Table 5. Most popular research focuses.

Period	High-Frequency Words (Valid Words and Weighted Percentage Higher than 0.5%)	Research Focuses
2010–2015	Enterprises, resources, industries, factors, allocation, productivity, economy, capital, China.	Enterprise productivity; TFP; distortion of resource allocation.
2016–2020	Resources, enterprises, economy, allocation, elements, industry, capital, industry, China.	Market distortion and misallocation of resources; influence of RM on TFP; misallocation of factors caused by industrial structure imbalance.

Figure 6. Word cloud for the 2010–2015 research papers.

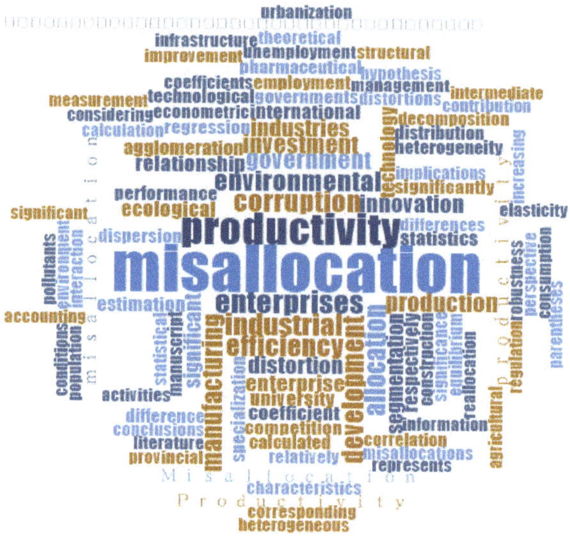

Figure 7. Word cloud for the 2016–2020 research papers.

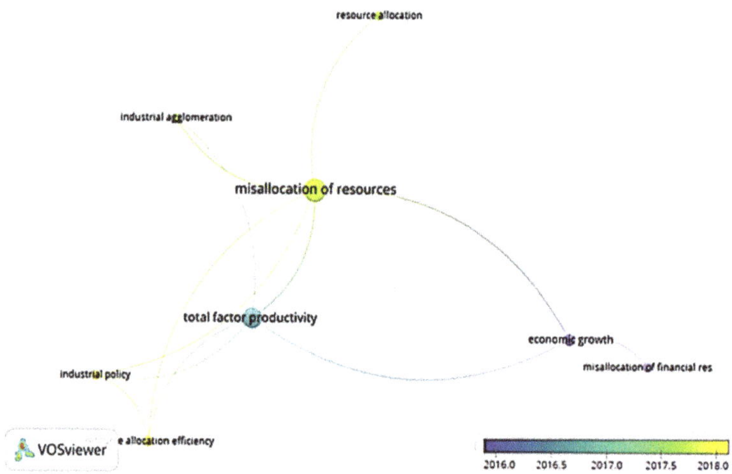

Figure 8. Keyword cluster analysis based on keyword classification.

The core issue is to manage the relationship between the government and the market in such a way that the market can play a decisive role in resource allocation and better reflect the government's role. Therefore, the most popular research focuses range from simply studying economic benefits, TFP, resource allocation, and other issues, to more in-depth research on the effect of market distortions on RM, as well as that of resource mismatching on TFP.

At the same time, there is an urgent need to strengthen the research and exploration of "extension misallocation". Both extensional RM defined by Banerjee and Moll (2010) or the net entry effect calculated by Brandt [50] are based on the difference in the literature between the TFP entering and exiting the enterprise. The effective dynamic replacement of enterprises should comprise three ways to realize the effective allocation of resources: (1) the entry of potentially high TFP companies and the exit of existing low TFP companies (selection effect); (2) after entering, the potentially high TFP companies achieve rapid growth of their own TFP and narrow the gap with incumbents or even surpass incumbents through acquired learning (learning effect); and (3) the process of enterprise entry and exit creates competitive pressure on the incumbents and forces them to improve their TFP (competitive effect).

These three effects together constitute the connotation of the process of optimal allocation of external resources. Distortion of the normal entry and exit process of enterprises will directly lead to the misallocation of extended resources. Therefore, it is necessary to overcome technical defects, exhaustively consider various distortive policies, further analyze the causes of RM and market distortion, and make a more accurate prediction of efficiency loss.

(3) Methods of misallocation degree measurement

RM leads to a loss of production efficiency. It is of theoretical and practical relevance to measure the degree of misallocation and its influence on productivity accurately [4]. It is necessary to clarify the mechanism and degree of efficiency loss caused by RM before attempting its rectification, while, a comprehensive understanding of the practical problems involved is needed to formulate policies to solve the RM problem, and thus help improve the economic benefits of enterprises and promote the economy and society's sustainable development. Table 6 shows the current methods used for measuring the efficiency loss caused by RM.

Table 6. Key literature for resource misallocation degree measurement and characterization.

	Research Direction	Research Connotation	Key Literature
Simple propo rtion method	TFP ratio of 90/10 enterprises	90/10 of all state-owned and above-scale enterprises have decreased and their average TFP increased year by year.	Huihua Nie [28]
	75/25 quantile difference	The degree of misallocation can be measured by productivity dispersion but may be interfered with by endogenous and selective change.	Puyang Sun [51]
Variable substi tution method	Use TFP to measure the degree of distortion of resource allocation across enterprises.		Klenow [24]
	Incorporate the cost of using elements into the analytical framework for studying misallocation.		Zhen Yang [29]
	Marginal product income of production factors as a tool to measure the allocation distortion.		Guan Gong [52]
Growth rate dec omposition method	Efficiency loss	Decompose TFP into technical efficiency, resource allocation efficiency, and economies of scale.	Ariel Pakes [53]
	Decompose the TFP growth rate into the industry's internal productivity growth effect, the positive effect of output share on productivity growth, and the resource reallocation effect.		Nordhaus [54]
	China's construction industry has a strong dependence on intermediate products. The introduction of intermediate input factors is more indispensable when studying the TFP and the growth of the specific sector.		Zhen Yang [29]

There are 67 reference points in total, including three second-level nodes: the variable substitution method, simple proportion method, and growth rate decomposition method. Of these, the variable substitution method is the main node with more related publications and 52 nodes. It contains the most reference points and occupies an important position in the Misallocation degree measurement and Characterization nodes. This, in turn, indicates that it is more suited for use when conducting a robustness test in RM measurement. The growth rate decomposition node contains the relatively small number of 10 reference points. In addition, the simple proportional method has a total of 5 reference points, with the least number of points and the least impact. This means that the simple proportional method is seldom used in the literature and is often ignored (or needs further study). A simplified study of problems related to resource allocation from the perspective of misallocation degree measurement and characterization can better reflect the extent of specific RM and effectively solve complex allocation problems.

Here, we considered the relationship of various dimensions and believe that misallocation degree measurement and characterization have a strong correlation with the sources and concepts and focused issues nodes. We conducted the measurement of the specific allocation of related resources and a quantitative study of their level of correlation (Table 7).

Table 7. Number of nodes at all levels and coding reference points included in the misallocation degree measurement and characterization node.

First Level Node	Secondary Nodes	Coding Reference Points
Misallocation degree measurement and characterization (3)	Simple proportional method	5
	Variant substitution method (main node)	52
	Growth rate decomposition method	10

The simple proportion method assumes a perfectly competitive market and uses productivity dispersion between firms to describe the degree of RM. The common indicator is the ratio of 90/10 firms' TFP. The greater the difference in TFP among firms, the more serious the degree of RM. This method has simple steps and is convenient to use for measurement. However, solely relying on the TFP ratio is insufficiently representative and may be distorted by endogeneity and selectivity biases.

With the gradual deepening of the analysis of corporate TFP and economic growth, a model of monopolistic competition was built with the variable substitution method used for its misallocation degree measurement and characterization. The TFP variance was used to measure the degree of distortion of resource allocation across companies. This method allowed us to demonstrate that the greater the difference in TFP between companies, the smaller is the total industry TFP. The assumption of constant returns to scale was gradually relaxed. The Levinsohn-Petrin semiparametric estimation method was used to estimate the cross-industry capital and labor output elasticity of China's construction industry. In addition, the marginal product returns of production factors were used as a tool for measuring the resource allocation distortion and the potential growth of manufacturing TFP after the Pareto improvement of a single factor allocation.

Furthermore, with the continuous improvement of data acquisition methods and the increasing availability of micro data, RM research has gradually moved from the macro to the micro level. At present, the general way of measuring RM-caused efficiency loss in many countries is the growth rate decomposition method. For instance, it is possible to decompose TFP growth into the industry's own TFP growth and the allocation effect of factors among industries. This operation allows the impact of structural changes in a country's production efficiency to be analyzed. It also enables the reallocation of resources between industries when we decompose it into the simply increasing industry share input effect and the factor price distortion effect.

Finally, with the continuous updating and maturity of theoretical models, the misallocation degree measurement and characterization have changed from a simple proportion to such other current mainstream methods as growth rate decomposition, DSEG, variable substitution, and the Aoki method. These methods not only measure RM between industries and within companies, but also can measure the impact on RM of different ownership systems, among regions, and even gradually measure the RM level of a whole national economy. All of these are relevant applications that can prevent negative effects on TFP and avoid hindering economic growth [55]. Therefore, the impact path of intermediate inputs should be considered when measuring construction industrial outputs. Intermediate inputs should also be included in the growth accounting framework when measuring the impact on the overall TFP. This will allow the impact of industrial TFP on overall TFP to be more accurately estimated, and the misallocation degree measurement and characterization to be continuously improved, to eventually solve specific RM-related problems.

3.4. Saturation Test

The saturation test involved randomly reserving 10 journal papers before coding, encoding the remaining 114 papers, and then recoding the 10 papers again. Upon performing this operation, it was found that all existing codes and categories coincided. That is, there were no new codes related to the study's topic, which was taken to mean that theoretical saturation had been achieved [56].

4. Discussion and Implications

4.1. Discussion

This study adopts a bottom-up qualitative research method with grounded theory to compare, analyze, condense, and summarize the main dimensions and hierarchical structure of RM concepts:

1. Current RM-related research involves a three-layer structural model. The first layer analyzes and describes the macro elements of RM. This layer contains 4 nodes in total: sources and concepts, misallocation degree measurement and characterization, focused issues, and research deficiencies. The second layer involves the study of intermediate elements and most of the RM direct elements. This contains 11 nodes: research beginning, the reason for the research, and the theoretical origin; the simple ratio method, the growth rate decomposition method, and the variable substitution method; the organization management, the upgrading of construction industry structure, and the effect analysis; and insufficient research on the mechanism of action and framework of analysis. Each node has a different influence and has also experienced a different amount of research activity
2. Focused issues and misallocation degree measurement and characterization are the "core categories" for exploring the RM problem, while sources, concepts, and research deficiencies are the "supported categories". The latter are not explained in this paper because their impact on RM is minimal. The research on the current and future trends of resource misallocation in the construction industry also has no core value.
3. With more RM-related research being developed and the updating and maturity of economic theoretical models, more methods for measuring RM have also been developed. These range from the simple proportional method to current mainstream methods (e.g., growth rate decomposition and variant substitution). This makes it possible to further analyze quantitatively the impact of RM changes in specific industries, including from BIM to extended reality (VR, AR, MR, Digital Twins, Data capturing technologies, etc.) in AEC industry. It also allows the factors that condition production efficiency between industries to be better understood.

4.2. Implications

This study develops a theoretical model to study the dimensions of the resource misallocation problem and shows that certain elements play a dominant role in the process (Figure 9).

More studies are urgently needed to explore the effects and influences of different market distortions on TFP. This will allow the implementation of more effective industrial policies and improved enterprise production efficiency while optimizing the country's economic structure. At the same time, it is important to identify effective methods for optimal resource allocation and reducing operating costs. For example, we suggest establishing market-oriented and rule-of-law mechanisms for the 'survival of the fittest' and match scientific technological progress paths to meet a substantial proportion of social production needs. Moreover, attracting investment that follows equal national treatment and pre-entry national treatment, and respects intellectual property rights and other principles, provides a significant way to develop the local economy, and provides development motivation and experience (such as capital, talents, technology, products, markets, and management methods for open areas).

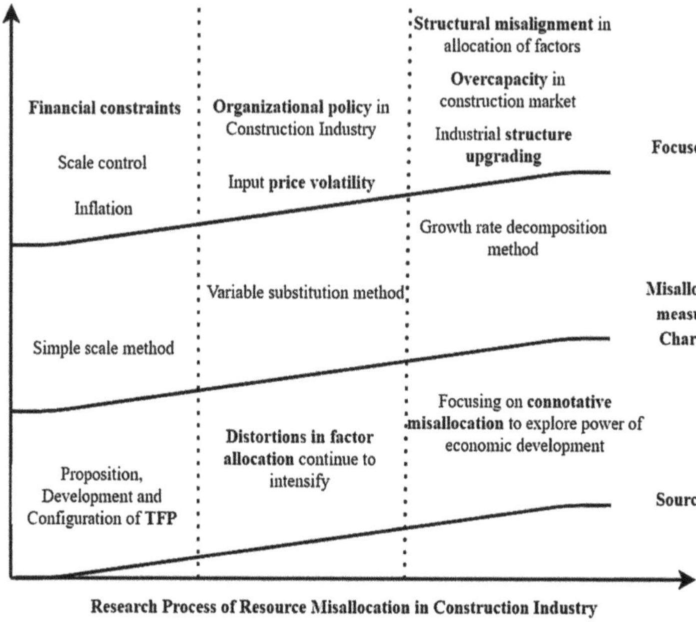

Figure 9. Dimensional model of resource misallocation.

Further improving management system design can also help achieve mutual benefits and complementary resources. When a higher resources efficiency is achieved, it will be possible to solve other core problems of the national economic system and promote the transformation of an economy from the traditional extensive growth model to a new sustainable version.

Overall, future studies would benefit from focusing on "core categories" and "supported categories". This can be achieved by concentrating on endorsing market-oriented reforms, transforming government functions, adjusting industrial structure, and implementing income distribution methods. At the same time, it is important to understand that RM is all-pervasive in many industries, companies, and national economies. Nevertheless, the value of marginal output—the increased revenue of a firm's additional use of a unit factor under conditions of perfect competition—still needs more in-depth empirical research on whole markets, industries, and companies from macro and micro perspectives to verify the conclusions of this study. This will allow the enrichment of the hierarchical dimensions and RM model structural elements proposed here.

5. Conclusions

RM-related research is at the core of economic theory, but usually involves regression analysis with a single explanatory variable. Consequently, there are many limitations, gaps, and deficiencies both in the RM analysis framework and its explanatory action mechanisms. For example, although studies have found that intermediate products are extremely important factor inputs in addition to capital, labor, and other basic factors, it is difficult to obtain data from intermediate inputs and purchased intermediate services from enterprises due to the data availability restrictions. Nevertheless, this influence path of intermediate input needs to be considered when calculating industry output, which means it is necessary for inputs to be also included in the RM growth accounting framework.

In this regard, measuring the impact of intermediate input itself on the overall TFP, for example, could more accurately predict the impact of the industry's TFP on the overall country's TFP.

Similarly, RM-related research in practice tends to explore the influence of individual factors in one dimension. That is, it neglects the potential mutual (interrelated) effects of these factors. Currently, the impact of market distortions (such as RM finance and labor) and changes in the process of TFP need to be resolved. The existence of RM is reflected in the excess of return on capital of individual companies by observing simple cross-sectional data at the micro level. But this excess of return on capital is the driving force behind most companies when innovating and improving productivity. Hence, RM's effect on the economy's overall productivity needs to be explored in both theory and practice. Additionally, errors of measurement and the existence of "externalities" in the production process may affect the estimation of efficiency loss. In particular, the (unobtainable) changes in the input of production factors will always lead to the inaccurate estimation of an enterprise's productivity. However, the literature has not yet considered any of this.

Finally, there is still a lack of countermeasures to solve RM-related problems. In real settings, concrete feasible improvement measures from governments or enterprises remain undeveloped. To address this gap, the cooperation between governments, scholars, disciplines, and institutions would be ideal—the aim being to build a high-level, comprehensive, and pioneering academic exchange platform to further develop more RM-related research with a special application to practical settings.

Author Contributions: J.Z. contributed to the study's conception, performed the experiment, F.D. and H.L. contributed significantly to analysis and manuscript preparation, F.D. performed the data analyses and wrote the manuscript together with M.S. and P.B.-P. helped perform the analysis with constructive discussions. All authors have read and agreed to the published version of the manuscript.

Funding: This research is supported by the National Social Science Fund projects (No. 20BJY010); National Social Science Fund Post-financing projects (No. 19FJYB017); Sichuan-Tibet Railway Major Fundamental Science Problems Special Fund (No. 71942006); List of Key Science and Technology Projects in China's Transportation Industry in 2018-International Science and Technology Cooperation Project (No. 2018-GH-006 and No. 2019-MS5-100); Emerging Engineering Education Research and Practice Project of Ministry of Education of China (No. E-GKRWJC20202914).

Conflicts of Interest: The authors declare that they have no known competing financial interests or personal relationships that could have influenced the work reported in this paper.

References

1. Eric, J.; Bartelsman, J.C.H.; Scarpetta, S. Cross-Country Differences in Productivity. *Role Alloc. Sel. Am. Econ. Rev.* **2013**, *103*, 305–334.
2. Caselli, F. Accounting for Cross-Country Income Differences. *LSE Res. Online Doc. Econ.* Available online: https://www.sciencedirect.com/science/article/abs/pii/S1574068405010099 (accessed on 4 August 2022).
3. Solow, R.M. Technical Change and Aggregate Production Function. *Rev. Econ. Stat.* **1957**, *39*, 312–320. [CrossRef]
4. Jing Hang, K.G.; Niu, M. Resource mismatch, capacity utilization and productivity. *Economics* **2021**, 20.
5. Loren Brandt, T.T.; Zhu, X. Factor market distortions across time, space and sectors in China. *Rev. Econ. Dyn.* **2013**, *16*, 39–58. [CrossRef]
6. Bai, Y. Research on the Transformation of my country's Industrial Economic Structure under the New Normal. *Trade Show Econ.* **2020**, *8*, 86–88. [CrossRef]
7. Song, J. An Analysis of the Causes and Countermeasures of Resource Misallocation in China's Economic Transformation. *J. Shandong Inst. Bus. Technol.* **2021**, *35*, 11. [CrossRef]
8. Khan, A.; Sepasgozar, S.; Liu, T.; Yu, R. Integration of BIM and Immersive Technologies for AEC: A Scientometric-SWOT Analysis and Critical Content Review. *Buildings* **2021**, *11*, 126. [CrossRef]
9. Mishlanova, M.; Patrina, T.Y.; Chekunova, A. Life cycle model material resources in the construction industry. *Bull. Belgorod State Technol. Univ. Named After. V. G. Shukhov* **2017**, *2*, 236–241. [CrossRef]
10. Anelauskas, J. Impact of Innovations on Human Resources in the Construction Industry. Available online: https://www.semanticscholar.org/paper/Impact-of-innovations-on-human-resources-in-the-Anelauskas/0b5439fd09c12159774662f91cb05c0a6b98b1cf. (accessed on 14 August 2022).

11. Fisunenko, N. Evaluation Market Investment Resources in Construction Industry. Available online: https://echas.vnu.edu.ua/index.php/echas/article/view/104 (accessed on 26 August 2022).
12. Piri, L.; Ghezavati, V.; Hafezalkotob, A. Developing a new model for simultaneous scheduling of two grand projects based on game theory and solving the model with Benders decomposition. *Front. Eng. Manag.* **2022**, *9*, 117–134. [CrossRef]
13. Richard, J.; Adams, P.S.; Huff, A.S. Shades of Grey: Guidelines for Working with the Grey Literature in Systematic Reviews for Management and Organizational Studies. *Int. J. Manag. Rev.* **2017**, *19*, 432–454. [CrossRef]
14. Li, Z. Analysis on the Application of Grounded Theory Method in Scientific Research. *Orient. Forum. J. Qingdao Univ.* **2007**, *4*, 90–94.
15. Pandit, N.R. The creation of theory: A recent application of the grounded theory method. *Qual. Rep.* **1996**, *2*, 1–15. [CrossRef]
16. Maocong Zhang, W.Z. On the Construction of my country's Crisis Education Content—Based on Nvivo Analysis of 32 Core Documents Since 2003. Courses. *Teach. Materials. Teach. Methods.* **2020**, *3*, 8.
17. Yifan Qin, F.L.; Li, J.; Wang, K.; Du, Z.; Xu, H. Investigation and Research on Doping Abuse in Adolescents of Physical Education College Students—A Qualitative Research Based on Grounded Theory. In Proceedings of the Collection of Abstracts of Papers of the 11th National Sports Science Conference, Nanjing, China, 1–3 November 2019; 2019; 10, p. 1.
18. Alessandro Pepe, L.A.; Veronese, G.; Glăveanu, V. Measuring Teacher Job Satisfaction: Assessing Invariance in the Teacher Job Satisfaction Scale (TJSS) Across Six Countries. *Eur. J. Psychol.* **2017**, *13*, 396–416. [CrossRef] [PubMed]
19. Lenzu, S.; Manaresi, F. Sources and implications of resource misallocation: New evidence from firm-level marginal products and user costs. *Quest. Di Econ. E Finanz* **2019**. [CrossRef]
20. Jeffers, J.R.; Bognanno, M.F.; Bartlett, J.C. On the demand versus need for medical services and the concept of "shortage". *Am. J. Public Health* **1971**, *61*, 46–63. [CrossRef]
21. Syrquin, M. Resource reallocation and productivity growth. *Econ. Struct. Perform.* Available online: https://www.sciencedirect.com/science/article/pii/B9780126800609500118 (accessed on 1 August 2022).
22. Jwa, S.H. Globalization and New Industrial Organization: Implications for Structural Adjustment Policies. In *Regionalism Versus Multilateral Trade Arrangements*; University of Chicago Press: Chicago, IL, USA, 1997; NBER Chapters 1997; pp. 313–344.
23. James, A.R.; Torvik, R.; Verdier, T. Political foundations of the resource curse. *J. Dev. Econ.* **2006**, *79*, 447–468. [CrossRef]
24. Hsieh, C.T.; Klenow, P.J. Misallocation and Manufacturing TFP in China and India. *Q. J. Econ.* **2009**, *124*, 1403–1448. [CrossRef]
25. Zhigang Yuan, D.X. Coordinating Urban and Rural Development: Coordinated Reallocation of Human Capital and Land Capital. *Res. Social. Chin. Charact.* **2011**, *2*, 31–42.
26. Yongwei Chen, W.H. Price distortion, factor mismatch and efficiency loss: Theory and application. *Economics* **2011**, *10*, 22.
27. Deming Luo, Y.L.; Shi, J. Factor market distortion, resource misplacement and productivity. *Econ. Res.* **2012**, *47*, 4–14+39.
28. Huihua Nie, R.J. Productivity and Resource Misplacement of Chinese Manufacturing Enterprises. *World Econ.* **2011**, *34*, 27–42.
29. Zhen Yang, Y.C. China's manufacturing resource misplacement and welfare loss measurement. *Econ. Res.* **2013**, *48*, 43–55.
30. Wang, S.; Sun, X.; Song, M. Environmental Regulation, Resource Misallocation, and Ecological Efficiency. *Emerg. Mark. Financ. Trade* **2021**, *57*, 410–429. [CrossRef]
31. Kong, Q.; Peng, D.; Zhang, R.; Wong, Z. Resource misallocation, production efficiency and outward foreign direct investment decisions of Chinese enterprises. *Res. Int. Bus. Financ.* **2021**, *55*, 101343. [CrossRef]
32. Buera, F.J.; Kaboski, J.P.; Shin, Y. Finance and Development: A Tale of Two Sectors. *Am. Econ. Rev.* **2011**, *101*, 1964–2002. [CrossRef]
33. Abhijit, V.; Banerjee, B.M. Why Does Misallocation Persist? *Am. Econ. J. Macroecon.* **2010**, *2*, 189–206.
34. Pan, Y. Focus on and solve the problem of financial resource mismatch from a strategic height. *Explor. Contention* **2016**, 93–98.
35. Asker, J.; Collard-Wexler, A.; Jan De, L. Productivity Volatility and the Misallocation of Resources in Developing Economies. CEPR Discussion Papers. Available online: https://www.nber.org/papers/w17175 (accessed on 21 August 2022).
36. Lewandowskakalina, M. Productivity Dispersion and Misallocation of Resources: Evidence from Polish Industries. *Economics.* Available online: https://ideas.repec.org/p/wse/wpaper/66.html (accessed on 5 August 2022).
37. Restuccia, D. Rogerson, Richard Misallocation and productivity. *Rev. Econ. Dyn.* **2013**, *16*, 1–10. [CrossRef]
38. Virgiliu Midrigan, D.Y.X. Finance and Misallocation: Evidence from Plant-Level Data. *Am. Econ. Rev.* **2014**, *104*, 422–458. [CrossRef]
39. Ziebarth, N.L. Are China and India Backwards? Evidence from the 19th Century U.S. Census of Manufactures. *Soc. Econ. Dyn.* **2013**, *16*, 138. [CrossRef]
40. Guner, N.; Ventura, G.; Xu, Y. Macroeconomic Implications of Size-Dependent Policies. *Rev. Econ. Dyn.* **2017**, *11*, 721–744. [CrossRef]
41. Barseghyan, L.; Dicecio, R. Entry costs, industry structure, and cross-country income and TFP differences. *J. Econ. Theory* **2009**, *146*, 1828–1851. [CrossRef]
42. Repetto, A.; Micco, A. Productivity, Misallocation and the Labor Market. *Economics.* Available online: https://ideas.repec.org/p/uai/wpaper/wp_020.html (accessed on 10 August 2022).
43. Bai, D. Distortion of industrial policy, market allocation and misallocation of regional resources. In Proceedings of the Modern Economic System and High-Quality Development-the 13th China Development Economics Annual Conference, Lanzhou, China, 5 March 2021; 2019; Volume 6, p. 1.
44. Jac, A.; Mg, B.; Al, C.; Fp, B. Environmental misallocation in the copper industry—ScienceDirect. *Resour. Policy* **2021**, *71*, 102003. [CrossRef]

45. Sahin, A.; Song, J.; Topa, G.; Violante, G.L. Mismatch in the Labor Market. Available online: https://users.nber.org/~{}confer/2011/EFGf11/Sahin_Song_Topa_Violante.pdf (accessed on 26 August 2022).
46. Sargent, T.; Manuelli, R. Instability, Misallocation and Productivity. In Proceedings of the 2013 Meeting Papers. Available online: https://ideas.repec.org/p/red/sed013/1043.html (accessed on 16 August 2022).
47. Xiaolan Deng, Z.Y. The Impact of Misallocation of Resources on China's Industrial Low-carbon TFP: An Empirical Study. *Financ. Sci.* **2014**, *5*, 74–83.
48. Ai Ma, J.Z.; Yu, J.; Xu, H. Does the mismatch of labor factors hinder the upgrading of the industrial structure? In Proceedings of the Modern Economic System and High-Quality Development-the 13th China Development Economics Annual Conference, Lanzhou, China, 5 March 2021; 2019; Volume 6, p. 1.
49. Wang, W.Z.S.; Niu, Z. Industrial Policy, Market Competition and Resource Mismatch. *Economist* **2014**, *9*, 22–32. [CrossRef]
50. Brandt, L.B.; Johannes, V.B.; Yifan, Z. Creative Accounting or Creative Destruction? Firm-Level Productivity Growth in Chinese Manufacturing. *J. Dev. Econ.* **2012**, *97*, 339–351. [CrossRef]
51. Puyang Sun, W.J.; Zhang, Y. Product Substitution and Productivity Distribution: An Empirical Study Based on Data from Chinese Manufacturing Enterprises. *Econ. Res.* **2013**, *48*, 30–42.
52. Guan Gong, G.H. Resource Allocation Efficiency and Total Factor Productivity of China's Manufacturing Industry. *Econ. Res.* **2013**, *48*, 4–15+29.
53. Ariel Pakes, S.O. A limit theorem for a smooth class of semiparametric estimators. *J. Econom.* **1995**, *65*, 295–332. [CrossRef]
54. Nordhaus, W.D. Productivity Growth and the New Economy. *Brook. Pap. Econ. Act.* **2002**, *2002*, 211–265. [CrossRef]
55. Yang, C. Research Progress and Frontier Analysis of Resource Mismatch. *China Sci. Technol. Resour. Guide* **2020**, *52*, 9. [CrossRef]
56. Barney, G.; Glaser, A.L.S.; Strutzel, E. The Discovery of Grounded Theory; Strategies for Qualitative Research. *Nurs. Res.* **1968**, *17*, 364.

buildings MDPI

Article

Barriers to Adopting Lean Construction in Small and Medium-Sized Enterprises—The Case of Peru

Cristian Huaman-Orosco [1], Andrews A. Erazo-Rondinel [2] and Rodrigo F. Herrera [3,*]

[1] Facultad de Ingenieria Civil, Universidad Nacional de Ingenieria, Lima 15333, Peru
[2] Facultad de Ingenieria, Universidad Continental, Huancayo 12001, Peru
[3] School of Civil Engineering, Pontificia Universidad Católica de Valparaíso, Av. Brasil 2147, Valparaíso 234000, Chile
* Correspondence: rodrigo.herrera@pucv.cl

Abstract: The competitiveness in the construction industry is leading companies to implement the Lean philosophy to improve project management and prepare to begin the adoption of Construction 4.0. However, this implementation generates several barriers, depending on the type of company and the country where it is implemented. For this reason, the following research aims to identify the barriers generated in implementing the Lean philosophy in Peru and SMEs. The following study began with a literature review, followed by a questionnaire, which six Lean experts evaluated. Then, 127 Peruvian professionals answered the survey, and the data were analyzed with RII and Exploratory Factor Analysis. As a result, the barriers to Lean Construction were grouped into four groups to analyze their impact on the industry: collaboration, education and implementation, organizational vision and system; Lean theory and philosophy; and Lean tools, coordination, and information flow. Barriers were also analyzed by project type, stage, and organization size to conclude that implementation barriers are directly related to company size. Finally, the obtained barriers are discussed, and Construction 4.0 is proposed to reduce their impact on construction projects. These results can be helpful for implementers to plan strategies, improve knowledge about Lean Construction implementation, and adopt new techniques that could help improve the construction industry's productivity.

Keywords: Lean Construction; lean implementation; barriers; construction 4.0; Peru

Citation: Huaman-Orosco, C.; Erazo-Rondinel, A.A.; Herrera, R.F. Barriers to Adopting Lean Construction in Small and Medium-Sized Enterprises—The Case of Peru. *Buildings* **2022**, *12*, 1637. https://doi.org/10.3390/buildings12101637

Academic Editors: Yongjian Ke, Jingxiao Zhang and Simon P. Philbin

Received: 19 August 2022
Accepted: 27 September 2022
Published: 9 October 2022

1. Introduction

The construction industry has lower productivity than other industries, such as manufacturing [1]. The temporality of projects, uniqueness of production, and the complex communication between the actors do not allow control over production, high rates of variability, unsatisfied customers, and waste of resources (workforce, activities, time, and money), resulting in projects with cost overruns and delays [2]. Lean Construction (LC) started its diffusion as a viable solution to the problems, achieving until today its implementation in building projects, road infrastructure, sanitary, mining, industrial plants, and energy and oil project [3].

Lean Construction applies the principles and tools of the Lean philosophy during the entire project lifecycle, from conception, execution, and commissioning [4]. Lean focuses on people as the main actor to improve the organization's management [2]. In addition, Lean encourages teamwork, improves communication, helps people focus on the value activities of the external and internal customer, knows the entire production flow, and identifies errors in the early stages [5]. As a result, Lean Construction has shown benefits of improving the project's overall productivity, reducing accidents, avoiding rework, satisfactory results for the customer, anticipating conflicts between project participants, and achieving project execution at the estimated cost and time [6].

The implementation of LC in projects is not simple. LC is not a set of tools or recipes; on the contrary, it is a construction management approach. LC implementation requires focusing on many factors. In addition, all organizations have multiple subcultures that generate a variety of conflicts of interest [7–9]. Moreover, integrating multiple cultures creates impressions among teams of activity overload, psychological fatigue, and increased resources to meet project goals [10]. These factors are known as barriers, resulting in the spread of lean with negative results and increased waste in different project stages [11].

Barriers to LC implementation are related to geographic factors of similar economic and cultural contexts [12–14]. These factors are common in most Latin American countries, with the limited implementation of LC in the private sector [15]. The need to implement LC in all sectors, with innovations in design, construction, technology, and materials, could help to efficiently optimize the use of resources (economic, materials, and others) to reduce the infrastructure gaps in about 42 million families [16].

Peru has not had a study identifying the main barriers professionals face. This problem generates concern because of the lack of a priori knowledge of the challenges that will be faced. Therefore, during the implementation, it generates excessive effort in the team, or they desist from using it in the project. Furthermore, Murguia [17] mentions that the company's size determines Lean implementation because it influences communication, complexity, maturity, and coordination effort. Therefore, this study aims to identify the main barriers to Lean implementation perceived by LC practitioners in Peru and to group them by company size. It performs a relationship and correlation analysis to find the most relevant factors for implementation.

SMEs are the companies that predominate in quantity and are the backbone of economic agility and job creation in any country [18]. SMEs are motivated by the positive benefits achieved by large organizations and adopt the lean philosophy [19–21]. In addition, SMEs are directly related to each other; e.g., small companies are active participants as subcontractors of medium and large companies [22]. However, the low Lean competence of small companies does not allow them to collaborate with large companies to transfer value directly to the customer (due to their agility in decision-making and direct dealings with the customer). Moreover, small companies can maintain competitiveness and adapt quickly to the strategies of large organizations [20,23]. For example, the investment of resources in implementation is a determining factor for decision-making in small companies and, to a lesser degree, in medium and large companies [11]. In addition, the maturity of LC implementation is different by the size of the organization; medium and large companies have more significant benefits in the long term by attracting talent, capabilities to negotiate with multiple suppliers, subcontractors, and rethinking strategies [24]. For the adoption of Lean in SMEs to be successful, it is necessary to manage implementation barriers, develop strategies to reduce their impact, and anticipate problems [11,25].

2. Literature Review

Increasing productivity in products or services through investment focused on technology has given good results in manufacturing, which is why gradually, it has been seeking to implement technology in the construction sector through the implementation of Construction 4.0 [26,27] and Lean Construction 4.0 [28]. However, the construction industry has temporary projects that do not justify personnel training and research in the long term [29]. Therefore, organizations have started to use Lean tools in specific projects [30]; some are successful, others do not obtain the expected results, and they return to traditional management systems [30]. Ballard et al. [31] suggest starting implementation by motivating project leaders, training all project members, implementing tools to form habits, forming a culture over time, and moving on to using the Lean philosophy for continuous improvement. Salvatierra et al. [32] suggested that it is necessary to maintain a balance between tools, culture, and philosophy, to maintain good Lean practices in the long term.

Organizations that implement Lean in their projects have challenges in adopting the philosophy of their workers, processes, and management. For example, in the United

States, there are challenges with low commitment from top management, low awareness of Lean practices, and poor communication between teams to share information [12]. On the other hand, the United Kingdom has the barriers of lack of knowledge of Lean by professionals, resistance to change, and difficulty in adopting the Lean culture [33]. In Colombia, there are barriers such as the perception of uncertainty to obtain positive results, the transparency of information between teams, and state laws that do not allow flexibility for Lean projects [34,35]. Finally, in Chile, the challenge is distributing information at all levels and implementing the project in the long term [36].

2.1. Global Barriers to LC Implementation

Lean Construction has been implemented in more than 48 countries for more than 20 years, and its evolution has been recorded in more than 1382 research papers [37]. Moreover, most publications are reported from the United States and Europe; however, the Lean system has spread and has a growing adoption in countries such as Brazil, Chile, and Peru. However, many organizations report that they cannot achieve the maximum benefit from the Lean philosophy due to multiple factors [30]. Globally, among the countries that have implemented Lean in construction, 110 barriers were identified, which are grouped into three categories: people (29%), including education, management, operation, and culture; production process (20%) focused on customer identification, management, planning and control; production management and logistics processes (51%) involving governance, long-term business philosophy, processes and system [13]. In the USA, it was identified that the main barrier is in people, especially in their educational and leadership training [12], as well as the scarce standardization of knowledge and Lean concepts, which do not agree among experts, creating learning complexity in beginner practitioners. Colombia has barriers such as resistance to change, high implementation costs, and information transparency manipulation [35,36]. Chile showed that the lack of support from top management and vertical information, which does not reach all levels of the organization, delays decision-making in implementation [38].

Moreover, short-term thinking does not allow solving problems entirely, creating vicious circles that repeat the same [32]. In Peru, the most relevant barriers are the lack of State policies to involve its institutions in Lean projects, the low collaboration between academia and industry, and a perceived high cost of implementation [39]. In addition, resistance to change [40] and problems in the construction stage are carried over from the design stage, making little or no improvement in the construction stage, affecting the degree of customer satisfaction, cost overruns, rework, and project stoppages [41].

2.2. Local Implementation Barriers

Barriers to Lean implementation have varied according to each country's socio-cultural conditions and technical capabilities [30]. Therefore, it was necessary to align the barriers from the literature and add the barriers perceived by the experts throughout their implementation experiences and to have a complete information base of the study phenomenon. The diversity of the experts in building projects (educational institutions, shopping malls, multi-family buildings, among others), infrastructure (road works, sanitation, railroads), energy, oil, industrial plants, and mining allowed adjusting the questionnaire to meet the objectives.

Most of the researchers and implementers evidenced that the main barrier to Lean is short-term thinking, so the good practices developed in the project are not managed to improve the next project [42]. The study focused on an exploratory study using Exploratory Factor Analysis, and to cover the study phenomenon; it was chosen to group the barriers in the sustainability model of Lean practices of the GEPUC triangle [32]. The GEPUC triangle is composed of culture, tools, and philosophy; the weakness of any of them leads to the implementation not obtaining the maximum benefit or falling over time. The expert interview and literature review could help adapt the barriers to the study phenomenon and achieve the study's objectives. In addition, the grouping in the GEPUC triangle could help to organize and maintain the balance between the barriers.

3. Research Methodology

The methodology of the study is shown in Figure 1. It includes the three stages, the tools used, and the activities of each process to achieve the study's objectives. The present study starts with a literature review followed by a valid survey design through semi-structured interviews with six Peruvian experts. The survey was designed with closed and semi-open questions, using the mixed method to understand the study phenomenon better. The figure below shows the process followed in the research.

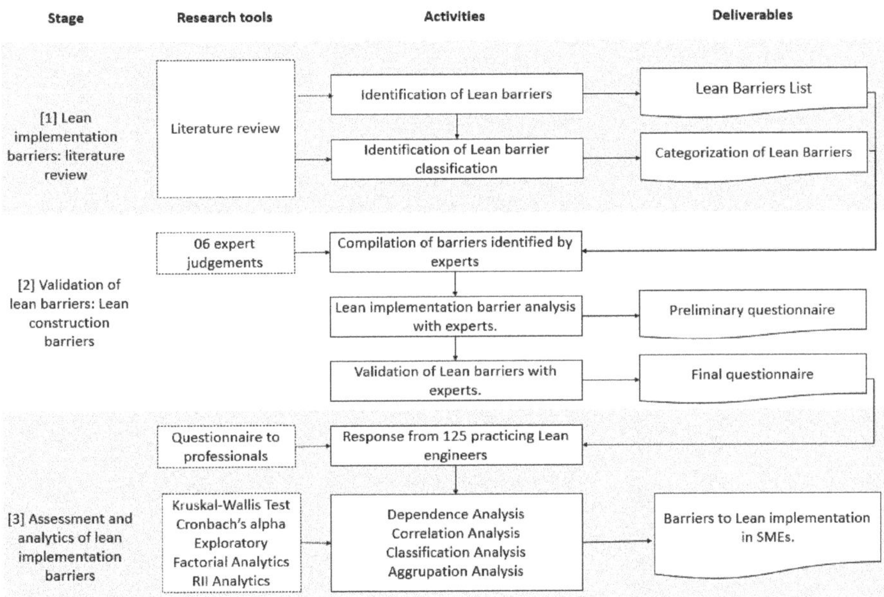

Figure 1. Research Methodology.

Through a literature review, the first stage identified the main barriers to implementation that the professionals have evidenced and obtained a list. Then, the best way to classify the barriers was reviewed, concluding by grouping them into three groups philosophy, tools, and Lean culture. On the other hand, implementation models were reviewed to be able to relate the barriers to the steps or raise barriers. Finally, libraries of Scopus, Web of Science, Google Scholar, International Group for Lean Construction (IGLC), ASCE, Lean Construction Journal, and the central repository of theses of Peru were consulted; comprised between the years 2000 to 2020, with keywords such as "Lean Implementation", "Lean challenges", "Lean barriers", and "Lean Construction".

In the second stage, researchers selected six experts from ten candidates with the following characteristics: (a) more than 12 years of experience as a Lean implementation leader and (b) experience in project diversity, university teaching, research, and Lean consultants. The general information of the experts is shown in Table 1. The general information of the experts is shown in Table 1.

Table 1. Information from lean experts.

Professional Expert	Academic Degree	Project Experience
Lean Expert 01	Civil Engineer with more than fifteen years of experience as a consultant, practitioner, teacher, and advanced instructor Lean.	Buildings, roadways, and energy infrastructure projects.
Lean Expert 02	Civil Engineer with more than twelve years of experience as a consultant, lean practitioner, and public sector consultant.	Railway, buildings, and hospital projects
Lean Expert 03	Civil Engineer with more than thirteen years of experience implementing lean, consultant, and advanced instructor lean.	Mining, buildings, infrastructure, and urban facilities projects
Lean Expert 04	Civil Engineer with more than ten years of experience implementing lean, production, planning, and design engineer	Education, hospital, commercial, and building projects.
Lean Expert 05	Civil Engineer with more than thirteen years of experience as a project manager, company owner, and advanced instructor lean.	Buildings, commercial, and roadways projects.
Lean Expert 06	Civil Engineer with more than twelve years of experience as a university teacher, investigator, and consultant.	Railway, buildings, commercial, sports infrastructure, and mining projects.

The questionnaire was evaluated and guaranteed with the opinion of six experts, who, based on their experience, suggested adding the barriers of: "Low capacity of people to identify waste" and "Lack of leadership and empowerment of people", which were searched in the literature, finding similarities in manufacturing. The questionnaire underwent three reviews by six experts, all with clear objectives and aligned to the study.

3.1. Research Design and Data Collection

The mixed method was used to take a holistic "snapshot" of the study phenomenon [43], simultaneously integrating qualitative and quantitative questions. The mixed method allows obtaining a wider range of perspectives of the problem in terms of frequency, generality, complexity, magnitude, and understanding. Quantitative to identify the company's size, years of experience, and frequency of use of good practices. Qualitative to describe their experiences, personal difficulties, or particular experiences. Integrating both methods allowed the questionnaire to be improved by the experts. After the pilot plan, new questions were identified and readjusted thanks to the corroboration of qualitative and quantitative data.

The survey followed a cross-sectional process to obtain the most significant amount of data and ease of remote response. The questionnaire had 40 open-ended and closed-ended questions. The barriers were evaluated with a Likert scale from 1 to 5 points, asking the respondents to assess which variables they most frequently experienced in their project and were evaluated as "Never = 1, Rarely = 2, Occasionally = 3, Frequently = 4, and very frequently = 5".

In addition, the questions of the questionnaire were changed to positive or neutral syntax, with the objective of not influencing the answers or sympathizing [44], avoiding having answers directed to the aim of the study. For example: what is the frequency you experience your team's reflections of the activities performed and suggestions for good practices? The respondent can choose to rate on a scale of 1 to 5. Therefore, the higher frequency of use is a common practice among professionals, and the lower use of practices causes barriers that prevent the development of the implementation.

The questionnaire was disseminated in Spanish through emails, social networks, and professional networks. Peruvian lean dissemination organizations and LC practitioner companies were invited to participate in the study. The survey was distributed virtually through Microsoft Forms. The evaluation survey can be found in supplemental material.

3.2. Population and Sample

The Last Planner System (LPS) is the commitment and variability control system for excellence [45]. Starting the implementation with LPS allows the control of setbacks and uncertainties, self-criticizing and enabling team members to identify problems and propose improvements [46,47]. Based on the premise, the study population was limited to all Peruvian professionals with more than two years of experience implementing LPS or participating in projects managed with LPS. The Peruvian Chapter of Lean Construction (LCI Peru) registers as of January 2021, a number of 1360 professionals registered and certified in its Lean Construction training program, which is composed of one year of theoretical training and requires the application of LPS for at least one project, which supports the certification and the ability to implement Lean Construction in more than 2 years.

In addition, the study used non-probabilistic and snowball sampling [48] to obtain as much data as possible under the conditions of the objectives of the study. The survey was disseminated from 18 february 2021 to 24 april 2021. A total of 127 professionals responded to the survey, and data from 125 professionals were used for the statistical analysis of the study. Unfortunately, one was discarded for not completing the questionnaire and the other for presenting strange variations in the correlation. The general data are shown in Table 2.

Table 2. Demographic profile of the interviewed.

	Demographic Characteristics	Frequency	Percentage
Experience	1–5 years	82	66.39%
	6–10 years	24	19.33%
	11–15 years	10	7.56%
	16–20 years	6	5.04%
	More than 20 years	2	1.68%
Experience working with lean	2–3 years	51	41.13%
	3–5 years	38	30.65%
	5–8 years	21	16.94%
	8–10 years	12	9.68%
	More than 10 years	2	1.61%
Enterprise type	Construction	92	73.95%
	Consulting and project supervision	9	7.56%
	Project formulation and design	8	6.72%
	Project logistics and maintenance	4	2.52%
	suppliers	2	1.68%
	other	9	7.56%
Project Type	Buildings	59	47.50%
	Infrastructure	37	30%
	Industrial plants	7	5%
	Energy and oil	6	4.17%
	Other	15	11.67%
Size of enterprise	micro (1 to 10 people)	27	22%
	small (10 to 50 people)	32	26%
	medium (50 to 250 people)	30	24%
	Large (more than 250 people)	35	28%

3.3. Validity Test and Data Analysis

The reliability of the data was determined by Cronbach's coefficient (α), whose values higher than 0.7 represent data with homogeneous distribution; therefore, they are reliable and solid data for the research [48].

The Exploratory Factor Analysis is composed of three main stages: (a) Assessment of the suitability of data for use in the EFA, the Kaiser–Maier–Olkin (KMO) measure, and Bartlett's test of sphericity were used to check the adequacy of the sample and the correlation between variables to test the adequacy of the sample and the correlation between variables. KMO values greater than 0.5 were accepted to perform EFA with data between 100 and 200. In order to reject the null hypothesis of Bartlett's test and show that the data are highly correlated, the matrix determinant is required to be less than 0.001 and have a significance of less than 0.05 [49–51]. (b) For extraction of factors or grouping of variables, Principal Component Analysis (PCA) generates a common grouping of data "commonality", whose commonality is accepted as EFA factors to maintain the greatest amount of information according to the objectives of the study, as well as to reduce the components as necessary [49,50,52]. (c) For rotation and interpretation of the grouping of variables, the orthogonal rotation Vari-max was used, which allows for a better and simple interpretation of the data; in addition, it allows to reduce the number of variables that have a high weight load in the factors, a recommended load is of 0.4 [52,53].

The data analysis was conducted with the statistical package for social sciences (SPSS) v25.0, reliability analysis, descriptive statistics, and Kruskal–Wallis tests (H) to differentiate the barriers by study groups. The Kruskal–Wallis test is a rank-based test for non-parametric variables, i.e., to compare more than two independent samples. The analysis starts with the hypothesis that the variables come from the same sample. If the Kruskal–Wallis statistic is significant ($p \leq 0.05$), the non-parametric multiple comparison tests indicate that the initial hypothesis is accepted; otherwise, it is assumed that the variables have independent provenances [54].

In addition, Microsoft Excel was used for the Relative Importance Index (RII) analysis to discuss priority barriers according to the segmented groups. RII allows the analysis to distribute the variance explained among multiple predictors to better understand the role played by each of them in a regression equation [55].

$$\boldsymbol{RII} = \frac{\sum w}{\boldsymbol{A.N.}}$$

where $\boldsymbol{w}$ is the frequency given to each factor by respondents (1 to 5); $\boldsymbol{A}$ is the highest frequency of each factor, 5 in this example; and $\boldsymbol{N}$ is the total number of respondents, 125 in this example.

4. Results

The literature review resulted in a list of lean implementation barriers, as shown in Table 3.

Table 3. List of lean implementation barriers.

Barrier	Code	Reference
Lack of collaborative work between academia and the construction industry	CEI1	[56,57]
Extensive duration of the Lean learning curve	CEI2	[58,59]
Lack of top management commitment to the implementation	CEI3	[47,60]
Lack of time to implement lean in ongoing projects	CEI4	[32,36]
Lack of up-front work planning and realistic scheduling using Lean tools	CEI5	[59,61]
People use tools without sustaining them with culture and philosophy	CEI6	[13,32,62]

Table 3. *Cont.*

Barrier	Code	Reference
Lack of improvement culture throughout the organization	CEI7	[38,63]
Inability to measure performance and team progress	CEI8	[64]
Lack of collaborative planning among project stakeholders	CEI9	[34,65]
The low ability of people to recognize waste	CEI10	[46,66]
Low knowledge in Lean in professionals who graduated from university	TFL1	[17,67]
Lack of knowledge of the fundamental purpose and rationale of Lean implementation	TFL2	[63,66]
Lack of knowledge and experience in the implementers	TFL3	[68]
Replicating another organization's lean strategy	TFL4	[58,69]
Low organizational commitment	TFL5	[13,69]
Lack of leadership and empowerment of people in the project	TFL6	[39]
Local and not global flow optimization	TFL7	[66,68]
Lack of centralized stored and shared information to generate continuous improvement cycle	HCF1	[13,66,70]
Lack of collaboration of all project stakeholders at all levels and early stages of design and production (suppliers, subcontractors, etc.)	HCF2	[65,71]
Wrong selection of Lean tools	HCF3	[72,73]
Lack of self-criticism to learn from mistakes and identify problems	HCF4	[74]
Lack of transparent information between team members and management, reducing reliability in Lean	HCF5	[75]
The difficulty of top management communication with improvement initiatives	HCF6	[12,32,72]
Lack of clear definition of scope, identification of value, and definition from the customer's point of view	VOS1	[63,69]
Lack of long-term thinking in the organization for the implementation of Lean	VOS2	[13,71]
Lack of information exchange between teams, suppliers, and subcontractors	VOS3	[13,76]
People in meetings do not respect the opinion of others and impose their ideas	VOS4	[77]

Table 4 shows the most relevant data from the semi-structured interviews with the experts. The interviews were semi-structured, opting for open and semi-open questions to obtain qualitative data and the most significant amount of information. In addition, the study phenomenon is relatively new in the country, so it is required to have a complete picture of the phenomenon [44].

Table 4. Results of interviews with experts.

Activity	Description
Collecting barriers identified by the experts	General data was obtained from the experts, implementation experiences, and barriers specific to project participation.
Analysis of the list of barriers by the experts	An interview was conducted with the experts, and each expert evaluated the barriers identified in the literature with a degree of priority. A total of 25 out of 68 barriers were used for the survey.
Validation of barriers with experts	The experts add barriers from their own experience, and two barriers are added in this process, ending the survey with 27 barriers.

4.1. Descriptive Analysis of the Barriers in SMEs

The reliability of the data is $\alpha = 0.924$, which allowed the following analyses to be performed. Table 5 shows the barriers to implementation in SMEs and is ranked according to the results of the RII. The CEI1 barrier "lack of collaborative work between academia and industry" has a more significant impact on SMEs compared to large companies; SMEs do not create links with academia; the reason is that they are focused on their processes and perceive these relationships as over costs. In addition, the projects they present are very temporary and have low investment, so they require high agility without losing resources [20].

Table 5. Barrier impact by project stage, type of project, and company size. Data sorted by RII.

Barrier	Construction	Design	Other Stages	Buildings	Infrastructure	Other Sectors	Long Enterprise	SMEs
CEI1	69.40%	56.90%	66.30%	67.90%	70.00%	58.60%	58.80%	70.90%
TFL1	60.60%	66.20%	56.30%	61.60%	60.80%	55.70%	50.30%	64.30%
CEI3	58.30%	58.50%	52.50%	55.90%	61.70%	51.40%	49.10%	60.70%
CEI2	57.90%	60.00%	57.50%	56.20%	61.70%	54.30%	53.30%	59.80%
CEI4	55.40%	55.40%	56.30%	54.90%	59.60%	44.30%	45.50%	59.10%
TFL2	55.20%	61.50%	52.50%	55.20%	59.20%	44.30%	47.90%	58.30%
CEI5	52.10%	66.20%	57.50%	55.90%	55.00%	44.30%	43.60%	58.00%
TFL3	54.20%	58.50%	48.80%	53.70%	57.10%	44.30%	43.00%	57.80%
HCF1	55.00%	56.90%	55.00%	56.50%	55.40%	48.60%	50.30%	57.00%
CEI6	54.40%	60.00%	48.80%	53.30%	57.50%	47.10%	47.30%	56.70%
TLF6	54.40%	53.80%	52.50%	52.70%	58.30%	45.70%	46.70%	56.70%
HCF2	55.80%	50.80%	45.00%	54.00%	56.30%	45.70%	47.30%	56.30%
CEI8	51.30%	50.80%	48.80%	50.20%	55.40%	38.60%	38.80%	55.20%
TFL4	53.10%	53.80%	51.30%	51.70%	55.40%	50.00%	47.90%	54.80%
HCF3	53.30%	55.40%	48.80%	53.30%	53.80%	48.60%	51.50%	53.50%
CEI7	51.00%	55.40%	50.00%	50.20%	53.80%	48.60%	45.50%	53.50%
TLF7	51.50%	58.50%	48.80%	50.50%	54.60%	48.60%	48.50%	53.00%
HCF5	51.70%	46.20%	42.50%	50.20%	50.80%	45.70%	42.40%	52.60%
TFL5	52.50%	49.20%	47.50%	50.20%	53.30%	51.40%	49.10%	52.40%
HCF6	50.40%	47.70%	46.30%	50.50%	50.80%	41.40%	43.60%	51.70%
HCF4	50.80%	53.80%	42.50%	50.20%	51.30%	45.70%	47.30%	51.10%
VOS2	46.50%	56.90%	47.50%	46.70%	50.00%	44.30%	38.20%	51.10%
VOS3	45.00%	61.50%	51.30%	45.10%	50.80%	47.10%	39.40%	50.40%
CEI9	46.90%	55.40%	45.00%	47.90%	50.00%	37.10%	39.40%	50.40%
VOS1	48.80%	52.30%	42.50%	45.70%	53.30%	42.90%	43.60%	50.00%
CEI10	45.80%	50.80%	37.50%	44.10%	49.20%	37.10%	37.60%	48.00%
VOS4	44.80%	46.20%	42.50%	43.80%	45.80%	44.30%	38.80%	46.70%

The TFL1 barrier, "low knowledge in professionals", has a significant impact on SMEs, i.e., the academy does not meet the competencies required by the companies to implement LC in their projects, and therefore, they have a long learning process in the members of their projects [78]. The large company has developed academies in its organization with customized knowledge, reducing the impact of this barrier [79].

From the table above, we can indicate that SMEs have three high-impact barriers, such as "Lack of collaborative work between academia and the construction industry" (CEI1), "Low knowledge in Lean in professionals graduated from the university" (TLF1), and "Lack of commitment of top management with the implementation" (CEI3) and are forcing the slow progress of Lean implementation in organizations.

4.2. *Difference in the Evaluation of the Barriers*

The Kruskal–Wallis test analysis was performed on the barriers to lean implementation grouped into (a) project stages (design, construction, and operation), (b) project types (buildings, infrastructure, energy, and mining), and (c) by company size.

The results and analysis of the Kruskal–Wallis tests determine that the barriers to implementation in the stages and types of projects are similar; therefore, their study is not significant. On the other hand, the result of the Kruskal–Wallis test of the SME groups and large companies had 76% of different barriers, with independent impacts and difficulties. Therefore, the study focuses on the barriers to lean implementation by company size, similar to other studies [18,22,80].

For the exploratory factor analysis (EFA), the KMO was greater than 0.5, and Bartlett's sphericity or correlation matrix was not an identity matrix (approx. Chi-square = 2755.743, *p*-value = 0.000). Therefore, the researchers can say that factor analysis is useful with these data, and the sample is adequate for analysis and representation. By using a loading of 0.4 and a Varimax rotation, the results of the EFA analysis showed that the barriers could be classified into four factors, as shown in Table 6.

Table 6. Exploratory factor analysis-rotated factor matrix for SMEs lean implementation barriers.

	Code	Factor			
		1	2	3	4
Collaboration, education, and implementation.	CEI1	0.541			
	CEI2	0.614			
	CEI3	0.641			
	CEI4	0.782			
	CEI5	0.728			
	CEI6	0.640			
	CEI7	0.635			
	CEI8	0.706			
	CEI9	0.702			
	CEI10	0.417			
Lean theory and Philosophy.	TFL1		0.602		
	TFL2		0.595		
	TFL3		0.605		
	TFL4		0.727		
	TFL5		0.608		
	TFL6		0.531		
	TFL7		0.414		
Lean tools, coordination, and information flow.	HCF1			0.755	
	HCF2			0.608	
	HCF3			0.825	
	HCF4			0.610	
	HCF5			0.589	
	HCF6			0.550	
Organizational vision and system.	VOS1				0.516
	VOS2				0.549
	VOS3				0.792
	VOS4				0.620

5. Discussion

The group of barriers "collaboration, education, and implementation" is perceived with more impact in large companies; respondents argue that it is because of the large number of people to train, adapting the areas of logistics, administration, resources, and others. In addition, the excessive effort to control many processes and people; in addition, it requires investment for massive training and multiple professionals with knowledge in Lean. On the other hand, the SMEs can start the implementation by hiring people with expertise in Lean and perform other functions simultaneously, which does not take them many resources. They have the facility to manage small groups of people. Similar results were obtained in [81] From the point of view of Construction 4.0, the use of BIM and

Artificial Intelligence [82] and Cloud Computing [83] can help us to improve the identification of waste and mitigate the barrier CEI10. Moreover, to promote greater collaboration and communication among project stakeholders, Construction 4.0 components such as BIM [82,84] and social media [85,86] reduce the impact of the barrier CEI9. In addition, the use of IoT (Internet of Things) [87], artificial intelligence [88], and data analysis [89,90] help us to improve lean planning in construction projects and hence mitigate the impact of the barrier CEI5.

The group of barriers to "organizational vision and system" is closely related to the lack of knowledge on the part of the clients, the permanence of the people during the implementation, and the company's offer. SMEs compete for the new market, reach directly to the customer and try to offer Lean-based services. However, customers are unaware of the lean philosophy and subcontractors and suppliers. SMEs significantly impact their barrier due to the high level of informality in their projects [91] and the hiring of professionals with low knowledge of Lean. As a result, they perceive the industry as non-productive [92,93]. In addition, large companies work with large investment projects; these projects are awarded in most cases by the government. However, the government has institutions that do not require the use of Lean. Likewise, when large companies use Lean, they cannot integrate state agencies due to their low knowledge, little knowledge of the project's value, and a low response to the acceptance of decisions [65]. Regarding this group of barriers, Construction 4.0 proposed the use of BIM [84] and augmented reality [94] to improve scope definition, customer value, and information sharing, thereby reducing the presence of VOS 1 and VOS 3 barriers. In addition, elements such as ECD (Common Data Environment) [84,95–97], cloud [83,84,98], data sharing [84,99], and social media [84,86,100] will help us to improve information sharing between teams, suppliers, and subcontractors, decrease fragmentation in the construction industry, helping the organization to combat the barrier VOS 3.

The group of barriers "Lean theory and philosophy" refers to the learning process of Lean, which should be formed from the university stage [101] and requires that leaders support the implementation by motivating the team, sustaining the practices, and supporting their learning [102]. It seems contradictory that SMEs have a more significant impact on these barriers concerning large companies, despite having few workers, and it should be easy to train, monitor performance, progress the team, and implement. However, the cause is that SMEs execute short-duration small projects with very temporary workers. Construction 4.0 can help us reduce this group's barriers by transferring knowledge using cloud computing [97].

The group of barriers "Lean tools, coordination, and information flow", SMEs tend to hide information to evade taxes and not abide by labor laws; this is a factor that overflows and affects professionals, obtaining a work culture of distrust between themselves and the other actors of the project. In addition, if the work team does not record data of problems and the best solution, they do not take advantage of solutions to similar problems and therefore do not have data support to select the right tool [103]. Likewise, the SME'S projects that are in progress do not have time to plan, coordinate and train the work teams; therefore, it requires quick actions of key tools of simple understanding for people. However, the teams are more focused on putting out fires, and the implementer has a false perception of low experience and knowledge.

6. Conclusions

The barriers of "lack of collaborative work between academia and the construction industry", "high cost of implementation", and "contracts do not require the use of Lean" are of direct impact on the implementation of Lean. Moreover, actions are required with solutions to improve the productivity of construction in general according to the size of the companies. Most studies explored barriers to LC implementation in large organizations. However, small and medium-sized companies are part of the world's economic engine, and their distinctive characteristics cause different impacts on LC implementation barriers.

Before the study, there was only evidence of barriers focused on Lean tools and no analysis of the type of companies. However, the study showed that the impact of Lean barriers in Peru is focused on the size of organizations. Moreover, this may help educators, implementers, and researchers to focus on adapting Lean thinking to the size of their organization, leaving aside the type of project (buildings, infrastructure, etc.) and the stages of the project (design, construction, operation, etc.). In addition, it is recommended that educators link curricula to the demands of companies and that companies invest in funded research to address the problems in their projects. Moreover, many professionals select incorrect tools when faced with a problem or otherwise work with Lean tools but are unaware of the theoretical underpinnings and are unable to obtain the full potential of Lean.

Finally, the study is biased toward the snowball sampling model, where the sample was selected conveniently to obtain results related to the study's objectives. Reducing the impact of barriers on LC implementers can help facilitators take precautions and rethink their implementation strategies. Moreover, for future research, consider strategies to reduce the impact of lean barriers, and create a roadmap. The roadmap must incorporate the lean principles, implementation barriers, construction 4.0, and lean maturity models; verify the implementation barriers through a lean maturity assessment in the organizations to corroborate the responses of the interviewees; and finally, perform a validation of the grouping of barriers through lean principles.

Supplementary Materials: The following supporting information can be downloaded at: https://www.mdpi.com/article/10.3390/buildings12101637/s1, Evaluation Survey.

Author Contributions: Conceptualization, C.H.-O. and A.A.E.-R.; Formal analysis, C.H.-O.; Methodology, A.A.E.-R. and R.F.H.; Supervision, R.F.H.; Writing—original draft, C.H.-O. and A.A.E.-R.; Writing—review & editing, R.F.H.. All authors have read and agreed to the published version of the manuscript.

Funding: This research received no external funding.

Institutional Review Board Statement: Not applicable.

Informed Consent Statement: Informed consent was obtained from all subjects involved in the study.

Data Availability Statement: The data presented in this study are available on request from the corresponding author. The data are not publicly available due to privacy.

Conflicts of Interest: The authors declare no conflict of interest.

References

1. Woetzel, J.; Mischke, J.; Barbosa, F.; Sridhar, M.; Ribeirinho, M.J.; Parsons, M.; Bertram, N.; Brown, S. *Reinventing Construction: A Route to Higher Productivity*; McKinsey Global Institute: London, UK, 2017.
2. Seed, W. *Transforming Design and Construction: A Framework for Change*, 1st ed.; Lean Construction Institute, Ed.; Lean Construction Institute: Arlington, VA, USA, 2017.
3. Erazo-Rondinel, A.A.; Huaman-Orosco, C. Exploratory Study of the Main Lean Tools in Construction Proojects in Peru. In Proceedings of the 29th Annual Conference of the International Group for Lean Construction (IGLC), Lima, Peru, 14–17 July 2021.
4. Koskela, L. *Application of the New Production Philosophy to Construction*; Stanford University: Stanford, CA, USA, 1992; Volume 72.
5. Modig, N.; Ahlstrom, P. *This Is Lean*; Morrison, J., Ed.; Rheologica Publishing: Bach, Switzerland, 2013; Volume 1.
6. McGraw Hill Construction Research y Analytics. *Lean Construction Leveraging Collaboration and Advanced Practices to Increase Project Efficiency*; Hill, M., Ed.; Dassault Systèmes: Vélizy-Villacoublay, France, 2013; Volume 3.
7. Salonitis, K.; Tsinopoulos, C. Drivers and Barriers of Lean Implementation in the Greek Manufacturing Sector. *Procedia CIRP* **2016**, *57*, 189–194. [CrossRef]
8. Liker, J. *The Toyota Way:14 Management Principles from the World's Greatest Manufacturer*; McGraw Hill: New York, NY, USA, 2004; Volume 1, ISBN 0070587477.
9. Kumar, A.R.; Subramanian, A.; Ware, B.F.; Fernandez, J.E. Lean Tools and Their Applications in an HPI Endeavor. In Proceedings of the 1st Annual World Conference of the Society for Industrial and Systems Engineering, Washington, DC, USA, 16–18 September 2012.
10. Rane, A.B.; Sunnapwar, V.K.; Rane, S. Strategies to Overcome the HR Barriers in Successful Lean Implementation. *Int. J. Procure. Manag.* **2016**, *9*, 223. [CrossRef]

11. Bhasin, S. Prominent Obstacles to Lean. *Int. J. Product. Perform. Manag.* **2012**, *61*, 403–425. [CrossRef]
12. Demirkesen, S.; Wachter, N.; Oprach, S.; Haghsheno, S. Identifying Barriers in Lean Implementation in the Construction Industry. In Proceedings of the 27th Annual Conference of the International Group for Lean Construction (IGLC), Dublin, Ireland, 3–5 July 2019; Volume 3.
13. Cano, S.; Delgado, J.; Botero, L.; Rubiano, O. Barriers and Success Factors in Lean Construction Implementation Survey in Pilot Contex. In Proceedings of the 23rd Annual Conference of the International Group for Lean Construction (IGLC), Perth, Australia, 29–31 July 2015.
14. Li, S.; Wu, X.; Zhou, Y.; Liu, X. A Study on the Evaluation of Implementation Level of Lean Construction in Two Chinese Firms. *Renew. Sustain. Energy Rev.* **2017**, *71*, 846–851. [CrossRef]
15. Martinez, E.; Reid, C.K.; Tommelein, I.D. Lean Construction for Affordable Housing: A Case Study in Latin America. *Constr. Innov.* **2019**, *19*, 570–593. [CrossRef]
16. Bouillon, C.P. *Room for Development*; Bouillon, C.P., Ed.; Palgrave Macmillan US: New York, NY, USA, 2012; ISBN 978-1-137-00563-2.
17. Murguia, D. Factors Influencing the Use of Last Planner System Methods: An Empirical Study in Peru. In Proceedings of the 27th Annual Conference of the International Group for Lean Construction (IGLC), Dublin, Ireland, 27 July 2019.
18. Singh, R.K. Developing the Framework for Coordination in Supply Chain of SMEs. *Bus. Process Manag. J.* **2011**, *17*, 619–638. [CrossRef]
19. Alaskari, O.; Ahmad, M.M.; Pinedo-Cuenca, R. Development of a Methodology to Assist Manufacturing SMEs in the Selection of Appropriate Lean Tools. *Int. J. Lean Six Sigma* **2016**, *7*, 62–84. [CrossRef]
20. Yadav, V.; Khandelwal, G.; Jain, R.; Mittal, M.L. Development of Leanness Index for SMEs. *Int. J. Lean Six Sigma* **2019**, *10*, 397–410. [CrossRef]
21. Yadav, V.; Jain, R.; Mittal, M.L.; Panwar, A.; Sharma, M.K. An Appraisal on Barriers to Implement Lean in SMEs. *J. Manuf. Technol. Manag.* **2018**, *30*, 195–212. [CrossRef]
22. Darcy, C.; Hill, J.; McCabe, T.; McGovern, P. A Consideration of Organisational Sustainability in the SME Context. *Eur. J. Train. Dev.* **2014**, *38*, 398–414. [CrossRef]
23. Dandage, R.V.; Mantha, S.S.; Rane, S.B.; Bhoola, V.; Ahmed, S.; Sobuz, M.H.R.; Dandage, R.V.; Mantha, S.S.; Rane, S.B.; Bhoola, V. Analysis of Interactions among Barriers in Project Risk Management. *J. Ind. Eng. Int.* **2018**, *14*, 153–169. [CrossRef]
24. Jones, D.T.; Womack, J.P. *Lean Thinking*, 1st ed.; Free Press: New York, NY, USA, 2003; ISBN 9788498751994.
25. Dora, M.; Kumar, M.; Gellynck, X. Determinants and Barriers to Lean Implementation in Food-Processing SMEs—A Multiple Case Analysis. *Prod. Plan. Control* **2016**, *27*, 1–23. [CrossRef]
26. Forcael, E.; Ferrari, I.; Opazo-Vega, A.; Pulido-Arcas, J.A. Construction 4.0: A Literature Review. *Sustainability* **2020**, *12*, 9755. [CrossRef]
27. Schönbeck, P.; Löfsjögård, M.; Ansell, A. Quantitative Review of Construction 4.0 Technology Presence in Construction Project Research. *Buildings* **2020**, *10*, 173. [CrossRef]
28. Hamzeh, F.; González, V.A.; Alarcon, L.F.; Khalife, S. Lean Construction 4.0: Exploring the Challenges of Development in the Aec Industry. In Proceedings of the 29th Annual Conference of the International Group for Lean Construction (IGLC), Lima, Peru, 14–17 July 2021; Alarcon, L.F., González, V.A., Eds.; pp. 248–283.
29. Salem, O.; Solomon, J.; Genaidy, A.; Minkarah, I. Lean Construction: From Theory to Implementation. *J. Manag. Eng.* **2006**, *22*, 168–175. [CrossRef]
30. Okere, G. Barriers and Enablers of Effective Knowledge Management: A Case in the Construction Sector. *Electron. J. Knowl. Manag.* **2017**, *15*, 85–97.
31. Ballard, G.; Kim, Y.W.; Jang, J.W.; Liu, M. *Roadmap for Lean Implementation at the Project Level University*; The Construction Industry Institute: Austin, TX, USA, 2007.
32. Salvatierra, J.L.; López, A.; Alarcon, L.F.; Velásquez, X. Lean Diagnosis for Chilean Construction Industry: Towards More Sustainable Lean Practices and Tools. In Proceedings of the 23rd Annual Conference of the Internatioanl Group for Lean Construction, Perth, Australia, 29–31 July 2015.
33. Sarhan, S.; Fox, A. Trends and Challenges to the Development of a Lean Culture Among UK Construction Organisations. In Proceedings of the 20th Annual Conference of the International Group for Lean Construction (IGLC), San Diego, CA, USA, 18–20 July 2012.
34. Forero, S.; Cardenas, S.; Vargas, H.; Garcia, C. A Deeper Look into the Perception and Disposition to Integrated Project Delivery (IPD) in Colombia. In Proceedings of the 23rd Annual Conference of the International Group for Lean Construction (IGLC), Perth, Australia, 29–31 July 2015; Volume 2015.
35. Castiblanco, F.M.; Castiblanco, I.A.; Cruz, J.P. Qualitative Analysis of Lean Tools in the Construction Sector in Colombia. In Proceedings of the 27th Annual Conference of the International Group for Lean Construction (IGLC), Dublin, Ireland, 27 July 2019; Volume 57.
36. Alarcón, L.F.; Diethelm, S.; Rojo, Ó. Collaborative Implementation of Lean Planning Systems in Chilean Construction Companies. In Proceedings of the 10th Annual Conference of the International Group for Lean Construction (IGLC), Gramado, Brazil, 10 August 2002.

37. Engebø, A.; Drevland, F.; Lohne, J.; Shkmot, N.; Lædre, O. Geographical Distribution of Interest and Publications on Lean Construction. In Proceedings of the 25th Annual Conference of the International Group for Lean Construction (IGLC), Heraklion, Greece, 8 July 2017.
38. Alarcón, L.F.; Diethelm, S. Organizing to Introduce Lean Practices in Construction Companies. In Proceedings of the 9th International Workshop on Lean Construction, Singapore, 6–8 August 2001; Volume 17.
39. Huaman-Orosco, C.; Erazo-Rondinel, A.A. An Exploratory Study of the Main Barriers to Lean Construction Implementation in Peru. In Proceedings of the 29th Annual Conference of the International Group for Lean Construction (IGLC), Lima, Peru, 14–17 July 2021.
40. Medina, A. Learning Through Failure: The Challenge of Lean Project Delivery from the Contractor's Perspective in Perú. In Proceedings of the 22nd Annual Conference of the International Group for Lean Construction (IGLC), Oslo, Norway, 25 June 2014.
41. Gutiérrez, F.M. Influence of Integrated Teams and Co-Location to Achieve the Target Cost in Building Projects. In Proceedings of the 28th Annual Conference of the International Group for Lean Construction (IGLC), Berkeley, CA, USA, 6–12 July 2020.
42. Lagos, C.I.; Herrera, R.F.; Alarcón, L.F. Contributions of Information Technologies to Last Planner System Implementation. In Proceedings of the 25th Annual Conference of the International Group for Lean Construction (IGLC), Heraklion, Greece, 10–12 July 2017; Volume 2, pp. 87–94.
43. Creswell, J.W. *Research Design: Qualitative, Quantitative and Mixed Methods Approaches*; Knight, V., Ed.; SAGE: Thousand Oaks, CA, USA, 2014; ISBN 9781452226095.
44. Meuser, M.; Nagel, U. *Research Methods Series*, 1st ed.; European Consortium for Political Research: Colchester, UK, 2014.
45. Ballard, G.; Tommelein, I.D. *2020 Current Process Benchmark for the Last Planner ®System of Project Planning and Control*; Berkeley university of California: Berkeley, CA, USA, 2021.
46. Cerveró-Romero, F.; Napolitano, P.; Reyes, E.; Teran, L. Last Planner System® and Lean Approach Process®: Experiences from Implementation in Mexico. In Proceedings of the 21st Annual Conference of the International Group for Lean Construction (IGLC), Fortaleza, Brazil, 21 July 2013; pp. 645–654.
47. Silvério, L.; Trabasso, L.G.; Pereira Pessôa, M.V. A Roadmap for a Leanness Company to Emerge as a True Lean Organization. *Concurr. Eng. Res. Appl.* **2020**, *28*, 3–19. [CrossRef]
48. Hogg, R.V.; Tanis, E.; Zimmerman, D. *Probability and Statistical Inference Ninth Edition*, 9th ed.; Deirdre Lynch, Ed.; Pearson: London, UK, 2015; ISBN 9780321923271.
49. Gaskin, C.J.; Happell, B. On Exploratory Factor Analysis: A Review of Recent Evidence, an Assessment of Current Practice, and Recommendations for Future Use. *Int. J. Nurs. Stud.* **2014**, *51*, 511–521. [CrossRef]
50. Watkins, M.W. Exploratory Factor Analysis: A Guide to Best Practice. *J. Black Psychol.* **2018**, *44*, 219–246. [CrossRef]
51. Shrestha, N. Factor Analysis as a Tool for Survey Analysis. *Am. J. Appl. Math. Stat.* **2021**, *9*, 4–11. [CrossRef]
52. Fabrigar, L.R.; Wegener, D.T.; MacCallum, R.C.; Strahan, E.J. Evaluating the Use of Exploratory Factor Analysis in Psychological Research. *Psychol. Methods* **1999**, *4*, 272–299. [CrossRef]
53. Kaiser, H.F. The Varimax Criterion for Analytic Rotation in Factor Analysis. *Psychometrika* **1958**, *23*, 187–200. [CrossRef]
54. Ostertagová, E.; Ostertag, O.; Kováč, J. Methodology and Application of the Kruskal-Wallis Test. *Appl. Mech. Mater.* **2014**, *611*, 115–120. [CrossRef]
55. Tonidandel, S.; LeBreton, J.M. Relative Importance Analysis: A Useful Supplement to Regression Analysis. *J. Bus. Psychol.* **2011**, *26*, 1–9. [CrossRef]
56. Alves, A.C.; Flumerfelt, S.; Kahlen, F.J. *Lean Education: An Overview of Current Issues*; Springer: Berlin/Heidelberg, Germany, 2016; pp. 1–179. [CrossRef]
57. Ahmed, S.; Hossain, M.M.; Haq, I. Implementation of Lean Construction in the Construction Industry in Bangladesh: Awareness, Benefits and Challenges. *Int. J. Build. Pathol. Adapt.* **2020**, *39*, 368–406. [CrossRef]
58. AlManei, M.; Salonitis, K.; Xu, Y. Lean Implementation Frameworks: The Challenges for SMEs. In Proceedings of the the 50th CIRP Conference on Manufacturing Systems Lean, Taichung City, Taiwan, 3–5 May 2017; Volume 63, pp. 750–755.
59. Koohestani, K.; Poshdar, M.; Gonzalez, V.A. Finding the Way to Success in Implementing Lean Construction in an Unfavourable Context. In Proceedings of the 28th Annual Conference of the International Group for Lean Construction (IGLC), Berkeley, CA, USA, 6–10 July 2020; pp. 373–384.
60. Torp, O.; Knudsen, J.B.; Rønneberg, I. Factors Affecting Implementation of Lean Construction. In Proceedings of the 26th Annual Conference of the International Group for Lean Construction (IGLC), Chennai, India, 18 July 2018; Volume 2, pp. 1261–1271.
61. Loosemore, M. Improving Construction Productivity: A Subcontractor's Perspective. *Eng. Constr. Archit. Manag.* **2014**, *21*, 245–260. [CrossRef]
62. Rahimian, F.P.; Goulding, J.; Akintoye, A.; Kolo, S. Review of Motivations, Success Factors, and Barriers to the Adoption of Offsite Manufacturing in Nigeria. *Procedia Eng.* **2017**, *196*, 512–519. [CrossRef]
63. Walter, R.; Weinmann, M.; Baier, C.; Oprach, S.; Haghsheno, S. A Requirement Model for Lean Leadership in Construction Projects. In Proceedings of the 28th Annual Conference of the International Group for Lean Construction (IGLC), Berkeley, CA, USA, 6–10 July 2020.
64. Aslam, M.; Gao, Z.; Smith, G. Exploring Factors for Implementing Lean Construction for Rapid Initial Successes in Construction. *J. Clean Prod.* **2020**, *277*, 123295. [CrossRef]

65. Gomez, S.; Ballard, G.; Naderpajouh, N.; Ruiz, S. Integrated Project Delivery for Infraestructure Projects in Perú. In Proceedings of the 26th Annual Conference of the International Group for Lean Construction (IGLC), Chennai, India, 18 July 2018.
66. Bashir, A.; Suresh, S.; Oloke, D.; Proverbs, D.; Gameson, R. Overcoming the Challenges Facing Lean Construction Practice in the UK Contracting Organizations. *Int. J. Archit. Eng. Constr.* **2015**, *4*, 10–18. [CrossRef]
67. Thomas, A.J.; Antony, J.; Francis, M.; Fisher, R. A Comparative Study of Lean Implementation in Higher and Further Education Institutions in the UK. *Int. J. Qual. Reliab. Manag.* **2015**, *32*, 982–996. [CrossRef]
68. Soren, W. Lean Construction with or Without Lean Challenges of Implementing. In Proceedings of the 22nd Annual Conference of the International Group for Lean Construction (IGLC), Oslo, Norway, 22 June 2014.
69. Kawish, S.E. Identifying and Prioritizing Barriers and Overcoming Strategies in Implementing Lean Construction Principles and Methods Whitin Transportation Projects. Master's Thesis, Michigan State University, East Lansing, MI, USA, 2017.
70. Nwaki, W.; Eze, E.; Awodele, I. Major Barriers Assessment of Lean Construction Application in Construction Projects Delivery. *CSID J. Infrastruct. Dev.* **2021**, *4*, 63. [CrossRef]
71. Shang, G.; Sui Pheng, L. Barriers to Lean Implementation in the Construction Industry in China. *J. Technol. Manag. China* **2014**, *9*, 155–173. [CrossRef]
72. Mejía-Plata, C.; Guevara-Ramirez, J.S.; Moncaleano-Novoa, D.F.; Londoño-Acevedo, M.C.; Rojas-Quintero, J.S.; Ponz-Tienda, J.L. A Route Map for Implementing Last Planner®System in Bogotá, Colombia. In Proceedings of the 24th Annual Conference of the International Group for Lean Construction (IGLC), Boston, MA, USA, 24 July 2016.
73. Albliwi, S.; Antony, J.; Lim, S.A.H.; van der Wiele, T. Critical Failure Factors of Lean Six Sigma: A Systematic Literature Review. *Int. J. Qual. Reliab. Manag.* **2014**, *31*, 1012–1030. [CrossRef]
74. Alarcón, L.F.; Diethelm, S.; Rojo, O.; Calderon, R.; Calderón, R. Assessing the Impacts of Implementing Lean Construction. In Proceedings of the 13th Annual Conference of the International Group for Lean Construction (IGLC), Sydney, Australia, 13 July 2008; Volume 23, pp. 26–33.
75. Liu, C.; González, V.A.; Liu, J.; Rybkowski, Z.; Schöttle, A.; Mourgues Álvarez, C.; Pavez, I. Accelerating the Last Planner System®(LPS) Uptake Using Virtual Reality and Serious Games: A Sociotechnical Conceptual Framework. In Proceedings of the 28th Annual Conference of the International Group for Lean Construction (IGLC), Berkeley, CA, USA, 6–12 July 2020.
76. Ivina, D.; Olsson, N. Lean Construction Principles and Railway Maintenance Planning. In Proceedings of the 28th Annual Conference of the International Group for Lean Construction (IGLC), Berkeley, CA, USA, 6–12 July 2020.
77. Abolhassani, A.; Layfield, K.; Gopalakrishnan, B. Lean and US Manufacturing Industry: Popularity of Practices and Implementation Barriers. *Int. J. Product. Perform. Manag.* **2016**, *65*, 875–897. [CrossRef]
78. Mossman, A. Why Isn't the UK Construction Industry Going Lean With Gusto? *Lean Constr. J.* **2009**, *5*, 24–36.
79. Izquierdo, J.L.; Cerf, M.; Gómez, S.A. Lean Construction Education: Basic Management Functions Workshop. In Proceedings of the 19th Annual Conference of the International Group for Lean Construction (IGLC), Lima, Peru, 13–15 July 2011.
80. Poshdar, M.; Gonzalez, V.A.; Antunes, R.; Ghodrati, N.; Katebi, M.; Valasiuk, S.; Alqudah, H.; Talebi, S. Diffusion of Lean Construction in Small to Medium-Sized Enterprises of Housing Sector. In Proceedings of the 27th Annual Conference of the International Group for Lean Construction (IGLC), Dublin, Ireland, 3 July 2019; pp. 383–392.
81. Khaleel, T.; Nassar, Y. Identification and Analysis of Factors Affecting Labour Productivity in Iraq. In *MATEC Web of Conferences*; EDP Sciences: Les Ulis, France, 2018; Volume 162, p. 02032. [CrossRef]
82. Akinade, O.O. Bim-Based Software for Construction Waste Analytics Using Artificial Intelligence Hybrid Models. Ph.D. Thesis, University of The West of England, Bristol, UK, 2017.
83. Bello, S.A.; Oyedele, L.O.; Akinade, O.O.; Bilal, M.; Davila Delgado, J.M.; Akanbi, L.A.; Ajayi, A.O.; Owolabi, H.A. Cloud Computing in Construction Industry: Use Cases, Benefits and Challenges. *Autom. Constr.* **2021**, *122*, 103441. [CrossRef]
84. Muñoz-La Rivera, F.; Mora-Serrano, J.; Valero, I.; Oñate, E. Methodological-Technological Framework for Construction 4.0. *Arch. Comput. Methods Eng.* **2021**, *28*, 689–711. [CrossRef]
85. Sarhan, J.G.; Xia, B.; Fawzia, S.; Karim, A.; Olanipekun, A.O.; Coffey, V. Framework for the Implementation of Lean Construction Strategies Using the Interpretive Structural Modelling (ISM) Technique. *Eng. Constr. Archit. Manag.* **2019**, *27*, 1–23. [CrossRef]
86. Ghani, N.A.; Hamid, S.; Targio Hashem, I.A.; Ahmed, E. Social Media Big Data Analytics: A Survey. *Comput. Hum. Behav.* **2019**, *101*, 417–428. [CrossRef]
87. Dave, B.; Kubler, S.; Främling, K.; Koskela, L. Opportunities for Enhanced Lean Construction Management Using Internet of Things Standards. *Autom. Constr.* **2016**, *61*, 86–97. [CrossRef]
88. Soman, R.K.; Molina-Solana, M. Automating Look-Ahead Schedule Generation for Construction Using Linked-Data Based Constraint Checking and Reinforcement Learning. *Autom. Constr.* **2022**, *134*, 104069. [CrossRef]
89. Lagos, C.I.; Alarcón, L.F. Assessing the Relationship between Constraint Management and Schedule Performance in Chilean and Colombian Construction Projects. *J. Manag. Eng.* **2021**, *37*, 04021046. [CrossRef]
90. Pérez, D.; Lagos, C.; Fernando Alarcón, L. Key Last Planner System Metrics to Assess Project Performance in High-Rise Building and Industrial Construction Projects. *J. Constr. Eng. Manag.* **2022**, *148*, 04021179. [CrossRef]
91. Cámara de Comercio de Lima. *El Impacto de la Informalidad*; Cámara de Comercio de Lima: Lima, Peru, 2017; p. 44.
92. Radhika, R.; Sukumar, S. An Overview of the Concept of Lean Construction and the Barriers in ITS Implementation. *Int. J. Eng. Technol. Manag. Res.* **2017**, *4*, 13–26. [CrossRef]

93. Oladiran Olatunji, J. An Investigation into the Usage of Lean Construction Techniques in Nigeria. *J. Constr. Proj. Manag. Innov.* **2017**, *7*, 1712–1725. [CrossRef]
94. Sacks, R.; Girolami, M.; Brilakis, I. Building Information Modelling, Artificial Intelligence and Construction Tech. *Dev. Built Environ.* **2020**, *4*, 100011. [CrossRef]
95. Radl, J.; Kaiser, J. Benefits of Implementation of Common Data Environment (CDE) into Construction Projects. In *IOP Conference Series: Materials Science and Engineering* ; IOP Publishing: Bristol, UK, 2019; Volume 471, p. 022021. [CrossRef]
96. Parn, E.A.; Edwards, D. Cyber Threats Confronting the Digital Built Environment. *Eng. Constr. Archit. Manag.* **2019**, *26*, 245–266. [CrossRef]
97. Sik, D.; Csorba, K.; Ekler, P. Implementation of a Geographic Information System with Big Data Environment on Common Data Model. In Proceedings of the 8th IEEE International Conference on Cognitive Infocommunications, CogInfoCom, Debrecen, Hungary, 11–14 September 2017; Volume 2018.
98. Vilutiene, T.; Hosseini, M.R.; Pellicer, E.; Zavadskas, E.K. Advanced BIM Applications in the Construction Industry. *Adv. Civil. Eng.* **2019**, *2019*, 6356107. [CrossRef]
99. Karaguzel, O.T.; Elshambakey, M.; Zhu, Y.; Hong, T.; Tolone, W.J.; das Bhattacharjee, S.; Cho, I.; Dou, W.; Wang, H.; Lu, S.; et al. Open Computing Infrastructure for Sharing Data Analytics to Support Building Energy Simulations. *J. Comput. Civ. Eng.* **2019**, *33*, 04019037. [CrossRef]
100. Mohamed, A.H. *Collaboration and Integration in Construction, Engineering, Management and Technology*; Springer: Berlin/Heidelberg, Germany, 2021; pp. 15–19. [CrossRef]
101. Shuman, L.J.; Besterfield-Sacre, M.; McGourty, J. The ABET "Professional Skills"—Can They Be Taught? Can They Be Assessed? *J. Eng. Educ.* **2005**, *94*, 41–55. [CrossRef]
102. Macomber, H.; Davey, C. *The Pocket Sensei*, 1st ed.; CreateSpace Independent Publishing, Ed.; Pemi River MEdia: Campton, UK, 2017.
103. González, V.A.; Senior, B.; Orozco, F.; Alarcon, L.F.; Ingle, J.; Best, A. Simulating Lean Production Principles in Construction: A Last Planner-Driven Game. In Proceedings of the 22nd Annual Conference of the International Group for Lean Construction (IGLC), Oslo, Norway, 22 June 2014; pp. 1221–1232.

buildings

MDPI

Article

Accelerating the Delivery of Low-Carbon Buildings by Addressing Common Constraints: Perspectives from High-Rise, High-Density Cities

Mohan M. Kumaraswamy, Nandun Madhusanka Hewa Welege * and Wei Pan

Department of Civil Engineering, The University of Hong Kong, Hong Kong, China
* Correspondence: nandun@hku.hk; Tel.: +852-67465223

Abstract: The delivery of low-carbon buildings (LCBs) in high-rise, high-density cities is still hindered by various common and interdependent constraints. However, a study that developed innovative strategies to address the common constraints to delivering LCBs focusing on traditional high-rise, high-density cities could not be identified in the previous literature. Therefore, this study aimed to identify potential strategies for accelerating the delivery of LCBs in high-rise, high-density cities by addressing relevant common constraints that were identified in recent studies. Accordingly, potentially relevant strategies were identified through eight semi-structured interviews with well-experienced experts in industry and academia. Consequently, 71 strategies were identified under six categories, i.e., policy implementation, building energy/carbon data utilisation, awareness raising/training, technology advancement, incentives, and organisational level commitments. This also required closer collaboration with different stakeholders/stakeholder classes in implementing these strategies, who were, therefore, also identified. An SNA-based analysis was also conducted to explore the connections between constraints and strategies. The strategies related to energy/carbon policy development, standardisation, codes and certifications, mandatory regulations, financial incentives, and technology adoption showed the ability to address a majority of the driving constraints related to policies and technologies. These study findings will assist policymakers and other relevant stakeholders in the arena of the project and asset management in accelerating the delivery of LCBs by adopting an innovative approach to prioritise potential strategies in order to suitably address and synergise the complex interdependencies among the constraints.

Keywords: constraints to low-carbon building; strategies to accelerate low-carbon building; high-rise, high-density cities

Citation: Kumaraswamy, M.M.; Hewa Welege, N.M.; Pan, W. Accelerating the Delivery of Low-Carbon Buildings by Addressing Common Constraints: Perspectives from High-Rise, High-Density Cities. *Buildings* **2023**, *13*, 1455. https://doi.org/10.3390/buildings13061455

Academic Editors: Simon P. Philbin, Yongjian Ke and Jingxiao Zhang

Received: 20 March 2023
Revised: 24 May 2023
Accepted: 26 May 2023
Published: 2 June 2023

1. Introduction

1.1. Carbon Emissions and Buildings in High-Rise, High-Density Cities

Buildings significantly contribute to global climate change. As a worldwide average, buildings consume 35–40% of the annual energy produced and emit 30–40% of annual carbon emissions [1,2] Ahmed et al. [3] emphasised that embodied carbon emissions of buildings account for about 11%, while operational carbon emissions from buildings account for about 28% of global carbon emissions annually. Accordingly, 'building stock' can be identified as an important focal point for initiating energy management and carbon emission reduction programmes [4].

Highly urbanised countries/regions with high-rise, high-density cities should implement effective policies and procedures to reduce energy consumption and carbon emissions from buildings since these highly dense cities consume about 80% of energy and emit about 75% of carbon in the country/region [5,6]. Nevertheless, more cities will grow vertically and become highly dense in future due to the rising building demand and migration of the population to urban areas [7]. UN DESA [8] predicted that about 68% of the global population will live in cities and urban areas by 2050.

Having identified the environmental threat from building stock due to the high energy consumption and carbon emissions, a majority of the countries with high-rise, high-density cities are continuously attempting to reduce the carbon emissions from their building stock [9,10].

Taking a few examples, Hong Kong initiated a 'climate action plan 2030+' to achieve the goals of the 'Paris Agreement' [11]. The UK government has initiated plans to achieve an 80% reduction in carbon emissions by 2050 compared to the emission levels in 1990 [12]. In Australia, several states have initiated plans to achieve the certifications of the Commercial Building Disclosure (CBD) programme and the National Australian Building Energy Rating System (NABRES) [13]. The 'national climate change plan of the UAE 2017–2050' also includes targets for reducing emissions from buildings [14]. 'Singapore green building masterplan' [15] targets to increase the percentage of green buildings up to 80% by 2030.

Despite the above initiatives, notable carbon emissions reduction from the buildings sector is not evident [16]. Moreover, about three billion m^2 of building floor area is being constructed yearly without complying with energy/carbon policies and procedures [17]. Indeed, databases clearly indicate, and scholars further highlight, how the rapid increase in carbon emissions and energy consumption from buildings in highly dense regions constitute a significant threat to the environment. This should raise a clarion call for the immediate attention of responsible stakeholders [10,16,18]. This situation demonstrates that the evolution of building carbon and energy reduction programmes is not keeping pace with the rapid growth in high-rise, high-density cities.

1.2. *Low-Carbon Buildings (LCBs)*

With increased rates of urbanisation, 'environmental sustainability' aspects are frequently incorporated into building construction, operations, and management activities throughout the world [19,20]. Accordingly, many scholars have emphasised that delivering LCBs is one of the highly supportive strategies to ensure the sustainability of buildings and the built environment [9,21].

LCBs are buildings which are specifically engineered with GHG reduction in mind. By definition, an LCB is a building which emits significantly less GHG than regular buildings [22]. An LCB typically includes energy-efficient features. However, a building with energy-efficient features does not necessarily mean that the building is an LCB [23]. The low-carbon concept covers a wide spectrum that transcends being just energy efficient. Accordingly, building orientation, structure, building envelope materials, window size, location and glazing, the efficiency of heating, ventilation and air conditioning (HVAC) systems, usage of materials with minimum GHG emissions while production, usage of onsite renewable energy, emission reduction in building systems, adapting low-carbon behaviour, reusing or recycling at the end of the lifecycle, compliance, minimising overall carbon footprint, etc. contribute to delivering an LCB.

To align with global carbon reduction efforts, all countries should implement efficient programmes to deliver LCBs. Further, yearly low-carbon intensive new building construction should improve up to 4 billion m^2 of floor area (the current level is around 250 million m^2 per year). Moreover, at least a 30–50% improvement in energy performance should be achieved through effective policies and technology adoption when renovating existing buildings [17].

Pan and Pan [9] emphasised the importance of delivering LCBs in high-rise, high-density cities/regions in order to minimise the carbon emissions from the building stock. Various action plans are being proposed and implemented for carbon emission reduction from buildings in most of the high-rise, high-density cities throughout the world, especially in countries with stable/developed economies [24]. However, these attempts could not achieve a noteworthy reduction in carbon emissions /energy usage in these cities [16,25]. This may arise from a number of constraints (related to policies and regulations, technologies, social conditions, geographical conditions, and financial status) that hinder the delivery of LCBs [26,27].

1.3. Constraints to Delivering LCBs and Strategies to Accelerate the Delivery of LCBs

Figure 1 shows a summary of constraints reported in the previous literature between 2011 and 2020. Accordingly, it is evident through the blue coloured lines that more studies reported the constraints related to financial level, policy/regulatory level, technology level, and knowledge. The number of reported constraints is also high for these categories.

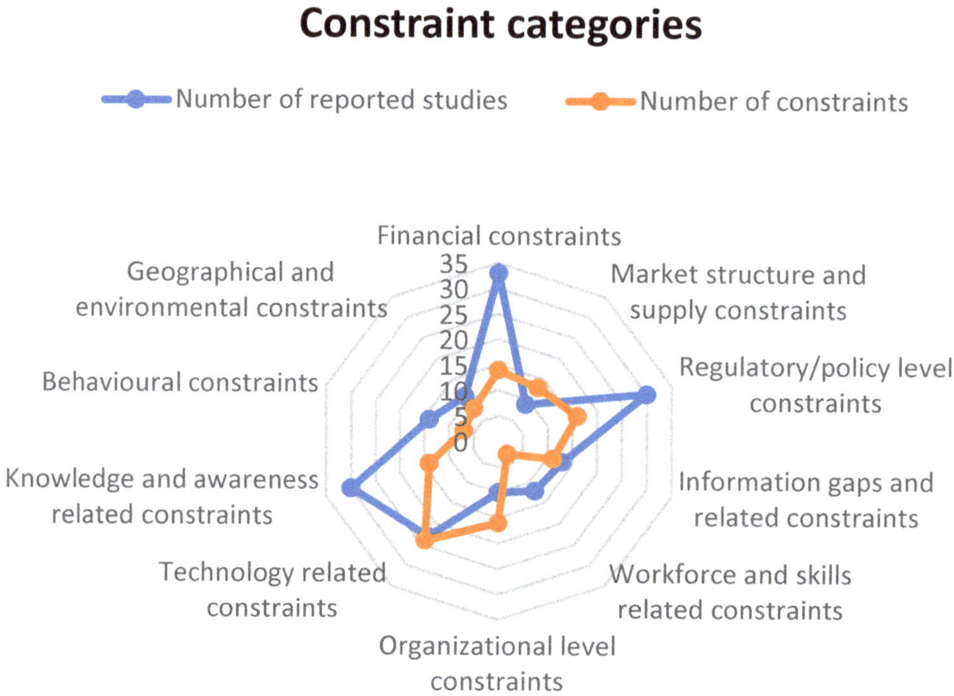

Figure 1. Constraint categories.

Many of these constraints show interdependencies and interrelationships among each other [28]. A clear identification and analysis of these interdependencies among constraints would be beneficial for relevant stakeholders to determine effective strategies to overcome the constraints [29]. Accordingly, developing strategies to address the driving constraints (significant constraints which generate other constraints) should be given a higher priority. Addressing the driving constraints upfront would ease the efforts in mitigating the dependent constraints [10,29].

Government involvement and policy intervention are regarded as essential and effective ways to promote building energy efficiency and low-carbon measures by overcoming common constraints [30–32]. Formulating policies for developing standards and designs of building materials as well as buildings, is crucial in accelerating the adoption of low energy/carbon buildings [31,33]. Design criteria should be duly supported by innovative and creative research methodologies and advanced technological support [33]. All future designs of buildings should ideally include the use of sustainable and energy-efficient materials. The development of such sustainable, eco-friendly, and energy-efficient building materials should be supported and promoted by the government, related institutions and departments [34,35]. Furthermore, effective collaboration among the industry, regulatory bodies, and academia should be maintained to facilitate the customised and effective research and development of feasible technological advancements [36]. Strict regulations should also be imposed on energy and carbon compliance documentation and reporting [34,37]. Furthermore, market-based incentive schemes are thought to be effective and cost-efficient instruments to support and enhance the low carbon and energy-efficient building investments [38,39].

Furthermore, carbon emissions and energy usage should be assessed throughout the life cycle of buildings [40]. Life cycle assessment should be supported by providing suitable calculation tools, technology, expertise, and training [33,39,41]. Relevant institutions, including government departments, should promote the new energy-efficient and low-carbon building development by setting examples by implementing pilot projects with measurable benefits [33]. Awareness raising, training, and skill development of the building professionals and the community, in general, should be considered as a long-term strategic approach to drive towards a sustainable built environment [42,43].

1.4. Research Gap and the Aim of the Present Study

Although many previous studies have identified the constraints to delivering LCBs from different dimensions and explored the strategies to accelerate the delivery of LCBs, a detailed exploration of strategies to address common constraints to delivering LCBs in high-rise, high-density cities could not be found in the previous literature. Although the constraints to delivering LCBs show complex interdependencies, none of the previous studies attempted to analyse these interdependencies and explore innovative methods to accelerate the delivery of LCBs by synergising with these interdependencies. Moreover, less attention has been paid in the literature to identifying and exploring the constraints and strategies for delivering LCBs by focusing on high-rise, high-density cities.

Therefore, this study aimed to identify the potential strategies to accelerate the delivery of LCBs by addressing the common and significant constraints to delivering LCBs in high-rise, high-density cities. The list of common and significant constraints to delivering LCBs in high-rise, high-density cities was adopted from a precursor study by the authors of this paper [29]. This precursor study explored the common constraints to Hong Kong, Singapore, Australia (Sydney and Melbourne), UAE (Dubai and Abu Dhabi), and Qatar (Doha). The present study summarised the findings of Madhusanka et al. [29] and identified suitable strategies to accelerate the delivery of LCBs by addressing the constraints. Moreover, the present study mapped and analysed the connections between constraints and strategies using an Interpretive Structural Modelling (ISM) based structure and a Social Network Analysis (SNA) based on a two-mode network. These analyses provided a clear view of the significance, centrality, and interdependencies of the identified strategies. Furthermore, the necessary involvement and collaboration of different stakeholders in implementing the identified strategies are also identified and discussed in the present study.

2. Methods

This section discusses the research methods chosen to suit this study. Accordingly, methods followed to identify and analyse the constraints, methods of identifying strategies and mapping with constraints, and the SNA approach utilised in this study are elaborated in Figure 2 and the sub-sections below. Some inputs to the present study were adopted from the precursor studies of the same authors of this paper. In Figure 2, the specific tasks covered in the present study are indicated by a dotted line.

2.1. Methods Used for Identifying and Analysing the Constraints

The list of common and significant constraints to delivering LCBs in high-rise, high-density cities and the interdependencies among these constraints were published in precursor studies by the authors of this paper [10,29]. The present study adopted a summary of these precursor studies and explored suitable strategies to accelerate the delivery of LCBs by addressing the constraints identified in the precursor studies. Madhusanka et al. [10] and Madhusanka et al. [29] identified the common and significant constraints to delivering LCBs from the perspectives of five high-rise, high-density regions through a comprehensive literature review followed by a questionnaire survey. Subsequently, the interdependencies among the identified constraints were analysed by using ISM and Matriced' Impacts Croise's Multiplication Appliquee a UN Classement (MICMAC) analysis approaches.

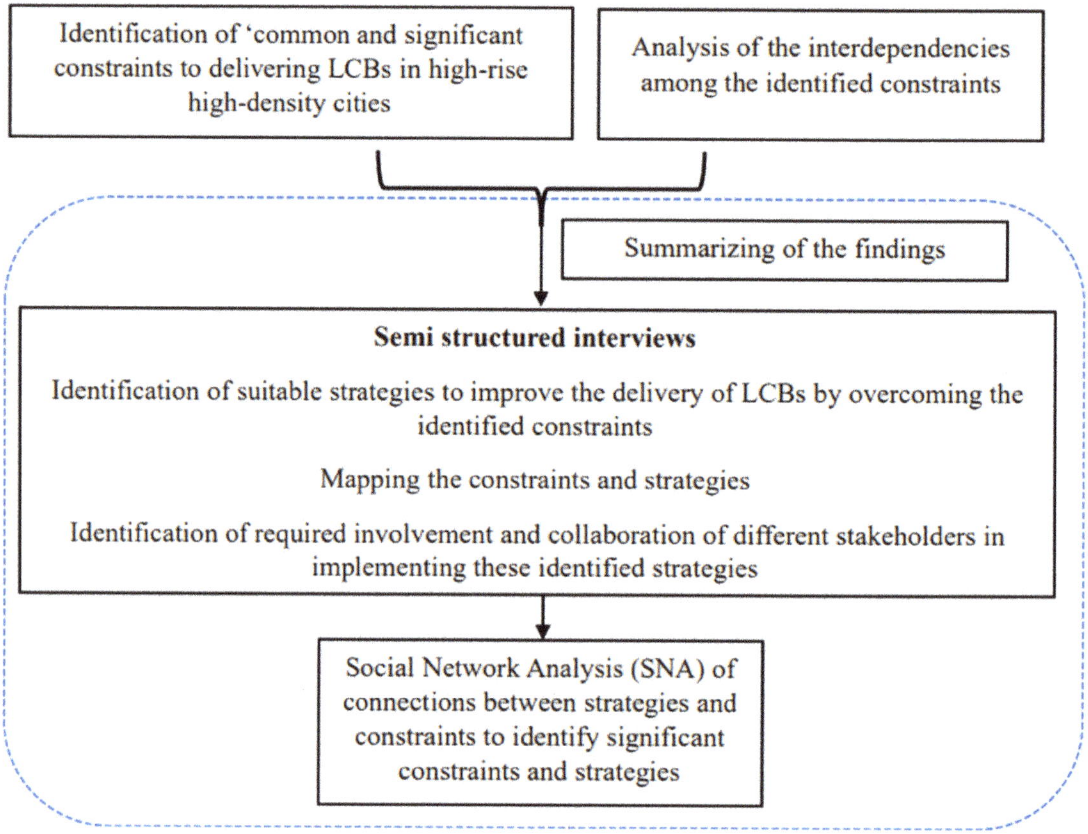

Figure 2. Research methods.

ISM approach is used to systematically reveal the relational links and interdependencies among a set of factors by placing the considered factors in a hierarchical model [44]. MICMAC analysis is used to categorise and prioritise a set of factors based on their driving/dependence nature [45]. Accordingly, these factors ('constraints' in the present study) can be categorised into four groups [46]: (1) autonomous barriers (weak driving power and weak dependence); (2) dependent barriers (weak driving potential but strong dependence); (3) linkage barriers (strong driving power and dependence); and (4) independent barriers (strong driving power but weak dependence).

2.2. *Methods Used for Identifying Strategies*

This step contributed to achieving the core target of this study. Accordingly, semi-structured interviews were conducted to identify suitable strategies to improve the delivery of LCBs by overcoming the identified constraints. A summary of the findings on constraints and their interdependencies was presented to each interviewee, who was asked to provide recommendations by referring to the summary of the findings where possible. The interview guidelines were sent to the interviewees beforehand to be prepared for the interview. Accordingly, the respondents were asked to state the constraints which could be addressed (or mitigated) through the strategies suggested by them. Furthermore, they were asked to indicate the required involvement and collaboration of different stakeholders in implementing these identified strategies.

In-depth semi-structured interviews were conducted with eight experts in the industry and academia, who were approached through personal contacts. Accordingly, three senior academics at different universities and in different countries (all of them were involved in research and teaching on low-carbon buildings, building energy efficiency, green buildings, and sustainable construction and built environment management), two senior facilities managers (who had relevant experiences in managing and carrying out facility management procedures in low-carbon/green buildings), a quality and compliance manager (involved in environmental and energy compliance in buildings), a government official (working in a senior management position at an energy/sustainability-related authority), and a senior quantity surveyor (who had experiences in constructing/costing the construction process of low-carbon buildings and green buildings) were interviewed. All the interviewees had more than 10 years of experience in relevant fields. Accordingly, a sample of experts with relevant exposure and experience covering different sectors could be approached for the semi-structured interviews.

The interviews were transcribed, and a written record was prepared. As there were only eight interviews, this task was performed manually without getting the support of any transcription software. Subsequently, the data were organised to easily understand the 'strategies to accelerate the delivery of LCBs by addressing the constraints', 'categories of the identified strategies', 'connections among strategies and the constraints', and 'required stakeholder involvement and collaboration in initiating the strategies'. Furthermore, the collected data from different interviewees were comparatively analysed, redundancies and repetitions were eliminated, and the list of strategies was finalised by utilising the most inclusive and appropriate suggestions of interviewees. Furthermore, detailed descriptions provided by the interviewees were utilised when interpreting and discussing the strategies. Accordingly, it was ensured that a clear and concise representation of the interviewees' opinions was incorporated in preparing the list of strategies, mapping the constraints with strategies, and discussing the findings. Furthermore, the identified strategies (through interviews) were further discussed with the literature support from several publications. The above steps were followed to ensure the reliability and validity of the collected data.

Methods Used for Social Network Analysis of Constraints and Strategies

SNA incorporates mathematical, statistical, and informational methodologies to analyse a specific set of linkages among a defined set of entities arranged into a network structure [47]. This study utilised an SNA two-mode (bipartite) network structure to map and analyse the connections among two sets. Accordingly, 'strategies' and 'constraints' were used as the two types of nodes in the social network. The links between strategies and constraints were identified through the semi-structured interviews, and the network of these identified links was developed through the UCINET social network software package.

The developed two-mode network was statistically analysed through the 'degree centrality' network analysis measure. This measure is suitable for analysing the nodes in a network in terms of their structural positioning and connectedness [47]. Further, this measure provides an indication of the significance of the nodes in terms of the number of connections.

Degree centrality (DC) indicates the number of connections of a node in the network. Accordingly, a node with a higher degree of centrality value has many direct links with other nodes in a network [48]. Thus, holding a centralised position in the network and acting as a hub [49]. In the present study, the DC of a strategy represents the number of constraints which can be addressed by the considered strategy. Similarly, the DC of a constraint represents the number of strategies which can contribute to addressing the considered constraint.

DC can be measured by using Formula (1).

$$DC(i) = \sum_{j}^{N} x_{ij} \quad (1)$$

Index *i* indicates the considered node for calculation; *j* refers to the nodes in the other category of the two-mode data set, and *x* represents the adjacency matrix. x_{ij} is the element of the *i*th row an *j*th column of *x*. Similarly, the DC of node type *j* can be calculated.

3. Results and Analyses

3.1. Constraints to Delivering LCBs and the Interdependencies among the Constraints

Figure 3 shows a summarised representation of the findings of the precursor study by the same authors [29]. Accordingly, the 21 common and significant constraints that were identified, are presented in a hierarchical structure of 12 levels (ISM hierarchical structure). The constraints placed at the bottom levels of the structure are significant and driving constraints which affect the other constraints in the middle and upper levels of the structure.

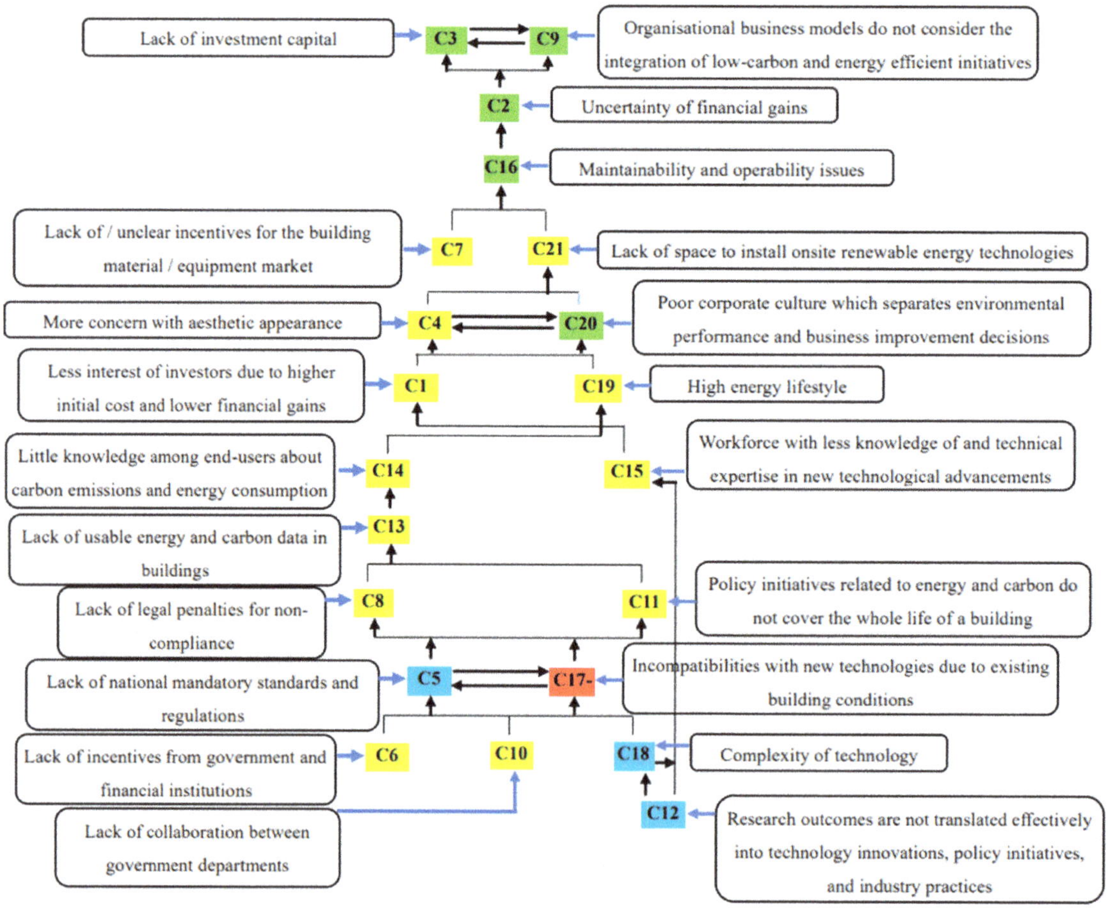

Figure 3. A summary of constraint interdependencies.

MICMAC technique and ISM approach are often used together to analyse the hierarchical structure among a set of barriers along with their driving/dependence nature [50,51]. Accordingly, constraints to delivering LCBs in high-rise, high-density cities were categorised and prioritised to identify the significant constraints which should be targeted when deciding strategies to accelerate the delivery of LCBs. According to the MICMAC analysis, the constraints were categorised into four different categories based on their driving and dependence nature. These four categories are highlighted by different colours in Figure 3, i.e., yellow—autonomous constraints with weak driving power and weak dependence, blue—independent constraints with strong driving power and weak dependence, red—linkage constraints with strong driving power and strong dependence, and green—dependent constraints with weak driving power and strong dependence.

According to the ISM hierarchical structure, the constraints C5 (lack of national mandatory standards and regulations), C6 (lack/unclear incentives towards owners and developers), C8 (lack of legal penalties due to non-compliance), C10 (lack of coordination, communication and collaboration between government departments to initiate building energy and carbon policy initiatives), C11 (policy initiatives related to energy and carbon do not cover the whole life of a building), C12 (research outcomes are not effectively translated into technology innovations, policy initiatives, and industry practices), C17 (incompatibilities with new energy efficient and low-carbon technologies due to existing building conditions), and C18 (complexity of technology) were identified in the bottom levels of the hierarchy (levels 9–12). Accordingly, these constraints can be identified as having a comparatively high driving power and significance. The MICMAC results also provide an indication of the significance of these constraints. Accordingly, it was evidently established that addressing these driving constraints upfront can effectively contribute to addressing and mitigating the other dependent constraints. Five of these eight significant constraints are related to the policy and regulatory level; one is related to research and the industrial applications of research outcomes, while the other two are related to technical constraints. Nevertheless, policy and regulatory actions will be the key to improving the practical applicability of research and also to introducing feasible low-carbon technologies to the building industry. Therefore, policy and regulatory sector stakeholders can be identified as having the highest influencing ability in overcoming the constraints of delivering LCBs.

3.2. *Strategies for Accelerating the Delivery of LCBs*

This section discusses the strategies to accelerate the delivery of LCBs by addressing the constraints identified through eight semi-structured interviews conducted with experts in industry and academia. Moreover, the required contributions of stakeholders in implementing the identified strategies are briefly discussed together with the identified strategies. Suitable strategies were identified to accelerate the development and delivery of LCBs under six categories. The discussions in this section are based on the findings from the in-depth semi-structured interviews.

Policy and regulatory constraints and technology-related constraints were identified as the most significant and driving constraints to delivering LCBs in the selected set of countries/regions. The interviewees also acknowledged that the policy and regulatory level strategic influences and effective technology advancements/adoption as the most significant driving forces in accelerating the delivery of LCBs. In addition to the identification of suitable strategies to accelerate the delivery of LCBs, the interviewees were requested to indicate the constraints which can be influenced through the identified strategies. Accordingly, Table 1 provides a summarised presentation of the strategies identified by the interviewees, and Figure 4 shows the connections among the identified strategies and the list of 21 'common and significant' constraints, as identified by the interviewees. Furthermore, Figure 4 provides an indication of the significance of the strategies based on their connections with driving, dependent, and intermediate constraints.

Table 1. Summary of identified strategies.

Categories	Strategies
P1	Decentralising the generation of electricity by proper urban planning
P2	Implementing regulations on phasing out the systems, equipment, and materials with high-carbon emissions and less energy efficiency
P3	Implementing national level realistic emission reduction policies and targets
P4	Integrated policy initiatives for urban development
P5	Mandating low-carbon retrofitting/refurbishment
P6	Standardising construction and building materials
P7	Implementing penalties, such as taxes for non-compliance, environmental tax, etc.
P8	Introducing city or state-based energy benchmarks and fines for not achieving the benchmarks
P9	Introducing regulations for building deconstruction and disposal
P10	Introducing/updating the mandatory building energy codes
P11	Labelling and certifying the new and existing buildings
P12	Labelling (energy/carbon) for building materials, equipment, and systems
P13	Introducing Life Cycle Analysis (LCA) regulations for building materials
P14	Mandating the required minimum energy/carbon performance levels of appliances and equipment
P15	Implementing regulations on thermal performance and embodied carbon emissions of building materials
P16	Ensuring proper inspection of major refurbishments
P17	Mandating structural provisions for onsite-renewable energy generation for new constructions
D1 (A,T)	Creating more opportunities for pilot projects, real-world examples, and sharing of performance data and evidences to build trust among investors
D2 (P)	Making energy audits compulsory for existing buildings and standardising the reporting procedures
D3 (P)	Mandating the disclosure of energy usage and carbon emissions
D4 (P)	Mandating LCA for major investments in the public sector
D5 (P,T)	Encouraging and supporting life cycle assessment (LCA) by providing relevant expertise and calculation tools
D6 (P)	Continuous improvement and reviewing of standards based on real world data
A1	Raising awareness of the general public through various media platforms
A2(O)	Raising end-user awareness, encouraging behavioural changes, and influencing changing their attitudes
A3	Including education on sustainability/emission reduction/energy efficiency to the school curriculums from primary education
A4	Training and certification of energy managers and energy auditors through a regulatory body
A5	Training relevant government sector officials
A6	Training organisations to develop sustainable business models
A7	Providing specialised training on relevant products, tools, systems and technologies
A8 (D)	Establishing databases/websites for relevant information sharing on energy-efficient and low-carbon initiatives (materials, equipment, technologies, suppliers, construction/installation methods/times/costs, energy/carbon performance data of materials/equipment/systems, payback information, important results of occupancy surveys/audits)
A9 (P)	Maintaining a pool of qualified energy and carbon auditors in the government sector
T1	Demonstrating, prototyping, and providing examples of developed advanced energy efficient materials, equipment, and technologies
T2	Exploring co-generation capacities and energy recovery opportunities
T3 (P)	Mandating the maintenance and servicing contracts with specialised and authorised service providers
T4 (P)	Ensuring the availability of globally applicable maintenance and servicing procedures for systems and equipment
T5 (P)	Assessing new constructions in terms of their potential for renewable energy generation and energy recovery
T6 (P)	Planning and implementing construction waste sorting and recycling facilities by the government or supporting the private sector to implement these facilities
T7 (P)	Implementing effective guidelines for building deconstruction and waste sorting
T8 (O)	Integration of BIM and utilisation of advanced tools at the design stage
T9 (P)	Validating the performances of new developments and evaluating the compliance with other requirements, such as safety and fire regulations

Table 1. *Cont.*

Categories	Strategies
T10 (P)	Conducting research on reusing/recycling materials and minimising the usage of important natural resources
T11 (O)	Adopting building management systems with integrated controls
T12 (O)	Ensuring timely standardised maintenance and servicing of high energy-intensive systems and equipment
T13 (P)	Obtaining international-level assistance for technology advancement for LCBs
T14 (P)	Industry, academia, and government collaborations in research and technology adoption
T15 (I,P)	Financing the conversion of research and development (on technical advancements of materials, products and equipment) into manufacturing (through tax incentives, providing grants, etc.)
T16 (P)	Providing suitable research grants to universities
T17 (P)	Ensuring the support of the government for manufacturers and producers (who develop advanced high-efficient technical solutions) to bare the risk of entering into new markets
T18 (O)	Obtaining the best performances of the systems through following proper (standardised) installation, maintenance, and operational procedures
T19 (P)	Promoting circular design approaches for buildings
I1	Offering attractive interest rates and longer tenure periods, incentives, such as grants, loans, and tax rebates, for low-carbon/energy efficient investments
I2 (A)	Raising awareness of financial institutions on financing low-carbon, energy efficient, and green building initiatives
I3	Offering incentives to encourage the building owners to consider more decentralised power generation options
I4 (P)	Establishing policies to finance the onsite renewable energy generation in buildings
I5	Providing attractive tariff rates for feeding power to the national grid through renewable energy generation in buildings
I6 (P)	Link financing procedures with the building energy/carbon certifications and standards
I7	Financing for low-carbon initiatives through energy service providers/companies (ESCOs) by initiating performance-based contracts
I8	Offering long-term warranties for the systems and equipment
I9 (T,P)	Offering incentives for manufacturing industries to invest in productive research and development
I10 (P)	Providing incentives, such as rebates for taxes and loans for energy efficient/low-carbon product development
O1	Establishing and following internal compliance procedures and energy/carbon policies in organisations
O2	Establishing an energy-efficient/low-carbon culture within organisations
O3	Developing organisational internal funding models to support investments in energy-efficient and low-carbon initiatives/procurement procedures to consider LCA, LCC
O4	Delegation of responsibilities for low-carbon, energy-efficient, and sustainability-related initiatives within organisations
O5	Generating, processing, and analysing the building-level data related to energy usage, carbon emissions, behaviour patterns, equipment/system performances, etc.
O6	Adopting sustainable maintenance procedures
O7	Incorporate energy saving and carbon reduction into their internal business models
O8 (D)	Determining building-specific initiatives (e.g., phasing out less efficient systems, demand management, load shedding, adjustment of operational procedures, investments in new energy/carbon efficient systems, adopting energy management systems) based on the generated data and information through building-level audits/surveys
O9	Implementing effective energy management procedures in buildings

In Table 1, 'P' indicates the strategies related to policies, regulations, standards, codes, and certifications; 'D' indicates the strategies related to the effective utilisation of building energy/carbon data; 'A' indicates the strategies related to awareness raising, training, and information dissemination; 'T' indicates the strategies for technology advancement; 'I' indicates the strategies related to providing incentives for low-carbon adoption, and 'O' indicates the organisational level strategies to accelerate the low-carbon adoption. Some strategies are difficult to mention under one specific category. Therefore, these strategies are mentioned (and numbered) under the most relevant category, and other relevant categories are mentioned within brackets. For example, 'D5 (P/T)' indicates that the relevant strategy relates to the data utilisation (D). Yet, it can also be related to policy and technology-related sectors.

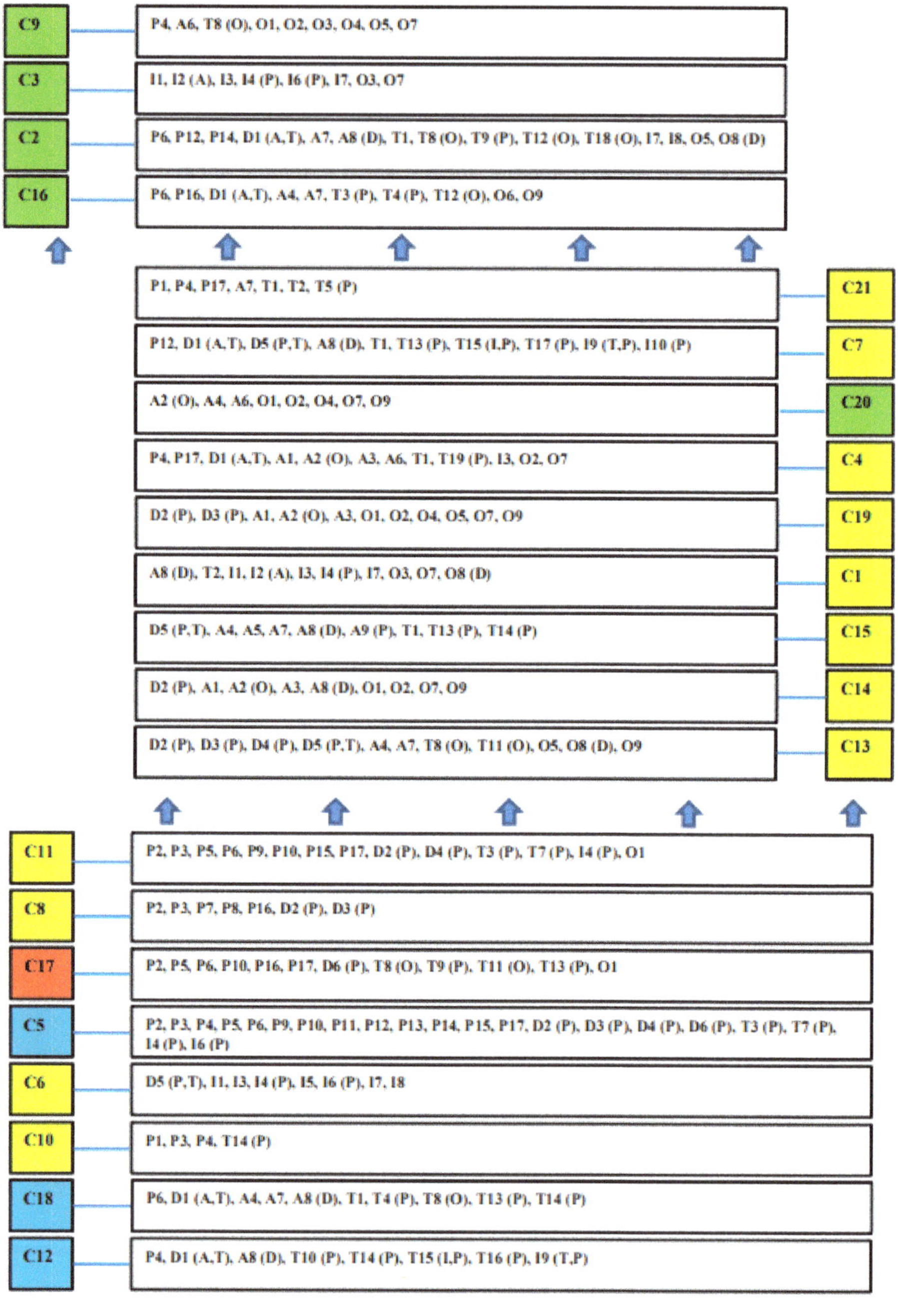

Figure 4. Connections between strategies and constraints.

Figure 4 is finalised with the 'summarised list of strategies' and 'connections between strategies and constraints', as identified by the interviewees. Moreover, the ISM and MICMAC findings of the precursor study [29] are incorporated into this figure to provide an indication of the significance of the constraints and strategies.

The constraints in Figure 4 above are coloured according to whether they are autonomous (yellow), independent (blue), linkage (red), and dependent (green) in nature (corresponding to the MICMAC analysis in Figure 3). Furthermore, the set of constraints in the bottom cluster (C11, C8, C17, C5, C6, C10, C18, and C12) are placed in levels 9–12 of the ISM hierarchical structure, indicating a higher driving power. Accordingly, the strategies targeting these driving constraints can also be considered significant driving strategies that could have a higher impact towards accelerating the delivery of LCBs by addressing the driving constraints. Accordingly, it is evident that the policy and regulatory level strategies, strategies to provide incentives, and strategies related to technology advancement are mostly linked with the driving constraints in Figure 4. The middle cluster in Figure 4 shows the connections of strategies towards the constraints placed in levels 4–8 of the ISM hierarchical structure (C21, C7, C20, C4, C19, C1, C15, C14, C13). Most of these constraints show both driving and dependence characteristics. The constraints placed in the upper cluster of Figure 4 show the highest dependent characteristics. Accordingly, these constraints are placed in levels 1–3 of the ISM hierarchical structure (C9, C3, C2, C16). The strategies related to the areas of knowledge and awareness, organisational commitments, and data utilisation are mostly influencing the constraints at middle and upper clusters. Accordingly, this implies that these strategies also depend on policy, technology, and incentive-related strategies.

3.3. SNA-Based Mapping and Analysis of Constraints and Strategies

In addition to the above analyses, a two-mode social network structure (Figure 5) was constructed to visualise the connections between the constraints and strategies. Accordingly, the red-coloured nodes show the identified 21 common and significant constraints to delivering LCBs in high-rise, high-density cities. Blue-coloured nodes show the identified strategies for accelerating the delivery of LCBs.

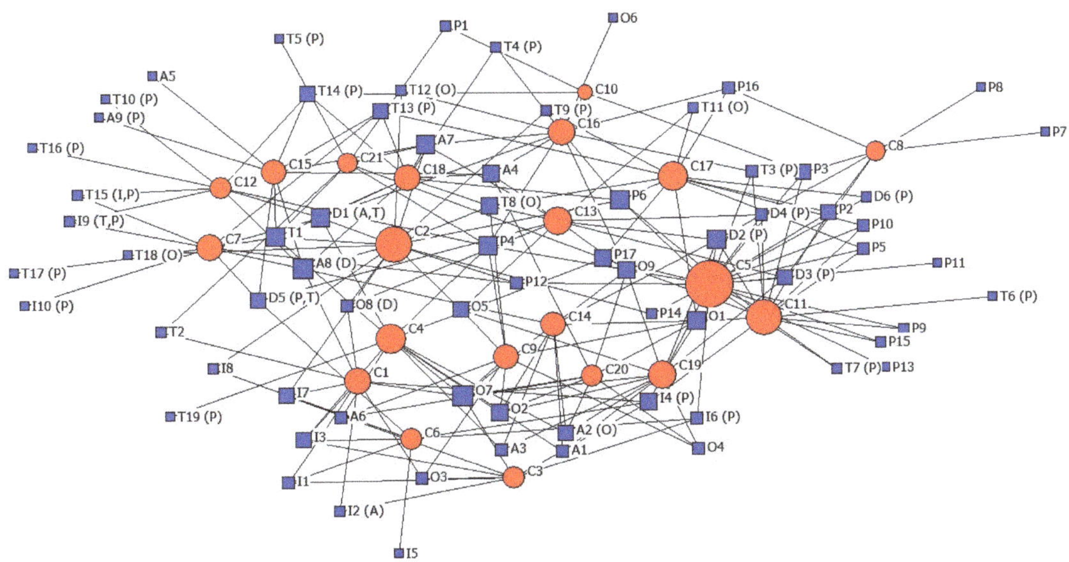

Figure 5. Two-mode social network of constraints and strategies.

The size of each node proportionally represents the 'degree centrality' value of the node. Furthermore, the significant nodes (which have a higher number of links) are clustered to the centre of the network structure, while the nodes with fewer connections are scattered around the network. Table 2 shows the degree centralities (DC) of 'strategies' and 'constraints'.

Table 2. Degree centralities of 'strategies' and 'constraints'.

Strategy ID	P1	P2	P3	P4	P5	P6	P7	P8	P9	P10	P11
DC	2	4	4	6	3	6	1	1	2	3	1
Strategy ID	P12	P13	P14	P15	P16	P17	D1 (A,T)	D2 (P)	D3 (P)	D4 (P)	D5 (P,T)
DC	3	1	2	2	3	5	6	6	4	3	4
Strategy ID	D6 (P)	A1	A2 (O)	A3	A4	A5	A6	A7	A8 (D)	A9 (P)	T1
DC	2	3	4	3	5	1	3	6	7	1	6
Strategy ID	T2	T3 (P)	T4 (P)	T5 (P)	T6 (P)	T7 (P)	T8 (O)	T9 (P)	T10 (P)	T11 (O)	T12 (O)
DC	2	3	2	1	1	2	5	2	1	2	2
Strategy ID	T13 (P)	T14 (P)	T15 (I,P)	T16 (P)	T17 (P)	T18 (O)	T19 (P)	I1	I2 (A)	I3	I4 (P)
DC	4	4	2	1	1	1	1	3	2	4	5
Strategy ID	I5	I6 (P)	I7	I8	I9 (T,P)	I10 (P)	O1	O2	O3	O4	O5
DC	1	3	4	2	2	1	6	5	3	3	4
Strategy ID	O (6)	O (7)	O8 (D)	O9							
DC	1	7	3	5							
Constraint ID	C1	C2	C3	C4	C5	C6	C7	C8	C9	C10	C11
DC	10	15	8	12	21	8	10	7	9	4	15
Constraint ID	C12	C13	C14	C15	C16	C17	C18	C19	C20	C21	
DC	8	11	9	9	10	12	10	11	8	7	

Accordingly, the strategies P4 (integrated policy initiatives for urban development), P6 (standardising construction and building materials), D1 (A,T) (creating more opportunities for pilot projects, real-world examples, and sharing of performance data and evidences to build trust among investors), D2 (P) (making energy audits compulsory for existing buildings, and standardising the reporting procedures), A7 (providing specialised training on relevant products, tools, systems and technologies), A8(D) (establishing databases/websites for relevant information sharing on energy-efficient and low-carbon initiatives (materials, equipment, technologies, suppliers, construction/installation methods/times/costs, energy/carbon performance data of materials/equipment/systems, payback information, important results of occupancy surveys/audits)), T1 (demonstrating, prototyping, and providing examples of developed advanced energy efficient materials, equipment, and technologies), O1 (establishing and following internal compliance procedures and energy/carbon policies in organisations), and O7 (incorporating energy saving and carbon reduction into their internal business models) show the highest number of connections (more than six) with constraints. This implies that these strategies could address many 'common and significant' constraints to delivering LCBs identified in this study.

Among the list of constraints, 21 strategies could contribute to addressing and overcoming constraint C5 (lack of national mandatory standards and regulations). C2 (uncertainty of financial gains) and C11 (policy initiatives related to energy and carbon do not cover the whole life of a building) are connected with 15 strategies, while the constraints C4 (more concern with aesthetic appearance) and C17 (incompatibilities with new technologies due to existing building conditions) are connected with 12 strategies. This indicates that the attention and commitment of multiple sectors are required to address and overcoming these constraints. Moreover, it can be highlighted that the centralised strategies with relatively higher degree centrality values in the SNA two-mode structure have the ability to address a majority of driving constraints (driving constraints could be identified from Figures 3 and 4). Once these driving constraints are effectively addressed and mitigated, it contributes to mitigating the connected dependent constraints (in upper levels of ISM structure) without much effort.

4. Discussion

This section discusses the strategies under the six main categories presented in Table 1. In addition to the identification of strategies and mapping them with constraints, the interviewees were asked to provide their views on the required involvement and collaboration of different stakeholders/stakeholder classes in implementing the identified strategies. Accordingly, the significant constraints, strategies to address the constraints and accelerate the delivery of LCBs, and the required involvement and collaboration of different stakeholders in implementing the strategies are discussed in detail in this section. This discussion is also supported by the literature where possible.

4.1. Formulating and Implementing Adequate Policies, Regulations, Standards, Codes, and Certifications

The lack of mandatory standards and regulations (C5) was identified as a common constraint in the considered countries and regions. Policy initiatives related to energy and carbon do not cover the whole life of a building (C11), and a lack of legal penalties due to non-compliance (C8) was also identified as a significant constraint hindering the effective delivery of LCBs. Accordingly, sufficient actions are required to strengthen the policies and regulations. Implementing national-level realistic emission reduction policies and targets for buildings and providing detailed step-by-step guidelines on moving towards the targets are needed [13], and this should be led by the government and standards and accreditation bodies. Academia and professional bodies should also be proactively involved in supporting policy development. Policy implementation needs to be accelerated by providing adequate incentives to the owners/clients, manufacturers, and contractors to adhere to the relevant guidelines. Policy implementation could start with putting mandates to retrofitting first and then moving into the areas, such as standardising construction and building materials, and then targeting zero environmental impact/zero-carbon buildings as long-term plans [33]. As an initial attempt, initiating regulations on phasing out the systems, equipment, and materials with high carbon emissions and less energy efficiency can be made. This can be achieved by introducing the minimum energy and carbon performance requirements for building retrofitting and refurbishment [52,53]. Energy efficient/low-carbon retrofitting policies can be developed specifically for cities/high-density urban areas, focusing on the facilities managed by the government sector. Subsequently, the best practices and procedures can be used as examples for the private sector to implement sustainable/green/low-carbon retrofitting and maintenance procedures. Implementing mandatory energy-efficient/low-carbon refurbishment requirements based on the age of the building (e.g., 25–30 years) could be pursued. Customised standards could be implemented for refurbishment. Proper inspection of the major refurbishments should be performed by the relevant regulatory authorities in the building sector. Market penetration of low-carbon/energy-efficient materials, equipment, and systems should be supported simultaneously through incentives (incentives are discussed in a separate section below).

Integrated and consistent policy development is another crucial requirement. Developing policies for multi-level governance in accelerating the development and delivery of LCBs in urban areas is a major required commitment of policy-makers and regulators [52]. Furthermore, integrated master plans should be developed for low-carbon high-rise, high-density city development. This initiative should be led by the government by involving the governmental authorities relevant to urban development, sustainability, energy, transportation, etc. The involvement and collaboration of owners/clients, contractors, NGOs, public and community bodies, and education and research institutes are also crucial in developing such a master plan [52]. The main responsibility in initiating policy initiatives is with the government, relevant departments, and authorities. Nevertheless, proper communication channels should be established with the academia, private sectors and industries, and public and civil societies to be successful in developing policies with the win–win scenarios. Furthermore, the governmental departments responsible for the environment, power generation and supply, industry relations, finance, buildings/built environment, and urban development should establish proper communication and collaboration channels with each other to achieve integrated approaches for decarbonising cities/regions as a whole. When it comes to high-rise, high-density cities, the main priority should be for buildings since they consume the majority of generated electricity and emit a significant amount of carbon. Lack of collaboration and coordination among relevant government sector authorities (C10) was also identified as a common and significant constraint to the considered set of countries/regions in the present study. Initiating the above actions could also help in enhancing collaboration among governmental entities, hence also facilitating innovative project management and asset management on building projects and project portfolios [54].

Government, with the collaboration of research entities, utility suppliers, industries, and building owners, should implement programmes to decentralise the generation of electricity through proper urban planning [52]. The buildings sector should be encouraged and supported for onsite renewable energy generation and feeding to the grid (net metering) based on the generation capacity. Mandates can be introduced to include structural provisions in buildings for onsite-renewable energy generation for the new constructions.

Government should also play a key role in providing policy support throughout the process of product development, starting from research and development, material extraction, processing, market penetration, financial incentives, and performance certification. There are multiple sectors which should be involved, such as researchers, manufacturers, financial institutions, and suppliers. Government agents should play the coordination role and driving role in this process. This should be performed until the technology or product is established in the market and gets matured/saturated in the market.

It is also possible to introduce city or state-based energy benchmarks and introduce fines and penalties for non-compliance [33,52]. Subsidies and tax incentives should be implemented prior to the implementation of penalties for non-compliance. Subsequently, penalties, such as taxes for non-compliance and environmental tax, can be introduced as penalty measures in the long run. Simultaneously, rewards or grants should be provided to the facilities with better energy and carbon performances.

4.2. *Effective Utilisation of Building Energy and Carbon Data*

Lack of usable energy and carbon data in buildings (C13) is another constraint commonly identified in the selected contexts. Effective policy interventions, skill development of professionals, and technological support are important in addressing this constraint. Mandatory energy audits for existing buildings should be supported by standardised reporting procedures and database maintenance [33]. The audits should be conducted by targeting different phases of the building life cycle. The operational phase audits should be conducted on a time scale. Audits can also be performed to identify the energy/carbon performances for mandatory compliance requirements. Accordingly, Gupta et al. [33] highlighted that life-cycle assessment (LCA) should be encouraged and supported by providing relevant expertise, regulatory support, comparative studies, and calculation tools. This can start with the public sector by mandating LCA for major investments.

Adequate training and certification programmes should be introduced for energy auditors and energy managers. Mandating the disclosure of energy usage and carbon emissions, including embodied carbon emissions, is an important step towards initiating a data-driven culture in decision-making for building energy efficiency improvement and carbon emission reduction. For this, energy/carbon performance data and audit results should be easily accessible to the relevant government sectors, academia and other relevant professionals at the project level, organisational level, and wider project portfolio levels. Continuous improvement and review of standards and regulations should be pursued based on the available real-world data. Standardisation and accreditation bodies and the governmental authorities are the primary responsible parties for this, while contractors, consultants, organisational-level energy managers, facility managers, and clients should actively engage and collaborate in arranging data generation and sharing at the project/organisational level.

Actual expenditures and payback period-related information on energy-efficient systems and equipment should be made available for the reference of investors and owners. Universities can focus on carrying out research projects and educational programmes to analyse the payback (return on investment) potentials of major energy-efficient installations and refurbishments. This helps to eliminate the uncertainty in energy saving potential/financial gains associated with low-carbon and energy-efficient investments in the buildings sector (C2). Furthermore, the information on actual construction/installation times, costs, and the amount of energy saved/reduction in carbon emissions should be published on easily accessible platforms (designated websites, media). Alam et al. [13] suggested creating websites to share the details of successful projects, awareness, training programmes, and research findings. This information is important in providing real-world evidence and examples for relevant stakeholders (owners, investors, material and equipment manufacturers, and relevant regulatory bodies) to make informed decisions on energy/carbon efficient investments and required improvements to the systems. Accordingly, Cristino et al. [55] also emphasised the importance of having a strong policy framework to disseminate relevant information on building energy-efficient technologies.

The data gathered through energy management systems (EMS) and building management systems (BMS) can be effectively utilised for decision-making on required improvements. These data could also effectively contribute towards deciding energy-saving strategies, evaluating compliance with policies and regulations, benchmarking, record keeping, and certifying. Therefore, the regulatory bodies should take appropriate measures to encourage/mandate the installation of these systems for new constructions.

4.3. Awareness Raising, Training, and Information Dissemination

It is important to develop and maintain proper databases of energy-efficient materials, equipment, technologies, and respective suppliers. This will create opportunities for comparison and selection of adequate materials, systems, and technologies. Moreover, databases should be developed with carbon and energy information on building materials, construction methods, construction projects, etc. Accordingly, new developments could benefit from past information.

Little knowledge among end-users about the consequences of their actions on carbon emissions and energy consumption (C14), high-energy lifestyle (C19), and more concern with aesthetic appearance (C4) are identified as constraints that are common to the selected countries/regions. End-user awareness raising, behavioural changes, and influencing changing their attitudes are important steps in addressing these constraints. Accordingly, Cristino et al. [55] also emphasised the importance of conducting education and training programmes for the occupants of the buildings and improving the awareness of occupants about the building energy-efficient programmes. Proper education should also be provided for occupants on changing their energy-use behaviour [53]. As an initial step, detailed occupancy surveys should be conducted to clearly identify the energy usage patterns and attitudes of the end-users and occupants. Subsequently, end-user-targeted awareness-raising programmes should be implemented to make them aware of the multiple benefits of low-carbon initiatives (the information gathered through the occupancy surveys could be used as the baseline for designing these awareness programmes in a particular country/region). Government and academia can collaboratively design uniform occupancy survey guidelines. Facility managers, energy managers (with the support of owners), and relevant government representatives can carry out the surveys while the relevant government sector authorities and academia can design suitable training and awareness-raising programmes. Furthermore, incentive schemes for adopting low-carbon/energy-efficient behaviours can be introduced at the organisational level.

Importantly, the overall awareness of the general public should be raised to self-motivate them in selecting low-carbon/sustainable alternatives available in the built environment sector. The public should be educated and encouraged to adopt sustainable behaviours. This is a long-term process since behaviour changes must be preceded by mindset changes. Therefore, this should start with primary education and continue through secondary and tertiary education, timely awareness programmes, training programmes, media support, etc.

A workforce with less knowledge and technical expertise in new technological advancements (C15) was identified as another significant constraint. Therefore, efficient training and accreditation systems should be implemented for the professionals engaged in the sectors of low-carbon construction, auditing, energy management, building maintenance and retrofitting. IEA [52] suggested implementing accreditation systems for professionals engaged in low-carbon construction and management as a feasible solution to enhance the quality, skills, and expertise of the workforce. Implementing specialised training on relevant products, tools, systems, and technologies is also an important step [26,33]. This will enable obtaining the best performances of these systems through proper installation, maintenance, operations, etc.

Government sector officials in relevant authorities and departments should also be provided with adequate training and accreditations. Cristino et al. [55] also emphasised the importance of providing adequate training to regulators and legislators. Accordingly, providing training and certifications and employing a qualified set of energy and carbon auditors in the government sector will be important to maintain the uniformity of energy/carbon auditing, comparison and further decision-making on policy initiatives for energy usage and carbon emission reduction in buildings. Accordingly, Alam et al. [13] highlighted the importance of developing a pre-qualified set of energy auditors. Moreover, training and certification of energy managers through a regulatory body should be conducted to facilitate the implementation of effective energy management procedures in buildings. Necessary training should be provided for organisations to support the development of their business models by including sustainability aspects in their operations and facilities.

4.4. Technology Advancement

Incompatibilities with new technologies due to existing building conditions (C17) and complexity of technology (C18) are identified as commonly prevailing constraints in the selected contexts. In order to adopt feasible technologies, the policy-makers and regulators should target obtaining international-level assistance for technology advancement for LCBs. This should be performed by considering the current social, economic, and technological status of the country.

Priority should be given to supporting the manufacturing of locally produced low-carbon building materials. Incentives should be given to relevant manufacturing industries to invest in productive research and development in developing low-carbon alternatives. Demonstration, prototyping, and examples of developed advanced energy-efficient materials, equipment, and technologies should be promoted through the sponsorship of the government [53,55]. This could help in building the trust of investors and clarifying the uncertainties associated with low-carbon and energy-efficient investments (C2).

Validating the performances of new developments and evaluating compliance with other requirements, such as safety and fire regulations, should be effectively supported and carried out with the involvement of relevant government authorities before the commercialisation of the materials and products. Furthermore, the cost-effectiveness of these alternatives should be assessed. For this, the manufacturers, relevant national-level laboratories and research institutions, such as universities, should work in collaboration. Initial market penetration of such innovative products should be supported with demonstrations, case studies, etc. Government, research institutions, academia, manufacturers, and the media should also work collaboratively to penetrate these markets with technological advancements. Government support is needed for the manufacturers during this transition to help bear the risk of entering into new markets. Financial institutions should also work collaboratively in providing incentives for promoting these technological advancements.

Useful research outcomes not being translated effectively into technology innovations, policy initiatives, and industry practices (C12) is also identified as a significant and common constraint. Research and development could be initiated by research institutions, such as universities or research and development sections of industries. The construction industry, government sector, and academia should collaboratively commit to addressing this constraint. Government should encourage and provide suitable research grants to universities and other institutions to initiate research centres and carry out research projects on developing cost-effective, low-carbon, and energy-efficient alternatives [33,55]. Research and development should be encouraged in exploring construction methods to minimise the usage of increasingly scarce natural resources, e.g., sand, and to increase the usage of recycled materials. Government should also provide incentives for industries to invest more in research and development.

Financing the conversion of research and development into manufacturing through tax incentives, providing grants, etc., should also be made with the support of government and financial institutions. Industry, academia, and government collaboration and information exchange should also be promoted through joint workshops, seminars, conferences, and exhibitions. Industries, energy generation, supplying sector, and the buildings sector should work collaboratively in researching and exploring co-generation capabilities of electricity and recovery of power through heat recovery at the industrial level. Government, academia, and industry-based research and development sectors should collaboratively target circular design approaches for innovative buildings, their materials, and equipment. This helps reduce the embodied energy/carbon of materials, increase the efficiency of construction, operation, and maintenance phases, and also increase the opportunities for reusing and recycling at the end of life.

Appropriate energy simulation, integration of BIM, and utilisation of advanced tools at the design stage effectively contribute to identifying energy-saving opportunities beforehand, thereby constructing a building with optimum energy efficiency and minimum carbon emissions [52]. Furthermore, all new constructions should be assessed in terms of their potential for renewable energy generation and energy recovery (solar electricity, solar thermal, geothermal energy, waste heat recovery, etc.). Accordingly, mandatory requirements should be introduced to plan and design buildings to ensure decentralised energy generation.

'Interoperability and compatibility' of building systems is another major concern (C17). For example, the daylighting control mechanism and HVAC system should have integrated controls. Fire safety systems and HVAC systems should also have proper integration. Therefore, adopting building management systems with integrated controls is important in enhancing the integration of different systems. The usage of different types of sensors and automated controllers and integrating them with building energy-management systems or building-management systems should be encouraged for all new constructions. This will ensure optimum energy utilisation, facilitate decision-making on maintenance, and also help optimise indoor environmental quality, while performance can also be monitored and logged for future decision-making.

Carrying out timely standardised maintenance and servicing of high energy-intensive systems and equipment is important to ensure their performance and efficiency. This can be performed by mandating maintenance and servicing contracts with specialised and authorised service providers. This also ensures the optimum energy/carbon performance of relevant systems and equipment. Availability of globally applicable maintenance and servicing procedures for the systems and equipment should be ensured at the point of installation [33].

Greater attention should also be paid to construction waste management, repurposing of buildings, recycling of materials, and reusing of materials [52]. Accordingly, the government should plan and implement construction waste sorting and recycling facilities or support the private sector to implement these facilities by providing adequate financial incentives and technical guidance. Development of effective guidelines for building deconstruction and waste sorting is another important initiative which could help in streamlining the recycling/reusing process.

4.5. Incentives for Low-Carbon Adoption

Incentives are important for material and equipment manufacturers, producers, and suppliers to survive and thrive in the market as well as for owners and investors to invest in LCBs. Lack of investment capital (C3), lack of incentives for the building material/equipment market (C7), and lack of incentives from the government and financial institutions towards owners and developers (C6) were identified as significant common constraints to LCB adaption in the selected countries and regions. Accordingly, government and financial institutions hold the primary responsibility for introducing adequate financial incentives. Government should initiate awareness raising of financial institutions on financing low-carbon, energy-efficient and green building initiatives [56]. Developing policies to finance onsite renewable energy generation in buildings is another important step towards delivering LCBs. Accordingly, Cristino et al. [55] highlighted the importance of developing a systematic method (pattern) for providing economic incentives.

Manufacturers of low-carbon materials and equipment should be supported with subsidies, such as rebates for taxes, loans, and incentives for research and development. Importers and distributors should also be supported by providing incentives for importing energy-efficient/low-carbon equipment and material by granting tax rebates. Implementing new technologies and developing new products/materials require significant capital and also require access to technology and expertise. More importantly, the manufacturers need a guarantee of the return on investment. Due to the many risks, private investors are less willing to pioneer such developments. Adequate incentives and support from government and financial institutions are required for manufacturers and investors until the end product saturates the market.

Banks can initiate providing financing support, such as mortgages, for buildings with high energy efficiency and less carbon footprint. Moreover, banks can offer attractive interest rates and longer tenure periods for loans which target low-carbon and energy-efficient investments. Furthermore, the incentives, such as grants, loans and tax rebates, can be provided based on the energy/carbon performance levels of developments [52]. A revolving loan fund mechanism can also be considered a feasible option for financing low-carbon and energy-efficient investments [13]. This provides the opportunity for the borrower to repay the loan through the cost savings achieved through the investments made for energy-efficient building retrofits. Additionally, the energy efficiency improvement/carbon emission reduction projects can be financed through the support of national or state level (local government level) budgets. Accordingly, governments could offer financial incentives for private developers to adopt low or zero-carbon initiatives for buildings [53]. UNECE [56] emphasised tax incentives and low-interest loans as the most appropriate methods to increase the viability of investing in energy-efficiency improvement projects. Furthermore, investors' trust can be enhanced, and the uncertainty of return on investment (C2) can be minimised by providing long-term warranties for energy-efficient /low-carbon systems and equipment.

Another feasible option is to link financing procedures with the building energy/carbon certifications and standards. More attractive incentives could be provided according to the ratings or certification levels. Providing attractive feed-in tariff rates for feeding power to the national grid through renewable energy generation in buildings also helps to encourage the building owners to consider more decentralised power generation options [52].

Financing can be made through an energy service provider/company (ESCO) for energy-efficient and low-carbon initiatives in buildings through performance-based contracts. This can be identified as a budget-friendly option for energy efficiency improvement. Accordingly, owners can enter into an energy performance-based contract with an ESCO to upgrade the building with energy-efficient features. The required capital for the improvements can be totally or partially from ESCOs. The relevant payments for the services and the investment of ESCOs are made based on the energy savings of the implemented retrofits and upgrades. In addition to the private sector, public sector facilities can also implement this energy-performance-based contract mechanism and help promote such initiatives in collaboration with ESCOs [52].

4.6. Organisational Level Commitments and Policy Initiatives

C9 (organisational business models are not considering the integration of low-carbon and energy-efficient initiatives) and C20 (a poor corporate culture which separates environmental performances and business improvement decisions) were identified as significant constraints common to the selected countries/regions. Not obtaining priority status for energy/carbon reduction at the organisational level (due to the primary focus on core-business activities), not having strong authority for decision-making and setting goals on energy efficiency improvement/carbon emission reduction, and the high initial cost for advanced energy/carbon efficient systems are major issues arising at the organisational level. Even though some organisations consider installing energy-efficient plants and equipment, there is a lack of attention to adopting continuous energy management approaches due to the additional workload and expenses. Mandates and sufficient incentives from the government sector and financial institutions are highly significant in directing organisations to adopt low-carbon approaches. In addition to the adoption of mandated compliance procedures, organisations and industries should also have proactive and self-motivated commitments to improving energy efficiency, reducing energy consumption, and minimising GHG emissions from their building stock. Accordingly, internal compliance procedures and energy/carbon policies should be developed by considering the nature of the core business and related procedures, for which organisations should incorporate energy saving and carbon reduction into their internal business models. Accordingly, Alam et al. [13] and Gupta et al. [33] also highlighted the importance of incorporating energy efficiency and carbon reduction into an organisation's internal economic model.

Appropriate delegation of responsibilities among the relevant employees (facility managers, energy managers, supervisors, and other responsible personnel on respective plants and systems) is an important step towards reducing operational stage energy usage and carbon emissions. This leads to proper management of buildings, their systems, and components, and, thereby, to achieving the optimum energy/carbon performances. Adopting sustainable maintenance procedures is another important step towards minimising operational carbon emissions from buildings. Furthermore, relevant responsible employees should take a proactive lead in establishing an energy-efficient/low-carbon culture within organisations. Gupta et al. [33] highlighted that the proper distribution of authority among the officers or managers helps in the timely implementation of energy and carbon-related policies and the proper utilisation of funds in an organisation.

Organisational internal funding models should be developed to support investments in energy-efficient and low-carbon initiatives. The organisational procurement procedures should also be developed to consider the options of life cycle cost analysis and life cycle environment impact analysis in procuring new systems and components [13].

Responsible personnel in building operations (facility managers, energy managers, maintenance engineers), with the guidance of their clients/owners, should target generating, processing, and analysing the building-level data related to energy usage, carbon emissions, behaviour patterns, equipment/system performances, etc. The services of energy services companies (ESCOs) or specialised consultants can also be obtained to effectively carry out these data generation and analyses. Subsequently, innovative building-specific initiatives (e.g., phasing out less efficient systems, demand management, load shedding, adjustment of operational procedures, investments in new energy/carbon efficient systems, adopting energy management systems) should be determined based on the generated data and information.

5. Conclusions

This study targeted to identify and explore suitable strategies for accelerating the delivery of LCBs by overcoming the common and significant constraints to delivering LCBs in high-rise, high-density cities.

Based on this study findings, strategies for accelerating the delivery of LCBs were identified and are recommended under six categories, namely, policy implementation, building energy/carbon data utilisation, awareness raising/training, technology advancement, incentives, and organisational level commitments. The strategies related to energy/carbon policy development, standardisation, codes and certifications, mandatory regulations, financial incentives, and technology adoption had the ability to influence a majority of the driving constraints. When these driving constraints are addressed or mitigated through these significant strategic approaches, it contributes to reducing the required efforts in addressing the intermediate and dependent constraints.

Policy and regulatory sector stakeholders were identified to have the highest influencing ability in overcoming the constraints of delivering LCBs. Other stakeholders were important in operationalising governmental actions. However, according to the overall discussions of stakeholder involvement and collaboration in accelerating LCB delivery, the stakeholders/stakeholder classes, contractors, consultants, financial sector, clients/owners, educational, training and research institutions, property and facility managers, manufacturers, accreditation and standardisation bodies, and energy and environmental service providers, were identified as having a significant influencing ability in implementing most of the strategies.

This study also adds to the SNA knowledge domain itself by extending the SNA two-mode applications to analyse connections of strategies and constraints to delivering LCBs. Moreover, the sequential ISM and SNA integrated methodological approach utilised in the present study contributes to the knowledge domains of respective research methods by providing prospective integrated analysis approaches for future researchers. Furthermore, the identified strategies provide important guidance to the relevant practitioners to determine suitable approaches to accelerate the delivery of LCBs by prioritising the required actions and identifying the responsible stakeholders/stakeholder classes in implementing these strategies. More importantly, this study provides an integrated, innovative approach to address the constraints by considering their driving and dependence characteristics. Accordingly, this study shows that an integrated systems approach, which considers the interdependencies among multiple sectors (e.g., policy level, project level, financing sector), is more suitable than traditional discrete/disconnected ad hoc attempts for addressing specific constraints in accelerating the delivery of LCBs.

Due to the difficulty in administering prolonged interviews and due to the time constraints, required stakeholder involvement and collaboration aspects were not discussed by specifically relating to each strategy. Therefore, future research can include a comprehensive stakeholder engagement and collaboration analysis for implementing the identified strategies. In order to fulfil the aim of the present study, the strategies for addressing the common constraints in delivering LCB to high-rise, high-density cities were identified based on data from a typical sample. Testing these strategies on yet another high-rise, high-density city would be a logical next step but may ideally require a longitudinal study in that city with close cooperation with all those involved. Such accessibility to the key stakeholders, longitudinal data availability, time, and resources were unavailable in this study for such a follow-up. However, such testing, followed by fine-tuning such strategies for specific cities/regions (as case studies) and then preparing city or region-specific strategic plans to accelerate the delivery of LCBs, are suggested as useful future research directions.

Author Contributions: Conceptualization, M.M.K., W.P. and N.M.H.W.; methodology, M.M.K., W.P. and N.M.H.W.; software, N.M.H.W.; validation, N.M.H.W.; formal analysis, N.M.H.W.; data curation, N.M.H.W.; writing—original draft preparation, M.M.K., N.M.H.W. and W.P.; writing—review and editing, M.M.K. and W.P.; supervision, W.P. and M.M.K.; project administration, M.M.K. and W.P. All authors have read and agreed to the published version of the manuscript.

Funding: This research received no external funding.

Data Availability Statement: Data available on request due to restrictions. The data presented in this study are available on request from the corresponding author. The data are not publicly available due to the restrictions applied when depositing the dataset to the research institution (University).

Conflicts of Interest: The authors declare no conflict of interest.

References

1. Intergovernmental Panel on Climate Change [IPCC]. Climate Change 2014 Synthesis Report Summary for Policymakers. 2014. Available online: https://www.ipcc.ch/pdf/assessmentreport/ar5/syr/AR5_SYR_FINAL_SPM.pdf (accessed on 8 January 2022).
2. Mardiana, A.; Riffat, S.B. Building Energy Consumption and Carbon dioxide Emissions: Threat to Climate Change. *J. Earth Sci. Clim. Chang.* **2015**. [CrossRef]
3. Ahmed, A.K.; Ahmad, M.; Yusup, Y. Issues, Impacts, and Mitigations of Carbon Dioxide Emissions in the Building Sector. *Sustainability* **2020**, *12*, 7427. [CrossRef]
4. Pérez-Lombard, L.; Ortiz, J.; González, R.; Maestre, I. A review of benchmarking, rating and labelling concepts within the framework of building energy certification schemes. *Energy Build.* **2009**, *41*, 272–278. [CrossRef]
5. Kennedy, C.; Cuddihy, J.; Engel-Yan, J. The Changing Metabolism of Cities. *J. Ind. Ecol.* **2007**, *11*, 43–59. [CrossRef]
6. Hunt, A.; Watkiss, P. Climate change impacts and adaptation in cities: A review of the literature. *Clim. Chang.* **2011**, *104*, 13–49. [CrossRef]
7. Saroglou, T.; Meir, I.; Theodosiou, T.; Givoni, B. Towards energy efficient skyscrapers. *Energy Build.* **2017**, *149*, 437–449. [CrossRef]
8. United Nations Department of Economic and Social Affairs [UN DESA]. 68% of the World Population Projected to Live in Urban Areas by 2050, Says UN. 2018. Available online: https://www.un.org/development/desa/en/news/population/2018-revision-of-world-urbanization-prospects.html (accessed on 23 January 2023).
9. Pan, W.; Pan, M. A dialectical system framework of zero carbon emission building policy for high-rise high-density cities: Perspectives from Hong Kong. *J. Clean. Prod.* **2018**, *205*, 1–13. [CrossRef]
10. Madhusanka, N.; Pan, W.; Kumaraswamy, M. Constraints to Delivering Low Carbon Buildings in High-Rise High-Density Cities. In Proceedings of the 37th Annual ARCOM Conference, Leeds, UK, 6–7 September 2021; pp. 804–813.
11. Hong Kong Environment Bureau [HKEB]. Energy Saving Plan for Hong Kong's Built Environment 2015–2025. 2015. Available online: http://www.enb.gov.hk/sites/default/files/pdf/EnergySavingPlanEn.pdf (accessed on 7 March 2022).
12. Jones, P.; Lannon, S.; Patterson, J. Retrofitting existing housing: How far, how much? *Build. Res. Inf.* **2013**, *41*, 532–550. [CrossRef]
13. Alam, M.; Zou, P.; Stewart, R.; Bertone, E.; Sahin, O.; Buntine, C.; Marshall, C. Government championed strategies to overcome the barriers to public building energy efficiency retrofit projects. *Sustain. Cities Soc.* **2019**, *44*, 56–69. [CrossRef]
14. Ministry of Climate Change and Environment [MOCCAE]. National Climate Change Planof the United Arab Emirates 2017–2050. 2021. Available online: https://www.moccae.gov.ae/assets/30e58e2e/national-climate-change-plan-for-the-united-arab-emirates-2017-2050.aspx (accessed on 13 October 2021).
15. Building and Construction Authority [BCA]. Green Building Masterplans. 2021. Available online: https://www1.bca.gov.sg/buildsg/sustainability/green-building-masterplans (accessed on 15 September 2021).
16. International Energy Agency [IEA]. Energy Efficiency 2020. 2020. Available online: https://iea.blob.core.windows.net/assets/59268647-0b70-4e7b-9f78-269e5ee93f26/Energy_Efficiency_2020.pdf (accessed on 4 April 2022).
17. International Energy Agency [IEA]. Estimated Global Buildings Construction in the Sustainable Development Scenario, 2018–2030. 2019. Available online: https://www.iea.org/data-and-statistics/charts/estimated-global-buildings-construction-in-the-sustainable-development-scenario-2018-2030 (accessed on 8 January 2022).
18. Regulatory Indicators for Sustainable Energy [RISE]. Indicators. 2021. Available online: https://rise.esmap.org/indicators (accessed on 4 May 2021).
19. Berke, P.R. Does Sustainable Development Offer a New Direction for Planning? Challenges for the Twenty-First Century. *J. Plan. Lit.* **2002**, *17*, 21–36. [CrossRef]
20. Chan, E.H.W.; Lee, G.K.L. Design-led sustainable urban renewal approach for Hong Kong. *HKIA J.* **2006**, *46*, 76–81.
21. Isiadinso, S.; Goodhew, S.; Marsh, J.; Hoxley, M. Identifying an appropriate approach to judge low carbon buildings. *Struct. Surv.* **2011**, *29*, 436–446. [CrossRef]
22. Bhatt, N. Low Carbon Building. Climate CoLab. 2016. Available online: https://www.climatecolab.org/contests/2016/buildings/c/proposal/1329602#:~:text=Low%2Dcarbon%20buildings%20(LCB),would%20qualify%20as%20a%20LCB (accessed on 25 October 2022).
23. Farhan, S.A.; Shafiq, N.; Azizli, K.A.M.; Jamaludin, N.; Gardezi, S.S.S. Low-Carbon Buildings: Renewable Energy Systems, Materials and Assessment Methods. *Aust. J. Basic Appl. Sci* **2014**, *8*, 260–263.

24. Pan, W.; Ning, Y. A socio-technical framework of zero-carbon building policies. *Build. Res. Inf.* **2015**, *43*, 94–110. [CrossRef]
25. Global Alliance for Buildings and Construction [GABC]. 2019 Global Status Report for Buildings and Construction: Towards a Zero-Emission, Efficient and Resilient Buildings and Construction Sector. 2019. Available online: https://www.worldgbc.org/sites/default/files/2019%20Global%20Status%20Report%20for%20Buildings%20and%20Construction.pdf (accessed on 10 September 2022).
26. Pan, W.; Pan, M. Opportunities and risks of implementing zero-carbon building policy for cities: Hong kong case. *Appl. Energy* **2019**, *256*, 113835. [CrossRef]
27. Pan, W.; Pan, M. Drivers, barriers and strategies for zero carbon buildings in high-rise high-density cities. *Energy Build.* **2021**, *242*, 110970. [CrossRef]
28. Mata, É.; Peñaloza, D.; Sandkvist, F.; Nyberg, T. What is stopping low-carbon buildings? A global review of enablers and barriers. *Energy Res. Soc. Sci.* **2021**, *82*, 102261. [CrossRef]
29. Madhusanka, N.; Pan, W.; Kumaraswamy, M. Constraints to low-carbon building: Perspectives from high-rise high-density cities. *Energy Build.* **2022**, *275*, 112497. [CrossRef]
30. Jiang, M.P.; Tovey, K. Overcoming barriers to implementation of carbon reduction strategies in large commercial buildings in China. *Build. Environ.* **2010**, *45*, 856–864. [CrossRef]
31. Baek, C.; Park, S. Policy measures to overcome barriers to energy renovation of existing buildings. *Renew. Sustain. Energy Rev.* **2012**, *16*, 3939–3947. [CrossRef]
32. Zhang, Y.; Wang, Y. Barriers' and policies' analysis of China's building energy efficiency. *Energy Policy* **2013**, *62*, 768–773. [CrossRef]
33. Gupta, P.; Anand, S.; Gupta, H. Developing a roadmap to overcome barriers to energy efficiency in buildings using best worst method. *Sustain. Cities Soc.* **2017**, *31*, 244–259. [CrossRef]
34. Reddy, B.S. Barriers and drivers to energy efficiency—A new taxonomical approach. *Energy Convers. Manag.* **2013**, *74*, 403–416. [CrossRef]
35. Andrić, I.; Koc, M.; Al-Ghamdi, S. A review of climate change implications for built environment: Impacts, mitigation measures and associated challenges in developed and developing countries. *J. Clean. Prod.* **2019**, *211*, 83–102. [CrossRef]
36. Madhusanka, N.; Pan, W.; Kumaraswamy, M. Social Network Analysis of Building Energy and Carbon Policy Networks in Developing Countries. *IOP Conf. Ser. Earth Environ. Sci.* **2020**, *588*, 022004. [CrossRef]
37. Iwaro, J.; Mwasha, A. A review of building energy regulation and policy for energy conservation in developing countries. *Energy Policy* **2010**, *38*, 7744–7755. [CrossRef]
38. Zhang, L.; Li, Q.; Zhou, J. Critical factors of low-carbon building development in China's urban area. *J. Clean. Prod.* **2017**, *142*, 3075–3082. [CrossRef]
39. Li, Y.; Zhu, N.; Qin, B. Major Barriers to the New Residential Building Energy-Efficiency Promotion in China: Frontlines' Perceptions. *Energies* **2019**, *12*, 1073. [CrossRef]
40. Pan, W.; Li, K.; Teng, Y. Briefing: Life-cycle carbon assessment of prefabricated buildings: Challenges and solutions. *Proc. Inst. Civ. Eng.—Eng. Sustain.* **2019**, *172*, 3–8. [CrossRef]
41. Yu, C.; Pan, W.; Zhao, Y.; Li, Y. Challenges for Modeling Energy Use in High-rise Office Buildings in Hong Kong. *Procedia Eng.* **2015**, *121*, 513–520. [CrossRef]
42. Biekša, D.; Šiupšinskas, G.; Martinaitis, V.; Jaraminienė, E. Energy efficiency challenges in multi-apartment building renovation in Lithuania. *J. Civ. Eng. Manag.* **2011**, *17*, 467–475. [CrossRef]
43. Shen, W.; Tang, W.; Siripanan, A.; Lei, Z.; Duffield, C.; Hui, F. Understanding the Green Technical Capabilities and Barriers to Green Buildings in Developing Countries: A Case Study of Thailand. *Sustainability* **2018**, *10*, 3585. [CrossRef]
44. Huang, J.; Tzeng, G.; Ong, C. Multidimensional data in multidimensional scaling using the analytic network process. *Pattern Recognit. Lett.* **2005**, *26*, 755–767. [CrossRef]
45. Sharma, H.D.; Gupta, A.D.; Sushil. The objectives of waste management in India: A futures inquiry. *Technol. Forecast. Soc. Chang.* **1995**, *48*, 285–309. [CrossRef]
46. Mandal, A.; Deshmukh, S. Vendor Selection Using Interpretive Structural Modelling (ISM). *Int. J. Oper. Amp; Prod. Manag.* **1994**, *14*, 52–59. [CrossRef]
47. Borgatti, S.P.; Everett, M.G.; Johnson, J.C. *Analyzing Social Networks*; SAGE: Los Angeles, CA, USA, 2013.
48. Loosemore, M. Social network analysis: Using a quantitative tool within an interpretative context to explore the management of construction crises. *Eng. Constr. Archit. Manag.* **1998**, *5*, 315–326. [CrossRef]
49. Wasserman, S.; Faust, K. *Social Network Analysis: Methods and Applications*; Cambridge University Press: Cambridge, UK, 1997.
50. Wang, G.; Wang, Y.; Zhao, T. Analysis of interactions among the barriers to energy saving in China. *Energy Policy* **2008**, *36*, 1879–1889. [CrossRef]
51. Movahedipour, M.; Zeng, J.; Yang, M.; Wu, X. An ISM approach for the barrier analysis in implementing sustainable supply chain management. *Manag. Decis.* **2017**, *55*, 1824–1850. [CrossRef]
52. Internatioanal Energy Agency [IEA]. Roadmap for Energy-Efficient Buildings and Construction in ASEAN Timelines and Actions towards Net Zero-Carbon Buildings and Construction. 2022. Available online: https://iea.blob.core.windows.net/assets/5255ea58-1fa7-4fb4-bca0-b32923e9184a/RoadmapforEnergy-EfficientBuildingsandConstructioninASEAN.pdf (accessed on 25 October 2022).

53. Ohene, E.; Chan, A.P.C.; Darko, A. Prioritizing barriers and developing mitigation strategies toward Net-Zero Carbon Building Sector. *Build. Environ.* **2022**, *223*, 109437. [CrossRef]
54. Madhusanka, N.; Kumaraswamy, M.; Pan, W.; Chan, S. Relationally Integrated Value Networks for Total Asset Management in Project Portfolios. *Infrastruct. Asset Manag.* **2022**, *9*, 123–134. [CrossRef]
55. Cristino, T.M.; Lotufo, F.A.; Delinchant, B.; Wurtz, F.; Faria Neto, A. A comprehensive review of obstacles and drivers to building energy-saving technologies and their association with research themes, types of buildings, and geographic regions. *Renew. Sustain. Energy Rev.* **2021**, *135*, 110191. [CrossRef]
56. United Nations Economic Commission for Europe [UNECE]. Overcoming Barriers to Investing in Energy Efficiency. United Nations New York and Geneva. 2017. Available online: https://unece.org/DAM/energy/se/pdfs/geee/pub/Overcoming_barriers-energy_efficiency-FINAL.pdf (accessed on 17 October 2022).

buildings

MDPI

Article

A Critical Scoping Review of Disability Employment Research in the Construction Industry: Driving Social Innovation through More Inclusive Pathways to Employment Opportunity

Susan Bailey [1], Phillippa Carnemolla [1], Martin Loosemore [1,*], Simon Darcy [2] and Shankar Sankaran [1]

[1] School of Built Environment, University of Technology Sydney, Sydney, NSW 2007, Australia
[2] UTS Business School, University of Technology Sydney, Sydney, NSW 2007, Australia
* Correspondence: martin.loosemore@uts.edu.au

Abstract: Innovation research in construction has almost exclusively focused on economic and technological innovation. In contrast, the emerging concept of social innovation has been largely ignored. This is despite the global growth of social procurement policies which incentivize construction firms to innovate in providing employment opportunities for equity-seeking groups. While there is an emerging body of research which is starting to explore innovative employment pathways into construction for certain equity-seeking groups such as women, refugees and Indigenous people, there has been relatively little research into employment pathways for people with a disability. Addressing this gap in research, this paper reports the results of a critical scoping review of Web of Science, Scopus, PubMed and Google Scholar publications on the employment of people with disability in construction. Using the Preferred Reporting Items for Systematic reviews (PRISMA-ScR) approach, extant research was mapped across seven themes of hiring disability practices. Results indicate that research into the employment of people with disability in construction internationally remains nascent with significant knowledge gaps compared to mainstream disability employment research. These key gaps include: barriers to employment based on the lived experiences of people with disability seeking employment in construction; the facilitation of cross-sector relationships with organizations that support people with disability into employment; the reduction of biases, ingrained stigmas and inequalities in recruitment practices for people with disability; and the role of informal norms and practices in undermining formal laws, regulations and policies designed to reduce barriers to employment. The scoping review also identifies a methodological gap in the research reviewed by highlighting the need for more construction research designs to include people with disability as prioritized research participants as well as research investigators and to adopt phenomenological and interpretive approaches which respect the lived experiences of people with a disability seeking work in the construction industry.

Keywords: disability; social innovation; corporate social responsibility; social value; employment; social procurement

Citation: Bailey, S.; Carnemolla, P.; Loosemore, M.; Darcy, S.; Sankaran, S. A Critical Scoping Review of Disability Employment Research in the Construction Industry: Driving Social Innovation through More Inclusive Pathways to Employment Opportunity. *Buildings* **2022**, *12*, 2196. https://doi.org/10.3390/buildings12122196

Academic Editor: Osama Abudayyeh

Received: 20 November 2022
Accepted: 8 December 2022
Published: 12 December 2022

1. Introduction

There has been a considerable amount of research into construction innovation going back more than four decades [1–3]. However, the vast majority of this research has focused on economic and technological innovations, whereas the concept of social innovation has received virtually no attention. In contrast, social innovation is a rapidly emerging field of research outside construction, encompassing business and management, sociology, economics, and other social science disciplines [4,5]. As an emerging area of innovation research, the concept of social innovation has many definitions. However, the Office of Economic Cooperation and Development [6] defines social innovation as the design and implementation of new solutions that imply conceptual, process, product, or organisational change, which ultimately aim to improve the welfare and wellbeing of individuals and communities. Although social innovations can be technological, many seek to address unresolved social needs with novel organisational solutions.

The neglect of social innovation in the field of construction is somewhat surprising given the growing focus on corporate social responsibility in the industry and the recent proliferation of social procurement requirements in many countries, which specifically weight social innovation in construction tender decisions [7,8]. In simple terms, social procurement involves construction clients requiring their construction supply chains to innovate in creating social value in the communities in which they build [9]. Social value can take many forms but social procurement policies tend to focus on creating employment and training opportunities for disadvantaged equity-seeking groups such as people with a disability, Indigenous people, refugees and migrants, ex-offenders and disengaged youth [10,11].

The construction sector is the world's largest employer but has struggled to increase the diversity of its workforce [12–14]. The long tradition of research in this area highlights that the industry has a strongly normalised view of an ideal construction industry employee (typically male and able-bodied) [15–17]. Barriers to more diverse employment are reinforced by narrow and exclusionary networks from which people are recruited into the construction industry, and negative stereotypes of those who lie outside the sector's institutionalized norms [8]. While diversity research in the field of construction management has focused on a wide range of excluded groups such as women [15], Indigenous Australians [18], culturally and linguistically diverse people [19] and refugees [20], research on the employment of people with disability has received relatively little attention. The term disability refers to a long-term physical, mental, intellectual, neurological or sensory impairment which can hinder [a person's] full and effective participation in society on an equal basis with others. Disability can be caused by genetic disorders, illnesses, workplace accidents, ageing, injuries or a combination of these factors and the way that people experience disability varies significantly depending on environmental factors such as community and employer attitudes, services and support available to them and personal factors such as an individual's determination and resilience in overcoming barriers.

The relative lack of research into the employment of people with disability in construction, compared to other equity-seeking groups is surprising given that the construction industry in many countries is facing calls to diversify its workforce to address severe skills shortages [21,22]. Furthermore, there is significant potential to increase the number of people with disability in the industry's workforce. For example, in the UK people with disability make up only 6% of the construction workforce [22] and in Australia it is around 8% [23] and these jobs are typically focused in low income, insecure and administrative type jobs which provide little opportunity for career progression compared to people without a disability. While there have been a small number of studies highlighting barriers to employment for people with disability in construction ranging from physical barriers and inaccessible workplace settings to negative attitudes and assumptions about higher costs, lower productivity and safety risks [10,12,24–27], there have been virtually no studies of pathways to more inclusive employment opportunities for people with disability in the construction industry. This contrasts starkly with the extensive body of research into disability employment outside construction which continues to identify entrenched personal and societal attitudes towards people with disability [28], a lack of knowledge of accommodations to enable people with disability to secure and maintain a career [29] and a lack of understanding of the diversity of disabilities [30]. Research also shows that those injured or disabled in the workplace face significant direct and indirect discrimination in maintaining their positions [31,32].

Given the above, there is a need for more research to understand and improve the opportunities for the employment of people with disability in construction. To address this gap in research, the aim of this paper is to present a scoping review of extant academic evidence relating to the inclusion of people with disability in the construction industry workforce. Specifically, this research addresses the question of what knowledge gaps exist in construction disability employment research, compared to Gewurtz et al.'s [29] review of mainstream disability employment research. The overall objective is to identify future research directions to advance this important yet under-researched area, enabling the construction industry to better harness the untapped potential benefits of a more diverse workforce [11,33].

As Munn et al. [34] notes, scoping reviews are especially useful when research is in an exploratory, disorganised and nascent state, as it is in the field of construction disability employment research. In contrast to systematic literature reviews, scoping reviews do not aim to explore a specific research question or testable hypothesis, but aim to provide an overview or map of the evidence in a particular field. To ensure that the results of this scoping review are valid and robust, this review employs the widely used Preferred Reporting Items for Systematic reviews and Meta-Analyses extension for Scoping Reviews (PRISMA-ScR) methodology [35]. Scoping reviews using PRISMA-Scr have been successfully applied across many built environment disciplines [36,37]. However, scoping reviews are less common in the field of construction management.

2. Method

Following the PRISMA-Scr method in Figure 1, the systematic literature review employed applied bibliometric analysis of relevant peer-reviewed research relating to the employment of people with disability in the construction industry. The searches were conducted in July 2021 and inclusion criteria related papers with key words: "disability" or "disabled" combined with "employment" or "recruitment" and "construction industry" or "construction sector". We excluded papers with keywords relating to 'safety' and 'injury' because while disability can arise from injury and safety risks in construction, such papers tend to focus on safety risks and laws and regulations rather than the forms of disability arising from them. We extracted data on: year of publication; journal/location of report; study design; study participants; main findings; sample size; key results.

While it is acknowledged that any bibliometric analysis is subject to the limitations of scientific research evaluation based on citations and potential biases towards certain types of publications in databases [38], this research focused explicitly on peer-reviewed journal articles published in the Scopus data base, ISI Web of Science (WoS), PubMed and Association of Researchers in Construction Management database. Peer-reviewed articles ensure a high degree of data integrity and are widely considered to encompass validated knowledge which has a more significant and reliable impact on a field's development than non-peer reviewed research [39]. In the review, duplicate citations were removed and we excluded protocol papers and commentaries with no reported results. The search was limited to literature written in English and was not restricted by date of publication. Eligible study designs included qualitative and quantitative methodologies as well as policy discussion. Whilst we did not include systematic literature reviews in the search criteria all systematic reviews resulting from our searches were analysed for relevant, empirical studies which were then assessed separately against our inclusion and exclusion criteria.

Based on the above criteria, 113 articles were initially identified. After removing duplicates and screening abstracts, we identified 91 articles that discussed disability in the construction sector but only 24 met the inclusion criteria and were included in this review. These articles were then coded using Gewurtz et al.'s [29] analytical framework which is based on a scoping review of mainstream disability employment literature outside construction which categorizes research into key seven themes, which are defined in Table 1. Gewurtz et al.'s [29] framework was chosen because it provides the most recent scoping review of the wider disability employment literature (not construction-specific) and thereby a useful point of comparison between our sector-specific findings and their findings in the wider field of disability employment research. While we acknowledge that the profile of mainstream disability research will have evolved since Gewurtz et al.'s [29] scoping review, the classification they produced remains the most up-to-date and valid.

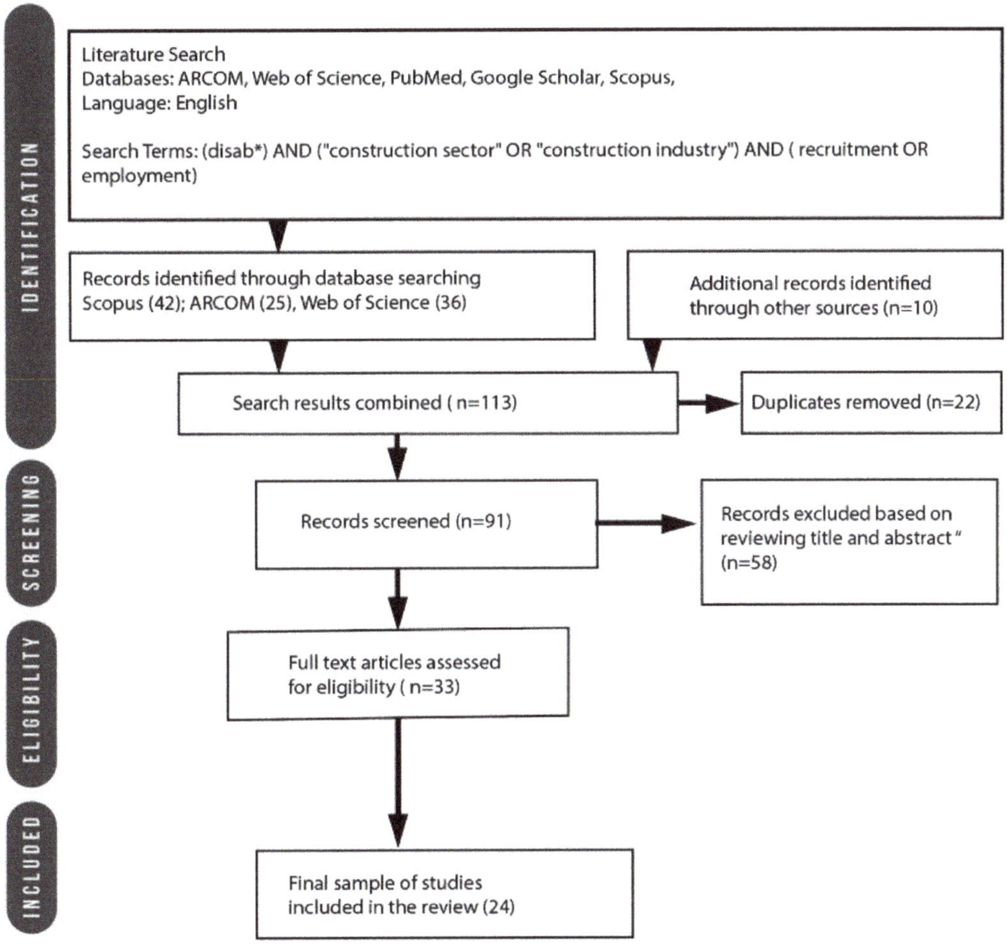

Figure 1. PRISMA Flow Chart. (* disability).

The coding process was conducted by a team of researchers from within and outside the construction industry to provide different perspectives on the data. This insider/outsider approach is widely used in psychology and social sciences research to provide different perspectives on data [40]. Comparing and cross-checking codes, categories, and themes between the researchers helped to minimized any potential disciplinary bias in the results. Instances of disagreement were resolved through discussion, a process which continued until 100% inter-rater agreement was achieved, providing a high level of 'fit' between the selection criteria and the articles and confidence in the validity of the coding process.

Table 1. Gewurtz et al.'s [1] seven themes.

Theme (Barriers and Enablers of Disability Employment)	Definition
Theme 1: Requirements versus Practice	Relates to the tensions and conflicts often found between actual disability employment practices and the laws, policies and rules that seek to protect or increase access to employment for people with disability but may also operate as barriers to employment.
Theme 2: Stigma or attitudinal barriers	Relates to the stigma and attitudinal barriers experienced by people with disability and the different ways in which stigma can act (directly and indirectly) as a barrier to employment and manifest and influence the hiring process through for example, instilling a sense of risk, fear, unpredictability, and avoidance behaviours that thwart hiring people with disability.
Theme 3: Disclosure	Relates to attitudes towards disclosure of disability, timing of disclosure, options faced by people with different disabilities, the various forms these disclosures can take and the employment consequences these choices have for the individuals concerned.
Theme 4: Accommodations	Relates to the legal requirement to offer and implement reasonable accommodations in the workplace which enable the employment of people with disability to perform their work with equal opportunity.
Theme 5: Relationship building	Relates to the idea that building relationships between employers and disability organisations that specialise in placing people with disabilities into jobs is critical to the hiring process, providing specialist advice to employers, addressing employer concerns, and showcasing opportunities and success stories.
Theme 6: Information and support to employers	Relates to the provision of information and support to employers in order to improve hiring practices and employment opportunities for people with disability. Types of information include: required accommodations and their cost, the impact of disability on job performance, the business benefits of hiring people with disability and education and support for employers.
Theme 7: Hiring practices that invite people with disability	Relates to the variety of approaches which normalise and support the hiring of people with disability, and provide equality of opportunity in applying for jobs for people with disability. Examples of such hiring practices include having disability recruitment plans, revamping job descriptions and hiring processes to minimise subjective bias, application forms that are available in a variety of formats to make them more accessible to applicants, etc.

3. Results

This section discusses the results in relation to our analytical framework which is based on a direct comparison to Gewurtz et al.'s [29] disability employment themes as discussed above and identified in the broader non-construction disability employment literature.

A comparison of studies in each of Gewurtz et al.'s [29] seven themes is illustrated in Figure 2 as a visualization device which has been applied in other scoping reviews such as [36].

In Figure 2, the height of each column illustrates the percentage of included studies that report on each theme in Table 1 rather than the number of studies, noting that an individual study can be reported across multiple themes. Interestingly, Figure 2 shows that the construction disability employment literature reflects different interests and priorities compared to mainstream disability employment literature reviewed by Gewurtz et al. [29]. In particular, the construction literature pays relatively more attention to the 'requirements versus practice', 'stigma' and 'accommodation' themes than mainstream disability employment literature. This may reflect the relatively high levels of psychological, cultural, procedural and environmental barriers that have been found to limit the employment of people with disability in the construction industry, and perceptions that people with a disability represent a risk rather than an asset who cannot be easily accommodated in the construction workplace [21,26,28]. In contrast, relatively less attention is paid to the 'disclosure', 'relationship building' and 'hiring practices' themes reflecting the immature state of corporate social responsibility reporting, human resource management and collaborative relationships with social economy organizations in the construction industry [8,12,41]. This lack of maturity of the sector, in terms of social models of disability, is also reflected in [26]'s study investigating enabling and disabling factors in the British and Dutch construction sectors.

The construction literature within each of these themes is discussed in more detail below to elaborate and explain these key differences.

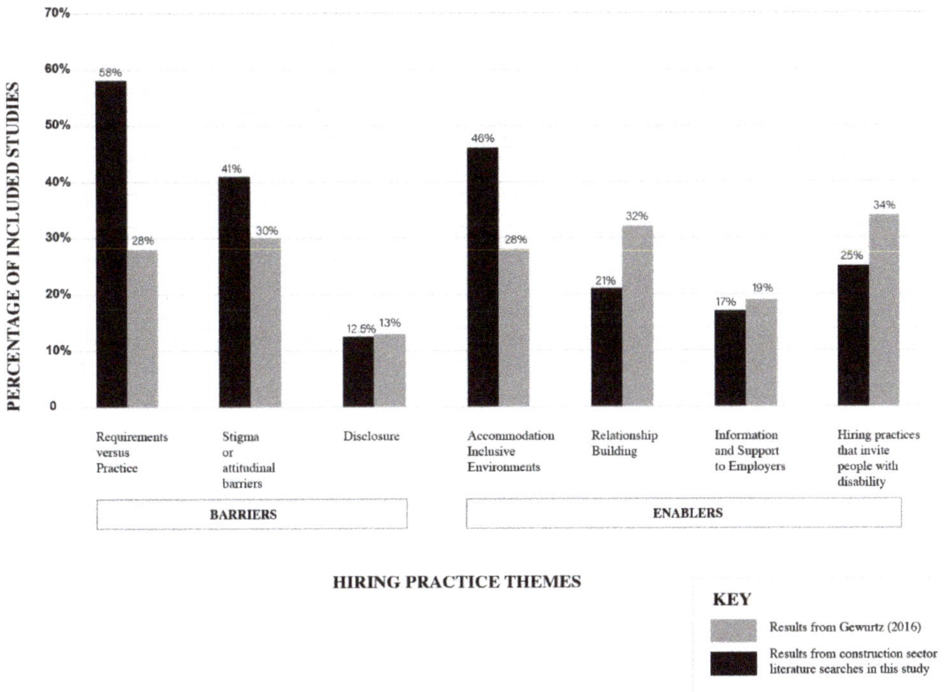

Figure 2. Research on employment of people with disability in construction mapped against Gewurtz et al.'s [29] themes.

3.1. Theme 1: Requirements Versus Practice

The literature included in this theme focusses largely on the laws, policies and rules that dictate anti-discrimination requirements within the hiring process and recognize the right of people with disability to access paid work compared to the actual hiring practices observed. Construction employment research contributes additional literature which identifies industry-specific barriers that limit the opportunity for people with disability to work in that sector in practice. For example, the traditionally homogenous (male and able-bodied) workforce of self-employed contractors, inflexible employment conditions and the practice of contractors to recruit tradesmen from their established, narrow social networks are barriers to the employment of people with disability in the construction industry in Britain, Holland and Australia [10,25,42]. The construction literature reports poor employment outcomes in practice. For example, Ref. [43]'s study reported low rates of participation in the UK construction workforce on the basis of gender, ethnicity and disability. Notably, although [43] made suggestions for increasing the representation of women and ethnic minorities, there were no suggestions for improving the employment participation rate for people with disability. Ref. [44]'s study focussed on professional transport employees and employee experiences of both invisible and visible disability types in a non-construction but comparable male-dominated industry. The employer's perspective on barriers to hiring people with disability was explored in [45]'s early Israeli research.

Only a small number of articles (all based on US and Canadian studies) demonstrate an effort to educate construction employers on their obligation not to discriminate in employment. For example, Ref. [46] described how the Americans with Disabilities Act (ADA) applies to real-life employment situations in the construction industry and explained the impact of the ADA on the hiring process. Ref. [47] developed a Construction Disabiity Management Maturity Model. Ref. [48] also delved into the implications of the ADA for the construction sector, more specifically focusing on the liabilities of engineers, as non-discriminatory employers as well as the designed outcomes of their work. Subsequent research by [49] examined cross-sector employer practices in relation to the ADA. Another study investigated the underemployment of people with disability in the construction industry in South Africa, and the role of that country's Employment Equity Act [50]. These research projects collectively demonstrate the importance of legislation in driving diverse employment practices in a compliance-based industry where competitive pressures and industry norms have not yet materialized to drive the employment of non-traditional workers [11]. Clarke and Gribling's [12]'s research reflects a new theme of social procurement research which is responding to the contractual imposition of disability employment requirements on construction supply chains. For example, Guimarães [51] explored the emergence of diverse employment requirements in the Swedish construction industry, as part of social procurement and as a tool to mitigate issues of exclusion on the job market. In Australia Loosemore et al [11] highlighted the value of cross-sector collaboration in implementing social procurement processes to find meaningful and sustainable work for people with disability (among others). This focus on collaboration across organisations (internally and externally) was reflected in Clarke and Gribling's [12] study of different strategies for reducing barriers to employment for people with disability, among other groups on the Heathrow Terminal 5 construction project in London. A group of studies of construction companies in Brazil focused on the higher rate of employment of people with physical disability under the employer quota obligations in that country [52,53]. Those studies investigated the types of accommodation for people with different types of disability working in different construction site roles in the context of the legal requirement to make reasonable accommodations in Brazil.

3.2. Theme 2: Stigma or Attitudinal Barriers

The literature in this theme includes research that explores or reports on stigma or attitudinal barriers to the employment of people with disability in the context of the hiring process and includes studies that explain the reasons behind the underemployment of people with disability. These barriers include the perceptions, assumptions, attitudes or beliefs that people with disability cannot work long hours, increase the cost of supervision and have health needs that will impact on productivity and absenteeism. Loosemore et al. [8,42] conducted research with Australian construction subcontractors, and reported that the main barriers to building a diverse and inclusive workforce include the perception of employers that marginalized groups (including people with disability) are a risk, not able to fit-in and not able to work effectively in the construction industry. People with disability suffered the third highest barriers of the six groups compared (women, refugees and migrants, Indigenous people, ex-offenders and disengaged youth) and a unique set of barriers which were different to the other groups, suggesting targeted employment strategies were necessary. This study also supported previous research which shows that many perceived barriers (such as higher costs) to employment are unfounded. For example, one study in Brazil compared the levels of absenteeism of people with disability to those without disability on construction worksites and concluded that the assumption or belief that people with disability have higher rates of absenteeism was not established [54]. Ref. [55]'s research into employment in the construction sector discussed the prejudice and discrimination experienced by people with disability in the UK. Clarke et al.'s [25] study explored the enablers and barriers to disability employment in the UK and Dutch construction sectors. The UK applies a capabilities approach that focuses on the individual's capability and the Dutch approach applies a social model of disability that views disability as socially constructed and, hence, focuses on the abilities of applicants.

Gewurtz et al.'s [29] review identified literature with a different focus, including strategies to reduce stigma like education and sharing success stories. Those authors cited research that found employers who had successfully employed a person with disability had a greater likelihood of employing other people with disability [56]. This finding is consistent with one construction study in the UK which concluded that UK contractors (among the top 100) were likely to continue to employ people who had acquired disability when employed [25], although this reflects the emphasize in construction research on people acquiring a disability by being injured in work. See for example, Clarke et al. [2] who reported regulatory initiatives in Britain and Holland for re-integrating injured employees into the construction workforce because of the high rate of injury and disability among that workforce.

3.3. Theme 3: Disclosure

The literature in this theme explores the person's experience or practice regarding the timing of disclosing disability when writing to apply for a job, during the interview or after employment. In contrast to Gewurtz et al.'s [29] review there were no construction studies regarding employee disclosure experiences or practices in the recruitment context. However, three studies discussed access to disclosed information [9,13,43] and three studies focused on the employment by construction companies of people with disability to meet employment quota obligations [52–54]. Briscoe [43] used UK Labour Force Survey data collected from people with and without disability to analyze the job/workforce share of people with disability (and other minority groups) working in the UK construction sector. There were insufficient statistics to provide an accurate understanding of the number of people with disability who work in construction because od low levels of disclosure and no information on workers who chose not to disclose their disability.

In contrast, the broader field of mainstream disability employment research provides insights into disclosure strategies of potential employees and the attitudes of employers in response [1,30]. Some of those studies reported potential employees resisting disclosure, the employee's choice or decision not to disclose invisible disability [57,58], the employer's response to disclosure in a cover letter [59,60] and the employer's negative response to late disclosure including disclosure at the end of the interview [58,61,62]. This is a gap for future construction disability employment research to address.

3.4. Theme 4: Reasonable Accommodations

The literature in this theme analyses the legal requirement and practices to offer and implement reasonable accommodations in the workplace, including during the hiring process [48]. Reasonable accommodations are referred to as reasonable adjustments under the law in some countries (like Australia) and some governments provide financial assistance for employers to make reasonable accommodations or adjustments to equipment or the work environment. Our scoping review highlighted a significant proportion of studies in this area. For example, a recent study by McCall and Simmons [17] explored the opportunities brought by new technologies to support more inclusive and productive workplaces in construction. A group of Brazilian studies investigated adaption and accommodation types for people with different disabilities to support them performing a range of construction labour roles on site [52–54]. Clarke and Gribling's [12] case study of Terminal 5 construction at Heathrow Airport documents the role of accommodations to retain workers on one of the largest construction sites in Europe while Newton and Ormerod [24] found that contractors in the UK construction sector were more likely to make the required adjustments to adapt workplaces for employees who acquired a disability if already employed.

In contrast to the construction literature, the mainstream disability employment literature reported by Gewurtz et al [29] notes the frequent lack of knowledge of employers of the requirement to provide reasonable accommodations and their lack of knowledge that accommodations are frequently nil or low cost [29]. The research also identified the belief of some employers that providing accommodations for the hiring process or the workplace creates financial and legal risks [29].

3.5. Theme 5: Relationship Building and Use of Disability Organisations

This theme includes studies that explore building relationships between employers and disability organizations that specialize in placing people with disability into jobs. We mapped three of the included articles to this theme [8,12,25]. However, we did not find the topic was prominent in the construction literature other than in relation to the recent emergence of social procurement as an approach to increase workforce diversity through new cross-sector collaborations between construction organizations and disability support organizations [9]. Interestingly, these collaborations were reported as immature and problematic due to different institutional drivers and constraints. This contrasts with Gewurtz et al.'s [29] review which revealed considerable emphasis in this area of research, leading to their conclusion that conclusion that "building relationships between community employers and disability organizations that specialize in placing people with disability into a job is critical to the hiring process" (p. 141). Indeed, Gewurtz et al. [29] identified that the relationship between the employer and these organizations was a factor associated with the likelihood of employment and appropriate supports in the workplace; and prospective employers could ask these organizations about disability and the employer's legal obligations to make accommodations. However, there were also concerns that some disability organizations do not provide a person-centered employment service, only help people with particular supports and not all disabilities, or give priority to the employers' interests [63]. This was seen as an important area of future research.

3.6. Theme 6: Information and Support to Employers

Research included in this theme reported on the provision of information and support for employers to improve their hiring practices and employment opportunities for people with disability. Loosemore et al. [8] reported on an innovative project-based intermediary in Australia which has been developed by a major contractor to reduce information asymmetries between the construction industry and the organizations which specialize in providing employment support for people with disability. It did this by providing important practical training, information and support to both job seekers and employers in the construction supply chain about the employment of marginalized groups like people with a disability. Collaboration between construction contractors and disability employment service providers was key to this support, reducing complexity and perceived risks for employers previously reluctant to employ people with disability. This research built on the collaborative theme of earlier research by [64]'s who reported on the opportunity for employers in the construction industry to employ and use the knowledge and expertise of people with disability across all aspects of the construction process including the design and planning stages of architectural projects. However, it should be noted that the intermediary analyzed by Loosemore et al. [8] was a unique single case study and the only one that could be found internationally by the authors. Innovation in this area therefore appears to be scant and Gewurtz et al. [29] also acknowledged the needs of employing organizations can best be met when the needs of the employer are understood based on a larger number of studies.

3.7. Theme 7: Hiring Practices That Invite People with Disability

The literature included in this theme reported on the importance of hiring practices that invite job seekers with disability to apply for jobs including organizational strategies to ensure the hiring process is accessible, organizational policies that support the employment of people with disability, and the broader study of organizational culture to build a culture of respect towards candidates and employees with disability. In our review, we identified six relevant articles in the construction sector, although the employment outcomes for those workers and their subsequent career trajectory is not reported. These papers included Loosemore et al.'s [8] case study of social procurement in Australia highlighted the importance of project-based intermediaries to bypass traditional recruitment processes which exclude people with disability being recruited into construction. Two other papers recommended changes in hiring practices to assist employers comply with the employment discrimination and accommodation provisions of the ADA when that US law was introduced [46,48]. Maroto and Pettinicchio's [65] chapter concluded that employer attitudes to hiring people with disability in the UK are improved when employers themselves have worked with a person with disability, and that it is a responsibility of the construction sector to support people with disability in pathways to employment (as well as supporting them when employed including career development opportunity). Clarke and Gribling [12] identified that existing recruitment practices in the construction sector in the UK were a major barrier to more inclusive recruitment of people in the community, from diverse backgrounds including people with disability. Ians et al.'s [57] survey of the top 100 UK contractors identified their lack of organisational policies and strategies of construction contractors to support people with disability through the hiring process, or provide evidence those contractors were complying with their employer obligations under the then UK Disability Discrimination Act 1995.

Overall, our analysis shows that while there is an established body of knowledge pointing to the role of organizational culture, strategies for inviting people with disability to apply for work and tools used to build and retain a diverse workforce (budget allocation, co-worker training and CEO commitment and leadership on diversity, etc.) [56,65,66], there has been little or no equivalent sector-specific research exploring how construction sector culture, organizational structures or other factors (job design, workforce planning, the managers or relationships that control the hiring process, recognition of jobs in technical and/or management roles, and allocation of jobs or even government policy, etc.) may be exclusionary, create work disincentives and contribute to current low levels of employment of people with disability [67].

There has also been little or no construction research studying hiring strategies and practices deployed for increasing the employment of people with disability in ways that can be made equitable whilst considering career access points, quality of work and career progression from the employee's perspective. Furthermore, while construction researchers have explored the employment barriers facing people with disability in construction, there is a clear contrast with the considerable research having been undertaken in other fields investigating strategies to navigate and reduce the barriers experienced by people with disability in gaining employment [30,65,68,69]. Another related gap in research which has received little attention in construction is the reporting of the effectiveness of these strategies in achieving improved employment outcomes for people with disability.

Connections have been made between technological advancement in the construction sector, and an increase in accessibility of construction workplaces and activities [17]. This opportunity to take advantage of technological and work changes has not yet been fully explored—one example being off-site construction. There is evidence within the literature that employers and agencies do not believe people with disability are suited for many construction professions [70]. There is also evidence of ableist practices within the sector that perceive people with disability as a homogenous group rather than a heterogenous group. An example of this is that the work environment for construction workers is often cited as a reason that all people with disability are less likely to be considered for work in the sector [25,43]. While there has been a considerable body of research into the benefits of offsite construction such as [71] there is a timely opportunity to address the gap in research of how off-site construction can support more inclusive employment, by better recognizing the diversity of disability and providing more-accessible work environments that can integrate assistive technology into construction work practice.

There is also a gap in construction research regarding disclosure, both from the perspective of people with disability and employers. Studies have identified that employees are reluctant to disclose any form of disability because of stigmas surrounding this and that employers are uncomfortable asking about an applicant's disability in the employment process making it hard to assess whether the person can do the job [30]. Employers seem to be uncertain about communicating necessary inherent requirement of jobs at the beginning of the application process and their right to ask all applicants how they would be able to perform the role.

Finally, the role of cross-sector collaboration, project-based intermediaries and organizational champions is also worthy of further investigation, especially within the context of emerging social procurement policies which mandate the employment of people with disability as part of project contractual requirements [8,51]. Such intermediaries are designed to encourage employment of people with disability within established supply chains. However, since the goals of social procurement is also to encourage supply chain diversity [72], the role of minority disability business and social enterprises merits particular attention in increasing employment of people with disability indirectly within the industry. Recent work on self-employment and entrepreneurship of people with disability, identified opportunity for the construction industry to draw on the skill and ability base of allied technical, professional and management businesses [73]. Questions revolve around whether they represent a sustainable long-term solution to diversity in construction or whether the focus should be on incumbent supply chains [74]. One of the differentiating characteristics of construction is its project-based production environment and the way that organizational champions and intermediaries can facilitate collaboration with specialist organizations which support people with disability could make a significant contribute to knowledge in this nascent field of research and practice.

4. Conclusions

The aim of this paper was to address the lack of social innovation research in construction relating the employment of people with a disability in the industry. The objective was to identify knowledge gaps in comparison to mainstream disability employment research and to clarify concepts to inform future research in this important yet under-explored area. Such research is important in informing social innovation practice, considering that the construction industry is experiencing severe skills shortages combined with a lack of diversity within the workforce population.

Our scoping review of the limited and fragmented academic evidence relating to the inclusion of people with disability in the construction industry workforce shows that it remains a relatively under-explored and under-theorized field and has identified significant gaps between research in construction disability employment compared with research in the broader field of mainstream disability employment. Overall, our analysis shows that while there is some research in construction relating to barriers to employment and stigmas surround people with disability, there is far less research relating to the enablers which can overcome these barriers. While there has been some research into accommodations which need to be made for employing people with a disability (especially around technological developments such as off-site prefabrication), there is a paucity of research in construction on relationship-building and cross-sector collaboration with support agencies, the role of social procurement and social enterprises and disability organizations in supporting disability employment, the provision of information to support employers to reduce ingrained stigmas and insights in to how to reduce biases and inequalities in highly traditional construction recruitment practices. In order to support social innovation in this area, construction research needs to balance its current emphasis on barriers (seeing people with disability as a risk) with equivalent research on solutions (seeing people with disability as an opportunity in a constrained labor market and because of their ability).

It has also become evident from our review that none of the extant research explores the employment cycle or addresses the project-based, itinerant and casual nature of construction employment and its impact on the quality work for people with disability. Therefore, further research is also needed to understand the informal factors influencing inclusive employment strategies and the identification of barriers within the workplace culture itself. It is clear from our review that despite the implementation of numerous 'formal' laws and regulations relating to inclusive employment for people with disability, there are numerous under-researched 'informal' and unwritten industry norms and practices which can potentially undermine the intent of these policies. This highlights the potentially valuable role of New Institutional Theory, employed successfully in construction gender studies and social procurement [11,14] in better understanding the interactions between formal and informal norms and practices in the construction industry which can undermine the intent of formal policies to employ people with disability. The results also reflect on the additional responsibility of the construction sector to engage with inclusive employment strategies given the high level of disabling injuries experienced by the construction workforce internationally.

The scoping review also identifies a methodological gap in the research reviewed by highlighting the need for more construction research designs to include people with disability as prioritized research participants as well as research investigators and to adopt phenomenological and interpretive approaches which respect the lived experiences of people with a disability seeking work in the construction industry. While the value of a phenomenological methodology has recently been acknowledged by in the area of housing design for neurodiversity [75], such research is rare in the field of construction, which has a long tradition of positivist research methodologies which tend to subjugate research participants. The lack of longitudinal research including success stories in construction employment and long-term employment outcomes in the construction industry are also methodological gaps for future construction researchers to address. The important role of these success stories is emphasized by Lundberg [76].

If the sector hopes to innovate in finding new employment pathways into construction for people with disability, it is critical that research investigates the perspectives of people with disability currently employed in the sector, and the large numbers of people who have been disabled as a result of working in the sector and who are working or no longer able to work in the sector. Understanding the how people with disability perceive the sector from the inside will help build successful recruitment strategies for people from the outside.

In conclusion, given the extent of the research gaps identified we recommend that there are twelve fundamental research questions that should be prioritized in finding innovative new pathways for people with a disability into construction, and thereby move this nascent field of social innovation research forward:

- What are the open employment opportunities in construction for people with disability? This includes exploration of the potential for types of roles, heterogeneous nature of disability, and in what parts of the construction sector?
- In what ways can the construction sector draw on open employment and disability employment services to improve the integration of recruitment of people with disability coming out of high school, TAFE and university education?
- What is the scope and quality of work that people with disability can be considered for in the sector?
- How can equality of opportunity in recruitment and career progression be supported for people with a disability?
- What are the barriers to the employment 'experienced' by people with disability in construction?
- What are the 'lived experiences' of people with disability when seeking work and in employment across all parts of the strategic human resource management process?
- How do employers improve the opportunities for construction employment for people with disability through innovative collaborative cross-sector partnerships and facilitative like project-based intermediaries?
- What is the role of self-employed professionals, social enterprises and disability enterprises in providing sustainable long-terms solutions to disability employment in construction?

- How does the industry reverse existing perceptions of risk associated with disability employment in the existing construction supply chain and make them a perceived opportunity for new labor supply?
- What role can off-site manufacturing play in providing more accessible workplace settings?
- How are the productivity benefits of disability assistive technology promoted to employers to challenge the conventional wisdom of lower productivity and higher costs employing people with disability?

Author Contributions: Conceptualization, S.B., P.C., M.L., S.D. and S.S.; methodology, S.B., P.C., M.L., S.D. and S.S.; validation, S.B., P.C., M.L., S.D. and S.S.; formal analysis, S.B., P.C., M.L. and S.D.; investigation, S.B. and P.C.; resources, S.B., P.C., M.L., S.D. and S.S.; data curation, S.B., P.C., M.L., S.D. and S.S.; writing—original draft preparation, S.B., P.C., M.L. and S.D.; writing—review and editing, S.B., P.C., M.L. and S.D.; project administration S.S.; funding acquisition, S.B., P.C., M.L., S.D. and S.S. All authors have read and agreed to the published version of the manuscript.

Funding: This research was funded by a University of Technology Social Impact seed grant.

Data Availability Statement: Not applicable.

Conflicts of Interest: The authors declare no conflict of interest.

References

1. Abbot, C.; Jeong, K.; Allen, S. The economic motivation for innovation in small construction companies. *Constr. Innov.* **2006**, *6*, 187–196. [CrossRef]
2. Abadi, A. A Study of Innovation Perception within the Construction Industry. Ph.D. Thesis, Department of Mechanical, Aerospace & Civil Engineering, University of Manchester, Manchester, England, 2014.
3. Adafin, J.; Wilkinson, S.; Rotimi, J.O.B.; MacGregor, C.; Tookey, J.; Potangaroa, R. Creating a case for innovation ac-celeration in the New Zealand building industry. *Constr. Innov.* **2022**, *22*, 185–204. [CrossRef]
4. Foroudi, P.; Akarsu, T.N.; Marvi, R.; Balakrishnan, J. Intellectual evolution of social innovation: A bibliometric analysis and avenues for future research trends. *Ind. Mark. Manag.* **2020**, *93*, 446–465. [CrossRef]
5. Martins, T.; Braga, A.; Ferreira, M.R.; Braga, V. Diving into Social Innovation: A Bibliometric Analysis. *Adm. Sci.* **2022**, *12*, 56. [CrossRef]
6. OECD. *Social Innovation*; The Organisation for Economic Co-Operation and Development: Paris, France, 2022. Available online: https://www.oecd.org/regional/leed/social-innovation.htm (accessed on 9 September 2022).
7. Watts, G.; Fernie, S.; Dainty, A. Paradox and legitimacy in construction: How CSR reports restrict CSR practice. *Int. J. Build. Pathol. Adapt.* **2019**, *37*, 231–246. [CrossRef]
8. Loosemore, M.; Alkilani, S.; Mathenge, R. The risks of and barriers to social procurement in construction: A supply chain perspective. *Constr. Manag. Econ.* **2019**, *38*, 552–569. [CrossRef]
9. Raiden, A.; Loosemore, M.; King, A.; Gorse, C. *Social Value in the Construction Industry*; Routledge: London, UK, 2019; ISBN 78-1-138-29510-0.
10. Loosemore, M.; Higgon, D.; Osborne, J. Managing new social procurement imperatives in the Australian con-struction industry. *Eng. Constr. Archit. Manag.* **2020**, *27*, 3075–3093. [CrossRef]
11. Loosemore, M.; Alkilani, S.; Murphy, R. The institutional drivers of social procurement implementation in Australian construction projects. *Int. J. Proj. Manag.* **2021**, *39*, 750–761. [CrossRef]
12. Clarke, L.; Gribling, M. Obstacles to diversity in construction: The example of Heathrow Terminal 5. *Constr. Manag. Econ.* **2008**, *26*, 1055–1065. [CrossRef]
13. Dainty, A.; Loosemore, M. *Human Resource Management in Construction: Critical Perspectives*; Routledge: London, UK, 2013.
14. Galea, N.; Powell, A.; Loosemore, M.; Chappell, L. Designing robust and revisable policies for gender equality: Lessons from the Australian construction industry. *Constr. Manag. Econ.* **2015**, *33*, 375–389. [CrossRef]
15. Galea, N.; Powell, A.; Loosemore, M.; Chappell, L. The gendered dimensions of informal institutions in the Australian construction industry. *Gender Work. Organ.* **2020**, *27*, 1214–1231. [CrossRef]
16. McCall, C.; Simmons, D.R. Exploring Disciplinary Technologies for Increased Accessibility in the Civil Engineering and Construction Industry: Starting the Conversation. In *Advances in the Human Side of Service Engineering. AHFE 2021*; Leitner, C., Ganz, W., Satterfield, D., Bassano, C., Eds.; Lecture Notes in Networks and Systems; Springer: Cham, Switzerland, 2021; Volume 266.
17. Carnemolla, P.; Galea, N. Why Australian female high school students do not choose construction as a career: A qualitative investigation into value beliefs about the construction industry. *J. Eng. Educ.* **2021**, *110*, 819–839. [CrossRef]
18. Denny-Smith, G.; Williams, M.; Loosemore, M. Assessing the impact of social procurement policies for Indigenous people. *Constr. Manag. Econ.* **2020**, *38*, 1139–1157. [CrossRef]

19. Al-Bayati, A.J. Satisfying the Need for Diversity Training for Hispanic Construction Workers and Their Supervisors at US Construction Workplaces: A Case Study. *J. Constr. Eng. Manag.* **2019**, *145*, 5019907. [CrossRef]
20. Loosemore, M.; Alkilani, S.; Hammad, A. 'Barriers to employment for refugees seeking work in the Australian con-struction industry: An exploratory study' in Engineering. *Constr. Archit. Manag.* **2021**, *38*, 552–569. [CrossRef]
21. Infrastructure Australia. *2021 Infrastructure Market Capacity Report*; Infrastructure Australia: Sydney, Australia, 2021.
22. Chartered Institute of Building. Future of Construction: Equality, Diversity and Inclusion. 2022. Available online: https://www.ciob.org/industry/policy-research/policy-positions/equality-diversity-inclusion (accessed on 3 March 2022).
23. Fritsch, J. *Yes, Wheelchair Users Can Work in Construction. Let's See This as an Opportunity*; The FifthEstate: Sydney, Australia, 2020. Available online: https://thefifthestate.com.au/innovation/building-construction/yes-wheelchair-users-can-work-in-construction-lets-see-this-as-an-opportunity/ (accessed on 2 September 2022).
24. Newton, R.; Ormerod, M. Do disabled people have a place in the UK construction industry? *Constr. Manag. Econ.* **2005**, *23*, 1071–1081. [CrossRef]
25. Clarke, L.; Van Der Meer, M.; Bingham, C.; Michielsens, E.; Miller, S. Enabling and disabling: Disability in the British and Dutch construction sectors. *Constr. Manag. Econ.* **2009**, *27*, 555–566. [CrossRef]
26. Quaigrain, R.A.; Winter, J.; Issa, M.H. A critical review of the literature on disability management in the construction industry. In Proceedings of the 30th Annual ARCOM Conference, Portsmouth, UK, 1–3 September 2014; Raiden, A., Aboagye-Nimo, E., Eds.; Association of Researchers in Construction Management: Portsmouth, UK, 2014; pp. 1121–1130.
27. Quaigrain, R.A.; Issa, M.H. Development and validation of disability management indicators for the construction industry. *J. Eng. Des. Technol.* **2018**, *16*, 81–100. [CrossRef]
28. World Health Organization. *World Report on Disability*; World Health Organization: Geneva, Switzerland; The World Bank: Washington, DC, USA, 2011.
29. Gewurtz, R.E.; Langan, S.; Shand, D. Hiring people with disabilities: A scoping review. *Work* **2016**, *54*, 135–148. [CrossRef]
30. Bonaccio, S.; Connelly, C.E.; Gellatly, I.R.; Jetha, A.; Ginis, K.A.M. The Participation of People with Disabilities in the Workplace Across the Employment Cycle: Employer Concerns and Research Evidence. *J. Bus. Psychol.* **2019**, *35*, 135–158. [CrossRef]
31. Darcy, S.; Taylor, T.; Green, J. 'But I can do the job': Examining disability employment practice through human rights complaint cases. *Disabil. Soc.* **2016**, *31*, 1242–1274. [CrossRef]
32. Winter, J.; Issa, M.; Quaigrain, R.; Dick, K.; Regehr, J. Evaluating disability management in the Manitoban construction industry for injured workers returning to the workplace with a disability. *Can. J. Civ. Eng.* **2016**, *43*, 109–117. [CrossRef]
33. Won, D.; Hwang, B.; Chang, J. Assessing the effects of workforce diversity on project productivity performance for sustainable workplace in the construction industry. *Sustain. Dev.* **2020**, *29*, 398–418. [CrossRef]
34. Munn, Z.; Peters, M.D.J.; Stern, C.; Tufanaru, C.; McArthur, A.; Aromataris, E. Systematic Review or Scoping Review? Guidance for Authors When Choosing between a Systematic or Scoping Review Approach. *BMC Med. Res. Methodol.* **2018**, *18*, 143. [CrossRef]
35. Page, M.J.; Moher, D.; McKenzie, J.E. Introduction to PRISMA 2020 and implications for research synthesis methodologists. *Res. Synth. Methods* **2021**, *13*, 156–163. [CrossRef] [PubMed]
36. Carnemolla, P.; Skinner, V. Outcomes Associated with Providing Secure, Stable, and Permanent Housing for People Who Have Been Homeless: An International Scoping Review. *J. Plan. Lit.* **2021**, *36*, 508–525. [CrossRef]
37. Koeman, J.; Mehdipanah, R. Prescribing Housing: A Scoping Review of Health System Efforts to Address Housing as a Social Determinant of Health. *Popul. Heal. Manag.* **2021**, *24*, 316–321. [CrossRef]
38. Campbell, D.; Pickard-Aitken, M.; Côté, G.; Caruso, J.; Valentim, R.; Edmonds, S.; Williams, G.T.; Macaluso, B.; Robitaille, J.-P.; Bastien, N.; et al. Bibliometrics as a Performance Measurement Tool for Research Evaluation: The Case of Research Funded by the National Cancer Institute of Canada. *Am. J. Eval.* **2010**, *31*, 66–83. [CrossRef]
39. Podsakoff, P.M.; MacKenzie, S.B.; Bachrach, D.G.; Podsakoff, N.P. The influence of management journals in the 1980s and 1990s. *Strat. Manag. J.* **2005**, *26*, 473–488. [CrossRef]
40. Hayfield, N.; Huxley, C.J. Insider and Outsider Perspectives: Reflections on Researcher Identities in Research with Lesbian and Bisexual Women. *Qual. Res. Psychol.* **2014**, *12*, 91–106. [CrossRef]
41. Wilkinson, A.; Johnstone, S.; Townsend, K. Changing patterns of human resource management in construction. *Constr. Manag. Econ.* **2012**, *30*, 507–512. [CrossRef]
42. Loosemore, M.; Reid, S. The social procurement practices of tier-one construction contractors in Australia. *Constr. Manag. Econ.* **2018**, *37*, 183–200. [CrossRef]
43. Briscoe, G. Women and minority groups in UK construction: Recent trends. *Constr. Manag. Econ.* **2005**, *23*, 1001–1005. [CrossRef]
44. Sang, K.J.; Richards, J.; Marks, A. Gender and Disability in Male-Dominated Occupations: A Social Relational Model: Gender and Disability in Male-Dominated Occupations. *Gend. Work. Organ.* **2016**, *23*, 566–581. [CrossRef]
45. Florian, V. Objective obstacles in hiring disabled persons–the employers' point of view. *Rev. Int. Rech. Readapt.* **1981**, *4*, 167–174.
46. Anderson, S.D.; Morgan, C.F.; Goel, M. Americans with Disabilities Act and Employment Practices in Construction. *J. Prof. Issues Eng. Educ. Pract.* **1994**, *120*, 360–378. [CrossRef]
47. Quaigrain, R.A.; Issa, M.H. Construction disability management maturity model: Case study within the Manitoban construction industry. *Int. J. Work. Health Manag.* **2021**, *14*, 274–291. [CrossRef]
48. Patterson, C. Engineers and ADA. *Civ. Eng.* **1995**, *65*, 73–75.

49. Bruyere, S.M. *Disability Employment Policies and Practices in Private and Federal Sector Organizations*; Cornell University Program on Employment and Disability School of Industrial and Labor Relations Extension Division: Ithaca, NY, USA, 2000. Available online: https://ecommons.cornell.edu/bitstream/handle/1813/89793/DA4_PDF1.pdf?sequence=1 (accessed on 10 September 2022).
50. Tshobotlwane, D.M.; Haupt, T.C.; Chileshe, N. An Empirical Study of the Factors Affecting the Employment of Disabled Persons within the South African Construction Industry. In Proceedings of the Annals of CIB W99 International Conference on Global Unity for Safety & Health in Construction, Beijing, China, 28–30 June 2006; pp. 457–468.
51. Troje, D.; Gluch, P. Populating the social realm: New roles arising from social procurement. *Constr. Manag. Econ.* **2020**, *38*, 55–70. [CrossRef]
52. Guimarães, B.; Martins, L.; Junior, B.B. Workplace Adaptation of People with Disabilities in the Construction Industry. *Procedia Manuf.* **2015**, *3*, 1832–1837. [CrossRef]
53. Martins, L.; Junior, B.B.; Guimarães, B. Including the people with disabilities at work: A case study of the job of bricklayer in civil construction in Brazil. *Work* **2012**, *41*, 4716–4721. [CrossRef] [PubMed]
54. Guimarães, B.; Junior, B.B.; Martins, L. Absenteeism of people with disabilities in the construction industry in Brazil. *Work* **2018**, *60*, 411–419. [CrossRef] [PubMed]
55. Ormerod, M.; Newton, R. Embracing diversity through the employment of disabled people. A missed opportunity. In *Managing Diversity and Equality in Construction: Initiatives and Practice*; Gale, A.W., Davidson, M.J., Eds.; Routledge: London, UK, 2006.
56. Wiggett-Barnard, C.; Swartz, L. What facilitates the entry of persons with disabilities into South African companies? *Disabil. Rehabil.* **2012**, *34*, 1016–1023. [CrossRef]
57. Jans, L.H.; Kaye, H.S.; Jones, E.C. Getting Hired: Successfully Employed People with Disabilities Offer Advice on Disclosure, Interviewing, and Job Search. *J. Occup. Rehabilitation* **2011**, *22*, 155–165. [CrossRef] [PubMed]
58. Dalgin, R.S.; Bellini, J. Invisible disability disclosure in an employment interview: Impact on employers' hiring deci-sions and vies of employability. *Rehabil. Couns. Bull.* **2008**, *52*, 6–15. [CrossRef]
59. Bishop, M.; Stenhoff, D.M.; Bradley, K.D.; Allen, C.A. The differential effect of epilepsy labels on employer perceptions: Report of a pilot study. *Epilepsy Behav.* **2007**, *11*, 351–356. [CrossRef]
60. Pearson, V.; Lo, E.; Yip, N. To tell or not to tell; disability disclosure and job application outcomes. *J. Rehabil.* **2003**, *69*, 35.
61. Gold, P.B.; Oire, S.N.; Fabian, E.S.; Wewiorski, N.J. Negotiating reasonable workplace accommodations: Perspectives of employers, employees with disabilities, and rehabilitation service providers. *J. Vocat. Rehabil.* **2012**, *37*, 25–37. [CrossRef]
62. Hebl, M.R.; Skorinko, J.L. Acknowledging one's physical disability in the interview: Does "when" make a difference? *J. Appl. Soc. Psychol.* **2005**, *5*, 2477–2492. [CrossRef]
63. Wren, M. *The Ten Demandments: How to Improve Employment Services for People with Disability*; Martin Wren: St Marys, Australia, 2016.
64. Ormerod, M.; Newton, R. Construction as a career choice for young disabled people: Dispelling the myths. *Constr. Manag. Econ.* **2013**, *31*, 928–938. [CrossRef]
65. Maroto, M.; Pettinicchio, D. Disability, structural inequality, and work: The influence of occupational segregation on earnings for people with different disabilities. *Res. Soc. Strat. Mobil.* **2014**, *38*, 76–92. [CrossRef]
66. Erickson, W.A.; von Schrader, S.; Bruyère, S.M.; VanLooy, S.A.; Matteson, D.S. Disability-Inclusive Employer Practices and Hiring of Individuals With Disabilities. *Rehabil. Res. Policy Educ.* **2014**, *28*, 309–328. [CrossRef]
67. Dainty, A.; Green, S.; Bagilhole, B. People and culture in construction: Contexts and challenges. In *People and Culture in Construction*; Routledge: London, UK, 2007; pp. 21–43.
68. Erickson, W.A.; von Schrader, S.; Bruyère, S.M.; VanLooy, S.A. The Employment Environment: Employer Perspectives, Policies, and Practices Regarding the Employment of Persons With Disabilities. *Rehabil. Couns. Bull.* **2014**, *57*, 195–208. [CrossRef]
69. Lindstrom, L.; Doren, B.; Miesch, J. Waging a Living: Career Development and Long-Term Employment Outcomes for Young Adults with Disabilities. *Except. Child.* **2011**, *77*, 423–434. [CrossRef]
70. Duncan, R.; Neale, R.; Bagilhole, B. Equal Opportunities and the Reconciliation of Work and Family within the Construction Process. *Management* **1999**, *1*, 171–180.
71. Alazzaz, F.; Whyte, A. Linking employee empowerment with productivity in off-site construction. *Eng. Constr. Arch. Manag.* **2015**, *22*, 21–37. [CrossRef]
72. Barraket, J.; Keast, R.; Furneaux, C. *Social Procurement and New Public Governance*; Routledge: London, UK, 2015.
73. Darcy, S.; Collins, J.; Stronach, M. Entrepreneurs With Disability: Australian Insights Through a Social Ecology Lens. *Small Enterp. Res.* **2022**, *29*, 1–25. [CrossRef]
74. Loosemore, M. Building a new third construction sector through social enterprise. *Constr. Manag. Econ.* **2015**, *33*, 724–739. [CrossRef]
75. Day, K.; Martel, A. Designing for Neurodiversity: Reimagining the Home for a COVID Normal Life. In Proceedings of the 37th Annual ARCOM Conference, Leeds, UK, 6–7 September 2021; Scott, L., Neilson, C.J., Eds.; Association of Researchers in Construction Management: Portsmouth, UK, 2021; pp. 67–76.
76. Lundberg, C.S. Is There an Ableist Paradox in Frontline Workers' Success Stories About Disability and Work Inclusion? *Scand. J. Disabil. Res.* **2022**, *24*, 1–14. [CrossRef]

buildings

MDPI

Article

Evaluation of Park Accessibility Based on Improved Gaussian Two-Step Floating Catchment Area Method: A Case Study of Xi'an City

Yunmei Li [1,*], Yuanli Xie [1,2,*], Shaoqi Sun [1] and Lifa Hu [1]

[1] College of Urban and Environmental Sciences, Northwest University, Xi'an 710127, China; sean20210511@gmail.com (S.S.); hlf_2019@163.com (L.H.)
[2] Shaanxi Key Laboratory of Earth Surface System and Environmental Carrying Capacity, Northwest University, Xi'an 710127, China
* Correspondence: liyunmei2020@163.com (Y.L.); xieyuanli@126.com (Y.X.)

Abstract: Park accessibility plays a critical role in evaluating the quality of park construction. However, the conventional accessibility model ignores non-spatial factors, so it is crucial to use more complex methods for evaluating park accessibility. This study aims to establish an improved Gaussian-based two-step floating catchment area method (iG2SFCA) based on Point of Interest (POI), population data and Baidu map, to measure the park accessibility of various travel modes (walking, riding and driving modes) in 5-min, 15-min and 30-min scenarios, and location quotient is used to assess spatial equity of parks. The results show that: (1) There are clear disparities between park supply and population demand at the street level. (2) iG2SFCA evaluates the level and attractiveness of the park comprehensively. It is more sensitive to identifying accessibility, which can lead to a more realistic assessment of Park accessibility. (3) Under the three modes of transportation, the accessible area of the park increases with time, and the accessibility difference between residential areas is the smallest under the 30-min scenario. Overall, accessibility of park is relatively high; however, there is an obvious tendency for the accessibility level to decrease from the park as the center. The areas with poor accessibility appear in the north and southeast of the research area. (4) There are regional variances in the spatial equity of parks within Xi'an 3 City, and the park configuration needs to be optimized. These findings can provide theoretical support for further optimizing the layout of park in Xi'an in order to improve the spatial equity of urban parks.

Keywords: park; accessibility; Gaussian based two-step floating catchment area (2SFCA) method; big data; Xi'an

Citation: Li, Y.; Xie, Y.; Sun, S.; Hu, L. Evaluation of Park Accessibility Based on Improved Gaussian Two-Step Floating Catchment Area Method: A Case Study of Xi'an City. *Buildings* **2022**, *12*, 871. https://doi.org/10.3390/buildings12070871

Academic Editors: Yongjian Ke, Jingxiao Zhang and Simon P. Philbin

Received: 16 May 2022
Accepted: 16 June 2022
Published: 21 June 2022

1. Introduction

As an important component of ecological products, urban parks serve not only ecological functions but also social functions, which can help cities enhance their environmental quality [1]. Numerous studies have proven that parks can improve physical health, reduce psychological stress [2–4], boost communication among residents [5], and improve residents' well-being [6]. In addition, as a third space, parks can meet the daily needs of inhabitants' for leisure activities, and positively promote the harmonious development of the urban social environment.

Park accessibility, a critical criterion for determining whether the layout of a park is balanced, is significant to the close integration of ecological civilization construction and urban development. Accessibility was originally presented as a notion in transportation geography, defined as the extent to which two nodes in a transportation network can communicate [7], and was later introduced into human geography [8]. Park accessibility refers to the proximity of a residential area to a park, in other words, residents' capacity to overcome travel expenses (time and distance) in order to visit the park [9].

A number of studies have been conducted on measurement methods of park accessibility. Early, measurement techniques included the ratio methods [10] and buffer analysis methods [8]. The computation approach is simple and intuitive, as well as easy to implement in urban planning. The widespread application of GIS encourages the growth of the cost weighted distance model [11], the minimal nearest neighbor distance method [12], and the network analysis method [13,14]. By establishing a particular resistance value, this approach may represent the cost of reaching the park from various destinations depending on the actual road network and can accurately portray the park's accessibility, assuming the essential data are complete and the computational capability permits. Due to the fact that the potential model and the two-step floating catchment area (2SFCA) take into account the supply capacity of the park and the demand of residents, it has become a widely used calculation method. However, there are still some issues in the 2SFCA: while using the binary division approach to determine the search range according to the threshold, the search beyond the domain is fully inaccessible; all supply points in the search domain have the same attraction; ignoring the distance attenuation influence on demand points [15]. A number of improved models have been proposed in response to these flaws: kernel density two-step moving search method (K2SFCA) [16], variable width search method (VFCA) [17], three-step floating catchment area method (3SFVA) [18], and G2SFCA [19]. Among them, the G2SFCA is a more scientific accessibility measurement method by introducing the Gauss attenuation function to fit the changing relationship between park attractiveness and distance, which conforms to the travel characteristics of residents and considers the supply scale and population demand of the park. However, it did not take into account non-spatial factors. Therefore, to overcome shortcomings in classic accessibility models, several scholars incorporated new criteria gauging a park's attractiveness. Dony [20] assessed park accessibility by looking at the park size and on-site amenities. Xing [21] incorporated park size and function into the 2SFCA model in order to provide a more accurate assessment of park accessibility. According to several surveys, even if it is not the closest park, residents choose high-quality parks under the influence of non-spatial factors [22–26]. However, previous research did not adequately consider the park's quality, research that incorporates park level and park attractions into accessibility models is rare.

Most previous studies depend on census data at a sub-district level [27–29], or divide the research region into grids and distribute the population to each grid evenly. In recent years, the use of big data to modify traditional processed data has emerged as a new trend in geographic research [30], allowing for a finer scale of research. Guo [31,32] uses mobile phone signaling data to estimate area population density, but the cost of acquiring data is too high. The data on the number of households by the Anjuke platform can reflect the static population distribution and the precise demand for parks. In addition, the majority of traffic trip data are derived from network analysis results, but it is difficult to collect complete basic data. At the moment, major map providers can improve route time predictions for multiple means of transport, which has been establish by relevant research as accurate and reliable [33,34].

Therefore, Xi'an was chosen as a research area, and the iG2SFCA was proposed in this study using the Baidu map route planning interface based on the POI and population data. This paper aims to: (1) Compare two methods before and after improvement to test the practicability of the iG2SFCA (2) explore the accessibility of the park in Xi'an in 5-min, 15-min, and 30-min scenarios, under three trip modes (walking, cycling, and driving mode); (3) study the spatial equality of park distribution in Xi'an. This paper is structured as follows: The following section introduces the study area and data preparation. The third part introduces the G2SFCA and explains the iG2SFCA. The fourth section illustrates the assessment results. Specifically, the comparison between two methods and the accessibility features in various modes. The final section summarizes and discusses the findings of this research.

2. Materials and Methods

2.1. Study Area

Xi'an, the only mega city in northwest China, is the capital city of Shaanxi province. It possesses distinct historical and cultural genes and is home to numerous heritage parks. The city was named "National Forest City" in 2016. By 2020, the permanent population had reached almost 12.96 million, with 7.61 million urban residents. There are 117 parks in all, covering a total area of 3490.53 hm^2. The built-up area has a green coverage rate of 39.32%. This study focuses on the area within the Xi'an City Ring Expressway, which encompasses six districts in the main city that serves as center area of the permanent population distribution.

2.2. Data and Preprocessing

In accordance with the definition in "Park Design Specification", the park data and the area of interest (AOI) in the Amap were collected and trimmed based on the 2 m image of Xi'an in 2019. There are 79 parks in all within the research area. Due to the length and narrowness of the Tang City Wall Heritage Park and the Ring Park, they were separated into sections in line with the main road, and serial numbers were applied to distinguish them so as to facilitate subsequent research. Ultimately, a total of 100 parks were recognized. According to the City Park Classification Standard DBJ61/T110-2015 and the green space system planning reports of Xi'an by urban planning department parks are divided into comprehensive park, theme park and community park, considering the grade, scale and facilities of parks. The spatial distribution of various parks is shown in Figure 1.

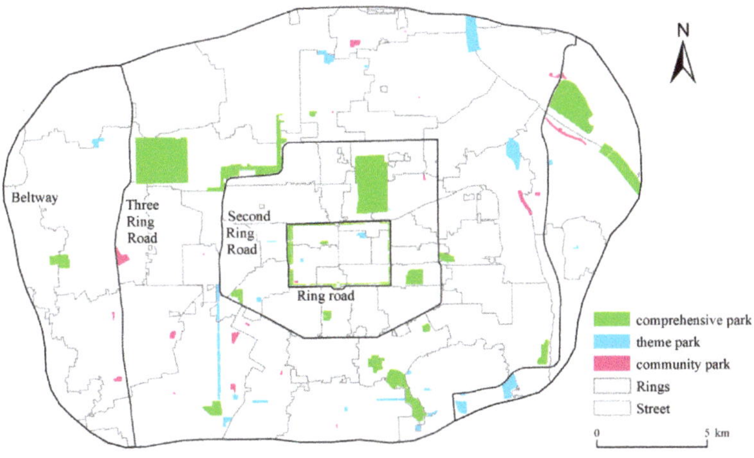

Figure 1. Distribution of parks within Xi'an Ring Expressway.

This project utilizes the Anjuke platform and python to collect residential district data within Xi'an City Ring Expressway. After data collection, 501 residential districts with no household data were deleted, with 4701 residential neighborhoods remaining as valid data. According to the 2020 statistical data of Xi'an (http://tjj.xa.gov.cn/tjnj/2020/zk/indexch.htm, accessed on 10 August 2021), the average permanent population of each household in Xi'an is 3.0, hence the total population is calculated by multiplying the number of households in the community by 3.0.

Due to the fact that different demographic groups favored different modes of transport [21], we analyzed the accessibility of the park using the three modes of transport—walking, cycling, and driving. Among them, the transit time from the residential area to the park is obtained entirely through the Baidu Map API interface [35], which involves taking the residential area as the starting point and the park as the destination, constructing the OD matrix, and then using python to request Baidu Maps' lightweight route planning service to obtain the transit time. This study recorded travel time in a week from 5 July 2021 to 11 July 2021. The travel time was gathered daily at 18:00, and the average time of seven days was used to determine the actual time taken from the residential area to the park.

2.3. Methods

2.3.1. G2SFCA Method

Dai [19] presented the 2SFCA approach for evaluating the accessibility of green space in Atlanta, Georgia, the United States. The fundamental calculating idea is divided into two steps:

Step 1: Calculate the supply–demand ratio. Determine a distance threshold for each park in the research area, calculate the demand within the threshold range, and multiply it by the Gaussian function, and divide the park supply by the result of multiplication to obtain the supply–demand ratio.

Step 2: Calculate accessibility. For each demand point, search all parks within the threshold range, multiply the park's supply–demand ratio by the Gaussian function, accumulate the results, and obtain the accessibility A_i of each demand point.

$$A_i = \sum_{j \in \{d_{kj} \le d_0\}} G(d_{kj}, d_0) R_j = \frac{S_j}{\sum_{k \in \{d_{kj} \le d_0\}} G\left(d_{kj}, d_0\right) P_k} \tag{1}$$

$$G\left(d_{kj}, d_0\right) = \begin{cases} \frac{e^{-\left(\frac{1}{2}\right) \times \left(\frac{d_{kj}}{d_0}\right)^2} - e^{-\left(\frac{1}{2}\right)}}{1 - e^{-\left(\frac{1}{2}\right)}}, & if \quad d_{kj} \le d_0 \\ 0, & if \quad d_{kj} > d_0 \end{cases} \tag{2}$$

In the formula: S_j is the supply of the park j, d_0 denotes the distance threshold, d_{kj} is the distance between the supply and the demand points, and P_k denotes the demand within the threshold range, which is frequently stated in terms of population. $G(d_{kj}, d_0)$ is the Gaussian attenuation function.

2.3.2. iG2SFCA Method

Although the G2SFCA assesses park accessibility from both supply and demand perspectives, there are still deficiencies. This paper is enhanced in two aspects as follows.

Firstly, consider the contrasts in park quality and park attractiveness factors. In addition to spatial features, park size, and nearby facilities also have a significant effect on park accessibility [10,36,37]. As a result, we assign attraction weights of 0.6, 0.4, and 0.2 to the three levels of classification. With an appropriate walking distance of 800 m [38], we take the normalized quantity of POI within the 800 m buffer outside the park as an indicator of attractiveness. The normalized quantity is calculated as the attractiveness factor. This article focuses on five types of POIs that are closely related to leisure and entertainment [39]: catering services, shopping services, sports and leisure services, scenic places, and scientific, educational, and cultural services; with a total of 77,089 POIs retrieved.

Secondly, the distance threshold is replaced with the time threshold. Due to the advancement of big data technology, it is possible to utilize a Baidu map to acquire travel time under multiple travel modes, which facilitates the analysis of the park's spatial accessibility and reveals the accessibility variations under different traffic modes [10]. Consequently, this study used three modes of transportation—walking, riding, and driving—to examine the accessibility change characteristics at various time thresholds.

The revised calculating formulas are as follows:

$$A_{ij} = \sum_{j \in \{t_{kj} \le t_0\}} G(t_{kj}, t_0) R_j = \frac{W_j S_j}{\sum_{k \in \{t_{kj} \le t_0\}} G\left(t_{kj}, t_0\right) P_k} \tag{3}$$

$$W_j = N_j \times A_j \tag{4}$$

$$G\left(t_{kj}, t_0\right) = \begin{cases} \frac{e^{-(\frac{1}{2})\times(\frac{t_{kj}}{t_0})^2} - e^{-(\frac{1}{2})}}{1-e^{-(\frac{1}{2})}}, if & t_{kj} \leq t_0 \\ 0, & if \quad t_{kj} > t_0 \end{cases} \tag{5}$$

In the formula, A_{ij} represents the accessibility of park j under traffic mode i, R_j represents the park's supply–demand ratio, S_j represents the area of park j, W_j represents the attractiveness of park j, expressed as the product of the normalized index N_j of the number of POIs within an 800 m buffer and the park's type weight A_j, P_k represents the population, t_0 represents the time threshold, t_{kj} represents the transit time between the park and the community, and $G(t_{kj}, t_0)$ is the Gaussian attenuation function.

2.3.3. Location Quotient

The location quotient is the ratio of the park area enjoyed by a street's population to the per capita park area in the research area, which can indicate the equity of street parks distribution [40]. Typically, the location quotient is compared to 1. When it is greater than 1, it implies that the street's park distribution level is greater than the average value of the study area. The greater the location quotient is, the greater the degree of street park distribution level is. The calculation formula for location quotient is:

$$LQ_i = (\frac{A_i}{P_i})/(\frac{A}{P}) \tag{6}$$

In the formula, LQ_i denotes the location quotient of street i, A_i and P_i denotes the park area and population of the street i respectively, and A, P respectively represent the total area and population of the park in the study area.

3. Results

3.1. Supply Analysis

Because some streets are divided by the ring expressway, this article only counts streets with an area of more than 50% in the ring expressway. There are 49 streets within the ring expressway, and parks are unevenly distributed among street (Table 1). In terms of quantity, there are nine streets with on parks situated around. Zhangbagou Street contains the largest number of parks compared to other streets, with a total of 14. This is because the Tang City Wall Heritage Park is scattered across the street in a long and narrow shape, split into many block parks by the road network. Due to the large area of Weiyang Palace Heritage Park, Daming Palace Heritage Park, and Chanba Wetland Park, the streets around these parks contain fewer number of parks, with a relatively high coverage rate of parks in contrast. Between West Third Ring Road and Ring Expressway, the parks are small and scattered due to spatial distribution.

3.2. Demand Analysis

Data from residential areas and data from streets within the ring expressway were connected and compared to investigate demand variances of each street (Figure 2). According to loop lines, the population is concentrated between the second and third ring roads. Meanwhile, the population within the ring road is relatively small, because of the implementation of the "Imperial City Rejuvenation" plan, which limits population growth within the ring road and adjusts the land use to reduce the height and density of buildings, transforming it into an urban functional area dominated by tourism and commerce. With the high density of firms in the high-tech zone, its rapid development has attracted an increasing number of talents. Moreover, communities in high-tech zone are primarily erected in recent years, with higher floors and larger capacities, resulting in a massive population in Yuhuazhai Street. On the other hand, the east area of the East Third Ring

Road is sparsely populated. Land of streets in northeast area is primarily for agricultural. Rural areas have a higher proportion of self-built residences, which is hard to quantify, with consequently lower population than other streets.

Table 1. Area table of some street Parks.

Street Name	Number of Parks (a)	Area of Park (km^2)	Area Ratio (%)	Street Name	Number of Parks (a)	Area of Park (km^2)	Area Ratio (%)
Weiyanggong Street	2	6.22	30.19	Shilipu Street	2	0.63	4.89
Baqiao Street	6	3.36	13.84	Dongguannanjie Street	1	0.51	19.47
Dayanta Street	7	1.75	20.91	Xiguan Street	6	0.42	8.20
Daminggong Street	1	1.65	15.68	Dengjiapo Street	2	0.40	2.86
Zhangbagong Street	14	1.21	4.78	Changyanbao Street	9	0.31	2.25
Xiwang Street	7	1.09	5.58	Xujiawan Street	3	0.30	3.67
Ziqianglu Street	1	0.95	39.92	Zhangjiabao Street	4	0.29	2.29
Taihualu Street	3	0.90	17.68	Tanjia Street	3	0.27	1.47
Qujiang Street	8	0.72	4.98	Hancheng Street	2	0.26	1.23
Sanqiao Street	4	0.71	2.23	Huanchengxilu Street	3	0.22	8.81

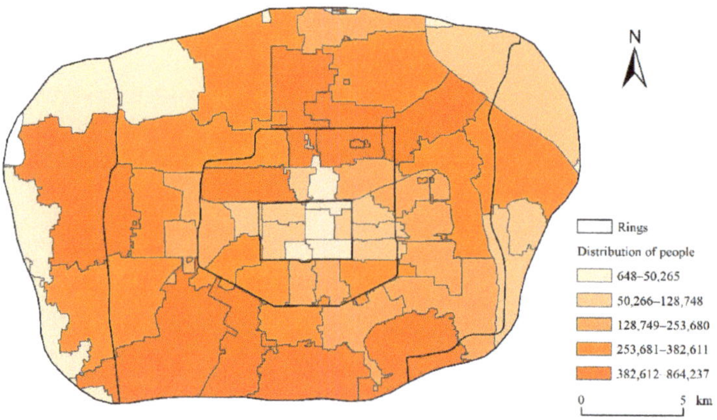

Figure 2. Population distribution in Xi'an Ring Expressway.

3.3. Comparing Analysis

Walking mode is chosen based on a 15-min threshold to compare the methods prior to and after the improvement. According to the statistical findings, significant differences between the maximum and minimum values of the G2SFCA and iG2SFCA are discovered by comparing the mean and standard deviation of the two methods reveals, which indicates that the distribution of park resources in the region is unbalanced. By comparing the mean and standard deviation of calculation results(Table 2), it is discovered that the mean value of iG2SFCA is less than that of G2SFCA. This is because the iG2SFCA considers the attractiveness of park type and the surrounding POI, and normalizes the number of POI (between 0–1), thus making the calculation result smaller.

Table 2. Comparison of accessibility calculation results.

	Maximum	Minimum	Mean	SD
G2SFCA	274.61	0.10	1.66	12.82
iG2SFCA	87.62	0.10	0.53	2.27

Normalized spatial distribution of the two approaches is shown in Figure 3. In terms of spatial distribution, the two approaches produce outcomes with a similar distribution of high values. Both methods identify that Chanba Wetland Heritage Park, Xi'an Park around city, and Qujiang Pond Heritage Park have a high degree of spatial accessibility. However, there are some discrepancies between the two outcomes. The results of the traditional method are obviously divergent, being unable to reflect the internal accessibility differences. Calculation results of the improved method, on the other hand, are more evenly distributed and exhibit obvious transitions, which can better reflect the internal variations among regions and are more sensitive to high value identification. Such as Xingqing Palace Park and Olympic Park, the iG2SFCA takes the park level and surrounding POI into account and determines that these two parks are high accessibility areas, which is more accurate than the traditional method.

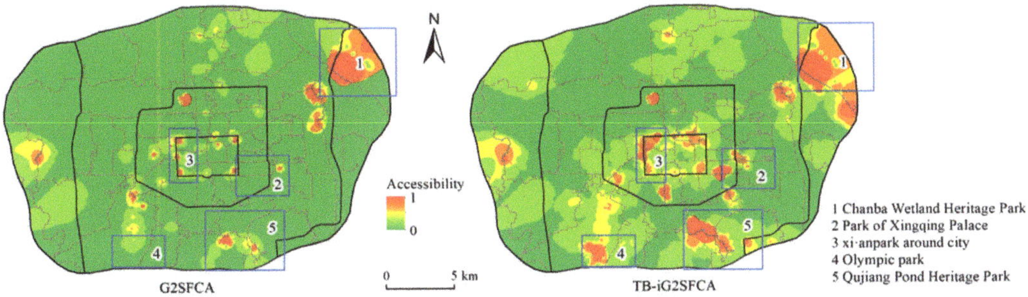

Figure 3. Comparison of spatial distribution of accessibility.

3.4. Accessibility Analysis

Given the various forms of transport, this article focuses on three frequently used travel modes: walking, cycling, and driving. The accessibility of 5-min, 15-min, and 30-min scenarios are calculated to provide a reference for inhabitants to choose their daily leisure travel mode. The mean and standard deviation of the accessibility level of each residential area for various means of transport are shown in Table 3. In the walking mode, the mean value of accessibility increases from 0.3 to 0.66 as travel time increases, and the reachable area continues to expand. In the cycling mode, the largest mean value is 0.92 in the 5-min scenario, and the SD steadily reduces from 5.60 to 0.76, demonstrating that riding is more convenient in the 5-min scenario, but the degree of convenience differs significantly across different residential communities. The mean value of residential community accessibility in driving mode is directly proportional to time, while the standard deviation is inversely proportional. When comparing the three modes of transport in a 30-min scenario, SD is the smallest at 1.99, 0.76, and 0.47. This demonstrates that as time passes, the accessibility gap between streets become smaller.

Table 3. Mean and standard deviation of accessibility level of residential area under different travelmodes.

Scenarios	Walking		Cycling		Driving		Multi-Mode	
	Mean	SD	Mean	SD	Mean	SD	Mean	SD
5-min	0.30	2.90	0.92	5.60	0.36	1.37	0.28	1.01
15-min	0.53	2.27	0.68	1.77	0.58	1.22	0.32	0.50
30-min	0.66	1.99	0.68	0.76	0.66	0.50	0.34	0.47

In order to compare the results from a spatial perspective, this study employs the kriging method for global analysis and categorizes the reachable areas according to the nature break to indicate the accessibility (Figure 4). Under three travel modes, the accessibility area of space expands with the growth of time. In walking mode, the area exhibits a point discrete distribution in 5-min scenario. These neighborhoods are centered along the Ring Road, near the Tang City Wall Heritage Park, and in the Qujiang neighborhood. Due to the large area of the Daming Palace, this region is inaccessible in a 15-min scenario. While in

30-min scenario, it becomes highly accessible. Under the 30-min scenario, the accessibility of most places remains 0, indicating that some residential communities are unable to use city parks effectively within 30-min. Since 18:00 is the evening rush hour in Xi'an, there may be traffic jams, and some roads sections cannot be driven when traveling by car. As a result, the riding reachable range is greater than the driving in the 5-min scenario. When the reach of the three modes of travel is compared, it is observed that walking accessibility can basically cover the area within the ring road during a 15-min scenario, while cycling and driving can effectively cover almost the entire area within the ring expressway.

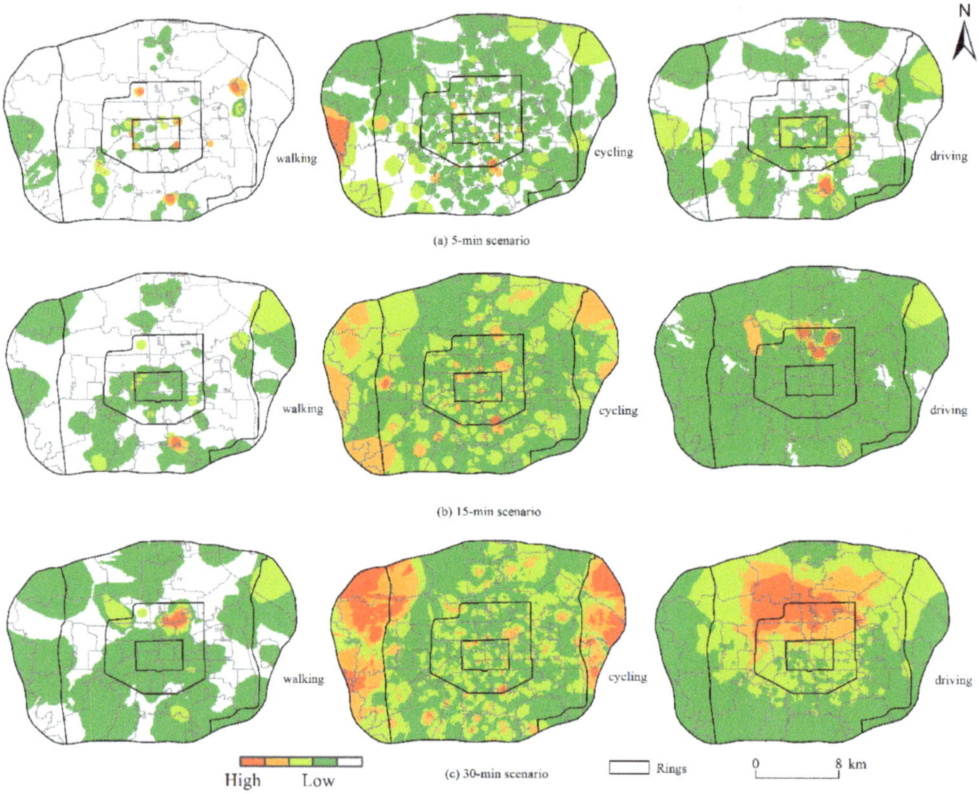

Figure 4. Spatial distribution of accessibility of three travel modes under different time thresholds.

Under the travel threshold of 5-min, the high value of accessibility is distributed around the ring road in walking mode. This is because Xi'an Park around city is situated along the city walls, which it is convenient and attractive to local inhabitants, transforming it into a favorable leisure and entertainment destination. High values of cycling and driving accessibility appear near Qujiang district. In recent years, Qujiang District has emphasized the development of cultural and tourism industries, resulting in the establishment of numerous parks with characteristics. These parks have comprehensive surrounding infrastructure, convenient transportation, thus strong attraction for residents. In addition, because the Qujiang District is predominantly a villa district with a small population, it is a high value area. From a 5-min to 30-min scenario, the high values of walking and driving accessibility have shifted from south to north, and the areas of Daming Palace Heritage Park and Baqiao Ecological Wetland Park have developed into new high-value center. Access to high values is evenly distributed in cycling modes, but as the time threshold increases, high-value gathering spots arise around the Third Ring Road's side. This is because there are numerous comprehensive parks on the east of the East Third Ring Road, and the populations in this area which means that demand is low. On the whole, the overall accessibility level within the Ring Expressway is reasonably good. Nonetheless, the southwest area is a low-value area. In comparison, riding accessibility is best in the 15-min scenario.

3.5. Spatial Equity Analysis

Accessibility is merely a spatial manifestation of the allocation of public service resources, while scholars and policy makers focus on the underlying spatial equity [39]. The degree of curvature of the Lorenz curve can indicate the equity of resource allocation. Figure 4 demonstrates the inequitable distribution of urban park resources (Figure 5). A total of 10% of the population enjoys only 0.2% of urban park resources, whereas 20% enjoy 1.04%.

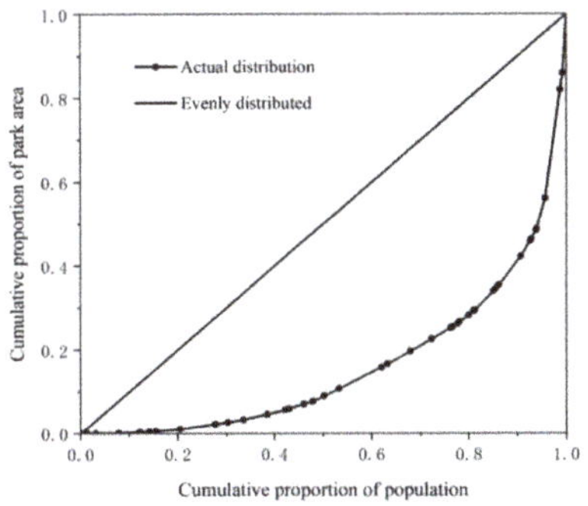

Figure 5. Lorentz curve of urban park resource allocation.

In order to investigate the spatial differentiation of equity of urban park resources allocation, this study calculates the location quotient of each street and categorizes the results into six groups (Figure 6, Table 4). In terms of the loop line, the most equitable route is located around the ring road. Additionally, there is significant disparity in the distribution of urban park resources between the second ring road and the third ring road. There are 16 streets that lack park distribution, which means that residents must cross roadways to use parks on daily basis. Due to the scarcity and small size of the parks, there are 16 streets with location quotient less than 0.2. The location quotient of 15 streets is above average. The greatest location quotient is in Baqiao Street which is ascribed to the street's national AAAA-level tourism attraction, Chanba National Wetland Park. Areas with large parks allow for high levels of urban park resource allocation per capita.

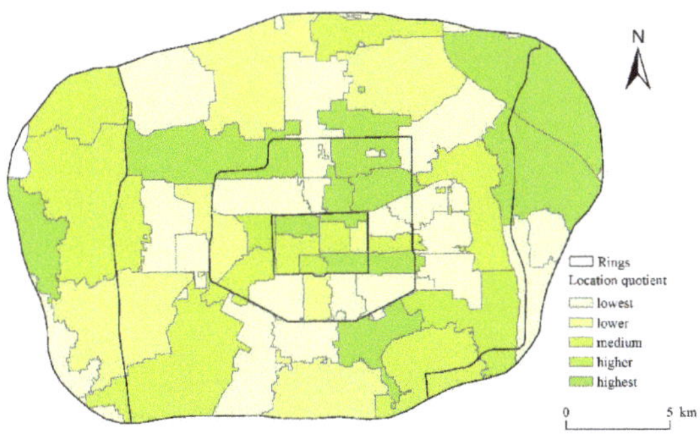

Figure 6. Street location quotient classification map.

Table 4. Street location quotient level table.

Grade	Location Quotient	Number of Streets	Ratio
Lowest	<0.2	17	34.6%
Lower	0.2–0.6	7	14.3%
Medium	0.6–1.0	10	20.4%
Higher	1.0–1.4	2	4.1%
Highest	>1.4	13	26.5%

4. Discussion

4.1. Methodological Contributions

Accessibility is influenced by both geographic and non-spatial variables, residents' preferences for parks may be altered by non-spatial factors and traffic pat-terns [22]. However, G2SFCA only evaluates the ratio between the supply and the demand, and the majority of studies focus on park accessibility via a single-mode [21]. Incorporating non-spatial criteria into the accessibility assessment of parks has therefore been enhanced by this study. The contributions of this paper are most evident in the following aspects: Firstly, the park level and POI are incorporated in the analysis of the attraction coefficient. Compared to the G2SFCA, the SD and mean of iG2SFCA are smaller, and the transition between high and low values is more natural, resulting in more realistic findings. In addition, all parameters are designed to be flexible. The park weight can be adjusted according to the change of facility type. The type of POI can be modified to consider other factors, such as adding the number of parking lots or giving higher weight to specific facilities. Therefore, this methodology can be used as a computational model for assessing the accessibility of other amenities, such as schools, medical centers, and shopping malls. Additionally, prior research indicates that assessing the accessibility of a park based just on supply and demand is not a reliable strategy [10,25]. Secondly, it is integrated with big data to collect population data, compensating for the inadequacy of statistics [30]. This research uses Baidu Maps API to determine the travel time between residential neighborhoods and parks, and incorporates it into an accessibility model that can reflect the real travel behavior of residents [15,34]. Different time thresholds have a considerable impact on the accessibility distribution features, as demonstrated by the findings. Thirdly, this research employs location quotients to identify locations that may be underserved by parks, thereby assisting policy-makers in optimizing the layout of urban parks. Although the focus of this study is on parks, the methodology can be used as a computational model for assessing the accessibility of other public amenities.

4.2. Implications for Urban Park Planning

The distribution of spatial equity is similar to the spatial distribution of 15-min walking accessibility, demonstrating that accessibility can represent spatial equity to a certain extent [8]. According to the research results of accessibility and spatial equity, combined with the actual condition of Xi'an, optimization ideas are offered to help improve unbalanced park services as much as possible: the low accessibility and poor spatial equity streets are mostly distributed between Ring Road and the Second Ring Road, limited by high-density development model and land resources, turning irregular or small vacant lots into mini-parks or pocket parks [41] can increase the supply of park and coordinate the relationship between nature and urban development, and create a high-quality road landscape that is both ecologically and culturally sensitive. Additionally, Xi'an has a large number of heritage parks. The characteristics of heritage parks should be fully exploited to create a comprehensive park that incorporates natural scenery, history, and culture. For strip park, existing ecological corridors should be utilized to maintain the natural wetland, divide multi-functional areas, and improve the park layout. Moreover, the attractiveness of the park is strongly influenced by non-spatial variables. Therefore, public service facilities (such as restaurants, cafes, gymnasiums, and other recreational areas) might be added to the park's vicinity to increase its appeal to the citizens [22].

4.3. Limitations and Future Research

This study also has certain drawbacks. Firstly, we generate the OD matrix using centroid coordinates of the park as end point, but, people enter the park when they arrive at the entrance [36]. In the subsequent study, the OD matrix can be generated by acquiring the coordinates of the park entrance as the end point of the trip.

In addition to distance and park attraction, individual socio-economic variations play a significant role in determining individuals' inclination to visit a park [42], this article makes no distinction between individuals' travel inclinations (e.g., income, ethno-racial characteristics, age, gender, and disability). Recent research has shown that people with low socioeconomic status have less access to parks [43,44]. However, few research has examined the disparities in park accessibility across Xi'an's various demographic groups. Consequently, future research might categories the population to investigate the environmental justice of the park in greater depth.

Thirdly, the accessibility of park during working days and on weekends also deserves further discussing. People have various preferences for parks on weekdays and weekends [45]. The prevalence of weekends encourages the practice of long-distance travel. This study focuses solely on the impact of various travel modes on park accessibility. Future researchers should describe accessibility factors at various times to reinforce the findings.

5. Conclusions

At the street level, the distribution of park supply and population demand in Xi'an 3 city is uneven. Zhangbagou Street Park has the largest number, including 14 parks, although there are still nine streets without parks. The relationship between population geographical distribution and policies is strong. The ring road has a small population density. However, the high-tech zone is densely populated due to company development policy support.

The G2SFCA underestimates regional accessibility. The iG2SFCA can provide a more realistic evaluation, yet the spatial pattern of accessibility under both models is comparable. Specifically, a very accessible gathering area is located next to the park. As distance increases, the level of accessibility eventually declines.

Under three modes of transport, the reachable range expands as the travel duration increases. In walking mode, the reachable area extends from the core of the Ring Road, Tang Great Wall Heritage Park and Qujiangchi Heritage Park area. In the 5-min scenario, the area of riding accessibility is the largest. The average values of park accessibility and SD decline as time passes. At 30-min, the park service gap between communities is at its smallest, but some areas remain unreachable in walking mode.

High accessibility areas differ with travel mode. In walking and driving modes, high-accessibility area moves northward. In riding mode, high value areas of accessibility are located on both sides of the east and west third ring roads. Under the 15-min walking condition, high values areas appear near Huancheng Park, Qujiangchi Heritage Park, and Daming Palace Heritage Park, which corresponds to Huang's research findings [46].

The distribution of park resources in the study area is inequitable. Due to the fact that 14.5% of the streets are not distributed with parks, the residents' demand for parks cannot be addressed within the streets. 33.3% of the streets have a location quotient of less than 0.2, while 31.2% having a location quotient of greater than 1, indicating those streets have a higher level of equality.

Author Contributions: Methodology, Y.L.; Data curation, Y.L. and S.S.; Writing—original draft preparation, Y.L.; writing—review and editing, Y.X. and L.H.; Visualization, Y.L. All authors have read and agreed to the published version of the manuscript.

Funding: This research received no external funding.

Data Availability Statement: The dataset utilized and/or analyzed during the present study are available on reasonable request from the corresponding author.

Conflicts of Interest: The authors declare no conflict of interest.

References

1. Chiesura, A. The Role of Urban Parks for the Sustainable City. *Landsc. Urban Plan.* **2004**, *68*, 129–138. [CrossRef]
2. Wood, L.; Hooper, P.; Foster, S.; Bull, F. Public Green Spaces and Positive Mental Health–Investigating the Relationship between Access, Quantity and Types of Parks and Mental Wellbeing. *Health Place* **2017**, *48*, 63–71. [CrossRef]
3. Brown, G.; Schebella, M.F.; Weber, D. Using Participatory GIS to Measure Physical Activity and Urban Park Benefits. *Landsc. Urban Plan.* **2014**, *121*, 34–44. [CrossRef]
4. Zhang, L.; Zhou, S.; Kwan, M.-P.; Chen, F.; Lin, R. Impacts of Individual Daily Greenspace Exposure on Health Based on Individual Activity Space and Structural Equation Modeling. *Int. J. Environ. Res. Public Health* **2018**, *15*, 2323. [CrossRef]
5. Moulay, A.; Ujang, N.; Said, I. Legibility of Neighborhood Parks as a Predicator for Enhanced Social Interaction towards Social Sustainability. *Cities* **2017**, *61*, 58–64. [CrossRef]
6. Tu, X.; Huang, G.; Wu, J. Review of the Relationship between Urban Greenspace Accessibility and Human Well-Being. *Shengtai Xuebao/Acta Ecol. Sin.* **2019**, *39*, 421–431.
7. Hansen, W.G. How Accessibility Shapes Land Use. *J. Am. Inst. Plann.* **1959**, *25*, 73–76. [CrossRef]
8. Nicholls, S. Measuring the Accessibility and Equity of Public Parks: A Case Study Using GIS. *Manag. Leis.* **2001**, *6*, 201–219. [CrossRef]
9. Rigolon, A. A Complex Landscape of Inequity in Access to Urban Parks: A Literature Review. *Landsc. Urban Plan.* **2016**, *153*, 160–169. [CrossRef]
10. Ibes, D.C. A Multi-Dimensional Classification and Equity Analysis of an Urban Park System: A Novel Methodology and Case Study Application. *Landsc. Urban Plan.* **2015**, *137*, 122–137. [CrossRef]
11. Kong, F.; Yin, H.; Nakagoshi, N. Using GIS and Landscape Metrics in the Hedonic Price Modeling of the Amenity Value of Urban Green Space: A Case Study in Jinan City, China. *Landsc. Urban Plan.* **2007**, *79*, 240–252. [CrossRef]
12. Akpinar, A. How Is Quality of Urban Green Spaces Associated with Physical Activity and Health? *Urban For. Urban Green.* **2016**, *16*, 76–83. [CrossRef]
13. Comber, A.; Brunsdon, C.; Green, E. Using a GIS-Based Network Analysis to Determine Urban Greenspace Accessibility for Different Ethnic and Religious Groups. *Landsc. Urban Plan.* **2008**, *86*, 103–114. [CrossRef]
14. Wang, S.; Wang, M.; Liu, Y. Access to Urban Parks: Comparing Spatial Accessibility Measures Using Three GIS-Based Approaches. *Comput. Environ. Urban Syst.* **2021**, *90*, 101713. [CrossRef]
15. Langford, M.; Higgs, G.; Fry, R. Multi-Modal Two-Step Floating Catchment Area Analysis of Primary Health Care Accessibility. *Health Place* **2016**, *38*, 70–81. [CrossRef]
16. Polzin, P.; Borges, J.; Coelho, A. An Extended Kernel Density Two-Step Floating Catchment Area Method to Analyze Access to Health Care. *Environ. Plann. B Plann. Des.* **2014**, *41*, 717–735. [CrossRef]
17. Luo, W.; Whippo, T. Variable Catchment Sizes for the Two-Step Floating Catchment Area (2SFCA) Method. *Health Place* **2012**, *18*, 789–795. [CrossRef]
18. Bissonnette, L.; Wilson, K.; Bell, S.; Shah, T.I. Neighbourhoods and Potential Access to Health Care: The Role of Spatial and Aspatial Factors. *Health Place* **2012**, *18*, 841–853. [CrossRef]
19. Dai, D. Racial/Ethnic and Socioeconomic Disparities in Urban Green Space Accessibility: Where to Intervene? *Landsc. Urban Plan.* **2011**, *102*, 234–244. [CrossRef]
20. Dony, C.C.; Delmelle, E.M.; Delmelle, E.C. Re-Conceptualizing Accessibility to Parks in Multi-Modal Cities: A Variable-Width Floating Catchment Area (VFCA) Method. *Landsc. Urban Plan.* **2015**, *143*, 90–99. [CrossRef]
21. Xing, L.; Liu, Y.; Liu, X. Measuring Spatial Disparity in Accessibility with a Multi-Mode Method Based on Park Green Spaces Classification in Wuhan, China. *Appl. Geogr.* **2018**, *94*, 251–261. [CrossRef]
22. Fan, Z.; Duan, J.; Lu, Y.; Zou, W.; Lan, W. A Geographical Detector Study on Factors Influencing Urban Park Use in Nanjing, China. *Urban For. Urban Green.* **2021**, *59*, 126996. [CrossRef]
23. Li, J.; Li, J.; Yuan, Y.; Li, G. Spatiotemporal Distribution Characteristics and Mechanism Analysis of Urban Population Density: A Case of Xi'an, Shaanxi, China. *Cities* **2019**, *86*, 62–70. [CrossRef]
24. Shi, P.; Xiao, Y.; Zhan, Q. A Study on Spatial and Temporal Aggregation Patterns of Urban Population in Wuhan City Based on Baidu Heat Map and POI Data. *Int. Rev. Spat. Plan. Sustain. Dev.* **2020**, *8*, 101–121. [CrossRef]
25. Wang, D.; Brown, G.; Liu, Y. The Physical and Non-Physical Factors That Influence Perceived Access to Urban Parks. *Landsc. Urban Plan.* **2015**, *133*, 53–66. [CrossRef]
26. Hu, S.; Song, W.; Li, C.; Lu, J. A Multi-Mode Gaussian-Based Two-Step Floating Catchment Area Method for Measuring Accessibility of Urban Parks. *Cities* **2020**, *105*, 102815. [CrossRef]
27. Li, Z.; Wang, Q.; Zhou, F.; Li, Y.; Han, X.; Mehmood, Q.; Cao, C.; Gu, F.F.; Han, M.; Chen, J. Integrating an Interferometric Synthetic Aperture Radar Technique and Numerical Simulation to Investigate the Tongmai Old Deposit along the Sichuan-Tibet Railway. *Geomorphology* **2021**, *377*, 107586. [CrossRef]
28. Wang, F.; Wang, K. Measuring Spatial Accessibility to Ecological Recreation Spaces in the Pearl River Delta Region: An Improved Two-Step Floating Catchment Area Method. *J. Spat. Sci.* **2018**, *63*, 279–295. [CrossRef]
29. Xiao, Y.; Wang, Z.; Li, Z.; Tang, Z. An Assessment of Urban Park Access in Shanghai–Implications for the Social Equity in Urban China. *Landsc. Urban Plan.* **2017**, *157*, 383–393. [CrossRef]

30. Lin-Lin, C.; Zhen-Wei, W.; Su-Feng, T.; Ya-Tong, L.I.; Meng-Yao, S.; Yu-Nian, Y. Evaluation of Eco-Environmental Quality in Mentougou District of Beijing Based on Improved Remote Sensing Ecological Index. *Chin. J. Ecol.* **2021**, *40*, 1177–1185. [CrossRef]
31. Guo, S.; Song, C.; Pei, T.; Liu, Y.; Ma, T.; Du, Y.; Chen, J.; Fan, Z.; Tang, X.; Peng, Y.; et al. Accessibility to Urban Parks for Elderly Residents: Perspectives from Mobile Phone Data. *Landsc. Urban Plan.* **2019**, *191*, 103642. [CrossRef]
32. Li, Z.; Fan, Z.; Song, Y.; Chai, Y. Assessing Equity in Park Accessibility Using a Travel Behavior-Based G2SFCA Method in Nanjing, China. *J. Transp. Geogr.* **2021**, *96*, 103179. [CrossRef]
33. Wang, F.; Xu, Y. Estimating O-D Travel Time Matrix by Google Maps API: Implementation, Advantages, and Implications. *Ann. GIS* **2011**, *17*, 199–209. [CrossRef]
34. Xu, M.; Xin, J.; Su, S.; Weng, M.; Cai, Z. Social Inequalities of Park Accessibility in Shenzhen, China: The Role of Park Quality, Transport Modes, and Hierarchical Socioeconomic Characteristics. *J. Transp. Geogr.* **2017**, *62*, 38–50. [CrossRef]
35. Su, S.; Li, Z.; Xu, M.; Cai, Z.; Weng, M. A Geo-Big Data Approach to Intra-Urban Food Deserts: Transit-Varying Accessibility, Social Inequalities, and Implications for Urban Planning. *Habitat Int.* **2017**, *64*, 22–40. [CrossRef]
36. Xing, L.; Liu, Y.; Liu, X.; Wei, X.; Mao, Y. Spatio-Temporal Disparity between Demand and Supply of Park Green Space Service in Urban Area of Wuhan from 2000 to 2014. *Habitat Int.* **2018**, *71*, 49–59. [CrossRef]
37. Xing, L.; Liu, Y.; Wang, B.; Wang, Y.; Liu, H. An Environmental Justice Study on Spatial Access to Parks for Youth by Using an Improved 2SFCA Method in Wuhan, China. *Cities* **2020**, *96*, 102405. [CrossRef]
38. Wang, N.; Du, Y. Resident Walking Distance Threshold of Community. *Transp. Res.* **2015**, *1*, 20–21.
39. Yu, K.; Duan, W.; Li, D.; Peng, J. Landscape Accessibility as a Measurement of the Function of Urban Green System. *Urban Plan.* **1999**, *8*, 7–10.
40. Tang, Z.; Gu, S. An Evaluation of Social Performance in the Distribution of Urban Parks in the Central City of Shanghai:From Spatial Equity to Social Equity. *Urban Plan. Forum.* **2015**, *02*, 48–56. [CrossRef]
41. Nordh, H.; Østby, K. Pocket Parks for People–A Study of Park Design and Use. *Urban For. Urban Green.* **2013**, *12*, 12–17. [CrossRef]
42. Gu, X.; Li, Q.; Chand, S. Factors Influencing Residents' Access to and Use of Country Parks in Shanghai, China. *Cities* **2020**, *97*, 102501. [CrossRef]
43. Fatima, K.; Moridpour, S.; Saghapour, T. Sustainable Transport Mobility for Older Adults: Influence of Neighbourhood Accessibility. *J. Transp. Health* **2021**, *22*, 101198. [CrossRef]
44. Harris, C.D.; Paul, P.; Young, R.; Zhang, X.; Fulton, J.E. Park Access among School-Age Youth in the United States. *J. Phys. Act. Health* **2015**, *12*, S94–S101. [CrossRef]
45. Bertram, C.; Meyerhoff, J.; Rehdanz, K.; Wüstemann, H. Differences in the Recreational Value of Urban Parks between Weekdays and Weekends: A Discrete Choice Analysis. *Landsc. Urban Plan.* **2017**, *159*, 5–14. [CrossRef]
46. Huang, X.; Li, H.; Zhang, Y.; Xie, L. Research and Evaluation of Accessibility of Xi'an City Park Green Space Based on Population Discretization and Transportation Network Analysis. *J. For. Environ. Sci.* **2014**, *29*, 98–102. [CrossRef]

buildings

MDPI

Article

A Real Estate Early Warning System Based on an Improved PSO-LSSVR Model—A Beijing Case Study

Lida Wang [1], Xian Rong [1,2], Zeyu Chen [1,*], Lingling Mu [3,*] and Shan Jiang [1]

1 School of Civil and Transportation Engineering, Hebei University of Technology, Tianjin 300401, China; 201711601010@stu.hebut.edu.cn (L.W.); xrong@hebut.edu.cn (X.R.); 18757587977@163.com (S.J.)
2 Hebei Civil Engineering Technology Research Center, Tianjin 300401, China
3 School of Economics and Management, Hebei University of Technology, Tianjin 300401, China
* Correspondence: czy19950719@163.com (Z.C.); lingmu1020@163.com (L.M.); Tel.: +86-177-9817-0019 (Z.C.); +86-022-6020-0385 (L.M.)

Abstract: The real estate market is vital for national economic development, and it is of great significance to research an early warning method to identify an abnormal status of the real estate market. In this study, a real estate early warning system based on the PSO-LSSVR model was created to train and test the indicator data of Beijing from 2000 to 2020, and to predict the early warning indicator of the Beijing real estate market from 2021 to 2030. The results showed that the warning status of the Beijing real estate market went from a fluctuation status to a stable "Normal" status from 2000 to 2020, and the warning status is expected to be more stable under a "Normal" status in the next decade under the same political and economic environment. The PSO-LSSVR model was found to have accurate prediction ability and demonstrated generalization ability. Furthermore, the warning status of the Beijing real estate market was analyzed in combination with national historical policies. Based on the results, this paper proposes policy recommendations to promote the healthy and sustainable development of the real estate market.

Keywords: real estate market; early warning system; PSO-LSSVR algorithm; sustainable development

Citation: Wang, L.; Rong, X.; Chen, Z.; Mu, L.; Jiang, S. A Real Estate Early Warning System Based on an Improved PSO-LSSVR Model—A Beijing Case Study. *Buildings* **2022**, *12*, 706. https://doi.org/10.3390/buildings12060706

Academic Editors: Yongjian Ke, Jingxiao Zhang and Simon P. Philbin

Received: 13 April 2022
Accepted: 19 May 2022
Published: 24 May 2022

1. Introduction

1.1. Background

Over the past few decades, the rapid development of China's real estate market has actively contributed to the rapid growth of China's GDP, especially after the 2008 global economic crisis. Moreover, the rapid development of the real estate industry has also led to the development of the construction industry, decoration, home appliances, furniture industry and so on. According to the China Statistical Yearbook 2021, the GDP of the real estate industry accounted for 7.3 percent of the country's total GDP, and the property-related construction industry accounted for 7.2 percent of the country's total GDP in the year 2020 [1]. On the other hand, the real estate industry has become an important part of fixed asset investments, where fixed asset investments are one of the troikas for economic growth (investment, export and consumption). Hence, the developing real estate industry has become one of the important economic pillars of China's economic development. However, the excessive development of any industry will bring negative effects on society.

First, when the real estate boom is excessively prosperous, there will be a large number of excess funds, bank loans and corporate funds flowing into the real estate market, which will lead to the phenomenon of "industry emptiness" (the decline in the development level of the primary and secondary industry lead to the unbalanced structural proportion of national economic demand and the continuous shrinkage of domestic investment, which, in turn, lead to the substantial reduction in employment opportunities and social problems, such as a rising unemployment rate) [2].

Second, the development of the real estate market is closely related to the macroeconomy. Excessive development of the real estate market promotes the investment flow into related industries, which leads to a shortage of resources supplied to other industries [3]. To solve such conditions, the government usually restrains the development of the real estate industry through credit regulation and land regulation policies, which leads to the rapid cooling of the real estate industry and economic recession [4]. The excessive development of the real estate industry and the drastic macro-control will lead to excessive fluctuations in the macroeconomy [5].

Last, but not least, the rapid development of the real estate industry has introduced a large amount of investment of capital and the public. Excessive investment will lead to the market price being greater than the actual price, which leads to economic bubbles in the real estate industry [6]. When the bubble bursts, the loans from financial institutions and banks to real estate enterprises will turn into bad debts, which will lead to the occurrence of an economic crisis and financial crisis in serious cases, and then lead to industrial depression, the sharp increase in unemployment and crime rate, and other social problems [7]. The economic depression caused by the bursting of the real estate bubble will bring enormous harm to the national development and living quality, as the U.S. real estate bubble burst in the late 1980s led to the bank crisis, and at the beginning of the 21st century, the U.S. real estate bubble burst led to the subprime mortgage crisis and the biggest financial crisis since 1930. The real estate bubble burst in Japan in the 1980s set off a decades-long economic recession. These classic real estate bubble bursts offer warnings for other countries' real estate developments.

In contrast, if the real estate industry appears excessively depressed, the public, capital and government will lose confidence in the real estate industry, which will further affect the development of the real estate industry and the development of the macroeconomy [8]. Moreover, the construction industry, which is closely related to the real estate industry, and the back-end industries will be seriously affected [9]. Therefore, it is necessary to establish an early warning system for the real estate market to prevent the bubble and further enormous harm to the national and social development. A real estate market early warning system can also help the urban government and supervisory agencies to reduce the risk of bubbles in advance [10]. Furthermore, a real estate market early warning system can provide a scientific basis for the government's macro-control and correctly guide investors away from irrational investment and consumption in the real estate market to promote the healthy and stable development of the market [11].

1.2. Literature Review

The real estate industry has the characteristics of information asymmetry, fixedness and transaction dispersion [12,13], and the effective balance of the general industrial structure and supply–demand relationship of the real estate industry cannot be achieved only through market forces [14]. Hence, research on the real estate market is complicated and diverse. With the continuous development of science and technology, more and more automatic methods were applied to the sustainable development of the real estate market, such as artificial intelligence [15] and machine learning [16]. However, from previous studies, we can still find some research ideas and theories, which are conducive to the comprehensive development of relevant research.

In a previous study, Pyhrr et al. observed the cyclical phenomenon of the Western real estate market since the 1960s and then researched the change laws and periodic mechanism of the real estate market [17,18]. With the in-depth study of the real estate cycle, scholars gradually realized the importance of early warning for the real estate market. In the 1990s, Nieboer et al. first established a real estate market early warning system based on the housing vacancy rate by studying the relationship between the housing vacancy rate and the trend of the real estate market in the Netherlands [19]. Subsequently, as a country with a well-developed real estate market, the United States found the real estate downturn signal of the Chicago community by observing the fluctuation and mechanism of the real estate cycle, and then detecting seven abnormal change indicators and establishing the original real estate early warning model [20]. With the proposal of the Case–Shiller (CS) housing price index, scholars used to research the relationship between the real estate market and macroeconomic phenomena by combining the Case–Shiller housing price index and early warning leading indicators [21,22]. Huang et al. adopted the deep learning method in the Time-Varying Parameter-Vector Autoregression (TVP-VAR) model to research the monitoring and early warning of systemic financial risk, where the results showed that the real estate market is closely related to the country's systemic financial risk [23].

In addition, scholars in different fields researched the methods and applications of early warning systems [24]. With the development of the research field, the concept and methods of early warning systems have been widely introduced into the aerospace study [25,26], ecological systems research [27], seismic research [28,29], transportation study [30,31], medical study [32,33] and real estate study [34].

With the rapid development of science and technology, more methods have been adopted in the research of early warning systems. Begusic et al. adopted the system dynamics to research the system risk in the financial market and real estate market, which was achieved by researching the spatio-temporal spillover effect of system information feedback to obtain early warning signals of the market [35]. Huang and Feng proposed an early warning method based on the monitor indicators and statistical method and simulated the policy regulation of the real estate market in Shenzhen by combining system dynamics [36]. Yang used computer technology based on the principle of the PROBIT model to design an early warning system to analyze risk verification in the real estate market and the results showed that the early warning system had an approximately 85% early warning rate [37]. Kholodilin and Michelsen used modern machine learning methods to forecast the real estate market bubbles in Germany and presented an early warning model to analyze the speculative risk in the real estate market; furthermore, they provided some market intervention suggestions for the German government [38]. Wang et al. analyzed the real estate market in Beijing by establishing an early warning model based on multi-class support vector machines and put forward policy suggestions for the healthy operation of the real estate market [34]. Based on previous research on the real estate market, it was found that the machine learning model has better processing capacity than the transmission statistical model for solving nonlinear problems [39,40]. In recent years, scholars established early warning index systems for the real estate market through different methods, and conducted early warning research through qualitative and quantitative methods, promoting the research of early warning index systems to gradually reach a mature stage of development.

Based on the relatively mature development of the real estate early warning indicators research, further research of indicator selection should take into account the market supply–demand relationship, internal coordination, external economic factors and so on. The methods for early warning systems were developed from basic index analysis research methods to statistical research methods and then to comprehensive early warning models. Moreover, with the in-depth research, the early warning research of the real estate market has changed from qualitative research to a combination of qualitative and quantitative research. With the rapid development of computer science and artificial intelligence, more and more theories and methods were introduced into the real estate early warning model, and new research ideas were developed [41]. However, in the current research on the real estate market early warning, the problems of a small sample, overall generalization ability and nonlinearity cannot be solved easily.

Due to the complexity of the real estate market system and the lack of historical data, there will be insufficient training, unstable performance and overlearning when the neural network is training. All of these factors will lead to unsatisfactory prediction results. As a result, Cortes [42] proposed the support vector regression (SVM), which has demonstrated excellent performance in solving small-sample, nonlinear, high-dimensional problems. That is why SVR, the regression version of SVM, is widely used in energy research [43]. SuyKens et al. [44] improved least-squares support vector regression (LSSVR) in 1999, the least-squares formulation of SVR, which has better generalization abilities and powerful computation. Chou et al. [45] found that LSSVR is more accurate than regression analysis and neural networks for building energy consumption. However, there are still disadvantages to LSSVR due to the choice of kernel and regularization parameters; therefore, particle swarm optimization (PSO) was introduced to improve the LSSVR. PSO-LSSVR was successfully employed in many fields, especially in the evaluation of engineering projects. For this reason, some scholars specifically analyzed and examined its operating principle [46,47]. In addition, the LSSVR model has the characteristics of a fast training speed, which is helpful to obtain a more effective regression model in a complex system. The PSO model has the characteristic of a fast convergence speed, which can further improve the accuracy of the prediction. Through continuous iteration of the PSO model, the most appropriate prediction parameters can be quickly found. The most important point is that the LSSVR model is suitable for exploring potential lows from historical data and predictions, while the PSO model has relatively low requirements for optimization functions; therefore, the PSO-LSSVR model is more suitable for the research of the real estate market with small data [48]. However, PSO-LSSVR in real estate market prediction has not been widely applied.

As the political, economic and cultural center of China, Beijing has a large population inflow and a high housing demand [49]. The real estate market in Beijing has the characteristics of a mass market, high land prices and high commercial housing prices, which are related to urban and economic sustainable development [50]. Based on the particularity and activity of the Beijing urban real estate market, it is vital to research the ability to provide an early warning of an abnormal status of the Beijing real estate market to ensure the stable and sustainable development of the Beijing real estate market. Furthermore, the sustainable development of the real estate market plays an important role in promoting the steady development of the urban economy and society. This study considered multiple indicators of the real estate market and adopted the PSO-LSSVR model to test and predict the early warning status of the Beijing real estate market. This study can make the real estate market early warning system research richer; furthermore, a PSO-LSSVR model can enrich the research on prediction methods more generally. The rest of this paper is arranged as follows. Section 2 establishes the PSO-LSSVR model and discusses the early warning system indicators. Section 3 presents the results of an empirical analysis of data of the Beijing real estate market from 2000 to 2020 and establishes the early warning system based on the analysis. Section 4 discusses the results of the Beijing real estate early warning system and presents early warning predictions for the Beijing real estate market for 2020, further providing policy implications. Section 5 concludes the paper.

2. Materials and Methods

In this section, the related theories and methods for the construction of the real estate market early warning system are introduced. The least-squares support vector regression (LSSVR) theory was used to train the sample data and make a prediction by finding the required parameters. Particle swarm optimization (PSO) theory was used to optimize the model parameters and find a combination of parameters with a higher accuracy via continuous iterations. To reflect the prediction mechanism of the PSO-LSSVR model, a flow chart was made, as shown in Figure 1.

The standardization method was used to make the values of the influencing factors dimensionless and convert the influencing factors from obeying the normal distribution to obeying the standard normal distribution, which could divide the predicted market status based on the standard normal distribution. The risk matrix method was used to analyze the situation of the real estate market, according to the market status of the different predicted factors, and provide an early warning of an abnormal market status.

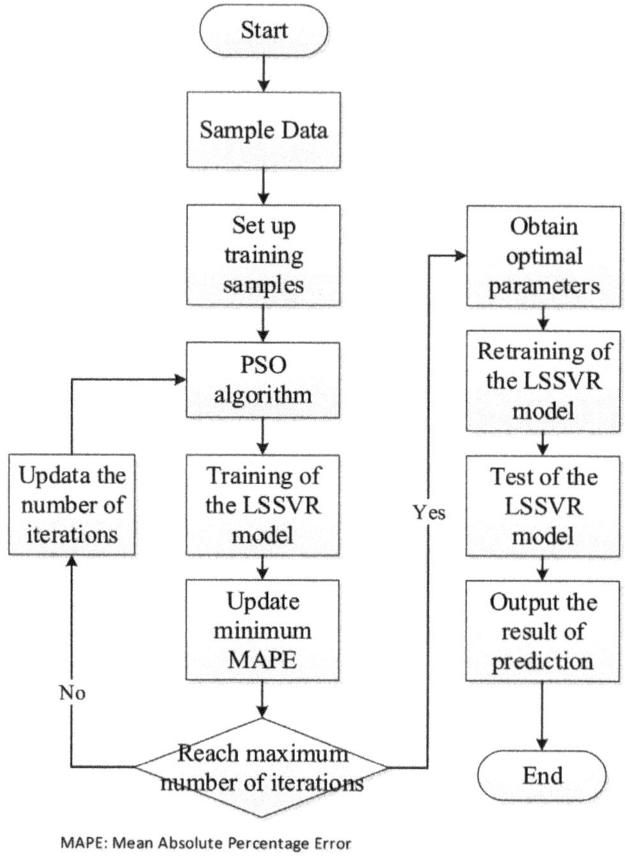

Figure 1. The operation mechanism for the PSO-LSSVR model.

2.1. Least-Squares Support Vector Regression (LSSVR) Theory

In this section, the principle of LSSVR is described briefly: Let $S = \{(x_k, y_k)|k = 1, 2, \ldots, n\}$ be a set of training sample points, where $x_k \epsilon R^n$ is the input vector, and $y_k \epsilon R^n$ is the output vector. With a nonlinear map $\varphi(\cdot)$ the sample from the original space R^{n} is mapped into the feature space $\varphi(x_k)$. The optimal decision function is defined as follows:

$$y(x) = \omega^T \cdot \varphi(x) + b \tag{1}$$

where ω is the weighted vector, $\varphi(x)$ is the kernel space mapping function and b is a constant. Usually, coefficients ω and b are obtained by minimizing the upper bound of generalization error.

According to the principle of structural risk minimization, the regression problem can be transformed into an optimization problem with constraints. The target function and constraint are formulated as follows:

$$\min_{\omega,b,e}(\omega, e) = \frac{1}{2}\omega^T\omega + \frac{1}{2}\gamma\sum_{k=1}^{n} e_k^2 \tag{2}$$

s.t.

$$y_k = \omega^T \cdot \varphi(x_k) + b + e_k$$

where $k = 1, 2, \ldots, n$; γ is the regularization parameter; and e_k is the slack variable.

To solve the problem, the Lagrange function is defined as follows:

$$L(\omega, b, e, \alpha) = \phi(\omega, e) - \sum_{k=1}^{n} \left\{ \alpha_k \left[\omega^T \cdot \varphi(x) + b + e_k - y_k \right] \right\} \tag{3}$$

where α_k is the Lagrange multiplier. According to the Karush–Kuhn–Tucker condition, the conditions for optimality are

$$\begin{cases} \omega = \sum_{k=1}^{n} \alpha_k \varphi(x_k) \\ \sum_{k=1}^{n} \alpha_k = 0 \\ \alpha_k = e_k \gamma \\ \omega^T \cdot \varphi(x_k) + b + e_k - y_k = 0 \end{cases} \tag{4}$$

Equation (4) can be transformed into linear equations:

$$\begin{bmatrix} 0 & 1 & \cdots & 1 \\ 1 & K(x_1, y_1) + \frac{1}{\gamma} & \cdots & K(x_l, y_l) \\ \vdots & \vdots & \ddots & \vdots \\ 1 & K(x_l, y_l) & \cdots & K(x_l, y_l) + \frac{1}{\gamma} \end{bmatrix} \begin{bmatrix} b \\ \alpha_1 \\ \vdots \\ \alpha_l \end{bmatrix} = \begin{bmatrix} 0 \\ y_1 \\ \vdots \\ y_l \end{bmatrix} \tag{5}$$

Once the values of α and b are found, the fitting function is represented as follows:

$$f(x) = \sum_{k=1}^{l} \alpha_k K(x, x_i) + b \tag{6}$$

where $K(x, x_i) = \varphi(x)^T \cdot \varphi(x_l)$ is the Kernel function, which expresses a nonlinear mapping from a low-dimensional to a high-dimensional space.

In this study, the radial basis function (RBF) was selected as the kernel function, which can be expressed as $K(x, x_i) = \exp\left(\frac{-\|x - x_i\|^2}{2\sigma^2} \right)$, where x is an m-dimensional input vector, x_i is the center of the ith RBF, σ^2 is the length of the Kernel function and $\|x - x_i\|$ is the norm of a vector, which is the distance between x and x_i.

It can be seen from the model that the LSSVR model transforms the original inequality constraints of the SVR model into equality constraints, which can improve the efficiency of operation, and further simplify the problem by solving equality constraints and least-squares problems.

2.2. Particle Swarm Optimization (PSO) Theory

The PSO, which simulates the hunting activities of birds and fish, is a random search algorithm that is able to solve the global optimal solution. The PSO is initialized with a random population of m particles, which is $X = \{X_1, X_2, \ldots, X_m\}$. Each particle is a point in a D-dimensional space and a feasible solution in the solution space. Particles change their position by moving in the solution space until arriving at the optimal solution. The position of the ith particle is $X_i = \{x_{i1}, x_{i2}, \ldots, x_{id}\}(i = 1, 2, \ldots, m)$ and the velocity of the ith particle is $V_i = \{v_{i1}, v_{i2}, \ldots, v_{id}\}(i = 1, 2, \ldots, m)$. $P_{best}(i)$ is the local best position, and G_{best} is the global best position. The position of the ith particle in the k_{th} iteration is computed using

$$x_{id}(k+1) = x_{id}(k) + v_{id}(k+1) \tag{7}$$

The velocity of the ith particle in the kth iteration is computed using

$$v_{id}(k+1) = w \cdot v_{id}(k) + c_1 r_1 \cdot (P_{best}(i) - x_{id}(k)) - c_2 r_2 \cdot (G_{best} - x_{id}(k)) \quad (8)$$

where r_1 and r_2 are two random numbers in a range of [0,1]; $c_1 = c_2 = 2$ are acceleration coefficients; and w is the inertia weight factor, which is determined using Equation (9).

$$w = w_{\max} - (w_{\max} - w_{\min}) \frac{n_i}{n_{\max}} \quad (9)$$

The velocity of each particle at the kth iteration depends on three components:

(1) The inertia term $w \cdot v_{id}(k)$ is affected by the constant inertia weight w and the previous step velocity term.
(2) The cognitive learning term $c_1 r_1 \cdot (P_{best}(i) - x_{id}(k))$ is the distance between the particle's best position so far found (called $P_{best}(i)$) and the particle current position (called $x_{id}(k)$).
(3) The social learning term $c_2 r_2 \cdot (G_{best} - x_{id}(k))$ is the distance between the global best position found thus far in the entire swarm (called G_{best}) and the particle's current position.

PSO implementation involves the following steps:

Step 1: Set the parameters of the PSO algorithm, such as the particle swarm size, inertial weight factor w, learning factors c_1 and c_2, and random numbers r_1 and r_2.

Step 2: Randomly initialize the velocity and position of the particles.

Step 3: Compute the fitness of the ith particle and obtain the local best position $P_{best}(i)$ and the global best position G_{best}.

Step 4: Update the local best position $P_{best}(i)$ and the global best position G_{best}.

Step 5: Update the velocity and position of the particles via Equations (7) and (8).

Step 6: If the maximum number of iterations is achieved or the optimal solution is found, then we may obtain the global best position; otherwise return to step 3.

Using different parameters as input combinations in LSSVR can produce a different model forecasting accuracy. Therefore, more importance is attached to the PSO-LSSVR model since it can select optimization parameters while modeling. The PSO algorithm optimizes the Kernel parameter σ^2 and regularization parameter γ in LSSVR, which can improve the prediction accuracy and reduce the uncertainty and randomness of the model results.

From the calculation process of the PSO model, the PSO model does not require the optimized function to have certain mathematical properties, such as being continuous, differentiable and derivable, which is especially suitable for small data samples, such as real estate markets.

2.3. Standardization Risk Matrix Method

The risk matrix method is a qualitative risk assessment analysis method that can comprehensively evaluate the risk of the possibility of danger and the severity of the injury [51]. Similarly, this method can be extended to all two-factor evaluation grading. This study divided the degree levels of the two factors at first and expresses them in the form of a matrix. According to the research needs, the result area is divided into different grades based on the effective diagonal line by using the risk matrix method for reference. When dividing the level of influencing factors, the influencing factors can be standardized. According to the distribution function of the standard normal distribution, the corresponding division points are determined, and the standardized influencing factors are divided based on the standard normal distribution.

The standardization risk matrix method implementation involves the following steps:

Step 1: Standardize the values of influencing factors according to historical data using the standardized formula.

$$x' = \frac{x - \mu}{\sigma} \sim N(0,1) \quad (10)$$

Step 2: According to the required classification number, find the corresponding classification points of the two influencing factors by the standard normal distribution table.

Step 3: Establish a risk matrix according to the levels of two factors.

Step 4: Classify the different units of the risk matrix using the diagonal lines.

Step 5: Substitute the standardized influencing factor values into the risk matrix and obtain the final grading.

Based on the research design, this study established an analytical framework to sort out selected research methods and research questions, as shown in Figure 2.

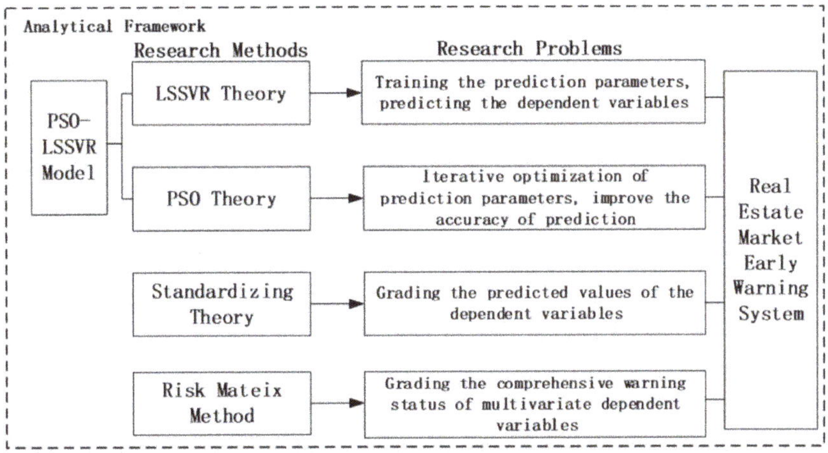

Figure 2. Analytical framework for the research methods and research problems.

3. Results

3.1. Choice of Urban Early Warning Indicators

The real estate market is a complex system, which is related to the constraints of urban development, urban residential economic behaviors and other fields; therefore, it is impossible to simply determine the relevant influencing indicators. In this study, considering the relationship between the real estate market and the national economy, the investment, gross domestic product (GDP), people's living standards and commercial housing prices were taken into account. On the other hand, the relationship between supply and demand in the real estate market is one of the key factors affecting the stable development of the real estate industry. Therefore, the sales of commercial housing, residential area per capital, Engels coefficient and housing income ratio were all taken into account. In addition, this study considered the coordination relationship within the real estate industry, extraction of construction area, completed area, residential and commercial housing construction investment ratio and other factors for analysis. Furthermore, the better-balanced development of measurement capabilities of relative indicators was considered rather than absolute indicators. Hence, most of the early warning indicators selected in this study were relative indicators. Furthermore, 16 indicators were initially selected as the early warning indicators for the urban real estate market from previous studies [52–63].

The indicator system of this study consisted of 2 variable types, 4 kinds of indicators and 16 sub-indicators, which can be seen in Table 1. To clearly show the details of the indicator system, the related factors and computation methods of the sub-indicators are listed below.

Table 1. Urban real estate market early warning indicators.

Types	Indicator	Sub-Indicators	Serial Number	References
Control variables	Relationship between the real estate market and the national economy	Real estate development investment level and urban economic development level	N1	[52]
		Real estate investment level	N2	[53]
		Real estate development self-raised financing level	N3	[54]
		Ratio of commercial housing price growth and urban residents' income growth	N4	[55]
		Real estate land area purchased level	N5	[56]
	Relationship between supply and demand	Commercial housing sales level	N6	[57]
		Land sales area level	N7	[54]
		Per capita residential area level	N8	[55,58]
		Commercial housing prices and urban residents' disposable income level	N9	[58,59]
	Inner relationship of the real estate industry	Residential investment level in commercial housing construction	N10	[52]
		Residential sales level	N11	[54]
		Residential area completion level	N12	[53]
		Proportion of new construction area of commercial housing	N13	[60,61]
		Construction and completed area ratio	N14	[62]
Dependent variables	Early warning indicators	Commercial housing price growth rate	N15	[55,61]
		Growth rate of the land area of commercial housing sales	N16	[56,63]

N1: Real estate development investment/GDP;

N2: Real estate investment/fixed-asset investment;

N3: Real estate development self-raised funds/real estate investment;

N4: Commercial housing price growth rate/urban residents' disposable income growth rate;

N5: Growth rate of real estate land area purchased/GDP growth rate;

N6: Commercial housing sales/real estate investment;

N7: Sales area/completed area;

N8: Per capita residential area/Engel coefficient;

N9: Commercial housing prices/urban residents' disposable income;

N10: Residential investment/commercial housing construction investment;

N11: Residential sales/all commercial housing sales;

N12: Completed residential area/completed area of all commercial housing;

N13: New construction area of commercial housing/construction area;

N14: Construction area/completed area.

For the dependent variables, the selection should consider the directness and speed of the reaction of factors. To reflect the situation of the real estate market comprehensively, it is necessary to analyze it from both the demand and supply aspects. From the demand aspect, housing consumption is one of the large types of asset consumption. Housing demand is highly sensitive to price, and the changes in housing prices are more likely to bring changes in the housing demand. Therefore, commercial housing price growth can be taken as one of the dependent variables to judge the status of the real estate market. From the supply aspect, the supply of land will largely determine the supply of the real estate market; therefore, the growth rate of the land area of commercial housing sales can also be taken as one of the dependent variables to judge the status of the real estate market.

The statistical data for the calculations were obtained from the China Statistical Yearbook (2001–2021) (National Bureau of Statistics), the China City Statistical Yearbook (2001–2021) (National Bureau of Statistics), the Beijing Statistical Yearbook, the China Entrepreneur Investment Club (CEIC) database and the China Real Estate Information

Corporation (CRIC) database [1,64–66], while the 2008–2020 statistical data came from the National Health and Family Planning Commission Migrant Population Service Centre. The data collected by these databases are highly accurate. As the capital and the political center of China, Beijing attaches great importance to the need for social stability. As a major factor affecting social stability, the housing problem has attracted the attention of the local government and society. Therefore, it is particularly necessary to establish an early warning system for the real estate market in Beijing. At the same time, Beijing is also one of the megacities with economic development in China, and the development of its real estate market is relatively thorough. However, China's real estate market originated in the 1990s of the last century, and various market mechanisms did not take shape until the 21st century, with the market gradually becoming mature. On this basis, according to the exploitability of data, the time frame of the research was determined as 2000–2020. The detailed data of early warning indicators for the Beijing real estate market from 2000 to 2020 are shown in Table 2.

Table 2. Early warning indicators data of the Beijing real estate market from 2000 to 2020.

Years	N1	N2	N3	N4	N5	N6	N7	N8	N9	N10	N11	N12	N13	N14	N15	N16
2000	0.16	0.4	0.25	−0.95	4.037	1.83	0.70	0.47	0.53	0.55	0.87	0.74	0.64	2.99	−0.13	0.76
2001	0.20	0.51	0.28	0.23	1.456	1.54	0.71	0.5	0.49	0.59	0.87	0.82	0.73	3.06	0.03	0.26
2002	0.22	0.55	0.29	−0.68	2.428	1.73	0.72	0.6	0.42	0.59	0.88	0.81	0.77	2.6	−0.06	0.42
2003	0.23	0.56	0.31	−0.05	0.67	1.58	0.73	0.66	0.38	0.53	0.88	0.8	0.81	2.7	−0.01	0.11
2004	0.24	0.58	0.29	0.49	1.625	1.68	0.81	0.72	0.35	0.53	0.87	0.78	0.76	2.45	0.07	0.30
2005	0.21	0.54	0.4	1.96	0.934	1.84	0.74	0.75	0.4	0.51	0.82	0.75	0.76	2.13	0.34	0.13
2006	0.21	0.51	0.32	1.52	−0.403	1.52	0.82	0.84	0.43	0.50	0.75	0.69	0.75	2.23	0.22	−0.07
2007	0.19	0.50	0.43	3.53	−0.68	1.09	0.75	0.86	0.54	0.50	0.73	0.64	0.74	2.35	0.40	−0.17
2008	0.16	0.50	0.49	0.55	−2.904	0.7	0.52	0.91	0.51	0.49	0.72	0.55	0.71	2.6	0.07	−0.39
2009	0.18	0.48	0.44	1.23	8.351	1.01	0.88	0.96	0.52	0.39	0.76	0.60	0.68	2.64	0.11	0.77
2010	0.19	0.53	0.61	2.89	−1.913	0.57	0.69	1.08	0.61	0.52	0.71	0.63	0.66	3.32	0.29	−0.31
2011	0.18	0.51	0.58	−0.39	−0.82	0.47	0.64	1.18	0.51	0.59	0.66	0.59	0.67	3.93	−0.05	−0.12
2012	0.17	0.49	0.51	0.09	3.275	0.62	0.81	1.18	0.46	0.52	0.74	0.64	0.65	3.86	0.01	0.35
2013	0.16	0.50	0.61	0.83	−0.188	0.55	0.71	1.3	0.45	0.50	0.69	0.63	0.65	3.55	0.09	−0.02
2014	0.17	0.49	0.46	0.17	−2.753	0.37	0.48	1.31	0.42	0.50	0.77	0.59	0.63	2.94	0.02	−0.23
2015	0.17	0.52	0.54	2.26	0.811	0.37	0.59	1.41	0.47	0.45	0.71	0.52	0.65	3.03	0.20	0.07
2016	0.15	0.47	0.49	2.56	0.848	0.41	0.70	1.51	0.52	0.48	0.61	0.53	0.58	3.25	0.21	0.08
2017	0.13	0.41	0.46	1.89	−4.545	0.23	0.60	1.61	0.56	0.46	0.74	0.41	0.59	4.98	0.17	−0.48
2018	0.12	0.46	0.4	0.70	−1.895	0.18	0.45	1.64	0.55	0.48	0.66	0.53	0.72	4.46	0.06	−0.20
2019	0.11	0.47	0.31	0.60	5.094	0.24	0.70	1.65	0.53	0.49	0.81	0.43	0.74	4.82	0.05	0.35
2020	0.11	0.47	0.36	1.98	1.65	0.25	0.63	1.55	0.54	0.57	0.83	0.63	0.79	5.93	0.05	0.03
Mean	0.17	0.50	0.42	1.02	0.72	0.89	0.68	1.08	0.49	0.51	0.77	0.63	0.70	3.32	0.10	0.08
SD	0.04	0.04	0.11	1.21	2.94	0.61	0.11	0.39	0.07	0.05	0.08	0.12	0.07	1.01	0.14	0.34
CV	0.21	0.09	0.27	1.19	4.10	0.69	0.16	0.36	0.14	0.10	0.11	0.19	0.09	0.30	1.33	4.33

In order to better analyze the data of indicators N1 to N16 from 2020 to 2020, the mean, standard deviation (SD) and coefficient of variation (CV) of the indicators are listed in Table 2. The CVs of N1, N2, N3, N7, N9, N10, N11, N12 and N13 were less than 0.3. which indicated that the changes in these indicators were stable. In addition, according to Rayda's criterion, $p\{\mu - 2\sigma \leq X_i \leq \mu + 2\sigma\} = 0.9544$, data with a difference of more than two standard deviations (SDs) from the mean can basically be regarded as abnormal data. Therefore, it can be seen that the value of N4 was abnormal in 2007 and that of N5 was abnormal in 2009. Among other indicators, except for 2008, the value of N6 had a relatively obvious downward trend, the value of N8 has a relatively obvious upward trend, and the value of N14 had a relatively obvious improvement after 2017.

3.2. Training and Testing of PSO-LSSVR

The PSO algorithm has the advantages of fast convergence speed, collaborative search and can easily leap over local optimal information. It can balance the global detection and local mining ability of the LSSVR algorithm and dynamically adjust the parameters of the LSSVR model linearly (or nonlinearly) according to the iterative process and particle flight. To balance the global search and convergence speed, in this study, the upper limit of the adjustment iterations was set at 300. Through further optimization of the LSSVM model, the accuracy of the prediction model was further improved. The iteration processes of the PSO model for indicators N15 and N16 are shown in Figures 3 and 4, respectively.

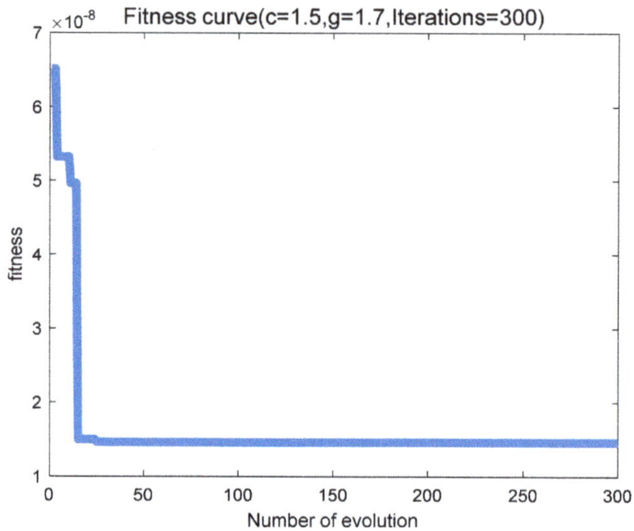

Figure 3. The iterative process of the PSO model for the N15 early warning indicator.

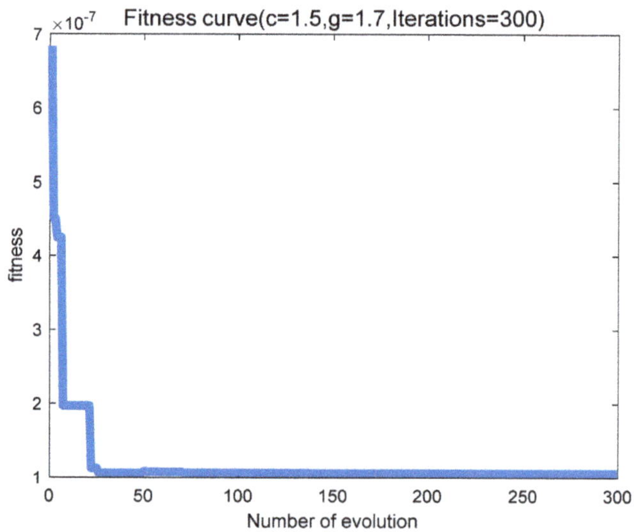

Figure 4. The iterative process of the PSO model for the N16 early warning indicator.

Figures 3 and 4 are the fitting curves of the indicators N15 and N16 in the training of the PSO-LSSVR model, respectively. It can be seen from the curve that the indicators N15 and N16 both tended to converge above the 300-fold fitting degree. It shows that the PSO model was better for the optimization of the LSSVR model. After the optimization of the PSO model, its prediction accuracy was greatly improved. Furthermore, the PSO-LSSVR model was used to train and test the indicators data of the Beijing real estate market, and Figures 5 and 6 were obtained for N15 and N16, respectively.

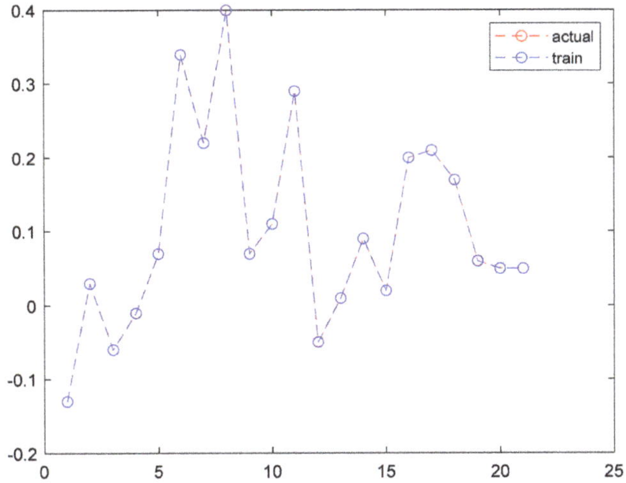

Figure 5. The training values and actual values of the N15 indicator.

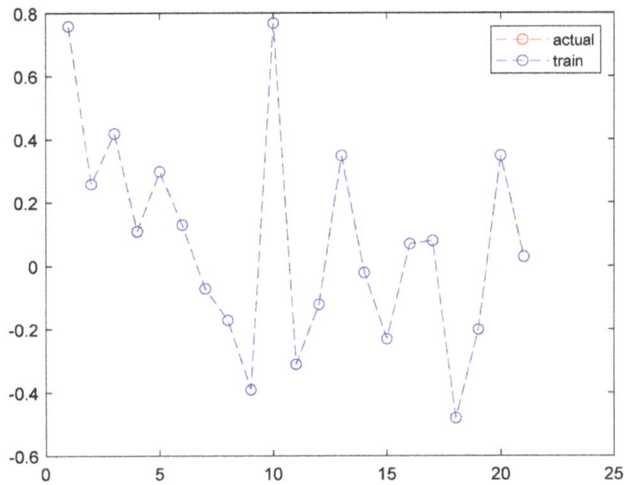

Figure 6. The training values and actual values of the N16 indicator.

Figures 5 and 6 are graphs that reflect the comparison between the training values and actual values obtained in the PSO-LSSVR model of the indicators N15 and N16, respectively. The blue points in the figure represent the results of the training values, and the red points represent the actual values. It can be seen from the figure that the training values obtained in the PSO-LSSVR model were highly consistent with the actual values. This situation showed that the prediction accuracy of the PSO-LSSVR model was very high, and the model could be used to predict the urban real estate market over the next ten years.

3.3. Prediction for Indicators N15 and N16 Based on PSO-LSSVR

In this study, the PSO-LSSVR model was constructed to predict the results of the early warning indicators N15 and N16 for the next ten years. First, the trend extrapolation method was adopted to extrapolate the results of the training indicators N1 to N14 over the next ten years. Then, the PSO-LSSVR model was used to predict the values for the next ten years, as shown in Table 3.

Table 3. The predicted values of the influencing factors N15 and N16 for the next 10 years.

Year	2021	2022	2023	2024	2025	2026	2027	2028	2029	2030
N15	0.0802	0.0675	0.0781	0.0838	0.0906	0.0940	0.0970	0.0985	0.0994	0.0998
N16	0.0686	0.0565	0.0730	0.0706	0.0747	0.0757	0.0766	0.0770	0.0772	0.0772

The values in Table 3 are collated into Figures 7 and 8 to show the numerical trend of indicators N15 and N16 for the next ten years more clearly.

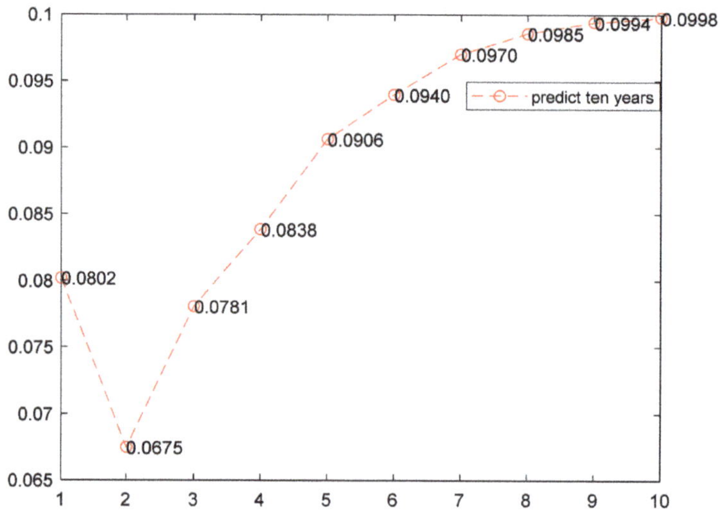

Figure 7. The prediction results of indicator N15 for the next 10 years.

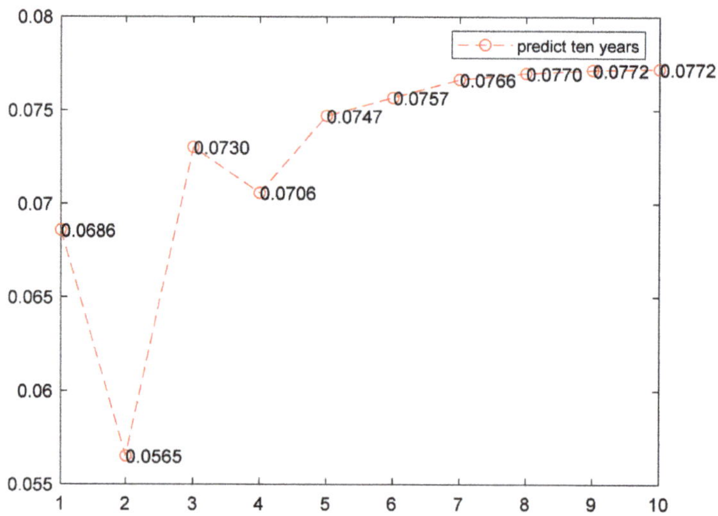

Figure 8. The prediction results of indicator N16 for the next 10 years.

Figures 7 and 8 respectively reflect the predicted values of the indicators N15 and N16 for the next ten years. It can be seen from the prediction results that the indicators N15 and N16 are expected to both show an overall upward trend for the next 7 years. However, in the fourth year, N15 is expected to continue its upward trend, while N16 is expected to decline. The following table reflects the predicted values of the influencing factors N15 and N16 for the next ten years.

3.4. Early Warning Degree of Beijing Real Estate Market

In this study, the status of the urban real estate market was divided into three categories, namely, "cold", "normal" and "hot", where the ranges of the factor standard values were used to determine the three categories. Considering the related research, the intervals for this study could be determined as one standard deviation above and below the 0; therefore, the state space could be divided into three intervals as $(-\infty,-1)$,$(-1,1)$,$(1,\infty)$ [34]. Subsequently, the three states of indicators N15 and N16 were divided into a matrix. According to the Guidelines for Tunneling Risk Management, the triangle of the risk matrix represents the line of two elements in the median status of one dimension and the extreme status of the other dimension represents one status. Since N15 and N16 are both factors reflecting the market conditions of the real estate market, there should not be a mixed "cold" and "hot" status. However, it is considered abnormal that two dependent variables exhibit diametrically opposite judgments; therefore, such states can be remarked as "abnormal". Based on the above analysis, the states where both factors were cold or there was a mixture of normal and cold were marked as "cold", the states where both factors were hot or there was a mixture of normal and hot were marked as "hot", the states where the two factors were a mixture of cold and hot were marked as "abnormal" and the remaining condition with both factors being normal was marked as "Normal". The results grading table was obtained as shown in Table 4 below.

Table 4. Two-factor status grading mark table.

Header	Cold	Normal	Hot
Cold	Cold	Cold	Abnormal
Normal	Cold	Normal	Hot
Hot	Abnormal	Hot	Hot

After standardizing the original data and prediction results of indicators N15 and N16 from 2000 to 2030 in Section 3.3, the results are shown in Tables 5 and 6. The indicator data column represents the raw data collected and the standardized value column represents the raw data normalized using the standardized formula $\bar{x} = \frac{x-\mu}{\sigma}$.

Table 5. The warning status of the Beijing real estate market from 2000 to 2020.

Year	N15			N16			Warning Degree
	Indicator Data	Standardized Value	Warning Degree	Indicator Data	Standardized Value	Warning Degree	
2000	−0.13	−1.71	Cold	0.76	2.02	Hot	Abnormal
2001	0.03	−0.54	Normal	0.26	0.54	Normal	Normal
2002	−0.06	−1.19	Cold	0.42	1.01	Normal	Cold
2003	−0.01	−0.80	Normal	0.11	0.09	Normal	Normal
2004	0.07	−0.26	Normal	0.30	0.67	Normal	Normal
2005	0.34	1.78	Hot	0.13	0.17	Normal	Hot
2006	0.22	0.87	Normal	−0.07	−0.44	Normal	Normal
2007	0.40	2.17	Hot	−0.17	−0.72	Normal	Hot
2008	0.07	−0.20	Normal	−0.39	−1.38	Cold	Cold
2009	0.11	0.07	Normal	0.77	2.05	Hot	Hot
2010	0.29	1.38	Hot	−0.31	−1.14	Cold	Abnormal

Table 5. *Cont.*

Year	N15			N16			Warning Degree
	Indicator Data	Standardized Value	Warning Degree	Indicator Data	Standardized Value	Warning Degree	
2011	−0.05	−1.14	Cold	−0.12	−0.59	Normal	Cold
2012	0.01	−0.68	Normal	0.35	0.81	Normal	Normal
2013	0.09	−0.09	Normal	−0.02	−0.29	Normal	Normal
2014	0.02	−0.64	Normal	−0.23	−0.92	Normal	Normal
2015	0.20	0.74	Normal	0.07	−0.04	Normal	Normal
2016	0.21	0.83	Normal	0.08	0.00	Normal	Normal
2017	0.17	0.49	Normal	−0.48	−1.65	Cold	Cold
2018	0.06	−0.30	Normal	−0.20	−0.84	Normal	Normal
2019	0.05	−0.37	Normal	0.35	0.80	Normal	Normal
2020	0.05	−0.39	Normal	0.03	−0.13	Normal	Normal
Mean	0.10	-	-	0.08	-	-	-
SD	0.14	-	-	0.34	-	-	-

Table 6. The predicted warning status of the Beijing real estate market from 2020 to 2030.

Year	N15			N16			Warning Degree
	Indicator Data	Standardized Value	Warning Degree	Indicator Data	Standardized Value	Warning Degree	
2021	0.08	−0.16	Normal	0.07	−0.03	Normal	Normal
2022	0.07	−0.26	Normal	0.06	−0.06	Normal	Normal
2023	0.08	−0.18	Normal	0.07	−0.02	Normal	Normal
2024	0.08	−0.14	Normal	0.07	−0.02	Normal	Normal
2025	0.09	−0.09	Normal	0.07	−0.01	Normal	Normal
2026	0.09	−0.06	Normal	0.08	−0.01	Normal	Normal
2027	0.10	−0.04	Normal	0.08	0.00	Normal	Normal
2028	0.10	−0.03	Normal	0.08	0.00	Normal	Normal
2029	0.10	−0.02	Normal	0.08	0.00	Normal	Normal
2030	0.10	−0.02	Normal	0.08	0.00	Normal	Normal

From Table 5, the warning statuses of the Beijing real estate market from 2000 to 2020 can be obtained. From 2000 to 2011, the warning statuses were chaotic, with the exception of the years 2001, 2003, 2004 and 2006. After the year 2011, the warning status gradually approached the "Normal" status, except for the year 2017, when it was "Cold".

From Table 6, the warning statuses of the Beijing real estate market from 2021 to 2030 can be obtained. The warning degrees of indicators N15 and N16 from 2021 to 2030 were all "Normal", and the warning statuses of the Beijing real estate market were also "Normal".

4. Discussion

4.1. The Validation of PSO-LSSVR Model

From the training of the PSO-LSSVR model on historical data of the Beijing real estate market, it can be seen that the training values had a high fitting degree to the historical data. From Figures 3 and 4, the results showed that the PSO optimization achieved the corresponding accuracy and stability quickly. The fitting degree illustrated that the PSO-LSSVR model had strong prediction accuracy and could effectively predict the information related to the real estate market.

4.2. Analysis of the Early Warning Status from 2000 to 2020

From the warning status from 2000 to 2020, it can be seen that the statuses of indicator N16 were relatively stable, where the warning statuses were basically in the "Normal" state, except for the fluctuation around 2010. For indicator N15, the warning statuses before 2012 were chaotic, with "Cold" or "Hot" states for many years, and after 2012, the statuses of indicator N15 tended to be stable. Reviewing the timing of the release and implementation of the relevant national real estate market regulation policies, it can be found that the General Office of the State Council published a *Notice on Promoting the Steady and Healthy Development of the Real Estate Market* in January 2010, which set out requirements and policies for housing price regulation and land supply management [67]. From the warning degrees of the Beijing real estate market, it can be seen that after a short policy lag period, the warning status tended to be stable.

4.3. Analysis of the Early Warning Status from 2021 to 2030

From the predicted early warning statuses of the Beijing real estate market from 2021 to 2030, there were slight fluctuations in the prediction results of both indicators N15 and N16. However, from the overall prediction results for the next decade, in the absence of major macro policy changes, the growth rates of N15 and N16 are expected to gradually slow down and eventually become stable. The warning degrees of the Beijing real estate market are expected to be in a stable "Normal" status.

On the other hand, the outbreak of COVID-19 at the end of 2019 had a great impact on the national economy and industry, including the real estate market. In fact, the slowdown of the material supply chain and the downturn of the consumer market is expected to lead to a certain slowdown in the growth rate of land sales and housing prices. It may lead to a "cold" status in the real estate market, and the government should put forward corresponding promotion strategies to manage the "cold" status of the real estate market during the epidemic.

5. Conclusions

In this study, a real estate early warning system was established based on the PSO-LSSVR method, which has good generalization ability and performance in the processing of small sample data. In the original sample data, the results for the years 2007, 2008, 2009 and 2010 were in an abnormal status because of frequent state policies regarding land supply, interest rates being reduced five times by the central bank in 2008, etc. Hence, the real estate market in Beijing showed continuous fluctuation from containment to stimulation and then to containment. By comparing the predicted values with the actual values, it was found that the prediction status of the model was very good; furthermore, the prediction values for N15 and N16 over the next five years were calculated using the prediction model combined with the trend extrapolation method. The results showed that the status of the real estate market in Beijing is expected to be "normal" from 2020 to 2024. The current policy theme of China's real estate market is to maintain the stability of the market. However, although China's real estate market is currently in a relatively stable state, it still needs to be constantly improved in terms of the system construction, financing channels and supervision system. In addition, it is necessary to emphasize the property of housing as consumer goods rather than investment goods, promote a people-oriented development policy, ensure the healthy and stable development of the real estate market, and prevent the formation of real estate market bubbles.

1. Balance regional housing supply and demand while promoting real estate marketization

The government's encouraging policies and reasonable planning for the real estate market are the premise to ensure its healthy and stable development. In the process of long-term planning of the urban real estate market, the development trend of population flow and the urbanization process should be analyzed, and the market demand should be predicted. Furthermore, a reasonable supply of land is vital for housing supply and demand; therefore, the urban government should plan reasonable land consolidation based on the market demand and expectations. In addition, the regularization management of the housing rental market is also one of the key points for promoting the regulation of urban housing supply and demand. Moreover, the reasonable supply of affordable housing is a key factor to guarantee the housing demand of middle- and low-income groups, and also a part of the balance between urban housing supply and demand.

2. Construct the long-term mechanism to promote the healthy development of the real estate market

From the analysis, the investment and the consumption of the real estate market are related to the development of the national economy. The government should balance the inner relationship through regulatory measures to prevent the market from overheating or overcooling. Furthermore, a long-term mechanism must be the foundation to ensure the healthy and sustainable development of the real estate market. The macroeconomic regulation and control strategies should ensure compliance with market rules. The government should also clarify the detailed issues of housing security based on specific laws. Moreover, a scientific and reasonable housing security system is vital, where the object of protection and its scope of protection should be explicit. The housing security system with diversified supply is a vital part of the housing structure system.

3. Expand financing channels in the real estate market and regulate financing behaviors in the real estate market

In order to better improve the mechanism of the urban real estate market, it is necessary to strengthen management from the perspective of financing. For affordable housing projects, the urban government should encourage financial organizations to provide multifaceted financing methods and encourage developers to build affordable housing by means of policy support. For the housing security system, it is important to enhance low-income people's ability to finance and pay by coordinating the housing security system with the housing provident fund, subsistence allowances for urban residents, unemployment insurance, old-age insurance and other social security systems. Simultaneously, the urban government should resolutely curb the purchase of houses for investment and speculation purposes to promote the steady and healthy development of the real estate market.

4. Strengthen supervision policies in the real estate market

To further strengthen the real estate market supervision, the government should standardize the market order, purify the market environment, curb unreasonable sales behaviors of developers and control property speculation by all parties to ensure the steady and healthy development of the real estate market. For the sale of commercial housing projects, all kinds of ways to collect deposits, advance payments and other charges, as well as the requirement of bank-verified certificates of buyers, should be strictly regulated and controlled. Real estate development enterprises should set reasonable prices according to market demand and standardize the price behavior of the commodity housing market. The provisions of the clearly marked price should be strictly implemented, and the sales price publicized should be completely consistent with the price department's record price.

As a complex industry, the development of the real estate market is associated with many related industries, such as the construction, decoration and materials industries. The real estate market as an indispensable part of urban development is also related to regional economic development, population flow, and social health and stability. In order to further promote the healthy and stable development of the real estate market, researchers should conduct a comprehensive system and analyze the dynamic trends to study the impact of relevant policy regulations on the trends of the real estate market.

Author Contributions: Methodology, L.M.; software, Z.C.; validation, L.W., X.R. and L.M.; formal analysis, Z.C.; investigation, L.W.; resources, L.M.; data curation, S.J.; writing—original draft preparation, L.W.; writing—review and editing, L.M.; supervision, X.R.; funding acquisition, L.M. and X.R. All authors have read and agreed to the published version of the manuscript.

Funding: This research was funded by "the National Natural Science Foundation of China (NSFC), grant number: 71904042"; "the Social Science Foundation of Hebei Province, grant number: HB17GL030"; and "the Social Science Development Research Project of Hebei Province, grant number: 20210101021".

Institutional Review Board Statement: Not applicable.

Informed Consent Statement: Not applicable.

Data Availability Statement: Not applicable.

Conflicts of Interest: The authors declare no conflict of interest.

References

1. Statistics Bureau of the People's Republic of China, China Statistical Yearbook. 2021. Available online: http://www.stats.gov.cn/ (accessed on 18 March 2022). (In Chinese)
2. Yan, J.Z.L. Empirical Researches on Macroeconomic Influence Factors in Real Estate Based on Data Mining (DM). *Agro Food Ind. Hi-Tech* **2017**, *28*, 2729–2732.
3. Kuil, P.; Wellner, K. The Imperatives of the Real Estate Industry and its Effects on Architecture and the Built Environment; Case Studies Presentation. In Proceedings of the 24th Annual European Real Estate Society Conference, Delft, The Netherlands, 30 June 2017.
4. Carrasco-Gallego, J.A. Real Estate, Economic Stability and the New Macro-Financial Policies. *Sustainability* **2020**, *13*, 236. [CrossRef]
5. Jian, C.; Cao, P.; Zhou, X.; Lai, K.K.; Chen, X.; Su, S. The Conductive and Predictive Effect of Oil Price Fluctuations on China's Industry Development Based on Mixed-Frequency Data. *Energies* **2018**, *11*, 1372.
6. Brzezicka, J. Towards a typology of housing price bubbles: A literature review. *Hous. Theory Soc.* **2021**, *38*, 320–342. [CrossRef]
7. Machaj, M. Can the Taylor Rule be a Good Guidance for Policy? The Case of 2001–2008 Real Estate Bubble. *Prague Econ. Pap.* **2016**, *25*, 381–395. [CrossRef]
8. Liow, K.H.; Zhou, X.; Ye, Q. Correlation Dynamics and Determinants in International Securitized Real Estate Markets. *Real Estate Econ.* **2015**, *43*, 537–585. [CrossRef]
9. Yan, X.; Lin, X. Business Expanding and Strategic Decision-Making Innovation for Real Estate Valuation Companies Based on Big-Data. In Proceedings of the 20th International Symposium on Advancement of Construction Management and Real Estate; Springer: Singapore, 2017; pp. 1057–1068.
10. Gu, H. Macroeconomic early warning research: Theory, method, history. *Econ. Theor. Econ. Manag.* **1997**, *4*, 1–7.
11. Xiao, L. Research on Urban Residential Area Early Warning System Based on Artificial Neural Network. *Agro Food Ind. HI-Tech* **2017**, *28*, 2250–2254.
12. Vergara-Perucich, J.; Aguirre-Nunez, C. Housing Prices in Unregulated Markets: Study on Verticalised Dwellings in Santiago de Chile. *Buildings* **2020**, *10*, 6. [CrossRef]
13. Echeverri, N.; Jylha, T.; Koppels, P. Searching for Flexibility in Corporate Real Estate Portfolio: Six Co-Working Strategies for User Corporations. *Buildings* **2021**, *11*, 115. [CrossRef]
14. Zhao, X.; Gao, Y. Impact of Transaction Characteristics on the Performance of Mergers and Acquisitions (M and A) in Chinese Real Estate Industry. *J. Tianjin Univ. (Soc. Sci.)* **2017**, *19*, 119–124.
15. Kang, J.; Lee, H.; Jeong, S.; Lee, H.; Oh, K. Developing a forecasting model for real estate auction prices using artificial intelligence. *Sustainability* **2020**, *12*, 2899. [CrossRef]
16. Yu, Y.; Lu, J.; Shen, D.; Chen, B. Research on real estate pricing methods based on data mining and machine learning. *Neural Comput. Appl.* **2021**, *33*, 3925–3937. [CrossRef]
17. Pyhrr, S.A.; Born, W.L.; Robinson, R.; Lucas, S.R. Real property valuation in a changing economic and market cycle. *Apprais. J.* **1996**, *64*, 14.
18. Pyhrr, S.A. A Computer Simulation Model to Measure the Risk in Real Estate Investment. *Real Estate Economics.* **2010**, *1*, 48–78. [CrossRef]
19. Nieboer, N.; Voogd, H. Housing vacancy and early warning systems. *Neth. J. Hous. Environ. Res.* **1990**, *5*, 237–249. [CrossRef]
20. Hoffman, A.V.; Belsky, E.S.; Lee, K. *The Impact of Housing on Community: A Review of Scholarly Theories and Empirical Research, Joint Center for Housing Studies, Graduate School of Design and John F. Kennedy School of Government*; Harvard University: Cambridge, MA, USA, 2006.
21. Chan, K.F.; Treepongkaruna, S.; Brooks, R.; Gray, S. Asset market linkages: Evidence from financial, commodity and real estate assets. *J. Bank. Financ.* **2011**, *35*, 1415–1426. [CrossRef]
22. Gonzalez, M.R.; Basse, T.; Kunze, F.; Vornholz, G. Early warning indicator systems for real estate investments: Empirical evidence and some thoughts from the perspective of financial risk management. *Z. Für Gesamte Versicher.* **2018**, *107*, 387–403. [CrossRef]
23. Huang, A.; Qiu, L.; Li, Z. Applying Deep Learning Method in TVP-VAR Model under Systematic Financial Risk Monitoring and Early Warning. *J. Comput. Appl. Math.* **2021**, *382*, 113065. [CrossRef]
24. Loris, A.B. A Financial Early Warning System for Over-the-Counter Broker-Dealers. *J. Financ.* **1976**, *31*, 1201–1217.
25. Caruzzo, A.; Belderrain, M.C.N.; Fisch, G.; Young, G.S.; Hanlon, C.J.; Verlinde, J. Modelling weather risk preferences with multi-criteria decision analysis for an aerospace vehicle launch. *Meteorol. Appl.* **2018**, *25*, 456–465. [CrossRef]
26. Burnaev, E.V. On Construction of Early Warning Systems for Predictive Maintenance in Aerospace Industry. *J. Commun. Technol. Electron.* **2019**, *64*, 1473–1484. [CrossRef]
27. Clements, C.F.; Mccarthy, M.A.; Blanchard, J.L. Early warning signals of recovery in complex systems. *Nat. Commun.* **2019**, *10*, 1681. [CrossRef] [PubMed]
28. Kong, Q.; Allen, R.M.; Schreier, L.; Kwon, Y.-W. MyShake: A smartphone seismic network for earthquake early warning and beyond. *Sci. Adv.* **2016**, *2*, e1501055. [CrossRef] [PubMed]

29. Li, Z.; Meier, M.-A.; Hauksson, E.; Zhan, Z.; Andrews, J. Machine Learning Seismic Wave Discrimination: Application to Earthquake Early Warning. *Geophys. Res. Lett.* **2018**, *45*, 4773–4779. [CrossRef]
30. Wang, J.; Wang, J. A New Early Warning Method of Train Tracking Interval Based on CTC. *IEEE Trans. Intell. Transp. Syst.* **2017**, *99*, 1–7. [CrossRef]
31. Wang, H.; Li, L.; Pan, P.; Wang, Y.; Jin, Y. Early warning of burst passenger flow in public transportation system. *Transp. Res. Part C Emerg. Technol.* **2019**, *105*, 580–598. [CrossRef]
32. Kalliomäki, J.; Kontula, T.; Kalliomäki, M.L.; Kivipuro, M.; Tirkkonen, J.; Solin, J.; Pauniaho, S.-L.; Huhtala, H.; Yli-Hankala, A.; Hoppu, S. The national early warning score (NEWS) could not predict secondary transportation to the tertiary hospital in Finnish prehospital care. *Resuscitation* **2016**, *106*, e69. [CrossRef]
33. Kovacs, C. Outreach and early warning systems for the prevention of intensive care admission and death of critically ill adult patients on general hospital wards. *Int. J. Nurs. Pract.* **2016**, *22*, 523–525. [CrossRef]
34. Wang, X.-J.; Zeng, G.-T.; Zhang, K.-X.; Chu, H.-B.; Chen, Z.-S. Urban real estate market early warning based on support vector machine: A case study of Beijing. *Int. J. Comput. Intell. Syst.* **2020**, *13*, 153–166. [CrossRef]
35. Begusic, S.; Kostanjcar, Z.; Kovac, D.; Stanley, H.; Podobnik, B. Information feedback in temporal networks as a predictor of market crashes. *Complexity* **2018**, *2018*, 2834680. [CrossRef]
36. Huang, F.; Feng, W. A system for early-warning and forecasting of real estate development. *Autom. Constr.* **2005**, *14*, 333–342. [CrossRef]
37. Yang, J. Design of early warning multimedia computer system for real estate market under PROBIT model. *Multimed. Tools Appl.* **2019**, 1–15. [CrossRef]
38. Kholodilin, K.A.; Michelsen, C. High Risk of a Housing Bubble in Germany and Most OECD Countries. *DIW Wkly. Rep.* **2019**, *9*, 265–273.
39. Kim, T.Y.; Oh, K.J.; Sohn, I.; Hwang, C. Usefulness of Artificial Networks for Early Warning System of Economic Crisis. *Expert Syst. Appl.* **2004**, *6*, 583–590. [CrossRef]
40. Zheng, Y.J.; Chen, S.Y.; Xue, Y. A Pythagorean-Type Fuzzy Deep Denoising Autoencoder for Industrial Accident Early Warning. *IEEE Trans. Fuzzy Syst.* **2017**, *25*, 1561–1575. [CrossRef]
41. Park, D.; Ryu, D. A Machine Learning-Based Early Warning System for the Housing and Stock Markets. *IEEE Access.* **2021**, *9*, 3077962. [CrossRef]
42. Cortes, C.; Vapnik, V. Support-vector network. *Mach. Learn.* **1995**, *20*, 273–295. [CrossRef]
43. Izadyar, N.; Ghadamian, H.; Chong, W.T.; Moghadam, Z.; Tong, C.W.; Shamshirband, S. Appraisal of the support vector machine to forecast residential heating demand for the District Heating System based on the monthly overall natural gas consumption. *Energy* **2015**, *93*, 1558–1567. [CrossRef]
44. Suykens, J.; Vandewalle, J. Least Squares Support Vector Machine Classifiers. *Neural Process. Lett.* **1999**, *9*, 293–300. [CrossRef]
45. Chou, J.; Ngo, N. Time series analytics using sliding window metaheuristic optimization-based machine learning system for identifying building energy consumption patterns. *Appl. Energy* **2016**, *177*, 751–770. [CrossRef]
46. Xue, X.; Xiao, M. Deformation evaluation on surrounding rocks of underground caverns based on PSO-LSSVM. *Tunn. Undergr. Space Technol.* **2017**, *69*, 171–181. [CrossRef]
47. Chen, Y.; Zhou, Z.; Chen, Q. The research and application of LS-SVM based on particle swarm optimization. In Proceedings of the IEEE International Conference on Automation and Logistics, Jinan, China, 18–21 August 2007; pp. 1115–1120.
48. Wang, B.; Tang, W.; Song, L.; Bai, G. PSO-LSSVR: A surrogate modeling approach for probabilitic flutter evaluation of compressor blade. *Structures* **2020**, *28*, 1634–1645. [CrossRef]
49. Sun, W.; Zheng, S.; Geltner, D.M.; Wang, R. The Housing Market Effects of Local Home Purchase Restrictions: Evidence from Beijing. *J. Real Estate Financ. Econ.* **2017**, *55*, 288–312. [CrossRef]
50. Li, Y.; Zhu, D.; Zhao, J.; Zheng, X.; Zhang, L. Effect of the Housing Purchase Restriction Policy on the Real Estate Market: Evidence from a Typical Suburb of Beijing, China. *Land Use Policy* **2020**, *94*, 104528. [CrossRef]
51. Degn Eskesen, S.; Tengborg, P.; Kampmann, J.; Veicherts, T.H. Guidelines for tunnelling risk management: International Tunnelling Association, Work Group No.2. *Tunn. Undergr. Space Technol.* **2004**, *19*, 217–237. [CrossRef]
52. Yin, S. Research on Early Warning of Real Estate Market in Harbin. Master's Thesis, Northeast Forestry University, Harbin, China, 2014.
53. Yi, X. China's Real Estate Market Overheating and Risk Early Warning. *Financ. Trade Economics* **2005**, *5*, 14–21.
54. Qian, S. Study on Real Estate Monitoring and Early Warning Model in Beijing Based on Neural Network. Master's Thesis, University of International Business and Economics, Beijing, China, 2009.
55. Han, L. Research on Urban Real Estate Early Warning System in China. Ph.D. Thesis, Sichuan University, Chengdu, China, 2004.
56. Shi, Y.; Wang, P. Research on Early-warning Index System and Integrated Warning of Real Estate. *Stat. Res.* **2011**, *28*, 16–21.
57. Wang, J. Research on Real Estate Market Early Warning System Based on Artificial Neural Network. Master's Thesis, Zhejiang University, Hangzhou, China, 2006.
58. Tajani, F.; Liddo, F.; Guarini, M.; Ranieri, R.; Anelli, D. An Assessment Methodology for the Evaluation of the Impacts of the COVID-19 Pandemic on the Italian Housing Market Demand. *Buildings* **2022**, *11*, 592. [CrossRef]
59. Yang, L. Research on Early Warning System of Real Estate Market. Master's Thesis, Wuhan, Wuhan University, 2005.
60. Ding, L.; Li, B. Technical Points and Process Design of Early Warning Control System in Real Estate Market. *Syst. Eng. Theory Pract.* **2002**, *4*, 58–62.

61. Ullah, F.; Sepasgozar, S. Key Factors Influencing Purchase or Rent Decisions in Smart Real Estate Investments: A System Dynamics Approach Using Online Forum Thread Data. *Sustainability* **2020**, *12*, 4382. [CrossRef]
62. He, F. Design and Application of Real Estate Early Warning System Based on BP Neural Network. Master's Thesis, Huazhong Agricultural University, Wuhan, China, 2011.
63. Su, X.; Qian, Z. State Intervention in Land Supply and Its Impact on Real Estate Investment in China: Evidence from Prefecture-Level Cities. *Sustainability* **2020**, *12*, 1019. [CrossRef]
64. Statistics Bureau of the People's Republic of China, China City Statistical Yearbook. 2021. Available online: https://www.stats.gov.cn/ (accessed on 18 March 2022).
65. China Entrepreneur Investment Club, China Entrepreneur Investment Club (CEIC) Database. 2021. Available online: https://www.ceicdata.com/zh-hans (accessed on 18 March 2022).
66. China Real Estate Information Corporation, China Real Estate Information Corporation (CRIC). 2021. Available online: https://cric.dichan.com/system.html (accessed on 18 March 2022).
67. Chang, L. Historical Review and Prospect of China's Real Estate Regulation Policy. *China Econ. Rev.* **2019**, *6*, 49–53.

buildings

MDPI

Article

The Influence Innovation Has on the Visual Appearance and Aesthetic Preference of Architectural Products

Mozammel Mridha [1,*], Blair Kuys [1] and Safia Najwa Suhaimi [2]

[1] School of Design and Architecture, Swinburne University of Technology, Melbourne, VIC 3122, Australia
[2] PDR-International Centre for Design and Research, Cardiff Metropolitan University, Cardiff CF14 3AT, UK
* Correspondence: mmridha@swin.edu.au

Abstract: The importance of innovation for architectural products in the building industry increases due to global competitive markets and users' increased value given to the visual aesthetics of products. Visual appearance is crucial in architecture and product design and influences users' product choices in many ways. Substantial research on innovation exists concerning users' purchasing, adopting, and recommending it to others, but little research investigates the link between product innovation and aesthetic preference. This gap in knowledge prompted our investigation. Quantitative analysis of a survey of 114 respondents from Australia was conducted in this study to examine whether innovation plays a significant role in perceiving the aesthetic preference of an architectural product more than other visual appearances. Standard multiple regression using SPSS V28 was applied for statistical data analysis. Results uncovered that innovation explained the highest percentage of variance in overall aesthetic preference, and the innovativeness of a product strongly influences the visual appearance and aesthetic preference. The findings of this study offer new insight into the level of innovation for new product development where visual appearance is of high importance.

Keywords: innovation; aesthetics; aesthetic preference; visual appearance; product design; architecture

Citation: Mridha, M.; Kuys, B.; Suhaimi, S.N. The Influence Innovation Has on the Visual Appearance and Aesthetic Preference of Architectural Products. *Buildings* **2023**, *13*, 19. https://doi.org/10.3390/buildings13010019

Academic Editors: Yongjian Ke, Jingxiao Zhang and Simon P. Philbin

Received: 28 October 2022
Revised: 16 December 2022
Accepted: 19 December 2022
Published: 22 December 2022

1. Introduction

Vision is considered the central of the senses [1]. Therefore, product preference is significantly governed by the visual properties or appearance of the product [2]. It can persuade product evaluations and choices by the users in quite a few ways [3–5]. Users' first impressions of a product can be determined by the product's design as the "look" of a product is very important, which elevates the product's value and looking at something beautiful is gratifying [6]. The aesthetic responses are very personal as they are mainly emotional or feelings [7] and may derive from seeing the product without considering utility [8]. It is recognised that response to design involves a full array of human responses as the design's sensory aspect is congruent with a product's visual appearance [9].

The aesthetic preference is fundamentally an aesthetic judgment made by an individual based on recognising the structure of the product, coherence, or order [10], including the individual's prior experience generated from inherent memory integration [11]. Reference [12] pointed out that aesthetic pleasure and aesthetic interest are basically two different responses to aesthetics in a positive way. The aesthetic response can also be generated from a comprehensive aesthetic preference judgement, evidenced by considerable empirical research on aesthetic preference [13]. These findings illustrate some fundamental inconsistencies in the literature. For example, several studies mentioned that 'fluency' prompts aesthetic preference, through which an observer can process an object [14]. In contrast, other researchers challenge this finding [15]. They found that a complex design can positively trigger aesthetic liking, which is difficult to process. Novelty also makes the processing less fluent but is associated with aesthetic preference [16].

Aesthetics are also proposed to be the 'soul' of innovation [17], including innovative product designs, and its importance can be further highlighted in its ability to influence users' acceptance of innovation. Previous studies have argued that there is a link between an individual's aesthetic expertise and their evaluation of innovative designs [18]. Individuals with expertise in aesthetics or art and design have a higher tendency to prefer innovative designs, as they are particularly "more sensitive to the changes of innovativeness, which presumably was due to higher cognitive design concepts" [18] (p. 617). As an extension of these studies, exploration has also been carried out to understand specific individuals' aesthetic preferences based on their background and expertise level. By classifying participants into categories of expertise level, it is suggested that there is a distinct difference in preference and evaluation of aesthetic appeal between experts in the design field compared to design novices [19]. This aligns with a previous study [18] that shows that individuals who are inclined and interested in aesthetics tend to respond more positively to innovative designs, while those without any interest are more inclined to choose more conventional designs.

This is contested by another study [20], arguing that the study [18] did not provide any empirical evidence to support their findings. In reference [20], researchers propose that "visual typicality in design is a more important criterion for design novices who are less sensitive to the aesthetic quality of design" (p. 528). Typicality is important for this group simply because their judgement is influenced by the ease of processing [20]. They prefer designs or stimuli that are easier to relate to based on their previous knowledge or memory than a more unusual, novel design that they cannot relate to or easily understand.

Considerable research is available on innovation concerning customers' purchasing, adopting, and recommending it to others, but little research investigated the link between product innovation and aesthetic preference. Therefore, to address the gap, if aesthetics is proposed to be the 'soul' of innovation as described by reference [17], this study aims to investigate whether innovation plays a significant role in perceiving the aesthetic preference of a product more than other visual appearances. Furthermore, this study views this issue from a psychological perspective rather than market and user research, which most other studies have done. Thus, the research questions were raised as follows:

- How do respondents evaluate the innovativeness of a product?
- Does the aesthetic preference of a product depend on its innovative appearance?

Visual stimuli that have received scholarly attention include but are not limited to sculptures, textures, faces, and geometrical shapes [21]. We used timber joints of the pagoda-style structure to represent an architectural product as visual stimuli to conduct the study. There were certain considerations for this choice. Firstly, we were concerned about the complexity of visual stimuli. By complexity, we mean the respondent's perception of the stimulus complexity in question. The study [12] shows that moderately complex stimuli is preferred over high or low level of complexity, which is supported by many other studies [22]. Reference [12] hypothesised the relationship between complexity and aesthetic preference as an inverted U shape, where stimuli with an intermediate level of complexity attain the highest preference. Timber joints used in the study are moderately complex. Secondly, we were concerned about visual working memory (VWM). Reference [23] identified that VWM lets people hold visual information in mind for a few seconds. A study [24] on the VWM capacity of simple and complex stimuli revealed that VWM is sensitive to the surface complexity of the stimuli, suggesting the heavier the information load, the lower the VWM capacity [25]. Therefore, we can anticipate that the perceptual limitation of complex stimuli can be compensated by viewing the stimuli for a longer time and thus allowing participants to view the stimuli as long as required. Finally, the present study uses timber joints as visual stimuli, as they combine both aesthetics and functionality. Unlike artwork that predominantly fulfils a hedonic need, timber joints provide practical motives that induce cognitive and affective aesthetic judgment. Therefore, a combination of the neural processing of aesthetic experience and emotional responses to visual stimuli can address the research gap identified in the present study.

2. Theoretical Background

2.1. Overview of Aesthetic Experience

According to reference [26], an aesthetic experience occurs in response to a visual encounter with any type of object, scene, or event. This encounter is not bound to the experience of encountering visual artworks. It can occur daily, for instance, when one appreciates the beauty of one's newly purchased decorative vase or a building product. Aesthetic experience can be further defined as a cognitive process influenced by a person's affective state that will lead to an aesthetic emotion [27]. Each aesthetic experience may be different depending on the state of visual processing. Because of this, aesthetic experiences are considered complex phenomena, and any gradual development or change must be considered when assessing the experience [28].

An aesthetic experience involves different stages of classifying, understanding, and cognitively mastering an artwork, or in this case, a stimulus [27]. This multi-stage process can also be referred to as aesthetic information processing [29]. At the first stage of aesthetic processing, a person generates a perceptual analysis of the stimulus, thus, creating the first impression of that stimulus. Reference [30] suggests that, in general, a person would spontaneously generate a global impression or the gist of the stimulus at first glance of a stimulus. This occurrence is an immediate awareness of the visual appearance; the gist is pre-cognitive in nature [30]. Reference [31] summarises the aesthetic-processing stages and states that the whole process starts with the participant's perceptual analysis of the stimulus, which is then compared with their previous personal encounters and experience. The stimulus will then be classified into a meaningful category that is later interpreted and assessed. This results in the final stage of aesthetic processing, namely, aesthetic judgement and aesthetic emotion.

2.2. Overview of Aesthetic Preference

To understand the aesthetic preference of individuals for a particular product, it is important to note that there are factors and principles that can be used to measure an aesthetic preference. The study [14] highlights possible factors that can "influence aesthetic judgments such as figural goodness, figure-ground contrast, stimulus repetition, symmetry and prototypicality" (p. 364). Reference [1] proposes that despite the differences in social settings, such as culture and time among individuals, it is possible to form a universal agreement on aesthetic pleasure. The properties of a designed product, such as having a balanced proportion or a familiar property that stimulates preference, have been studied and measured to produce a universal agreement that can represent the aesthetic preference of the majority of individuals for a particular product.

Studies have experimented with using product elements and properties to measure aesthetic preference. One such study [32] measures a product's aesthetic preference by testing product angularity as a specific element. Their experiment shows a preference for arrays of circles and hexagons when it comes to angularity. Another similar study examining the difference in preference rate between curved and sharp objects suggests that sharp objects or sharp-angled contours induced lower preference in participants [33]. Some studies look into an individual's preference for physical elements of products, such as its physical form and shape (for example, [34]). The evaluation of other product elements and the global perception of a product itself also differs between design beginners and design experts. According to reference [35], beginners have a higher probability of using the level of novelty to indicate a product's apparent usability compared with design professionals. This means they perceive a novel-designed product as more usable than a designed product with typical features.

2.3. Aesthetics and Innovation

Linking aesthetics and innovation focuses mostly on 'soft' innovations, which are products that have a strong aesthetic component [36]. The author, in reference [36], also argues that most soft innovations are regarded best as new differentiated products. Because products that are primarily the same can be ascertained individually based on performance and aesthetic appeal, they are preferred by users differently due to different tastes or preferences.

A report from the Organization for Economic Co-operation and Development (OECD) [37] defines product innovation as "the introduction of a good or service that is new or significantly improved with respect to its characteristics or intended uses. This includes significant improvements in technical specifications, components and materials, incorporated software, user-friendliness or other functional characteristics" [37] (p. 48). We claim that aesthetic innovation plays a vital role in product characteristics and intended uses because nowadays, people are more concerned about the look, feel, and functionality of a product due to aesthetics being an essential element in our society [38]. As a result, practitioners comprehend the significance of aesthetic design in user choice [39], and many industries experience aesthetic innovation when the visual attributes of a product ascertain novelty [40].

Supporting this, we refer to another form of innovation listed in the OECD report [37] (p. 49), known as marketing innovation. "A marketing innovation is implementing a new marketing method involving significant changes in product design or packaging, product placement, product promotion or pricing" (p. 49). Despite the fact that the OECD's introduction of the marketing innovation concept has made it accepted that innovation no longer necessitates a change in product performance or functionality, and that innovation can exhibit just an aesthetic transformation as opposed to a functional change, none of the previously defined types of innovation encompasses soft innovation as defined above [37].

3. Methods

3.1. Participants

Australians 18 years of age and above were eligible to participate in the survey. Participants under 18 years of age and participants with limited or no capacity or authority to give voluntary and informed consent were excluded from the survey. An experience management company, Qualtrics©, recruited a total of 114 participants. Qualtrics© organised financial rewards for the participants for their contribution to the survey.

3.2. Procedures

An online survey was conducted in line with the National Statement on Ethical Conduct in Human Research, outlining the ethics protocol and approvals. All participants' data were anonymous, and they were informed about the intention of this study and gave implied consent. Participants were presented with five images of timber joints of the pagoda-style structure (Figure 1), one after another. Each image carried several visual appearance questions, and participants were asked to rate their agreement with the statements on a 7-point Likert scale (from 'Disagree' to 'Agree'). The online survey was prepared in English.

Figure 1. Stimuli were shown to the respondents.

3.3. Stimuli

The experimental stimuli were computer-generated visual images of a post and roof beam detail of five timber joints for a pagoda-style structure (Figure 1). The experimental stimuli were created to understand better visual appearance's contribution to determining aesthetic preference. Therefore, stimuli ranged from simple and straightforward (Joint 1) to intricate and highly decorative (Joint 5). All timber joints' images were taken from the same viewpoint and in a similar setting for consistency. Although the setting and the viewpoint for all stimuli were controlled, other visual features which might affect participants' responses were outside the experimental control. Furthermore, participants were not given any indication of performance or other product specifications.

3.4. Variables

Previous research confirms that visual appearance is a key component in product design and influences users' product preference or choice in many ways [1,3–5]. Since the aim of this study is to examine if innovation plays a significant role in perceiving the aesthetic preference of a product more than other visual appearances, the overall aesthetic preference of the timber joints has been considered as the dependent variable, while 'Joint appears innovative' is the independent variable in this study. We incorporated several controls in our study to account for other factors that are likely to influence the result. Therefore, we added three visual properties of the timber joint (i.e., long-lasting, functional, and strong) as a controlled variable (Table 1).

Table 1. Definition and description of all variables used in the study.

Variable	Definition	Description
Overall aesthetic preference	Dependent variable	Cognitive and affective response to timber joints
Innovative	Independent variable	The joint appears novel and aesthetically preferred
Long-lasting	Control variable Visual properties of the stimuli (timber joints) in the study	Joint appears durable and will last for a long time
Functional		Joint appears that it is designed to be practical and useful rather than purely attractive
Strong		Joint appears that it is able to withstand force, pressure, and/or wear

3.5. Statistical Analysis

The major objective of data analysis in this study was to investigate whether innovation plays a significant role in perceiving the aesthetic preference of a product more than other visual appearances. The SPSS (Statistical Package for the Social Sciences) version 28 program was used for data analysis. Before commencing data analysis, data were checked for possible errors. Furthermore, all assumptions of the statistical method applied for this study were validated and met before any analysis. Due to the nature of the investigation, 'standard multiple regression' analysis was employed for this purpose, which can identify the most important explanatory variables and determine the relative importance of these variables.

4. Results

4.1. Description of the Sample

The sample's most significant proportion was young females from Australian and indigenous backgrounds in full-time employment with a bachelor's degree and above. The sample population does not indicate any dominant occupation category. This study provides a non-biased analysis of the general public, as this group will interact with the final product outcome. The following figure (Figure 2) shows the frequency distribution of the sample population in percentage.

4.2. Checking the Assumptions of Multiple Regression

Before analysis, the data were verified, assessing whether they were suitable for analysis concerning multicollinearity outliers, normality, linearity, and homoscedasticity.

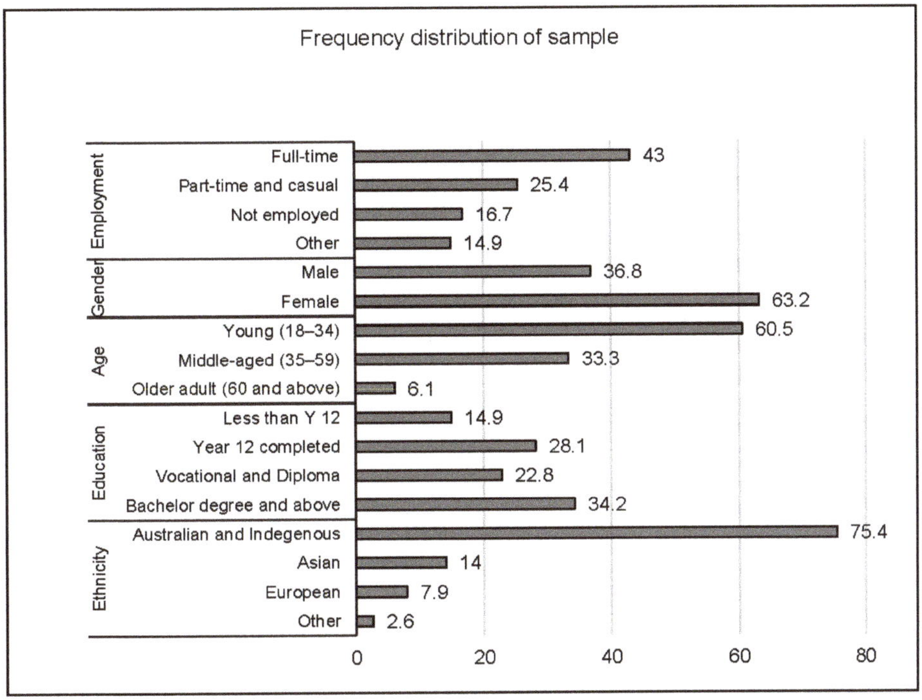

Figure 2. Characteristics of the respondents: Gender, age, education, ethnicity, and employment.

4.2.1. Multicollinearity

The correlation between independent variables (i.e., the visual appearance of timber joints) with the dependent variable (i.e., overall aesthetic preference) was checked. The preferred bivariate correlation value between the independent and dependent variable is above 0.3 [41]. Collinearity diagnostics were also performed to check further multicollinearity that might not be evident in the correlation matrix. Two values were scrutinised at Tolerance and VIF (Variance inflation factor). The tolerance values for independent variables were above 0.10; the multicollinearity assumption was not violated. These were also supported by the VIF values, as all values are well below the cut-off of 10.

4.2.2. Outliers, Normality, Linearity, Homoscedasticity, Independent of Residuals

Assumptions about outliers, normality, linearity, homoscedasticity, and independent of residuals were verified by inspecting the normal P-P plot of the regression standardised residual and the scatter plot of standardised residuals. In the normal P-P plot, all points are laid in a reasonably straight diagonal line from bottom left to top right, suggesting that there is no major deviation from normality. Additionally, residuals are roughly rectangularly distributed in the scatter plot, with most of the scores concentrated along with the 0 points. The presence of an outlier was checked from the scatter plot (i.e., cases that had standardised residual of more than 3.3 or less than −3.3) and none were found. The presence of outliers was also checked by inspecting the Mahalanobis distance. According to reference [41], the critical value is 18.47 for four independent variables. Three cases above that value were observed, which were slightly above the critical value and hence retained. The maximum value for Cook's distance was also verified to check whether these cases had an undue influence on the result of the model. The maximum value for the Cook's distance was 0.084, suggesting that there was no major problem in the data as reference [42] recommended that a value above 1 would be a potential problem.

4.3. Model Evaluation

In this section, the evaluation of the regression model is discussed. There are five models developed from five stimuli (see Figure 1). Table 2 represents the model summary and statistical significance of all five models. The R square value of the models in Table 2 varies between 0.461 and 0.550. This implied that the models explained a minimum of 46% (Joint 1) and a maximum of 55% (Joint 2) of the variance in overall aesthetic preference. This is quite a respectable result. The statistical significance of the result was also assessed in Table 2. The F-ratio in the ANOVA tests whether the overall regression model is a good fit for the data. The table shows that the independent variables statistically significantly predict the dependent variable, F (4, 109) $p = (0.001) < 0.05$ (i.e., the regression model is a good fit for the data).

Table 2. Summary (b) of the model predicting overall aesthetic preference and statistical significance of the models.

Model Summary [b]			**ANOVA**			
Joint		Variance Explained [a] (R Square)	df		F	Significance
1		46.1%	Regression	4	23.31	<0.001
			Residual	109		
2		55.0%	Regression	4	33.35	<0.001
			Residual	109		
3		46.1%	Regression	4	23.31	<0.001
			Residual	109		
4		52.9%	Regression	4	30.57	<0.001
			Residual	109		
5		51.3%	Regression	4	28.68	<0.001
			Residual	109		

[a] Predictors: the joint appears appropriate for the overall structure, the joint looks difficult to manufacture, the joint appears to be long lasting, the joint appears innovative, the joint appears strong, the joint appears functional.
[b] Dependent variable: overall aesthetic preference.

4.4. Variable Evaluation

The models were developed according to the types of timber joints (five in total), and each model was composed of four visual appearance components. Each component was evaluated to ascertain which of the variables contributed to the prediction of overall aesthetic preference. Table 3 illustrates information about the independent variable and how it affects the dependent variable.

Table 3. Coefficients ([a]) of aesthetic preference of timber joints.

Joint		Visual Appearance	B	Beta	Sig	Part
1		Strong	0.187	0.207	0.095	0.118
		Functional	0.071	0.074	0.547	0.042
		Long-lasting	0.121	0.127	0.205	0.090
		Innovative	0.347	0.390	<0.001	0.308
2		Strong	−0.078	-0.070	0.588	−0.035
		Functional	0.287	0.218	0.071	0.117
		Long-lasting	0.407	0.325	0.002	0.203
		Innovative	0.408	0.363	<0.001	0.253
3		Strong	0.187	0.207	0.095	0.118
		Functional	0.071	0.074	0.547	0.042
		Long-lasting	0.121	0.127	0.205	0.090
		Innovative	0.347	0.390	<0.001	0.308
4		Strong	0.000	0.000	0.998	0.000
		Functional	0.433	0.392	<0.001	0.225
		Long-lasting	0.119	0.112	0.372	0.059
		Innovative	0.330	0.305	0.001	0.216
5		Strong	0.293	0.263	0.032	0.145
		Functional	0.119	0.104	0.365	0.061
		Long-lasting	−0.031	−0.028	0.818	−0.015
		Innovative	0.483	0.469	<0.001	0.334

[a] Dependent variable: overall aesthetic preference.

4.5. Significance

The following independent variables have a statistically significant impact on the outcome variable (overall aesthetic preference) according to joint type:

Joint 1: The joint appears innovative

Joint 2: The joint appears innovative; the joint appears to be long-lasting

Joint 3: The joint appears innovative

Joint 4: The joint appears innovative; the joint appears functional

Joint 5: The joint appears innovative; the joint appears strong

It is evident from the above that "The joint appears innovative" has a statistically significant impact ($p < 0.001$) on overall aesthetic preference for all joint types.

4.6. Unstandardised Coefficients

Unstandardised coefficients (B) indicate how much the dependent variable varies with an independent variable when all other independent variables are held constant. Since 'The joint appears innovative' is statistically significant across all joint types, we look at its influence on overall aesthetic preference. Table 3 also indicates that if 'The joint appears innovative' index increases by a value of 1, we observe 0.347, 0.408, 0.347, 0.330, and 0.483 units increase on the dependent variable for Joint 1 to 5, respectively. So, the more a respondent perceives the timber joint to look innovative, the joint becomes more aesthetically pleasing and is hence preferred. Clearly, the effect of this independent variable is more pronounced on Joint 5. However, the 'confidence interval' for Joint 5 indicates that there is a 95 percent chance that the actual value of the unstandardised coefficient is between 0.291 and 0.675.

4.7. Standardised Coefficients

Table 3 illustrates the contribution of each independent variable included in the model contributed to the prediction of the dependent variable. The beta value in this table is the standardised coefficient. These values for each of the different variables have been converted to the same scale for comparison. Therefore, the higher the beta value, the stronger the unique contribution to explaining the dependent variable. For Joint 1, as seen in Table 3, 'The joint appears innovative' had the largest beta coefficient of 0.39. Therefore, 'The joint appears innovative' caused the strongest unique contribution to explaining 'Overall aesthetic preference' when all other variables in the model were controlled. This was followed by 'The joint appears strong' (0.21), 'The joint appears long lasting' (0.13), and 'The joint appears functional' (0.07) and made the least contribution to predicting overall aesthetic preference. From Table 3, Joints 2, 3, and 5 can be explained in the same fashion where 'The joint appears innovative' had the largest beta coefficient. However, for Joint 4, we see that 'The joint appears functional' had the largest beta coefficient (0.39), followed by 'The joint appears innovative' (0.31).

Further potential information generated from Table 3 is the part correlation coefficient. It shows how much of the total variance in the independent variable is uniquely explained by that variable, and how much R square would drop if it was not included in the model. For Joint 1, 'The joint appears innovative' had a part correlation value of 0.308. This value was squared and multiplied by 100 to ascertain the percentage of variance [41]. The new value came out as 9.48. This represented that the component uniquely explained 9.5% of the variance in overall aesthetic preference (Table 4). Table 4 also explains which of the variables included in the models contribute more to the prediction of overall aesthetic preference for other joints. For example, 'The joint appears innovative' makes the strongest unique contribution to explain the overall aesthetic preference for Joints 2, 3, and 5. It explains 6.4%, 9.5%, and 11.1% of the variance in overall aesthetic preference for Joints 2, 3, and 5, respectively. Joint 4 illustrates a slightly different scenario. 'The joint appears functional' makes the strongest unique contribution to explaining 5.6% of the variance in overall aesthetic preference. However, 'The joint appears innovative' trailed by just 1%. 'The joint appears strong' explained variance in overall aesthetic preference for Joints 1, 3, and 5. However, Joint 5 (2.1%) was only statistically significant. The only other statistically significant variable is 'The joint appears to be long lasting', which explained a 4.12% variance in overall aesthetic preference for Joint 2.

Table 3 also shows two exceptional cases where for Joint 2, a 1 unit increase in 'The joint appears strong' is associated with a 0.078 unit decrease in 'overall aesthetic preference' and Joint 5, a 1 unit increase in 'The joint appear long-lasting' is associated with a 0.031 unit decrease in 'overall aesthetic preference'. However, both cases were statistically insignificant and omitted from further investigation.

Table 4. Percentage of variance explained by visual appearance components in overall aesthetic preference for the timber joints.

	Joint	Visual Appearance (Standardised Beta Coefficients)
1		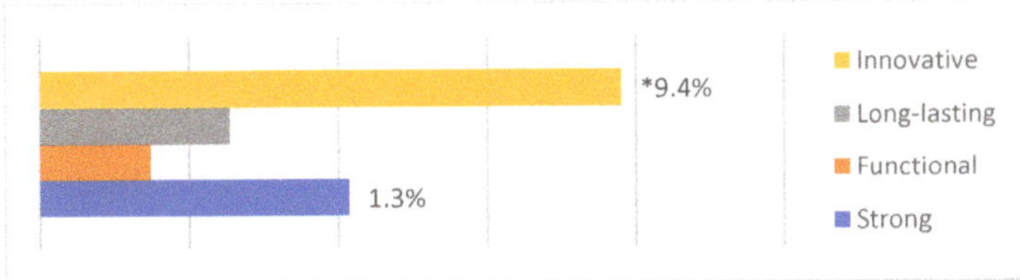
2		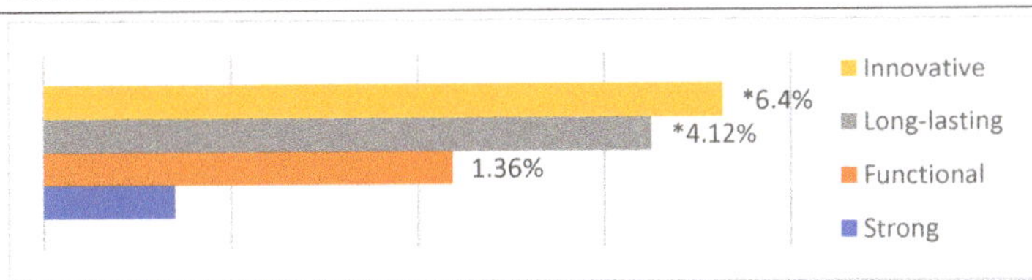
3		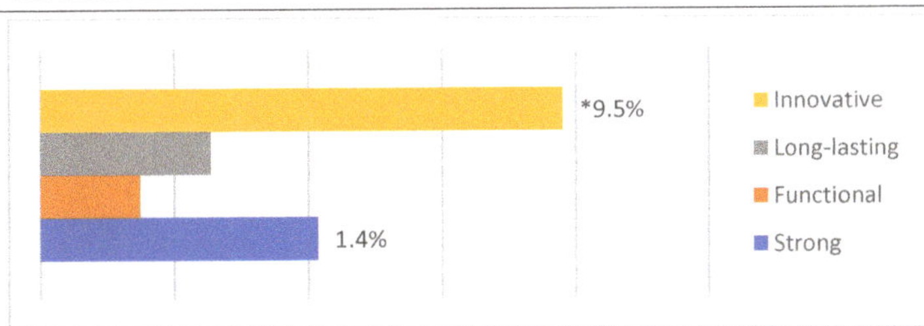
4		
5		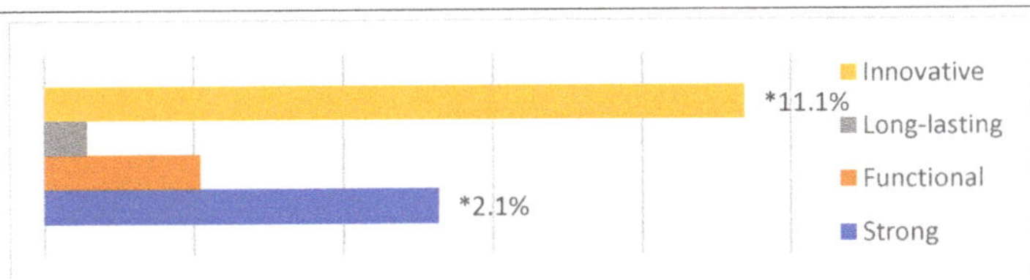

* $p < 0.001$

5. Discussion

In this study, we examined whether innovation plays a significant role in perceiving the aesthetic preference of a product more than other visual appearances. Our findings suggest several theoretical and practical implications.

5.1. Theoretical Implications

5.1.1. Innovation and Aesthetic Preference

The results section shows that the independent variable significantly predicts the dependent variable (overall aesthetic preference). To further investigate whether a product's aesthetic preference depends on its innovative appearance, we look at Table 4, which shows which of the variables incorporated in the models contributes more to predicting overall aesthetic preference. Except for Joint 4, innovation uniquely explained the highest percentage of the variance in overall aesthetic preference. Although Joint 4 was a bit exceptional, where 'Functional' took the lead, the difference was only by 1%.

We mentioned earlier reference [1] findings that a product's visual properties are vital in determining product preference. In an innovation, the first thing users notice is the product's visual appearance, i.e., aesthetics [43]. The effect of visual complexity was examined in several early studies [44,45]. A medium level of complexity (which all stimuli in this study fall into) was often preferable [27] due to the arousal potential resulting from visual stimulation. Since innovative appearance is a visual property, we contend that aesthetic preference depends on the innovative appearance of a product.

5.1.2. Innovation and Aesthetic Experience

The results section finds that 'Joint appears innovative' statistically impacts overall aesthetic preference for all joint types. Let us explore respondents' aesthetic experiences when viewing the stimuli to record aesthetic preferences. Stimuli were shown chronologically. Therefore, respondents observe Joint 1 first and Joint 5 last. From the literature, we know that aesthetic experience occurs in response to a visual encounter with any type of object, scene, or event [26]. Reference [46] pointed out that when a user sees a product, one of the first responses is aesthetic perception, which is closely related to visual information. Hence, reference [47] argues that overall, it significantly impacts the perception of a product. During the observation, stimuli 1 appeared as something new for many respondents, and they rated Joint 1 as significantly innovative. Having this experience when they observed stimuli 2, which is almost similar to stimuli 1 except for the introduction of a small protrusion of timber beam, they scored a little low on innovation. The introduction of capital and elaborate rounded protrusion of stimuli 3 was a substantial departure from stimuli 2. As a result, respondents rated it as significantly innovative (even higher than stimuli 1). Respondents rated stimuli 4 low in terms of innovation due to the exact reason for stimuli 2. Stimuli 5 was significantly different from previous stimuli. The timber column took the shape of the Greek Corinthian order. As rated by the respondents, the flute on the shaft, volute, and acanthus leaf on the capital made it the most innovative (Tables 3 and 4). According to reference [48], a product's outer form can affect customer perceptions in quite a few ways, (i) by accentuating or concealing different factors of technology that are introduced by innovation, (ii) by providing visual cues for product interpretation, and (iii) by triggering sensory experiences, which influence cognition and emotion. Therefore, we argue that products perceived as innovative provide observers with initial cues that trigger various cognitive and emotional responses that underlie their assessment of aesthetic preference.

5.1.3. Aesthetics and functionality

Reference [8] pointed out that a product's aesthetic value may relate to the pleasure of seeing the product without considering utility. The finding is in accordance with reference [6], who mentioned that a user might prefer a product entirely based on its 'look' as looking at something beautiful is satisfying. The functionality of Joint 4 uniquely explains the highest percent of the variance. The result is different from other joints. Although functionality reflects the users' perceptions of a product's ability to fulfil its purpose [49]. However, according to reference [50], aesthetics are significantly more important than functionality for product appreciation and observation, which is also supported by reference [51], which argues that visual appeal is more important than functionality. Since the variance difference between 'innovation' and 'functional' is only 1%, this deviation can be ignored and will not impact the overall empirical premise of the study.

5.2. Practical Implications

In practice, this can be used for designing and developing a new product that includes products from the architectural and building industry, where innovativeness is considered a condition for generating public preference that promotes product success. Notably, the study consents to new thoughts and debates on how innovation should be defined, evaluated, and construed, including product innovation's role in professional practice and design-related research areas.

6. Conclusions

The research presented throughout this paper revealed that a product's innovativeness strongly influences the visual appearance and aesthetic preference. The study also has some strengths. First, our samples were randomly selected across all of Australia, ensuring variability in the population. Second, our study investigated product innovation and aesthetic preference from a psychological perspective rather than market and user research. Third, our findings uniquely contribute to design research and design practice.

Reference [52] identified that there is a disconnect of belief on what is new and innovative between marketers and users. There may be a number of factors for this disengagement but what attracts a user to a new product is the visual aesthetic design [43]. Users increasingly value the visual aesthetics of product design [53,54]; however, there have been small attempts to ascertain how innovative visual aesthetics influence the perceptions of novelty and product assessments [43]. The visual appearance is critical to the product's user response and success [3]. As we pointed out in the introduction section, that response to design involves a full array of human responses as the design's sensory aspect is congruent with a product's visual appearance [9]. As a result, the product's visual appearance as perceived by the users is characteristically based on users' cognitive and affective responses. The aesthetic properties of objects can activate a multifaceted combination of secondary emotional and cognitive responses, which according to Csikszentmihalyi and Robinson [55] (p. 18), is "the aesthetic experience". So, a user may positively evaluate a product if product innovation prompts a positive emotional response through its aesthetic properties.

The survey of 114 respondents revealed that the joints' innovativeness explained the highest percentage of variance in overall aesthetic preference. This displays that the visual complexity and appeal (aesthetics) tend to be more important than a product's performance (functionality) upon first observation. Furthermore, the study shows a link between cognitive and emotional responses in their assessment of aesthetic preference. Therefore, we expect this study's findings to offer new insight into the design process of new product development, not only in the architectural and building industry, but in a holistic context for architectural and product design in general.

7. Limitations and Future Research

This study highlights innovation as an important aspect of visual appearance to determine aesthetic preference among respondents living in Australia; however, it has several limitations which suggest useful guidance for future research. First, our study design means causality cannot be concluded. There could be source bias as both exposure and outcome measures are self-reported. We encourage future research to go further with larger sample sizes and use control and experimental groups to reduce bias and further extend our understanding. Second, the respondents could only respond to the selected visual appearance questions. No open-ended questions were included to give respondents an opportunity to include other aspects of their lives that contribute to aesthetic preference. Thus, we invite studies examining whether other aspects of life (e.g., sociodemographic characteristics) may influence aesthetic preference. Third, it is difficult to claim that the study represented the complete breadth and all types of constructs and variables as a predictor of aesthetic preference. Therefore, it is difficult to assess the generalisability of the findings to projects other than those in the sample without the benefit of further research that includes a comprehensive range of predictors of aesthetic preference.

Author Contributions: Conceptualisation, M.M.; methodology, M.M.; validation, M.M. and B.K.; formal analysis, M.M.; investigation, M.M., B.K. and S.N.S.; resources, M.M., B.K. and S.N.S.; writing—original draft preparation, M.M.; writing—review and editing, M.M., B.K. and S.N.S.; visualisation, B.K.; supervision, M.M. and B.K.; project administration, B.K.; funding acquisition, B.K. All authors have read and agreed to the published version of the manuscript.

Funding: This research received no external funding.

Institutional Review Board Statement: The study was conducted in accordance with the Declaration of Helsinki, and approved by Swinburne University of Technology Human Ethics Committee (Ref: R/2019/209, 18/09/2019) for studies involving humans.

Informed Consent Statement: Informed consent was obtained from all subjects involved in the study.

Data Availability Statement: The data presented in this study are available on request from the corresponding author. The data are not publicly available due to approved ethical restrictions.

Acknowledgments: The authors would like to thank Australian manufacturer Timberfy, who worked with the research team to manufacture the pagoda for one of Melbourne's largest cemeteries. The authors would also like to thank the research participants for their time in completing the survey.

Conflicts of Interest: The authors declare no conflict of interest.

References

1. Hekkert, P.; Leder, H. Product Aesthetics. In *Product Experience*; Schifferstein, H.N.J., Hekkert, P., Eds.; Elsevier: San Diego, CA, USA, 2008; pp. 259–285.
2. Ceballos, L.M.; Hodges, N.; Watchravesringkan, K. Consumer preference and apparel products: Investigating the role of the Centrality of Visual Product Aesthetics concept. *Int. J. Fash. Des. Technol. Educ.* **2021**, *14*, 325–337. [CrossRef]
3. Bloch, P.H. Seeking the ideal form: Product design and consumer response. *J. Mark.* **1995**, *59*, 16–29. [CrossRef]
4. Garber, L., Jr.; Burke, R.R.; Jones, J.M. *The Role of Package Color in Consumer Purchase Consideration and Choice*; Marketing Science Institute: Cambridge, MA, USA, 2000.
5. Veryzer, R.W., Jr. The Place of Product Design and Aesthetics in Consumer Research. In *Advances in Consumer Research*; Kardes, F.R., Provo, M.S., Eds.; UT: Association for Consumer Research NA: Seattle, WA, USA, 1995; Volume 22, pp. 614–645.
6. Creusen, M.E.H.; Schoormans, J.P.L. The Different Roles of Product Appearance in Consumer Choice. *J. Prod. Innov. Manag.* **2005**, *22*, 63–81. [CrossRef]
7. Bamossy, G.; Scammon, D.L.; Johnston, M. A Preliminary Investigation of the Reliability and Validity of an Aesthetic Judgment Test. In *Advances in Consumer Research*; Bagozzi, R.P., Tybout, A.M., Eds.; Association for Consumer Research: Ann Arbor, MI, USA, 1983; pp. 685–690.
8. Holbrook, M.B. Some Preliminary Notes on Research in Consumer Esthetics. In *NA-Advances in Consumer Research*; Olson, J.C., Ed.; Association for Consumer Research: Ann Arbor, MI, USA, 1980; Volume 7, pp. 104–108.
9. Crilly, N. Product Aesthetics: Representing Designer Intent and Consumer Response. Ph.D. Thesis, University of Cambridge, London, UK, 2005. Available online: https://dl.icdst.org/pdfs/files/52725a8fcd239b24698050ed22a5f343.pdf (accessed on 9 August 2022).
10. Desmet, P.M.A.; Hekkert, P. Framework of product experience. *Int. J. Des.* **2007**, *1*, 57–66.
11. Leder, H.; Nadal, M. Ten years of a model of aesthetic appreciation and aesthetic judgments: The aesthetic episode—Developments and challenges in empirical aesthetics. *Br. J. Psychol.* **2014**, *105*, 443–464. [CrossRef]
12. Berlyne, D.E. *Aesthetics and Psychobiology*; Appleton-Century-Crofts: New York, NY, USA, 1971.
13. Graf, L.K.M.; Landwehr, J.R. Aesthetic pleasure versus aesthetic interest: The two routes to aesthetic liking. *Front. Psychol.* **2017**, *8*, 1–15. [CrossRef]
14. Reber, R.; Schwarz, N.; Winkielman, P. Processing fluency and aesthetic pleasure: Is beauty in the perceiver's processing experience? *Personal. Soc. Psychol. Rev.* **2004**, *8*, 364–382. [CrossRef]
15. Landwehr, J.R.; Labroo, A.A.; Herrmann, A. Gut liking for the ordinary: Incorporating design fluency improves automobile sales forecasts. *Mark. Sci.* **2011**, *30*, 416–429. [CrossRef]
16. Hekkert, P.; Snelders, D.; van Wieringen, P.C. Most advanced, yet acceptable: Typicality and novelty as joint predictors of aesthetic preference in industrial design. *Br. J. Psychol.* **2003**, *9*, 111–124. [CrossRef]
17. Godoe, H. Innovation theory, aesthetics, and science of the artificial after Herbert Simon. *J. Knowl. Econ.* **2012**, *3*, 372–388. [CrossRef]
18. Leder, H.; Carbon, C.C. Dimensions in Appreciation of Car Interior Design. *Appl. Cogn. Psychol.* **2005**, *19*, 603–618. [CrossRef]
19. Person, O.; Snelders, D. Initial Findings on Design and Product Category Expertise in Aesthetic Evaluation. In Proceedings of the 4th Nordic Design Research Conference, Helsinki, Finland, 29–31 May 2011; pp. 1–5.

20. Creusen, M.E.H.; Snelders, D. Visual typicality and novelty as joint predictors of aesthetic preference: The influence of design expertise and product category interest. In *E-European Advances in Consumer Research*; Bradshaw, A., Hackley, C., Maclaran, P., Eds.; Association for Consumer Research: Duluth, MN, USA, 2011; Volume 9, pp. 528–529.
21. Cheung, M.C.; Law, D.; Yip, J.; Wong, C. Emotional Responses to Visual Art and Commercial Stimuli: Implications for Creativity and Aesthetics. *Front. Psychol.* **2019**, *10*, 1–10. [CrossRef] [PubMed]
22. Van Geert, E.; Wagemans, J. Order, Complexity, and Aesthetic Appreciation. *Psychol. Aesthet. Creat. Arts* **2020**, *14*, 1–20. [CrossRef]
23. Luck, S.J.; Vogel, E. The capacity of visual working memory for features and conjunctions. *Nature* **1997**, *309*, 279–281. [CrossRef] [PubMed]
24. Alvarez, G.A.; Cavanagh, P. The capacity of visual short term memory is set both by visual information load and by number of objects. *Psychol. Sci.* **2004**, *15*, 106–111. [CrossRef] [PubMed]
25. Eng, H.Y.; Chen, D.; Jiang, Y. Visual working memory for simple and complex visual stimuli. *Psychon. Bull. Rev.* **2005**, *12*, 1127–1133. [CrossRef] [PubMed]
26. Palmer, S.E.; Schloss, K.B.; Sammartino, J. Visual aesthetics and human preference. *Annu. Rev. Psychol.* **2013**, *64*, 77–107. [CrossRef]
27. Leder, H.; Belke, B.; Oeberst, A.; Augustin, D. A model of aesthetic appreciation and aesthetic judgments. *Br. J. Psychol.* **2004**, *95*, 489–508. [CrossRef]
28. Verhavert, S.; Wagemans, J.; Augustin, M.D. Beauty in the blink of an eye: The time course of aesthetic experiences. *Br. J. Psychol.* **2018**, *109*, 63–84. [CrossRef] [PubMed]
29. Markovic, S. Components of aesthetic experience: Aesthetic fascination, aesthetic appraisal, and aesthetic emotion. *I-Perception* **2012**, *3*, 1–17. [CrossRef]
30. Locher, P.J. The Aesthetic Experience with Visual Art at First Glance. In *Investigations into the Phenomenology and the Ontology of the Work of Art*; Springer: Cham, Switzerland, 2015; pp. 75–88. [CrossRef]
31. Hekkert, P. Design aesthetics: Principles of pleasure in design. *Psychol. Sci.* **2006**, *48*, 157–172.
32. Silvia, P.J.; Barona, C.M. Do People Prefer Curved Objects? Angularity, Expertise, and Aesthetic Preference. *Empir. Stud. Arts* **2009**, *27*, 25–42. [CrossRef]
33. Bar, M.; Neta, M. Humans prefer curved visual objects. *Psychol. Sci.* **2006**, *17*, 645–648. [CrossRef]
34. Silvera, D.H.; Josephs, R.A.; Giesler, R.B. Bigger Is Better: The Influence of Physical Size on Aesthetic Preference Judgments. *J. Behav. Decis. Mak.* **2002**, *15*, 189–202. [CrossRef]
35. Mugge, R.; Schoormans, J.P.L. Product design and apparent usability. The influence of novelty in product appearance. *Appl. Ergon.* **2012**, *43*, 1081–1088. [CrossRef]
36. Stoneman, P. Soft innovation: Changes in product aesthetics and aesthetic products. In Proceedings of the Royal Economic Society Annual Conference, Coventry, UK, 17–19 March 2008.
37. OECD. *The Measurement of Scientific and Technological Activities: Proposed Guidelines for Collecting and Interpreting Technological Innovation Data*, 3rd ed.; Commission Eurostat: Paris, France, 2006.
38. Postrel, V. *The Substance of Style: How the Rise of Aesthetic Value Is Remaking Commerce, Culture, and Consciousness*; HarperCollins: New York, NY, USA, 2004.
39. Swan, K.S.; Kotabe, M.; Allred, B.B. Exploring Robust Design Capabilities: Their Role in Creating Global Products, and Their Relationship to Firm Performance. *J. Prod. Innov. Manag.* **2005**, *22*, 144–164. [CrossRef]
40. Alcaide-Marzal, J.; Tortajada-Esparza, E. Innovation Assessment in Traditional Industries: A Proposal of Aesthetic Innovation Indicators. *Scientometrics* **2007**, *72*, 33–57. [CrossRef]
41. Pallant, J. *SPSS Survival Manual: A Step by Step Guide to Data Analysis Using SPSS*, 3rd ed.; Allen & Unwin: Crows Nest, NSW, Australia, 2007.
42. Tabachnick, B.G.; Fidell, L.S. *Using Multivariate Statistics*; Pearson Education: Boston, MA, USA, 2007.
43. Goode, M.R.; Dahl, D.W.; Moreau, C.P. Innovation aesthetics: The relationship between category cues, categorisation certainty, and newness perceptions. *J. Prod. Innov. Manag.* **2013**, *30*, 192–208. [CrossRef]
44. Berlyne, D.E. *Studies in the New Experimental Aesthetics*; Wiley: New York, NY, USA, 1974.
45. Frith, C.D.; Nias, D.K.B. What determines aesthetic preferences? *J. Gen. Psychol.* **1974**, *91*, 163–173.
46. Ulrich, K.T. Design: Creation of Artifacts in Society. 2011. Available online: https://ssrn.com/abstract=1951106 (accessed on 2 August 2022).
47. Mata, M.P.; Ahmed-Kristensen, S.; Shea, K. Implementation of design rules for perception into a tool for three-dimensional shape generation using a shape grammar and a parametric model. *J. Mech. Des.* **2018**, *141*, 011101. [CrossRef]
48. Rindova, V.; Petkova, A. When Is a New Thing a Good Thing? Technological Change, Product Form Design, and Perceptions of Value for Product Innovations. *Organ. Sci.* **2007**, *18*, 217–232. [CrossRef]
49. Homburg, C.; Schwemmle, M.; Kuehnl, C. New product design: Concept, measurement, and consequences. *J. Mark.* **2015**, *79*, 41–56. [CrossRef]
50. Haug, A. A framework for the experience of product aesthetics. *Des. J.* **2016**, *19*, 809–826. [CrossRef]
51. Han, J.; Forbes, H.; Schaefer, D. An exploration of how creativity, functionality, and aesthetics are related in design. *Res. Eng. Des.* **2021**, *32*, 289–307. [CrossRef]

52. Gourville, J.T. Eager sellers stony buyers: Understanding the psychology of new product adoption. *Harv. Bus. Rev.* **2006**, *84*, 98–106.
53. Bloch, P.H.; Brunel, F.F.; Arnold, T.J. Individual differences in the centrality of visual product aesthetics: Concept and measurement. *J. Consum. Res.* **2003**, *29*, 551–565. [CrossRef]
54. Page, C.; Herr, P.M. An investigation of the processes by which product design and brand strength interact to determine initial affect and quality judgments. *J. Consum. Psychol.* **2002**, *12*, 133–147. [CrossRef]
55. Csikszentmihalyi, M.; Robinson, R.E. *The Art of Seeing*; J. Paul Getty Museum: Malibu, CA, USA, 1990.

MDPI
St. Alban-Anlage 66
4052 Basel
Switzerland
Tel. +41 61 683 77 34
Fax +41 61 302 89 18
www.mdpi.com

Buildings Editorial Office
E-mail: buildings@mdpi.com
www.mdpi.com/journal/buildings

www.ingramcontent.com/pod-product-compliance
Lightning Source LLC
Chambersburg PA
CBHW061511220926
51850CB00004B/228
9783036581149